PSYCHIATRY - THEORY, APPLICATIONS AND TREATMENTS

NEUROBIOLOGY OF POST-TRAUMATIC STRESS DISORDER

PSYCHIATRY - THEORY, APPLICATIONS AND TREATMENTS

Additional books in this series can be found on Nova's website under the Series tab.

Additional E-books in this series can be found on Nova's website under the E-books tab.

PSYCHIATRY - THEORY, APPLICATIONS AND TREATMENTS

NEUROBIOLOGY OF POST-TRAUMATIC STRESS DISORDER

LEO SHER
AND
ALEXANDER VILENS
EDITORS

Nova Science Publishers, Inc.
New York

For permission to use material from this book please contact us:
Telephone 631-231-7269; Fax 631-231-8175
Web Site: http://www.novapublishers.com

Additional color graphics may be available in the e-book version of this book.

Library of Congress Cataloging-in-Publication Data

Neurobiology of post-traumatic stress disorder / editors, Leo Sher and Alexander Vilens.
p. ; cm.
Includes bibliographical references and index.
ISBN 978-1-61470-593-2 (softcover)
1. Post-traumatic stress disorder--Pathophysiology. 2. Neuropsychiatry. 3. Stress (Psychology)--Physiological aspects. I. Sher, Leo. II. Vilens, Alexander.
[DNLM: 1. Stress Disorders, Post-Traumatic--physiopathology. 2. Stress Disorders, Post-Traumatic--therapy. 3. Diagnostic Techniques, Neurological. 4. Nerve Net. 5. Stress Disorders, Post-Traumatic--diagnosis. WM 170 N491 2010]
RC552.P67N475 2010
616.85'21--dc22

2010015618

Published by Nova Science Publishers, Inc. † New York

Contents

Preface

Post-Traumatic Stress Disorder (PTSD) is a common and severe psychiatric disorder precipitated by exposure to a psychologically distressing event. PTSD is associated with significant morbidity and mortality and is characterized by the presence of three distinct, but co-occurring, symptom clusters. Research evidence suggests that PTSD has a neurobiological basis. Current research on the neurobiology of PTSD include the utilization of functional brain imaging; molecular genetic research; the incorporation of cross-system research including neuroendocrine, neurochemical, and neuroimmunological systems. This book examines the neurobiological basis of PTSD and the future research goals in regards to these findings.

Chapter I-Advances in neuroimaging permit *in vivo* investigation of the structural and functional neurobiology of post-traumatic stress disorder (PTSD). Recent reviews of structural and functional neuroimaging in PTSD are discussed together with recent studies. The significance of the findings is limited by the disparate groupings of PTSD trauma exposures, making generalization difficult, and this, combined with the heterogeneity inherent in the existing diagnostic criteria, renders the often conflicting data confusing. There has been an emphasis on studies of cortical structures, without adequate consideration of neural circuit models implicating other brain structures. There has been a move towards investigating subsets of symptoms, such as hyperarousal and dissociation, particularly in functional imaging studies. Overall, neuroimaging research in PTSD faces challenges for the future. The key improvements lie in: the definition and classification of trauma exposures, as they are not all equivalent in nature or indeed comparable; the study of restricted phenotypic subsets of symptoms or endophenotypes, such as hyperarousal and dissociation which may have different neural substrates; adequate study designs for power and control of confounders; and more focused research based upon targeted investigation of neural networks putatively involved in the processing and re-experiencing of trauma.

Chapter II- Magnetic resonance imaging (MRI) studies have demonstrated changes in brain morphology in PTSD, including smaller volume of the hippocampus, anterior cingulate cortex and morphologic alterations in the corpus callosum. This paper reviews the recent progress of structural and functional neuroimaging studies in PTSD. MRI volumetric studies have shown smaller hippocampal volume in men and women with chronic PTSD from early abuse or combat; however, these results have not been replicated in children with PTSD.

Neuroimaging studies demonstrated functional and morphologic abnormalities in ACC of the patients with PTSD. Brain functional studies using positron emission tomography (PET) or functional magnetic resonance imaging (fMRI) showed decreased function in ACC. In addition, morphometric MRI studies demonstrated smaller volume of the ACC in comparison to non-PTSD subjects. Alterations in this brain region may underlie changes in emotion that are a part of PTSD. Some studies of children with PTSD showed smaller corpus callosum area in comparison to healthy subjects. Findings of smaller corpus callosum size in children with PTSD related to maltreatment have led to the hypothesis that maltreatment is associated with a reduction in corpus callosum size.

Chapter III- Voxel based morphometry (VBM) methods are used to detect the difference in brain structures between the posttraumatic stress disorder (PTSD) patients and the normal controls. Standard VBM method can detect the difference of the gray matter or white matter densities while the optimized VBM method can detect the difference of gray matter or white matter volumes in the whole brain. The experimental results show that for the patient group, gray matter density or volumes significantly increased in right frontal lobe, middle frontal gyrus, vermis, left caudate and parietal lobe compared with the normal controls. However, in the left frontal lobe and middle frontal gyrus, gray matter density significantly decreased. There is no significant difference in white matter between the two groups. These results are consistent with those studies based on the fMRI. It not only provides the evidence for further study of the pathogeny in PTSD but also validates the efficiency of the VBM methods for detecting the difference in the whole brain structure.

Chapter IV- Post-traumatic stress disorder (PTSD) is classified as an anxiety disorder. However, clinical impressions and preclinical work suggest that PTSD may be also associated with dysfunctional reward processing. Here we highlight three lines of research evidence from our laboratory to further support this association. First, PTSD patients' self-reports on a wheel of fortune-type task suggest lower subjective experience of expectancy and satisfaction with receipt of monetary reward. Second, patients with PTSD displayed less motivational effort to extend the viewing time of attractive female faces on a computer key pressing procedure. Third, the abovementioned wheel of fortune-type game administered during functional magnetic resonance imaging was associated with smaller activations of the brain reward/motivational regions in the PTSD group. Further research is needed to unravel the primary versus secondary nature of the observed deficits as well as the place of the reward-enhancing behavioral and pharmacotherapeutic interventions within the PTSD treatment armamentarium.

Chapter V- This chapter summarises current findings on the processing and representation of rewards in PTSD. Behavioural studies using money and attractive faces as rewards point to changes particularly in the evaluation and the motivational impact of rewards. In accordance with these findings, neuroimaging results showed decreased responsiveness in the nucleus accumbens and mesial prefrontal cortex – central structures in the reward pathway – during the processing of gain feedback. Related to these results, reward function has also been demonstrated to be affected by depression, a disorder that is often co-morbid with PTSD, and stress exposure. The common ground of these findings and their implications are discussed to inspire further research on this new topic.

Chapter VI- Extant research suggests genetic and biological factors shape individuals' responses to trauma and play a role in determining who develops posttraumatic stress disorder (PTSD). However, the role of genetic and biological risk factors in the etiology of PTSD has been largely neglected by many trauma researchers. The aim of this chapter is to review current evidence for the role of genetic and biological risk factors in PTSD etiology in a way that is broadly accessible to the field of traumatic stress. In order to accomplish this aim, we first operationalize our definition of risk factors and discuss the specific challenges to identifying biological risk factors for PTSD. Next, we review the evidence for genetic risk factors for PTSD. Then, the authors review the evidence for other biological risk factors, with a specific emphasis on neurocognitive factors. Finally, the authors conclude by reflecting on future research needed to better understand the role of genetic and biological risk factors in PTSD etiology.

Chapter VII- Traumatic events during early life may affect the neural systems associated with emotional regulation, depending on exposure period, and lead to altered sensitivity to stress later in life. The authors elucidated whether early postnatal stress alters the neural circuits associated with fear conditioning and fear extinction during the postadolescent period in rats using electrophysiological approaches combined with behavioral analysis. Aversive stimuli during the second postnatal week (during postnatal day (PND) 14 to PND 18; 2W-FS) caused a marked reduction in fear-related freezing behavior during exposure to contextual fear conditioning, presumably because of disturbance of the retention of fear memory, an effect that was attributable to changes in synaptic transmission in the hippocampal CA1 field. Aversive stress exposure during the third postnatal week (during PND 21 to PND 25; 3W-FS group) impaired the extinction of context-dependent fear memory evaluated by the expression of freezing behavior, an effect that was associated with dysfunction of the hippocampal-medial prefrontal cortex neural pathway. Thus, the altered behavior observed in adulthood may likely be the result of neurodevelopmental perturbations elicited by early life stress, suggesting that a "critical period" may exist for neural circuits involved in emotional expression. Furthermore, we found that extinction deficits, i.e., behavioral and synaptic dysfunction observed in the 3W-FS group was abolished by repeated treatment with the partial *N*-methyl-D-aspartate receptor agonist D-cycloserine. Considering the similarities between exposure psychotherapy in humans and fear extinction training in animals, both pharmacological interventions and psychotherapy targeting the neural circuits underlying fear memory may have significant clinical value in the treatment of anxiety-related disorders, such as posttraumatic stress disorder (PTSD). A better understanding of the neural mechanisms of extinction deficits induced by early postnatal stress may help clarify the pathophysiology of PTSD in which fear extinction is compromised.

Chapter VIII- Many men suffering from stress, including post-traumatic stress disorder (PTSD), report sexual dysfunction, which is traditionally treated *via* psychological counseling. Recently, we identified a gastrin-releasing peptide (GRP) system in the lumbar spinal cord that is a primary mediator for male reproductive functions. To ask whether acute severe stress could alter the male-specific GRP system, we used a single-prolonged stress (SPS), a putative rat model for PTSD. Exposure of male rats to SPS decreases both the local content and axonal distribution of GRP in the lower lumbar spinal cord and results in an attenuation of penile reflexes *in vivo*. Remarkably, pharmacological stimulation of GRP receptors restores penile reflexes in SPS-exposed males, and induces spontaneous ejaculation in a dose-dependent manner. Furthermore, although the level of plasma testosterone is normal one week after SPS exposure, there is a significant decrease in the expression of androgen receptor protein in this spinal center, which may make the spinal center less responsive to circulating androgens. The authors conclude that the spinal GRP system for male reproductive function is vulnerable to stress, which may provide new insights into clinical treatment of erectile dysfunction triggered by stress and psychiatric disorders.

Chapter IX- Sleep disturbances are common complaints among patients with posttraumatic stress disorder (PTSD), as well as intrusive recollection, avoidance of disturbing stimuli, difficulty in concentration, and depressive symptoms. Patients with PTSD frequently report difficulty falling or staying asleep and recurrent nightmares related to the traumatic events. These sleep disturbances are part of the diagnostic criteria for PTSD and are included in both the re-experiencing (Criterion B) and increased arousal (Criterion D) symptom cluster in the DSM-IV-TR. Despite these sleep complaints, polysomnographic studies have not provided consistent evidence of sleep disturbances, especially regarding non-rapid eye movement (REM) sleep. However, the characteristic sleep disturbances, in particular the repetitive, stereotypical nightmares may be relatively specific to PTSD. The presence of nightmares in PTSD has been hypothesized as a dysfunction of REM sleep mechanisms and investigators have reported various types of REM sleep abnormalities. Treatment studies also have shown that antidepressants, such as nefazodone, trazodone, fluvoxamine, and image rehearsal therapy are beneficial in the treatment of PTSD-related sleep disturbances as well as core symptoms of this anxiety disorder; however, these sleep disturbances may persist and impair the quality of life in PTSD patients for several decades. It is important, therefore, to understand the characteristics of sleep disturbances in PTSD and these treatments. In this review, the authors discuss the clinical features and objective findings of sleep disturbances in PTSD and these treatments based on the published literature, including our findings.

Chapter X- Since Posttraumatic Stress Disorder (PTSD) was recognized as an anxiety disorder in childhood, a limited number of studies investigated the neuroendocrine parameters underlying the clinical manifestations of the disease. In contrast to the majority of adult studies, where low or normal peripheral cortisol concentrations are reported in individuals with PTSD, cortisol is frequently found high in pediatric PTSD patients, whereas high catecholaminergic activity has been reported in both age groups. Developmental stage of the individual, and time since the trauma occurred, seem to be crucial variables in explaining discrepancies between pediatric and adult studies.

Longitudinal psychobiological studies in children, after motor vehicle accidents, have shown that initially elevated cortisol concentrations gradually normalize while norepinephrine concentrations become greater, revealing a progressive interrelation between these two main hormones of the stress system that may underlie the maintenance of the disorder. Thus, the authors hypothesized, that low cortisol levels, together with the high catecholamines in adults with chronic PTSD may reflect a late event in the natural history of the disorder, months or years after the trauma.

Chapter XI- Posttraumatic stress disorder (PTSD) has been related to a number of functional and structural brain alterations. This chapter aims to review the current understanding of the role of functional brain asymmetries for negative (PTSD) and positive posttraumatic outcomes (posttraumatic growth) and for the successful treatment of PTSD. Theoretically, this chapter is based on theories of brain asymmetry and emotion which propose that asymmetries of brain activation are related to certain features of human emotion (e.g. valence, approach or withdrawal tendencies, arousal). These theories are supported by asymmetries found in the alpha frequency of the electroencephalogram (EEG). Recent findings of EEG alpha asymmetry in patients with PTSD are reviewed and implications for its theoretical understanding, treatment and further research are discussed.

Chapter XII- The diagnosis of PTSD dates back to the 80s and is characterized by the observation that as a consequence of a series of traumatic events, it is possible to develop a psychiatric syndrome. Recent epidemiological data show a correlation between stress-related disorders, as in the case of PTSD, and substance abuse disorders. A large majority of drug dependent patients have been reported experiencing one or more PTSD criterion of traumatic experiences. For a better understanding of this comorbidity, it is important to consider shared neurobiological risk factors and relevant stressful life events as possible vulnerability conditions to both PTSD and substance use disorders. The main reason given by subjects with PTSD for the abuse of psychotropic drugs, including illicit drugs, prescription drug and alcohol seems to be obtaining temporary relief and improve impulse control. Yet, the onset of drugs use may also anticipate the development of PTSD syndrome, determining an increased individual susceptibility to PTSD. Common neurotransmitter abnormalities i.e. low plasma levels of serotonin, elevated plasma levels of catecholamines, a similar derangement of the HPA axis, a generalized over-activity of the sympathetic nervous system and morphological and functional alterations of the central nervous system have been reviewed in this chapter.

Chapter XIII- Posttraumatic Stress Disorder and Alzheimer's Disease have many similarities and only some differences. The main similarities are that both disorders have as risk factor a severe traumatic event (100% in PTSD and 80% in AD), the prevalence F/M ratio of PTSD is the same as in AD, (2:1), stress hormones and other neurotransmitters are involved in both diseases, the areas of brain which are involved in both disorders are about the same (hippocampus, amygdale, prefrontal cortex, anterior cingulated gyrus) and finally the areas that are activated in fMRI are about the same in both conditions. Perhaps there are some differences in genes, but this area needs more research and the age of onset is the point which is different in vast majority of patients with AD, because the mean age in AD patients is 75 years old. This is the reason which leads to different therapeutic strategies.

Chapter XIV- Recent reported findings indicate that stress experiences are related to psychological and neurobiological processes that may have lasting consequences and significantly influence brain functions. Stress also causes atypical memory consolidation leading to traumatic memory that in patients with PTSD has certain specific psychological and neurophysiological characteristics. Together these findings suggest that specific memory impairment and intrusive autobiographical memories of stressful experiences, and increased sensitivity to stress that occur in PTSD are related to dissociation. Cognitive and emotional dysregulation related to traumatic stress and dissociation is also likely linked to deficits in inhibitory functions and increased limbic excitability that may lead to temporo-limbic seizure-like activity. These findings strongly suggest that stress-activated limbic kindling may be involved in the pathogenesis of PTSD and may explain efficacy of antiepileptic drugs in the treatment of PTSD. Future research needs to determine whether patients with PTSD who have heightened level of psychosensory seizure-like symptoms are more likely to benefit from treatment interventions that use antiepileptic drugs.

Chapter XV- This chapter reviews human and animal literature on the psychological and biological consequences of maintaining and intense, prolonged fear response. This type of psychological trauma alters neural regulatory mechanisms in the CNS that renders fear- and anxiety-related circuitry hyper-responsive such that individual overreact to relatively mild stressors. The metabolic challenge associated with maintaining a state of hyperactivity engages a compensatory signaling mechanism involving adenosine. The authors review evidence from the learned helplessness rat model of PTSD and comorbid depression that enhanced brain adenosine signaling mediates the transition from a state of anxiety and agitation to a depression-like state termed conservation-withdrawal. The authors also present new evidence that the conservation-withdrawal aspects of helplessness are mediated in the striatum where adenosine activates specific A_{2A} receptors to uncouple dopamine's motivational signal from ongoing behavior.

Chapter XVI- The psychological consequences of trauma exposure are well recognized throughout history. The study of post-traumatic stress disorder (PTSD) dates back more than 100 years. Before 1980, post-traumatic syndromes were recognized by different names, including railway spine, shell shock, traumatic (war) neurosis, concentration-camp syndrome, and rape-trauma syndrome. The symptoms described in these syndromes overlap considerably with what we now recognize as PTSD. Individuals exposed to severe traumatic stress often develop comorbid PTSD and major depressive disorder (MDD). The author developed a new concept suggesting that that some or all individuals diagnosed with comorbid PTSD and MDD have a separate psychobiological condition that can be termed "post-traumatic mood disorder" (PTMD). This concept is supported by a significant number of clinical, translational and other studies suggesting that patients suffering from comorbid PTSD and MDD are different clinically and biologically from individuals with PTSD alone or MDD alone. Patients with comorbid PTSD and MDD are characterized by greater impairment compared to individuals with PTSD alone or MDD alone. Neurobiological evidence supporting the concept of PTMD includes the findings related to dopaminergic, serotonergic, and hypothalamic-pituitary-adrenal axis function and other observations. It is crucial to develop interventions to prevent and to treat PTMD and measures to prevent and to treat suicidal ideation and behavior in individuals with PTMD.

Chapter XVII- Post-traumatic stress disorder (PTSD) is a common psychiatric illness with significant morbidity and impact on function. Several classes of pharmacological agents, including antidepressants, anticonvulsants, atypical antipsychotics, and several novel agents, have been investigated for the treatment of PTSD. This chapter reviews the published literature with an emphasis on randomized controlled trials. The evidence for benefit as monotherapy appears to be strongest for antidepressants, specifically SSRIs and SNRIs, with some benefit from trycyclics and MAOIs. Promising preliminary results have been reported for mirtazapine, prazosin and some anticonvulsant such as lamotrigine. Atypical antipsychotics seem to be most useful as adjunctive agents. Limited efficacy data exists on the efficacy of the combination of medication and psychotherapy but this strategy would likely benefit refractory illness. The general dearth of large, well-designed RCTs on treatments for PTSD must be noted.

Chapter XVIII- Aberrant glutamatergic transmission has been implicated in the pathophysiology of anxiety disorders. Alterations in NMDA receptor (NMDAR) expression, function, and binding are induced in animal models of fear and stress. NMDARs have a central role in mediating the acquisition and extinction of learned fear responses. Consequently, this has led to investigations examining the therapeutic utility of compounds targeting the NMDAR for the treatment of anxiety syndromes. High-affinity open channel blockers of the NMDAR promote anxiolytic effect in preclinical rodent studies; however the psychotomimetic effects induced by these antagonists have hampered their applicability in clinical settings. Instead, low-affinity NMDAR inhibitors and agonists of the NMDAR glycine binding site have demonstrated therapeutic potential without significant adverse effects. Furthermore, NMDAR glycine site activators have been shown to be effective in accelerating fear extinction in animal studies, and recent clinical trials have supported their efficacy in enhancing exposure therapy in patient with anxiety syndromes. Thus, an improved understanding of the molecular mechanisms underlying stress, fear, and extinction has led to the discovery of highly promising treatment interventions involving the NMDAR.

Chapter XIX- Extinction is the learned inhibition of the expression of previously acquired memories. It is particularly useful for the treatment of fear motivated disorders, particularly PTSD. It relies on the interlinked electrophysiological and molecular activity of the ventromedial prefrontal cortex, the hippocampus, the lateral or basolateral amygdala and the entorhinal cortex. For some memories it may involve in addition other brain sites, like the insular cortex.

Chapter XX-The newly discovered endocannabinoid system (ECS) and the ongoing research into its role in the central nervous system (CNS) has important ramifications in the understanding and treatment of posttraumatic stress disorder (PTSD). While the connection with this system and PTSD is in its early stages, and only one small clinical study involving patients with PTSD being treated with an endocannabinoid agonist has been published to date, [1] nonetheless emerging investigations into the endocannabinoid system are promising and warrant a dialogue and speculation about the potential benefits of endocannabinoid ligands in the treatment of PTSD. This chapter will consider three areas related to the endocannabinoid system and how it may be connected to the field of posttraumatic stress disorder. First will be an overview of the recent discovery of endocannabinoid system. Next, research studies pointing to potentially interesting relationships between cannabinoids, endocannabinoids and PTSD will be discussed. Finally, clinical issues and questions arising out of actual field experience using nabilone, a synthetic cannabinoid, will be presented. This latter area will be of interest to clinicians who may consider using a synthetic cannabinoid for the first time with a PTSD patient. It will be based on the above mentioned clinical trial which used a synthetic cannabinoid in managing treatment resistant PTSD nightmares which was carried out by this author

Chapter XXI- The experience of surgery can produce both physical and psychological stress. Awareness during anesthesia is a major concern of patients preoperatively. While its occurrence is relatively rare, awareness increases the likelihood of negative post-surgical outcomes, including Posttraumatic Stress Disorder (PTSD). By using an animal model of PTSD called stress-enhanced fear learning (SEFL), the authors have studied how inhaled anesthetics suppress PTSD-like symptomotology. Sufficient concentrations of the inhaled anesthetics isoflurane and nitrous oxide suppress PTSD-like symptomotology, as measured by SEFL. However, their mechanism of action is unclear. The authors have examined the role of *N*-methyl-D-aspartate receptors (NMDA-R) previously, and the authors examine the role of γ-aminobutyric acid type A receptors (GABAA-R) in this chapter. The results collectively show that neither receptor system alone suppresses SEFL, suggesting that multiple receptor systems underlie the ability of inhaled anesthetics to suppress SEFL.

Chapter XXII- This chapter reports on the symposium, "Translational neuroscience may inform treatment development for posttraumatic stress disorder" (Chair – Leo Sher, M.D., Co-Chair – Alexander Neumeister, M.D.) that was held on June 28, 2009 during the 9th World Congress of Biological Psychiatry (June 28-July 2, 2009, Palace of Congresses (Le Palais des Congrès de Paris), Paris, France). Compelling animal and human data indicate that stress compromises neurotrophic functions producing neural atrophy and reducing neuroplasticity resulting in dysfunction in brain circuitry associated with PTSD. This problem has been identified as the most important priority for studies of the neurobiology of PTSD. This symposium was the first translational interdisciplinary attack on this problem, from molecular neuroscience to treatment and targets major gaps in our understanding of the neurobiology, neural circuitry, phenomenology, and treatment of PTSD. Four research works were presented at the symposium: "Early postnatal stress: Impact on neural circuits and relevance to psychophysiology" by Machiko Matsumoto (Japan), "Long term alterations in brain and neurobiology in PTSD" by Eric Vermetten (Netherlands), "Mechanisms of resilience: Lessons from PET imaging" by Alexander Neumeister (USA) and "Neurobiology of suicidal behavior in PTSD and implications for suicide prevention" by Leo Sher (USA).

In: Neurobiology of Post-Traumatic Stress Disorder
Editors: L. Sher and A. Vilens, pp. 1-20
ISBN: 978-1-61470-593-2

Chapter I

Structural and Functional Neuroimaging in PTSD: A Neurobiological Update

***Jeffrey Chee Leong Looi*[1,2], *Marco Pagani*[3], *Davide Nardo*[4], *Beverley Raphael*[1], *Lars-Olof Wahlund*[2]**

[1]Australian National University Medical School, Canberra, Australia
[2]Karolinska Institute, Huddinge, Stockholm, Sweden;
[3]Karolinska Hospital, Solna, Stockholm, Sweden & Institute of Cognitive Sciences and Technologies, Rome, Italy
[4]Sapienza University of Rome, Italy

Abstract

Advances in neuroimaging permit *in vivo* investigation of the structural and functional neurobiology of post-traumatic stress disorder (PTSD). Recent reviews of structural and functional neuroimaging in PTSD are discussed together with recent studies. The significance of the findings is limited by the disparate groupings of PTSD trauma exposures, making generalization difficult, and this, combined with the heterogeneity inherent in the existing diagnostic criteria, renders the often conflicting data confusing. There has been an emphasis on studies of cortical structures, without adequate consideration of neural circuit models implicating other brain structures. There has been a move towards investigating subsets of symptoms, such as hyperarousal and dissociation, particularly in functional imaging studies. Overall, neuroimaging research in PTSD faces challenges for the future. The key improvements lie in: the definition and classification of trauma exposures, as they are not all equivalent in nature or indeed comparable; the study of restricted phenotypic subsets of symptoms or endophenotypes, such as hyperarousal and dissociation which may have different neural substrates; adequate study designs for power and control of confounders; and more focused research based upon targeted investigation of neural networks putatively involved in the processing and re-experiencing of trauma.

Cotidie damnatur qui semper timet.
"The man who is in fear is constantly condemned."
Publilius Syrus (c. 100 B.C.E.)

Introduction

This chapter aims to provide a critical update on the status of structural and functional neuroimaging studies of PTSD as a means of studying the *in vivo* neurobiology of the disorder. As such, we have focused upon studies from the last decade and we are summarizing from number of comprehensive reviews that have been published recently on structural [1,2] and functional imaging [3,4]. The overall impression in this field is that there is much of interest, but equally much more to be done, both theoretically and methodologically, to advance our understanding of the *in vivo* structural neurobiology of PTSD.

We begin with an introduction to some of the theoretical and methodological issues in neuroimaging of PTSD. The first section of the chapter summarizes the findings from structural (MRI) and functional magnetic resonance imaging (fMRI). The second section summarizes radioisotope functional neuroimaging, Single Photon Emission Computed Tomography (SPECT) and Positron Emission Tomography (PET). Finally, we summarize our findings and draw conclusions.

Theoretical and Methodological Issues

Characterization of PTSD

A key issue in such correlative brain imaging studies is the selection of study groups, and in particular, diagnostic criteria. Most studies have accepted diagnoses for PTSD, as defined via the DSM-III-R or DSM-IV [5,6]. Surprisingly, within reviews of these studies, there has been little discussion of the diagnostic criteria and their suitability for characterizing the disorder from a neuropathologic and hence, *in vivo* neuroimaging viewpoint.

Whilst different types of stressors, exposures and impacts may result in different presentations, the field has apparently assumed that these processes drive the same neurobiological processes. In the main, exploration to date has focused upon the cerebral cortex, in what has been termed a cortico-centric model of cognition [7]. This differs from the understanding of cognitive processing as being segregated into multiple, parallel distributed networks. For example, visual components of experience, arguably including trauma, are processed via different circuitry than auditory and cognitive components being routed via different pathways through the basal ganglia [7]. However, recent structural and functional studies have begun to investigate networks of distributed cognition, particularly in relation to subsets of symptoms in PTSD.

Methodological Issues

The key methodological issues we have identified in this review are:

1. Insufficient characterization of the trauma exposure, which is significant and may partly explain conflicting results from non-equivalent exposures or reactions to those being compared;
2. Evolution of diagnostic criteria;
3. Impact of number of stressors and exposures;
4. Specificity of symptom constellations (such as hyperarousal and dissociation) within these criteria;
5. Severity of symptomatology;
6. Studies of inadequate statistical power to examine group differences;
7. Variation in imaging methodology, even within modality and technological innovation with the passage of time, compromising comparisons from different periods;
8. Lack of accounts of treatment effects;
9. Variation in candidate neuroanatomical structures and networks being studied.

The trauma exposures for criteria A1 and A2 of the accepted DSM-IV diagnostic criteria for PTSD remain controversial because of the inherent subjectivity of these experiences. That is, what is perceived as constituting actual or threatened death or serious injury (A1); and intense fear, horror or helplessness (A2); is subjective and difficult to standardize. In children, this assessment is even more problematic.

Finally, the issue of the nature of the threat: to life (e.g. serious assault) or integrity (e.g. sexual abuse), may vary in experience or meaning. Similarly, the evolution of the vague diagnostic criterion A from: "The person has experienced an event that is outside the range of usual human experience and that would be markedly distressing to almost anyone," in DSM-III-R adds further complexity to neuroimaging studies; yet this variation in definition has received little comment in existing reviews.

Indeed, as is common in psychiatric research, the disorder is assumed to exist in some reified form based upon diagnostic criteria. For example, Karl et al. [1] structural MRI review includes: 6 studies on combat related trauma, 4 of childhood sexual abuse, with a variety of other exposures from domestic violence, experiences of police officers and cancer survivors to be considered among the 21 groups included in the meta-analysis. Can such diverse experiences be considered equivalent? In relation to neuroimaging studies, are they necessarily processed by the same neural circuitry? This would seem unlikely.

We have not been the only group to propose that grouping all PTSD subjects together is questionable. For example, Lanius et al., [3], has observed that doing so may interfere with our understanding of post-traumatic symptomatology. In raising this criticism, we accept existing cohorts have varied exposures and indeed have had to resort to combining subjects with different exposures to increase the power of such studies.

However, a more logical, parsimonious, approach would be to classify types of trauma exposure, document subsequent psychological sequelae and investigate structural correlates of the same. The logical comparison, which has laudably been adopted in most extant studies, is to compare those with the same exposure who developed PTSD with those who did not.

One prism which could be used to focus such research may to investigate more restricted subsets of phenotypic symptomatology and behaviour, known as *endophenotypes*. Thus, clusters of symptoms, such as dissociative phenomena, traumatic re-experiencing, and hyperarousal might usefully be investigated for their structural and functional correlates, rather than the whole ensemble of PTSD diagnostic criteria. In addition, then we might usefully investigate the correlation of the severity of symptomatology with structural and functional imaging.

When we examine studies purporting to identify structural or functional changes in the brain of persons suffering from PTSD, we find that the numbers of subjects in the studies are relatively small. We accept that enrolling subjects with specific traumatic experience is very difficult, and acknowledge the further challenge of finding those exposed to the same experience that did not develop PTSD. Nonetheless, studies of less than fifteen patients, by our calculations, would require a very large effect size to show significance and, perhaps, overestimate random effects as difference, especially if significance is set at $p<0.05$.

Whilst acknowledging the informative work of previous reviewers, we also question the utility of comparing research using different diagnostic criteria and different image modalities for meta-analysis. Depending on the segmentation methods used, there may be considerable variation in delineation of brain structures or identification of activity due to different MRI acquisition and SPECT/PET protocols. Each study can contribute a piece to solving the puzzle through how they fit together, but perhaps not all pieces are equally useful.

This brings us to a general problem of brain imaging, and that is of adequate hypotheses, for adequate designs, preceding acquisition of images. It seems to us there has been a dearth of coordinated approaches to examining the neural circuitry implicated in PTSD. By this, we mean specific investigations based upon some putative circuitry or functional neuroanatomy of PTSD [8]. Indeed Lanius et al. [3] has proposed decomposing the PTSD symptoms into two clusters which could be examined separately, from the perspective of different exposure at least, functional imaging: hyperarousal and dissociative patterns and endophenotypes, which may have different neural circuit substrates. These patterns were previously proposed by Bremner [9]. To this could be added traumatic re-experiencing, in the form of flashbacks, studied in recent functional imaging symptom-provocation studies. That is, focus should be upon endophenotypes of PTSD, such as dissociation, re-experiencing and hyperarousal, which may have their basis in particular brain circuitry. To this, we may add that memory disorders, such as re-experiencing may also occur. For example, some of the authors of this chapter have investigated the structure of the caudate in frontostriatal circuits postulated to be involved in re-experiencing of trauma-related experience [10].

Having raised these methodological issues, we now turn to the summaries of recent research, acknowledging and drawing upon previous reviews [1-4] for this update.

Structural MRI Studies

Recent Reviews

Karl et al., [1] – Meta-analysis of Structural Brain Abnormalities in PTSD

Karl et al., [1] reviewed 21 MRI studies from 1990 to 2005 via meta-analysis. Their inclusion criteria were relatively rigorous, so their review is a good base to build upon for an update. Therefore, we have summarized directly from their review in this section. However, we add the reservations, noted above, that not all trauma exposures are equivalent, although symptoms resulting may be similar; and the variation in diagnostic criterion A from DSM-III-R to DSM-IV remains problematic for those classified as having PTSD. The issues addressed in inclusion criteria for the meta-analysis were: inclusion of DSM-III-R or DSM-IV PTSD compared to non-PTSD exposed or controls; sufficient methodological specification (sample, MRI methodology); and sufficient statistical testing.

However, the issue of endophenotypic characterization, in contrast to the use of the extant criteria was not discussed, nor was the issue of heterogeneity of trauma exposures explored. Thus, the studies summarize findings from disparate trauma exposures and populations, using a broad classification of PTSD which may obscure differences due to differential pathophysiology of the endophenotypes of hyperarousal, re-experiencing or dissociation.

Hippocampal Volumetric Studies

This was the most frequently investigated area. The meta-analysis found that PTSD subjects had smaller bilateral hippocampal volumes than healthy controls, whilst meta-analysis of exposed non-PTSD persons compared to those with PTSD was only significant for left hippocampal volume [1]. Interesting methodological factors were moderators: MRI acquisition protocols and delineation of hippocampal boundaries (mostly manual in these studies). Only the studies using high resolution MRI and correction for whole brain volume showed differences for non-PTSD exposed compared to those exposed with PTSD in the left hippocampus; whilst the findings for PTSD versus controls were robust even with low resolution and no brain volume correction. There were indications that studies using the alveus/fornix and superior colliculi/basilar artery to trace the hippocampus coronally found significantly smaller bilateral hippocampal volume in PTSD compared to controls, in contrast to studies using mamillary bodies/fornix as a boundary. Other significance moderators were age, severity and gender, as might be expected. Overall, the effect size for the hippocampal volume reductions in PTSD were generally between 0.0 and 0.5, with the percentage of reduction generally between 0-20%, which is quite a large reduction in volume. Examining the breakdown by trauma type, there seems a trend toward combat trauma resulting in larger changes, followed by childhood sexual abuse, accidents and interpersonal violence. Also, trauma severity may be proportional to the extent of the violence directed towards an individual.

Amygdala

Left amygdala volume was found to be smaller in PTSD subjects compared to healthy controls (n=320) and trauma-exposed non-PTSD subjects (n=213) with an effect size between 0.1-0.4. However, Karl et al., [1] noted there was a significant sample heterogeneity in all these groups.

Corpus Callosum

The corpus callosum (n=221) was primarily investigated in children and adolescents and was found to be smaller in those with PTSD, with an effect size circa 0.4.

Caudate

No significant group differences were found in caudate nucleus volumes in a heterogenous group of children, adolescents and adults (n=281).

Frontal Lobes

The tracing or delineation of the frontal lobes remains technically challenging due to variations in the morphology of these regions across individuals. In children and adolescents (n=223), the prefrontal cortex was significantly smaller in PTSD compared to healthy controls and non-PTSD exposed, albeit with a small effect size c. 0.3-0.4.

Cavum Septum Pellucidum

In a very small meta-analysis of a heterogenous grouping of adults and children (n=63), no significant differences were found across PTSD compared to healthy controls and non-PTSD exposed.

Jackowski et al. [2] – Pediatric Neurostructural Findings in PTSD

In a brief non-systematic review, Jackowski et al. [2] have summarized some of the relevant findings in pediatric populations. They proposed that the failure to find significant volumetric alterations in the hippocampus in children and adolescents with PTSD may represent neuro-maturational differences. That is, early life stresses may have an impact that is not evident until later development. Whilst they also emphasize the corpus callosum differences identified in the Karl et al. [1] meta-analysis, the authors of the pediatric review speculate these changes may be due to either developmental differences and/or selective vulnerability of callosal regions in critical periods of development to trauma. Generalized reduction in brain volume, reduced frontal lobe, prefrontal cortex, right temporal lobe, and white matter volumes has been found in pediatric populations with PTSD. In addition, bilateral ventricular enlargement and increased prefrontal CSF volume has been found. These finding appear isolated to children, and the authors suggested generalized atrophy or impaired brain development as mechanisms.

Recent Structural MRI Studies since the Karl et al. [1] Meta-analysis

There have been interesting developments, some of which have used new MRI acquisition and analysis methods.

Anterior Cingulate Cortex

Using diffusion-tensor analysis, Abe et al. [11] studied 9 victims of the Tokyo subway Sarin attack with PTSD compared with 16 controls, and found a significant increase in fractional anisotropy in the left anterior cingulum underlying the region in which the same group found a volumetric reduction in the left anterior cingulate grey matter. This structural finding supports the hypothesis of neural circuit dysfunction, with the suggestion that there is increased connectivity and thus, hyper-responsivity of the circuits traversing this region. Why this should result in volumetric reduction of the adjacent grey matter is more problematic. Researchers from the same group have used voxel-based morphometry and found that there were significant grey matter reductions in right hippocampal cortex, pregenual anterior cingulate cortex and bilateral insular regions in combat-exposed twins with (n=18) and without PTSD (n=23) [12]. In another study on the same cohort, using manual segmentation of the cerebellum, no significant differences in the volume of the cerebellar vermis were found [13].

Other Cortical Regions

Two recent studies have found changes, albeit in opposite directions, in pediatric PTSD and in adults with PTSD. Geuze et al. [14] compared 25 male veterans with PTSD with 25 without PTSD, matched for age, year and region of deployment; using a cortical thickness algorithm to find that veterans with PTSD had reduced thickness in bilateral superior and middle temporal gyri, and left superior temporal and inferior frontal gyri. However, cortical thickness was not correlated with memory measures. The pediatric study of Carrion et al. [15] used semi-automated segmentation and voxel-based morphometry in 24 children with PTSD compared to 24 age and gender matched controls. The pediatric study found increased gray matter volume in the bilateral inferior and superior prefrontal cortex, with reduction in the pons and posterior cerebellar vermis. The VBM study showed increased gray matter density in ventral prefrontal cortex. Again, the pediatric studies are intriguing, with evidence of some possible developmental hypertrophic differences which may precede volume loss in adult life.

Caudate Nucleus

Looi et al. [10] assessed 36 subjects aged less than 65 recruited from transport workers in Stockholm who reported having been unintentionally responsible for a person-under-the-train accident or among employees having experienced an assault in their work (1999–2001) between 3 months and 6 years before MRI scanning. After adjustment for the covariates (age, sex, intracranial volume, years since trauma, and number of trauma episodes), there was a significant difference in raw right caudate nucleus volume between subjects with PTSD (n=19) compared with those without PTSD (n=17). Volume of the left caudate nucleus was not significantly different between the PTSD and no PTSD groups. The right caudate volume in the PTSD group was 9% greater compared with the no PTSD group, superimposed upon a baseline asymmetry, indicating possible involvement of frontostriatal circuitry in PTSD.

Conclusion: Structural MRI Findings

In general, structural MRI findings have been restricted to the ensemble of the entire PTSD criteria, rather than endophenotypic subsets, and this, together with trauma exposure heterogeneity, renders the findings somewhat questionable in their specificity for PTSD. In addition, there has been a focus upon the cortex as structural basis for PTSD, but it is not necessarily clear how specific the cortical changes are to PTSD. Indeed hippocampal volumetric loss has been reported in depression, dementia and schizophrenia, and may merely be a general stress response, rather than specific to PTSD. Similarly, the findings for the amygdala, corpus callosum and developmental abnormalities might represent some generalized, non-specific structural basis for vulnerability to stress. The more specific findings relating to key structures in cortico-subcortical circuits relevant to processing trauma, such as the anterior cingulate and caudate, both components of frontostriatal circuits potentially serving as a basis for endophenotypes, are step forward to more focused, parsimonious research.

Functional MRI Studies

We have synthesized findings here derived from the reviews of Lanius et al. [3] and Francati et al. [4], and followed this with an update of recent studies since these reviews.

The group led by Lanius has reported a number of fMRI findings in PTSD [3,4]. Using trauma scripts in PTSD subjects with sexual assault/abuse and motor vehicle accidents (n=9) paired with controls, decreased activation was found in the thalamus, anterior cingulate and medial frontal gyrus in PTSD subjects with hyperarousal and reliving/re-experiencing [16]. They subsequently presented findings with increased activation in the medial frontal gyrus and anterior cingulate gyrus, and activation in superior and middle temporal gyri, inferior frontal gyrus, parietal and occipital lobes in 7 subjects with dissociative PTSD responses compared to 9 controls [17]. There have been interesting findings supportive of circuit dysfunction, with enhanced patterns of right hemisphere connectivity in the posterior cingulate gyrus, parietal and occipital lobes and in the caudate nucleus [18]. Other clusters of symptoms may have different patterns, as the same group found activation in bodily state networks in PTSD subjects with dissociative symptoms such as the superior temporal gyrus, right middle frontal gyrus, right insula and cuneus, and left parietal lobe [19].

Other groups have investigated different stimuli [4]. Exposure to emotional face representations has been used in studies of PTSD in comparison with healthy controls. Rauch et al. [20] presented such images to combat veterans with PTSD, finding enhanced amygdala responsivity; whilst Shin et al. [21] also found enhanced amygdala responsivity with decreased medial prefrontal cortex activity in civilians with PTSD. Emotionally-valenced words have also been used as stimuli [4], with findings of amygdalar activation [22] and underactivity of rostral anterior cingulate [23].

fMRI Findings since Lanius et al. [3]; Francati et al. [4]

Recent fMRI studies have explored either the dissociative or hyperarousal clusters of symptomatology, indicating more focussed and parsimonious approaches are being adopted to PTSD.

Another study from the Lanius [24] group compared 17 women with chronic early life trauma exposure or those who were adults when exposed to trauma with PTSD to 15 controls using fMRI. They demonstrated increased connectivity in healthy controls of the posterior cingulate/precuneus to the right amygdala, hippocampus and parahippocampal gyrus in comparison to those with PTSD. The authors concluded that the default process of spontaneous self-reflection and monitoring was dysfunctional in PTSD and thus, may result in dissociative symptomatology.

Felmingham et al. [25] investigated the function of arousal networks in 11 subjects with PTSD and 11 matched controls on an oddball task that involved responding to non-trauma-related auditory target tones embedded in low frequency background tones and also measured skin conductance response. There was evidence of greater activity in PTSD subjects compared to controls on background tasks in greater dorsal anterior cingulate, supramarginal gyrus and hippocampus. However, when skin conductance responses were evident, showing arousal, the PTSD subjects showed reduced ventral anterior cingulate activity, which due to implication in fear-extinction processes, suggests faulty fear-extinction may occur in PTSD.

Conclusion: Functional MRI Findings

In general, functional MRI findings have shown more specific findings relating to key structures in circuits relevant to processing trauma, focusing on components of neural circuits potentially serving as a basis for endophenotypes, such as hyperarousal and dissociation. For example, dissociation may arise from faulty self-reflection and monitoring processes, whilst hyperarousal may result from faulty fear-extinction. In addition, memory disturbances, such as traumatic re-experiencing, which have been investigated in some functional imaging studies, may also serve as a useful endophenotype. Thus, hyperarousal, dissociation and re-experiencing may arise from dysfunction in different neural circuits, resulting in specific patterns.

Functional Imaging: SPECT and PET

Over the past decade, it has become increasingly clear that a number of specific brain structures play a key role in the generation of PTSD symptoms. These structures are involved in emotional, memory, linguistic, visuospatial and motor processing, all of which might be affected in PTSD. Proposed neural correlates of psychotherapy have also been investigated, revealing neurobiological effects on brain functions.

Functional studies by single photon emission computed tomography (SPECT) and positron emission tomography (PET) can now reliably detect changes in cerebral blood flow (CBF) and metabolism patterns, suggesting a specific role for each of the brain areas in various components of emotional processing.

Functional mapping of brain by PET is considered to be superior to SPECT due to: the applicability of positron-emitting “bioisotopes”, i.e. oxygen-15 and fluorine-18, allowing investigation of both brain CBF and metabolism; the relatively good spatial resolution (4-7 mm); and the ability to perform quantitative assessment [26]. Nevertheless, its availability is restricted by the heavy on-site capital investment required to build an in-house cyclotron facility, even if currently several centres perform the studies with commercially available positron emitters. Due to this, the number of PET cameras remains relatively small, as compared to the numerous SPECT facilities, making SPECT the most commonly used technique especially in routine examinations. SPECT imaging allows measurements of CBF also reflecting neuronal activity. The maximum spatial resolution of present high-resolution SPECT cameras is of the order of 7-9 mm. SPECT is a qualitative method, reflecting the distribution of radiotracers, and whose results are usually expressed in semi-quantitative values. However, SPECT has the advantage compared to PET, allowing for experiments to be performed in ideal psychological conditions in quiet environments outside the camera gantry. In fact, due to the characteristics of SPECT, image acquisition can be commenced some hours after administration whilst still depicting the radiopharmaceutical brain distribution at the moment of injection.

Both SPECT and PET have shown in PTSD regional cerebral blood flow (rCBF) changes during trauma recall. In this respect symptom provocation paradigms are an extremely useful and powerful way of delineating the functional anatomy of traumatic memories that characterize PTSD. Alterations of local activations at specific tasks indicate dysfunctions of neural processing. Autobiographical trauma-script exposure [27-29] or audio and visual trauma-related stimuli [30,31] are a valid approach to elicit rCBF changes in PTSD and the improvements in both technical capabilities and methodology have made neuroimaging studies particularly suitable in investigating *in vivo* the neurobiology of emotions.

Following symptom provocation in PTSD patients, rCBF was found to be either increased [32-34], or decreased [33- 35] within hippocampus, amygdala, medial pre-frontal cortex (mPFC), including orbito-frontal (OFC) and anterior cingulate (ACC) cortices, cerebellum, as well as temporal and posterior cingulate cortices. These initial findings resulted in the formulation of neurocircuitry models of PTSD that emphasize the functional relationship between a triad of brain structures: amygdala, ventro-medial prefrontal cortex (vmPFC) and hippocampus [32,36-41]. The findings from studies performed by some of our authors [42] suggest an involvement of these brain structures in PTSD, and are consistent with the results of previous functional neuroimaging studies based on symptom provocation paradigms investigating rCBF response to different stressors.

However, other structures have been shown to be involved in PTSD such as the thalamus [43-45], insula [41,46,47], Broca's area [48-50], ACC [8,30,51,52] parietal lobes [53-56].

Pagani et al. [27] investigated 20 transport workers in Stockholm who reported having been unintentionally responsible for a person-under-the-train accident or among employees having experienced an assault in their work (1999–2001) between 3 months and 6 years before MRI scanning, symptomatic for PTSD and compared their CBF to a group of 27 drivers undergoing the same trauma without developing PTSD. Multivariate analysis identified the right hemisphere as the brain region showing statistical differences between the CBF in the two groups.

The right hemisphere integrates sensory modalities, processes nonverbal emotional communication and seems to be strongly connected with the amygdala, from which receives incoming stimuli of fear and hostility, and the relative regulation of autonomic and hormonal responses. It is also responsible for intrusive emotional memory component of PTSD and autobiographical memories [57].

The left hemisphere generates symbolic representations by categorizing stimuli and personal experiences into novel images and symbols and labels perceptions.

From the analysis of the regions involved in PTSD it appears that a large part of the limbic system (hippocampus, fusiform gyrus, amygdala and nucleus accumbens, lentiform nucleus, anterior cingulate and orbitofrontal cortex) plays a key role in the regulation of emotions and storage and retrieval of memories. The reported dysfunctions in governing cortices and medial prefrontal cortex are likely related to alterations in planning, execution, inhibition of responses and extinction of fear resulting in motor responses, and in effects on peripheral, sympathetic and cortisol systems. Furthermore in PTSD patients changes in prefrontal cortex, essential for encoding and retrieving verbal memories, might result in the difficulties in cognitively restructuring traumatic experiences.

Considering the key role of this structure in PTSD, it is worth noting the amygdala has been suggested to be mainly activated during recognition and induction of emotions by visual stimuli rather that during the reaction to recalled stimuli [58]. Furthermore due to the finest spatial resolution of PET and to the small dimensions of amygdala the detected functional blood flow and metabolism changes in this structure are reliably described only in PET studies.

In general, in PTSD patients, limbic hyperactivation is paralleled by higher cortical hypofunction [59] resulting in a lack of inhibition of reaction to fear from amygdala and lack of adequate attenuation of peripheral sympathetic and hormonal responses to stress. It has been proposed that such hyperperfusion and hyperactivity of limbic and paralimbic regions is related to stress-induced long-term potentiation between amygdala and periacqueducal grey through N-methyl-D-aspartate (NMDA)-mediated pathway, once a sufficient amount of glutamate is released following stressing events [60]. In PTSD, psychological hyperactivation of an overlearned survival response with intrusive phenomena representing an active reworking of trauma memories at cognitive level has also been hypothesized by [61].

The critical involvement of limbic system has been postulated to be connected to the fear-related stimuli and to the emotional responsiveness to the retrieved traumatic experience elicited by symptom provocation. It is worth noting that chronic PTSD is often associated with long-term pharmacological treatment and/or alcohol and substance abuse further affecting brain structures and function and in case confounding the results of the investigations.

Following the predominance of PTSD studies performed in the past on veterans and abused women and children, there has been a recent focus on traumas more related to the daily life and to societal problems. In this respect, the choice of the control group is a critical step of the global analysis in neuroimaging. In PTSD subjects that were exposed to the same trauma as patients without suffering any symptom are likely the best ones to be compared to the study group since the CBF distribution differences found in group comparisons would be completely related to the disorder itself and not confounded by possible group and trauma discrepancies nor biased by other variables.

Lindauer et al. [28] studied a group of 30 traumatized police officers with and without PTSD that underwent SPECT scanning after neutral and trauma scripts. Within the PTSD group, rCBF was higher in reaction to the neutral scripts in comparison with the trauma scripts in the ventral part of the insula. The insula has an important role in body representation, pain experience and subjective emotional experience. Alterations in left insular activity (increased or decreased) have been found in healthy volunteers with induced anxiety [62], subjects with a variety of anxiety disorders [32] and affective disorder patients with anxiety symptoms [63]. In a PET study, correlating regional cerebral perfusion with flashback intensity (a novel and interesting approach to severity of presentation) following an auditory script of their traumatic event [64], the investigators scanned a group of chronic PTSD subjects with a history of flashbacks. They detected the involvement of a number of brain areas, including bilateral insula and somatosensory regions, but no gradient of dysfunction based upon severity of symptoms.

As observed in a study performed on a group of civilian trauma survivors by Bonne and co-workers [65], rCBF distribution was higher in subjects with PTSD than in trauma-exposed healthy subjects in several brain regions, including a region extending from inferior parietal cortex (BA 40) to the post-central gyrus (BAs 1, 2, and 3).

Other neuroimaging studies highlighted the role of the parietal lobes, which are thought to be implicated in visuo-spatial information processing [53-55] as well as in temporal and spatial orientation [56].

A previous study by Liberzon et al. [56] with combat veterans who developed PTSD reported increased subcortical activity centred over the thalamus, as compared to cortical regions, while presenting combat sound stimuli. Conversely, a group of PTSD patients in resting state showed decreased rCBF in the right thalamus as compared to a group of healthy subjects [66]. Thalamus has been reported to play an important role in temporal binding, especially in the unity of perception [43-45] and in the generation of dissociative symptoms in PTSD, characterized by a significant alteration of the relative distribution of CBF between cortex and thalamus during flashbacks [67].

Posterior cingulate, parietal and motor cortices are functionally related to prefrontal cortex and mediate cognitive functions of the visuospatial processing, critical in extreme stress situations. The lentiform nucleus, consisting of the putamen and the globus pallidus, was found by Lindauer et al. [28] to show a decreased 99mTc-HMPAO uptake distribution possibly resulting in reduced motor activities and fewer active coping reactions [68].

The cerebellar involvement in PTSD has been described in two studies [65,69]. Cerebellum is involved in the autonomic regulation of both cardiovascular and skin responses and emotional behaviour as well as in motor conditioning response. Patients with cerebellar dysfunction have a diminished heart rate reduction upon fear conditioning [70] and an abnormal startle response in PTSD [71]. This might result in abnormal delay in habituation, increased heart rate variability, exaggerated startle response and sleep abnormalities [65]. Importantly, hese findings suggest that the methodological use of cerebellum as reference region for data normalization in PTSD might be inappropriate. The involvement in PTSD of regions processing learning, preparation and execution of motor tasks results in an increased basal level of anxiety and arousal through a continuous preparatory motor activation [65].

Nucleus accumbens, also known as ventral striatum, was found to be activated in two studies [29,33]. It receives inputs from amygdala, prefrontal cortex and subiculum and projects to basal forebrain, diencephalon and rostral brainstem. Its functional-anatomic location makes it involved in creating appropriate emotional responses to incoming stimuli and sensitive to aversive incoming signals. In turn, it modulates the activity in downstream brain structures affecting in PTSD the motor aspect of adaptive responding resulting in exaggerated reactions to incoming information (i.e. hyperirritability, hypersensitivity and hyperexcitability).

Among the four SPECT studies we are aware of in which the effect of therapy on PTSD has been investigated by neuroimaging (see Table 4) two dealt with eye movement desensitization and reprocessing [EMDR] [27,72]; one with selective serotonin reuptake inhibitor [SSRI] [73]; and one with cognitive restructuring therapy [74].

EMDR is an eclectic therapy method utilizing, among other techniques, relaxation exercises, safe place exercises, cognitive restructuring, future projections and imaging of the trauma combined with positive sensory stimulation. Cognitive restructuring was an exposure-based therapy consisting of an introductory session, an anamnesis session and three restructuring sessions. The SSRI citalopram was administered for 8 weeks at 20 mg/day (first two weeks) and 40 mg/day (last 6 weeks). The therapy delivered in three studies [27,73,74] lasted 8 weeks only and was a little longer in the Lansing et al. [72] study (on average about 10 weeks).

The traumatic events were very heterogeneous across the EMDR treatment studies, but the designs of the studies included an individual trauma script during SPECT in three out of four investigations [27,73,74]. As for the effect of therapies on CBF, both increases and decreases were reported. Probably due to the large variety of traumatic events and therapies the patients underwent to, the regional changes were distributed in almost all cortex. In general in PTSD and in anxiety disorders deactivations are considered to be related to symptom relief and to the normalization of reciprocal associative circuitry regulation and of hyperreactivity in emotional and memory disturbances. Activations were seen as an effect of therapies in improving negative symptoms as depression [27,72] or as a better inhibition of feed-back processes related to amygdala activity [74].

Conclusion: Functional Imaging –SPECT and PET

Overall, the findings for functional imaging with SPECT and PET demonstrate some interesting data, but there are limitations in common with structural MRI and to a lesser extent, functional MRI. The majority of the studies have used symptom-provocation models to elicit trauma-related responses to map regional cerebral blood flow changes. These studies show functional activation in hippocampus, amygdala and ventromedial prefrontal cortex to symptom provocation. Apart from such presumably direct activation, it has been observed there may also be relative hyperactivity of the limbic system, with cortical hypofunction. A number of other discrete brain regions have been implicated, both cortical and subcortical, including the cerebellum.

However, the emphasis on symptom provocation in SPECT and PET to date would appear to restrict these findings to the domain of hyperarousal PTSD symptomatology, in contrast to the fMRI findings for possible dissociative symptomatology noted above.

Treatment studies have been much more heterogenous in relation to exposures and the interpretation of increased or decreased activity in the context of symptomatic improvement remains problematic.

Overall Conclusion

In this brief update on the structural and functional imaging of PTSD we have summarized recent reviews and studies from the last few years, and conclude with some suggestions for future research. The increasing sophistication of studies is evident in: the extensive exploration of different symptom subsets that may constitute endophenotypes, such as hyperarousal, re-experiencing and dissociation; consideration of moderating factors; inclusion of larger numbers of subjects and controls; and, exploration of the role of both cortical and subcortical neural networks implicated in processing trauma.

More emphasis should be placed on classifying the trauma exposures into different subgroups, such as combat trauma, sexual abuse, physical abuse, domestic violence etc. These exposures are neither similar in nature, homogenous or functionally equivalent in their impact on persons, and thus should not be conflated. Perhaps separate studies of different types of trauma should be specifically designed, based upon the expected symptom profile from previous studies.

We suggest that rather than studying the entire cluster of symptoms as a whole, a more productive approach to investigation of PTSD may be to concentrate on subsets of clinical features that may serve as endophenotypes, predicated upon disturbances in neural circuitry, such as hyperarousal, re-experiencing and dissociation. The rationale for this has been established in our summary of imaging, in that different brain structures are implicated in the neural circuitry of different trauma responses, and such studies need to include subcortical as well as cortical structures. This will also help avoid the problem of evolving diagnostic criteria, and perhaps specific symptom cluster-related diagnostic criteria should be developed for researchers.

Further improvements will result from grouping homogeneous trauma exposures (by appropriate selection) and recruiting larger numbers of subjects, to permit studies with greater power to demonstrate differences. Controls exposed to the same trauma, but without persistent symptoms, are necessary to act as satisfactory controls, and it is even more preferable to have healthy controls without trauma exposure for further comparison.

Combinations of structural and functional imaging methods on larger, well-characterized samples will help advance understanding of structural-functional correlations. Longitudinal studies, involving follow-up of symptomatology and outcomes will add to our understanding of structural-functional correlations in the neuroimaging of PTSD.

Dulce bellum inexpertis.
"War is sweet to those who have not experienced it."
Pindaros (c. 522-443 B.C.E.)

Author Contributions

JCLL wrote the first draft of the sections on structural and functional MRI, introduction, conclusions; and was the lead editor for the chapter. MP and DN wrote the first draft of the sections on functional imaging. All authors contributed to the editing of the chapter.

Acknowledgments

JCLL partly self-funded travel and accommodation costs for some of the research described in this chapter, with additional support from the Canberra Hospital Specialists Private Practice Trust Fund and A.C.T. Health.

References

[1] Karl A, Schaefer M, Malta LS, Dorfel D, Rohleder N, Werner A: A meta-analysis of structural brain abnormalities in PTSD. *Neurosci Biobehav Rev* 2006; 30:1004-1031.

[2] Jackowski AP, de Araujo CM, de Lacerda ALT, Mari J deJ, Kaufman J: Neurostructural imaging findings in children with post-traumatic stress disorder: brief review. *Psychiatry and Clinical Neurosciences* 2009; 63:1-8.

[3] Lanius RA, Bluhm R, Lanius U, Pain C: A review of neuroimaging studies in PTSD: heterogeneity of response to symptom provocation. *J Psychiatr Res* 2006; 40:720-729.

[4] Francati V, Vermetten E, Bremner JD: Functional neuroimaging studies in posttraumatic stress disorder: review and current findings. *Depression and Anxiety* 2007; 24:202-208.

[5] American Psychiatric Association: *Diagnostic and statistical manual of mental disorders* DSM-III-R (3rd ed.) Washington D.C.: American Psychiatric Association, 1987.

[6] American Psychiatric Association: *Diagnostic and statistical manual of mental disorders* DSM-IV-TR (4th ed.) Washington D.C.: American Psychiatric Association, 2000.

[7] Koziol LF, Budding DE: *Subcortical structures and cognition.* New York, NY, Springer, 2009.

[8] Liberzon I & Sripada CS: The functional neuroanatomy of PTSD: A critical review. *Prog Brain Res* 2008; 167:151-69.

[9] Bremner JD: Acute and chronic responses to psychological trauma: where do we go from here? *Am J Psychiatry* 1999; 156:349-351.

[10] Looi JCL, Maller J, Lindberg O, Liberg B, Hogberg G, Pagani M, Botes L, Engman E-L, Zhang Y, Svensson L, Wahlund L-O: Caudate volumes in public transportation workers exposed to trauma in the Stockholm train system. *Psychiatry Research Neuroimaging* 2009; 171:138-143.

[11] Abe O, Yamasue H, Kasai K, Yamada H, Aoki S, Iwanami A, Ohtani T, Matsumi Y, Kato N, Ohtomo K: Voxel-based diffusion tensor analysis reveals aberrant anterior cingulum integrity in posttraumatic stress disorder due to terrorism. *Psychiatry Research Neuroimaging* 2006; 146:231-242.

[12] Kasai K, Yamasue H, Gilbertson MW, Shenton ME, Rauch SL, Pitman RK: Evidence for acquired pregenual anterior cingulate gray matter loss from a twin study of combat-related posttraumatic stress disorder. *Biol Psychiatry* 2008; 63:550-556.

[13] Levitt JJ, Chen QC, May FS, Gilbertson MW, Shenton ME, Pitman RK: Volume of the cerebellar vermis in monozygotic twins discordant for combat exposure: lack of relationship to post-traumatic stress disorder. *Psychiatry Research Neuroimaging* 2006; 148:143-149.

[14] Geuze E, Westenberg HGM, Heinecke A, de Kloet C.S., Goebel R, Vermetten E: Thinner prefrontal cortex in veterans with posttraumatic stress disorder. *Neuroimage* 2008; 41:675-681.

[15] Carrion VG, Weems CF, Watson C, Eliez S, Menon V, Reiss AL. Converging evidence for abnormalities of the prefrontal cortex and evaluation of midsagittal structures in pediatric posttraumatic stress disorder: an MRI study. *Psychiatry Research Neuroimaging 2009*; 172:226-234.

[16] Lanius RA, Williamson PC, Densmore M, Boksman K, Gupta M, Neufeld RW, et al. Neural correlates of traumatic memories in posttraumatic stress disorder: a functional MRI investigation. *American Journal of Psychiatry* 2001;158:1920–2.

[17] Lanius RA, Williamson PC, Boksman K, Densmore M, Gupta M, Neufeld RWJ, et al. Brain activation during script-driven imagery induced dissociative responses in PTSD: a functional MRI investigation. *Biological Psychiatry* 2002;52:305–11.

[18] Lanius RA, Williamson PC, Densmore M, Boksman K, Neufeld RW, Gati JS, et al. The nature of traumatic memories: a 4.0 tesla fMRI functional connectivity analysis. *American Journal of Psychiatry* 2004;161(1):36–44.

[19] Lanius RA, Williamson PC, Bluhm RL, Densmore M, Boksman K, Neufeld RW, Gati JS, Menon RS. Functional connectivity of dissociative responses in posttraumatic stress disorder: a functional magnetic resonance imaging investigation. *Biological Psychiatry 2005*;57(8):873–84.

[20] Rauch SL, Whalen PJ, Shin LM, McInerney SC, Macklin ML, Lasko NB, Orr SP, Pitman RK: Exaggerated amygdala response to masked facial stimuli in posttraumatic stress disorder: A functional MRI study. *Biol Psychiatry* 2000; 47:769–776.

[21] Shin LM, Wright CI, Cannistraro PA, Wedig MM, McMullin K, Martis B, Macklin ML, Lasko NB, Cavanagh SR, Krangel TS, Orr SP, Pitman RK, Whalen PJ, Rauch SL: A functional magnetic resonance imaging study of amygdala and medial prefrontal cortex responses to overtly presented fearful faces in posttraumatic stress disorder. *Arch Gen Psychiatry* 2005; 62:273–281.

[22] Protopopescu X, Pan H, Tuescher O, Cloitre M, Goldstein M, Engelien W, Epstein J, Yang Y, Gorman J, LeDoux J, Silbersweig D, Stern E: Differential time courses and specificity of amygdala activity in posttraumatic stress disorder subjects and normal control subjects. *Biol Psychiatry* 2005; 57:464–473

[23] Shin LM, Whalen PJ, Pitman RK, Bush G, Macklin ML, Lasko NB, Orr SP, McInerney SC, Rauch SL: An fMRI study of anterior cingulate function in posttraumatic stress disorder. *Biol Psychiatry* 2001; 50:932–942.

[24] Bluhm RL, Williamson PC, Osuch EA, Frewen PA, Stevens TK, Boksman K, Neufeld RWJ, Theberge J, Lanius RA: Alterations in default network connectivity in posttraumatic stress disorder related to early-life trauma. *J Psychiatry Neurosci* 2009; 34:187-194.

[25] Felmingham KL, Williams LM, Kemp AH, Rennie C, Gordon E, Bryant RA: *Anterior cingulate activity to salient stimuli is modulated by autonomic arousal in posttraumatic stress disorder.* Psychiatry Research: Neuroimaging doi:10.1016 /jpscychresns .2008.12.005

[26] Pagani M, Stone-Elander S, Larsson SA: Alternative PET with non-conventional positron emitters: effects of their physical properties on image quality and their potential clinical application. *Eur J Nucl Med* 1997; 24: 1301-1328.

[27] Pagani M, Högberg G, Salmaso D, Tärnell B, Sanchez-Crespo A, Soares J, Åberg-Wistedt A, Jacobsson H, Hällström T, Larsson SA, Sundin Ö: Regional cerebral blood flow during auditory recall in 47 subjects exposed to assaultive and non-assaultive

trauma and developing or not posttraumatic stress disorder. *Eur Arch of Psychiatry Clin Neurosci* 2005; 255:359-365.

[28] Lindauer RJ, Booij J, Habraken JB, Uylings HB, Olff M, Carlier IV, den Heeten GJ, van Eck-Smit BL, Gersons BP: Cerebral blood flow changes during script-driven imagery in police officers with posttraumatic stress disorder. *Biol Psychiatry* 2004; 56:853-61.

[29] Pavic L, Gregurek R, Petrovic R, Petrović D, Varda R, Vukusić H, Crnković-Marković S: Alterations in brain activation in posttraumatic stress disorder patients with severe hyperarousal symptoms and impulsive aggressiveness. *Eur Arch Psychiatry Clin Neurosci* 2003; 253:80-3.

[30] Liberzon I, Taylor SF, Amdur R, Jung TD, Chamberlain KR, Minoshima S, Koeppe RA, Fig LM: Brain activation in PTSD in response to trauma-related stimuli. *Biol Psychiatry* 1999; 45:817-826.

[31] Zubieta JK, Chinitz JA, Lombardi U, Fig LM, Cameron OG, Liberzon I: Medial frontal cortex involvement in PTSD symptoms:A SPECT study. *J Psychiatry Res* 1999; 33:259–64.

[32] Rauch SL, Savage CR, Alpert NM, Fischman AJ, Jenike MA: The functional neuroanatomy of anxiety: a study of three disorders using positron emission tomography and symptom provocation. *Biol Psychiatry 1997*; 42: 446-52.

[33] Bremner JD, Narayan M, Staib LH , Southwick SM, McGlashan T, Charney DS: Neural correlates of memories of childhood sexual abuse in women with and without posttraumatic stress disorder: *Am J Psychiatry* 1999; 156:1787-95.

[34] Shin LM, McNally RJ, Kosslyn SM, Thompson WL, Rauch SL, Alpert NM, Metzger LJ, Lasko NB, Orr SP, Pitman RK: Regional cerebral blood flow during script-driven imagery in childhood sexual abuse-related PTSD: A PET investigation. *Am J Psychiatry* 1999; 156: 575-84.

[35] Bremner JD: Neuroimaging Studies in Post-traumatic Stress Disorder. *Curr Psychiatry Reports* 2002; 4:254–63.

[36] Rauch SL and Shin LM: Functional neuroimaging studies in posttraumatic stress disorder. *Ann NY Acad Sci* 1997; 821: 83-98.

[37] Shin LM, Shin PS, Heckers S, Krangel TS, Macklin ML, Orr SP, Lasko N, Segal E, Makris N, Richert K, Levering J, Schacter DL, Alpert NM, Fischman AJ, Pitman RK, Rauch SL: Hippocampal function in posttraumatic stress disorder. *Hippocampus* 2004; 14:292-30.

[38] Gilboa A, Shalev AY, Laor L, Lester H, Louzoun Y, Chisin R, Bonne O: Functional connectivity of the prefrontal cortex and the amygdala in posttraumatic stress disorder. *Biol Psychiatry* 2004; 55:263-72.

[39] Rauch SL, van der Kolk BA, Fisler RE, Alpert NM, Orr SP, Savage CR, Fischman AJ, Jenike MA, Pitman RK: A symptom provocation study of posttraumatic stress disorder using positron emission tomography and script-driven imagery. *Arch Gen Psychiatry.* 1996; 53:380-7.

[40] Pitman RK, Shin LM, Rauch SL: Investigating the pathogenesis of posttraumatic stress disorder with neuroimaging. *J Clin Psychiatry* 2001; 62(Suppl 17):47-54.

[41] Liberzon I, Phan KL: Brain-imaging studies of posttraumatic stress disorder. *CNS Spectr* 2003; 8:641-50.

[42] Pagani M, Högberg G, Salmaso D, Nardo D, Sundin O, Jonsson C, Soares J, Aberg-Wistedt A, Jacobsson H, Larsson SA, Hällström T: Effects of EMDR psychotherapy on 99mTc-HMPAO distribution in occupation-related post-traumatic stress disorder. *Nucl Med Commun* 2007; 28:757-65.

[43] Lanius RA, Williamson PC, Densmore M, Boksman K, Gupta MA, Neufeld RW, Gati JS, Menon RS: Neural correlates of traumatic memories in posttraumatic stress disorder: a functional MRI investigation. *Am J Psychiatry* 2001; 158:1920-2.

[44] Lanius RA, Williamson PC, Hopper J, Densmore M, Boksman K, Gupta MA, Neufeld RW, Gati JS, Menon RS: Recall of emotional states in posttraumatic stress disorder: an fMRI investigation. *Biol Psychiatry* 2003; 53:204-10.

[45] Llinas RR, Leznik E, Urbano FJ: Temporal binding via cortical coincidence detection of specific and non-specific thalamocortical inputs: a voltage-dependent dye-imaging study in mouse brain slices. *Proc Nat Acad Sci* USA 2002; 99:449-454.

[46] Liberzon I, Britton JC, Phan KL: Neural correlates of traumatic recall in posttraumatic stress disorder. *Stress* 2003; 6:151-6.

[47] Liberzon I, Martis B: Neuroimaging studies of emotional responses in PTSD. *Ann N Y Acad Sci* 2006; 1071:87-109.

[48] Petersen SE, Fox PT, Posner MI, Mintun M, Raichle ME: Positron emission tomography studies of the processing of single words. *J Cogn Neurosci* 1989; 1:153-170.

[49] Paulesu E, Frith CD, Frackowiak RSJ: The neural correlates of the verbal component of working memory. *Nature* 1993; 362:342-5.

[50] McGuire PK, Silbersweig DA, Murray RM, David AS, Frackowiak RSJ, Frith CD: Functional anatomy of inner speech and auditory verbal imagery. *Psychol Med* 1996; 26:29-38.

[51] Yamasue H, Kasai K, Iwanami A, Ohtani T, Yamada H, Abe O, Kuroki N, Fukuda R, Tochigi M, Furukawa S, Sadamatsu M, Sasaki T, Aoki S, Ohtomo K, Asukai N, Kato N: Voxel-based analysis of MRI reveals anterior cingulate gray-matter volume reduction in posttraumatic stress disorder due to terrorism. *Proc Nat Acad Sci* 2003; 100:9039-43.

[52] Bush G, Luu P and Posner MI: Cognitive and emotional influences in anterior cingulate cortex. *Trends Cogn Sci* 2000; 4:215–222.

[53] Coull JT, Frith CD: Differential activation of right superior parietal cortex and intraparietal sulcus by spatial and nonspatial attention. *Neuroimage* 1998; 8:176-187.

[54] Bushara KO, Weeks RA, Ishii K, Catalan MJ, Tian B, Rauschecker JP, Hallett M: Modality-specific frontal and parietal areas for auditory and visual spatial localization in humans. *Nat Neurosci* 1999; 2:759-766.

[55] Harris IM, Egan GF, Sonkkila C, Tochon-Danguy HJ, Paxinos G, Watson JD: Selective right parietal lobe activation during mental rotation: a parametric PET study. *Brain* 2000; 123:65-73.

[56] Liberzon I, Taylor SF, Fig LM, Koeppe RA: Alteration of corticothalamic perfusion ratios during a PTSD flashback. *Depression and Anxiety* 1996-1997; 4:146-150.

[57] Freeman TW, Roca V: Gun use, attitudes toward violence, and aggression among combat veterans with chronic posttraumatic stress disorder. J Nerv Ment Dis 2001; 189:317-20.

[58] Damasio AR, Grabowski TJ, Bechara A, Damasio H, Ponto LL, Parvizi J, Hichwa RD: Subcortical and cortical brain activity during the feeling of self-generated emotions. *Nature Neurosci* 2000; 3:1049-56.

[59] Garcia R, Vouimba RM, Baudry M, Thompson RF: The amygdala modulates prefrontal cortex activity relative to conditioned fear. *Nature* 1999; 402(6759):294-6.

[60] Hull AM: Neuroimaging findings in post-traumatic stress disorder. *Br J Psychiatry* 2002; 181:102-10.

[61] Silove D: Is posttraumatic stress disorder an overlearned survival response? An evolutionary-learning hypothesis. *Psychiatry* 1998; 61:181-90.

[62] Chua P, Krams M, Toni I, Passingham R, Dolan R: A functional anatomy of anticipatory anxiety. *Neuroimage* 1999; 9:563-71.

[63] Osuch EA, Ketter TA, Kimbrell TA, George MS, Benson BE, Willis MW, Herscovitch P, Post RM: Regional cerebral metabolism associated with anxiety symptoms in affective disorder patients. *Biol Psychiatry* 2000; 48:1020-3.

[64] Osuch EA, Benson B, Geraci M, Podell D, Herscovitch P, McCann UD, Post RM: Regional cerebral blood flow correlated with flashback intensity in patients with posttraumatic stress disorder. *Biol Psychiatry* 2001; 50:246-53.

[65] Bonne O, Gilboa A, Louzoun Y, Brandes D, Yona I, Lester H, Barkai G, Freedman N, Chisin R, Shalev AY: Resting regional cerebral perfusion in recent posttraumatic stress disorder. *Biol Psychiatry* 2003; 54:1077-86.

[66] Kim SJ, Lyoo IK, Lee YS, Kim J, Sim ME, Bae SJ, Kim HJ, Lee JY, Jeong DU: Decreased cerebral blood flow of thalamus in PTSD patients as a strategy to reduce re-experience symptoms. *Acta Psychiatr Scand.* 2007;116: 145-53.

[67] Krystal JH, Bennett AL, Bremner JD, Southwick SM, Charney DS:. Toward a cognitive neuroscience of dissociation and altered memory functions in post-traumatic stress disorders. In: D.S. Charney, M.J. Friedman and A.Y. Deutch, Editors, *Neurobiological and Clinical Consequences of Stress: From Normal Adaptation to PTSD,* Lippincott-Raven Publishers, Philadelphia (1995).

[68] LeDoux JE, Gorman JM: A call to action: overcoming anxiety through active coping. *Am J Psychiatry* 2001; 158:1953-5.

[69] Chung YA, Kim SH, Chung SK, Chae JH, Yang DW, Sohn HS, Jeong J: Alterations in cerebral perfusion in posttraumatic stress disorder patients without re-exposure to accident-related stimuli. *Clin Neurophysiol* 2006; 117:637-42.

[70] Maschke M, Schugens M, Kindsvater K, Drepper J, Kolb FP, Diener HC, Daum I, Timmann D: Fear conditioned changes of heart rate in patients with medial cerebellar lesions. *J Neurol Neurosurg Psychiatry* 2002; 72:116-8.

[71] Cohen H, Benjamin J, Geva AB, Matar MA, Kaplan Z, Kotler M: Autonomic dysregulation in panic disorder and in post-traumatic stress disorder: application of power spectrum analysis of heart rate variability at rest and in response to recollection of trauma or panic attacks. Psychiatry Res 2000; 96:1-13.

[72] Lansing K, Amen DG, Hanks C, Rudy L: High-resolution brain SPECT imaging and eye movement desensitization and reprocessing in police officers with PTSD. *J Neuropsychiatry Clin Neurosci* 2005 Fall; 17:526-32.

[73] Seedat S, Warwick J, van Heerden B, Hugo C, Zungu-Dirwayi N, Van Kradenburg J, Stein DJ: Single photon emission computed tomography in posttraumatic stress disorder before and after treatment with a selective serotonin reuptake inhibitor. *J Affect Disord* 2004; 80:45-53.

[74] Peres JF, Newberg AB, Mercante JP, Simão M, Albuquerque VE, Peres MJ, Nasello AG: Cerebral blood flow changes during retrieval of traumatic memories before and after psychotherapy: a SPECT study. *Psychol Med* 2007; 37:1481-91.

In: Neurobiology of Post-Traumatic Stress Disorder
Editors: L. Sher and A. Vilens, pp. 21-30

ISBN: 978-1-61470-593-2

Chapter II

Magnetic Resonance Imaging (MRI) Studies in PTSD

Noriyuki Kitayama
Kitayama Clinic, Tokyo, Japan; Showa University School of Medicine, Tokyo, Japan

Abstract

Magnetic resonance imaging (MRI) studies have demonstrated changes in brain morphology in PTSD, including smaller volume of the hippocampus, anterior cingulate cortex and morphologic alterations in the corpus callosum. This paper reviews the recent progress of structural and functional neuroimaging studies in PTSD. MRI volumetric studies have shown smaller hippocampal volume in men and women with chronic PTSD from early abuse or combat; however, these results have not been replicated in children with PTSD. Neuroimaging studies demonstrated functional and morphologic abnormalities in ACC of the patients with PTSD. Brain functional studies using positron emission tomography (PET) or functional magnetic resonance imaging (fMRI) showed decreased function in ACC. In addition, morphometric MRI studies demonstrated smaller volume of the ACC in comparison to non-PTSD subjects. Alterations in this brain region may underlie changes in emotion that are a part of PTSD. Some studies of children with PTSD showed smaller corpus callosum area in comparison to healthy subjects. Findings of smaller corpus callosum size in children with PTSD related to maltreatment have led to the hypothesis that maltreatment is associated with a reduction in corpus callosum size.

Introduction

Stress plays a critical role in the development of considerable number of psychiatric diseases. Brain areas responsible for memory (e.g., hippocampus and amygdala) and regulating emotion (e.g., anterior cingulate cortex) play an important role in the stress response. Hormones such as cortisol and adrenaline, which are released during stress, bathe these brain areas and modulate their function, bringing the organism back to a similar state as during prior times of danger, when there was a similar stress-induced outpouring of these stress hormones. In this way, the stress response mobilizes brain systems that are critical for survival [1].

With excessive or repetitive stress, the individual can develop long-term changes in these same brain systems that mediate memory and the stress response. Individuals exposed to repeated stress may develop dysfunction in their stress response systems, and can lose the capacity to properly adapt to new stressors. Stress responses that are useful for short-term survival can be at the expense of long-term function. For instance, excessive levels of cortisol result in a gastric ulcers, osteoporosis, cardiovascular disease, diabetes, and asthma. Stress also impairs the immune system, which increases the risk of developing infections and other medical disorders [1].

Posttraumatic Stress Disorder (PTSD), which is characterized by re-experiencing symptoms, emotional numbing, and increased arousal following exposure to a traumatic event, is a highly prevalent disorder, affecting 8% of the population by some estimates [2]. Understanding the neurobiology of PTSD is important for the development of new treatments for this disorder.

Magnetic resonance imaging (MRI) studies have demonstrated changes in brain morphology in PTSD, including smaller volume of the hippocampus [3], anterior cingulate cortex [4, 5] and morphologic alterations in the corpus callosum [6, 7].

This paper reviews the recent progress of structural and neuroimaging study in PTSD using MRI.

Hippocampus

One area of intensive investigation has been related to the hippocampus, a brain area involved in learning memory, and probably emotional regulation.

Studies in animals suggest that stress results in structural changes in the hippocampus [8, 9]. Psychosocial stress in monkeys [10, 11] and tree shrews [12] resulted in decreased dendritic branching and /or neuronal loss in the CA3 region of the hippocampus. It was demonstrated in various animal species and across studies that direct glucocorticoid exposure resulted in decreased dendritic branching [13, 14, 15], and alterations in synaptic terminal structure [16, 17]. Stress also results in an inhibition of neuronal regeneration [18] within the hippocampus. Mechanisms proposed for these findings have included stress induced elevations in glucocorticoids [19, 20, 21, 22] and/or decreased brain derived neurotrophic factor (BDNF) [23]. These findings are consistent with the hypothesis that stress is associated with hippocampal damage.

Stressors were also associated with deficits in memory function that are mediated by the hippocampus. Stress was associated with both deficits in working memory tasks, as well as an inhibition of long-term potentiation, which is dependent on the N-methyl-D-aspartate (NMDA; excitatory amino acid) receptor (highly concentrated in the hippocampus) felt to represent a model for memory at the molecular level [24, 25].

Table 1. Effect Size (Hedges' *gu*) Estimates for Hippocampal Volume

PTSD vs Healthy Controls	Hedges' *gu*	95% CI (lower, Upper)
Left hippocampus	-0.568*	-0.813, -0.322
Right hippocampus	-0.491*	-0.735, -0.247
PTSD vs Traumatized Controls		
Left hippocampus	-0.732*	-1.18, -0.287
Right hippocampus	-0.560*	-0.998, -0.122

*$p<0.002$

The first report regarding structural changes in the human central nervous system in PTSD showed smaller hippocampal volume as demonstrated with magnetic resonance imaging (MRI) volumetrics [26]. This study involved subjects with a history of Vietnam combat-related traumatic stress and the diagnosis of PTSD and suggested the possibility that exposure to extreme stressors may reduce the volume of the hippocampus in subjects with PTSD.

Since then, MRI volumetric studies have shown smaller hippocampal volume in men and women with chronic PTSD from early abuse or combat; however, these results have not been replicated in children with PTSD [26, 27, 28, 29, 30, 31, 32, 33, 34, 35, 36, 37, 38, 39, 40, 41, 42, 43, 44]. Therefore, we performed a meta-analytic study that combines all of the known studies to better estimate the role of hippocampal structural changes in adult subjects with chronic PTSD. This search identified 9 studies published between 1995 and 2003 [26, 27, 28, 31, 32, 42, 44, 45, 46]. This meta-analysis of published magnetic resonance imaging (MRI) volumetric studies in the field of PTSD showed that adult subjects with chronic PTSD have smaller hippocampal volume in comparison to both healthy subjects and traumatized subjects without PTSD (Table 1). PTSD subjects showed smaller volumes for both the left and the right hippocampus, both for subjects with PTSD related to combat exposure and with childhood abuse, and for both men and women. Smaller hippocampal volume was not seen in children with PTSD, or subjects with new onset PTSD. These studies suggest that smaller hippocampal volume may be related to illness duration or aging. Findings from animal studies in fact show that early life stress may not have an immediate effect on the hippocampus, but may only manifest during the adult phase of development [45].

Several studies also found lower concentrations of N-acetylaspartate (NAA), a marker of neuronal integrity, in the hippocampi of individuals with PTSD [39, 47, 48, 49]. Findings of lower NAA suggest that volumetric changes are associated with changes in neuronal structure.

Anterior Cingulate Cortex

The medial prefrontal cortex, a brain area which makes up a part of the frontal lobe, plays an important role in the negative feedback regulation of hypothalamic-pituitary-adrenal (HPA) activity during physiologic and behavioral stress. In recent animal studies, repeated restraint stress or corticosterone injections induced apical dendritic reorganization by reducing total branch number of pyramidal neurons in the anterior cingulate cortex (ACC) [50, 51, 52]. Furthermore, neuroimaging studies demonstrated functional and morphologic abnormalities in ACC of the patients with PTSD. Brain functional studies using positron emission tomography (PET) or functional magnetic resonance imaging (fMRI) showed decreased function in ACC [53, 54, 55, 56, 57]. These results suggest a possibility that a failure of medial prefrontal/anterior cingulate activation with re-experiencing of the trauma is hypothesized to represent a neural correlate of the failure of extinction seen in PTSD. The brain has the capacity for plasticity in the aftermath of traumatic stress. In addition, morphometric MRI studies demonstrated smaller volume of the ACC [5, 58] in comparison to non-PTSD subjects. However, volumetric studies of ACC in PTSD using MRI have had conflicting findings; one study using the conventional manual tracing method found significant volume reduction in pregenual ACC but not in dorsal ACC (Brodmann's area [BA] 24) [4], and the other using voxel-based analysis found a gray-matter volume reduction in dorsal ACC [58]. Furthermore, there was no difference in ACC volume between acute PTSD and healthy subjects but was found shape differences in a recent study [59]. Kitayama et al examined the effects of early childhood abuse and PTSD on ACC volume using the conventional manual tracing method in morphologic MR images and reported that right dorsal ACC volume in PTSD, including both BA 24 and 32, was significantly smaller than in non-PTSD ([4], Figure 1). On the other hand, there is considerable variability of pattern and asymmetry in human cingulate and paracingulate sulci [60]. In addition, this study does not address whether the changes are stress-related or are related to a pre-existing difference in brain structure. These questions can be addressed in longitudinal studies. Furthermore, these results, interpreted in conjunction with other functional and structural imaging studies, suggest that PTSD is associated with altered function and structure of the ACC. Alterations in this brain region may underlie changes in emotion that are a part of PTSD.

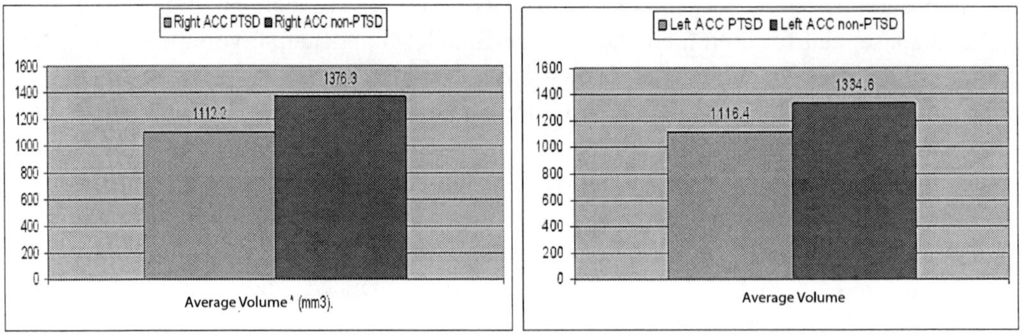

Figure 1. Whole ACC volume (BA24 and BA32) in subjects with abuse-related PTSD and non-PTSD. Whole ACC volume was significantly reduced on the right side in PTSD compared with non-PTSD (* $P < 0.05$).

Corpus Callosum

Some studies of children with PTSD showed smaller corpus callosum area in comparison to healthy subjects [7, 61]. Since neuronal fibers connect the left and right hemispheres of the brain and maintain a somatotopic arrangement [62], morphological abnormalities in specific corpus callosum areas are thought to be associated with abnormalities in corresponding cortical regions [63]. Findings of smaller corpus callosum size in children with PTSD related to maltreatment have led to the hypothesis that maltreatment is associated with a reduction in corpus callosum size. It is also possible that a pre-existent smaller volume of the corpus callosum predisposes the individual to develop PTSD after traumatic events. There is only one study which showed smaller size of the corpus callosum in adults with PTSD; this study was performed in a heterogeneous sample with mixed gender and a variety of traumas [64]. Some studies of normal human subjects have shown gender differences [65] while others have not [66]. Kitayama et al measured the midsagittal area of the corpus callosum as well as sub-regions of the corpus callosum in 9 female subjects with abuse-related PTSD and 9 healthy female subjects [7] As the results, no differences were found in total area of the corpus callosum or in individual sub-regions, but the sub-region / total area ratio was significantly smaller in posterior mid-body in PTSD compared with the healthy subjects (Table 2). These results suggest that relatively smaller areas of the posterior mid-body of the corpus callosum are associated with childhood abuse related PTSD in adults; these findings are consistent with findings in children with abuse-related PTSD. Additionally, these findings suggest that PTSD in both adults and children is associated with alterations in brain structure.

Table 2. Corpus Callosum Measures of Abuse-related PTSD and non-PTSD Controls

	Absolute Area (SD) mm^2		Percentage of Sub-region/ Total area Ratio (SD) ×100 %		*P*	
	PTSD (N = 9)	non-PTSD (N = 9)	PTSD (N = 9)	non-PTSD (N = 9)	Abso -lute	Ratio
Region 1 (Rostrum)	27.6 [12.7)	22.9 (9.7)	4.4 (1.8)	3.7 (1.5)	0.39	0.41
Region 2 (Genu)	125.3 (29.7)	122.6 (13.5)	19.8 (2.9)	20.0 (1.2)	0.80	0.86
Region 3 (Rostral body)	68.4 (18.5)	61.9 (18.5)	10.9 (2.8)	9.9 (2.5)	0.47	0.44
Region 4 (Anterior midbody)	79.6 (15.6)	84.9 (12.6)	12.6 (1.0)	13.8 (1.8)	0.43	0.095
Region 5 (Posterior midbody)	72.4 (19.1)	79.4 (9.6)	11.3 (1.3)	12.9 (1.5)	0.34	0.027 *
Region 6 (Isthmus)	65.0 (12.3)	63.7 (11.6)	10.3 (9.5)	10.4 (1.7)	0.83	0.95
Region 7 Splenium)	195.4 (40.2)	180.6 (26.3)	30.8 (1.7)	29.3 (2.6)	0.37	0.17
Total area	633.6 (117.1)	616.0 (56.4)	-	-	0.69	-

*P < 0.05

Conclusion

This paper reviewed the recent progress of structural and neuroimaging study in PTSD. MRI study is one of the most useful methods to detect morphologic and functional brain changes in PTSD. It is also an important method for the development of new treatments for this disorder.

Acknowledgment

The author is most grateful to J. Douglas Bremner, M.D. for his generous provision of advice and supports for the author's every MRI study and publication regarding PTSD.

References

[1] Kitayama N, Bremner JD: Stress in dementia and depression, in Cognition and Mood Interactions. Edited by Miao-Kun Sun. New York, *Nova Science Publishers, Inc.,* 2005, pp 149-160.

[2] Kessler RC, Sonnega A, Bromet E, Hughes M, Nelson CB: Post-traumatic stress disorder in the national comorbidity survey. *Arch Gen Psychiatry* 1995; 52: 1048-1060.

[3] Kitayama N, Vaccarino V, Kutner M, Weiss P, Bremner JD: Magnetic resonance imaging (MRI) measurement of hippocampal volume in posttraumatic stress disorder: a meta-analysis. *J Affect Disord* 2005; 88: 79-86.

[4] Kitayama N, Quinn S, Bremner JD: Smaller volume of anterior cingulate cortex in abuse-related posttraumatic stress disorder. *J Affect Disord* 2006; 90: 171-174.

[5] Rauch SL, Shin LM, Segal E, Pitman RK, Carson MA, McMullin K, Whalen PJ, Makris N: Selectively reduced regional cortical volumes in post-traumatic stress disorder. *Neuroreport* 2003; 14: 913-916.

[6] De Bellis MD, Keshavan MS, Clark DB, Casey BJ, Giedd JN, Boring AM, Frustaci K, Ryan ND: Developmental traumatology part II: brain development. *Biol Psychiatr* 1999; 45: 1271-1284.

[7] Kitayama N, Brummer M, Hertz L, Quinn S, Kim Y, Bremner JD: Morphologic alterations in the corpus callosum in abuse-related posttraumatic stress disorder: a preliminary study. *J Nerv Ment Dis* 2007; 195:1027-9.

[8] McEwen BS, Angulo J, Cameron H, Chao HM, Daniels D, Gannon MN, Gould E, Mendelson S, Sakai R, Spencer R, Woolley C: Paradoxical effects of adrenal steroids on the brain: protection versus degeneration. *Biol Psychiatry* 1992; 31: 177-199.

[9] Sapolsky RM: Why stress is bad for your brain. *Science* 1996; 2-3: 749-750.

[10] Sapolsky RM, Uno H, Rebert CS, Finch CE: Hippocampal damage associated with prolonged glucocorticoid exposure in primates. *J Neurosci* 1990; 10: 2897-2902.

[11] Uno H, Tarara R, Else JG, Suleman MA, Sapolsky RM: Hippocampal damage associated with prolonged and fatal stress in primates. *J Neurosci* 1989; 9: 1705-1711.

[12] Magarinos AM, McEwen BS, Flugge G, Fuchs E: Chronic psychosocial stress causes apical dendritic atrophy of hippocampal CA3 pyramidal neurons in subordinate tree shrews. *J Neurosci* 1996; 16: 3534-3540.
[13] Packan DR, Sapolsky RM: Glucocorticoid endangerment of the hippocampus: tissue, steroid and receptor specificity. *Neuroendocrinology* 1990; 51: 613-618.
[14] Virgin CE, Taryn PTH, Packan DR, Tombaugh GC, Yang SH, Honer HC, Sapolsky RM: Glucocorticoids inhibit glucose transport and glutamate uptake in hippocampal astrocytes: implications for glucocorticoid neurotoxicity. *J Neurochem* 1991; 57: 1442-1428.
[15] Woolley CS, Gould E, McEwen BS: Exposure to excess glucocorticoids alters dendritic morphology of adult hippocampal pyramidal neurons. *Brain Res* 1990; 531: 225-231.
[16] Magarinos AM, McEwen BS: Stress-induced atrophy of apical dendrites of hippocampal CA3c neurons—involvement of glucocorticoid secretion and excitatory amino acid receptors. *Neuroscience* 1995; 69: 89-98.
[17] Magarinos AM, Verdugo JM, McEwen BS: Chronic stress alters synaptic terminal structure in hippocampus. *Proc Natl Acad Sci USA* 1997; 94: 14002-14008.
[18] Gould E, Tanapet P, McEwen BS, Flugge G, Fuchs E: Proliferation of granule cell precursors in the dentate gyrus of adult monkeys is diminished by stress. *Proc Natl Acad Sci USA.* 1998; 95: 3168-3171.
[19] Armanini MP, Hutchins C, Stein BA, Sapolsky RM: Glucocorticoid endangerment of hippocampal neurons in NMDA-receptor dependent. *Brain Res* 1990; 532: 7-11.
[20] Lawrence MS, Sapolsky RM: Glucocorticoids accelerate ATP loss following metabolic insults in cultured hippocampal neurons. Brain Res 1994; 646: 303-306.
[21] Sapolsky RM: Glucocorticoid toxicity in the hippocampus: synergy with an excitotoxin. *Neuroendocrinology* 1986; 43: 440-446.
[22] Sapolsky RM, Packan DR, Vale WW: Glucocorticoid toxicity in the hippocampus: in vitro demonstration. *Brain Res1988*; 453: 367-371.
[23] Nibuya M, Morinobu S, Duman RS: Regulation of BDNF and trkB mRNA in rat brain by chronic electroconvulsive seizure and antidepressant drug treatments. *J Neurosci* 1995; 15: 7539-47.
[24] Diamond DM, Branch BJ, Fleshner M, Rose GM: Effects of dehydroepiandosterone and stress on hippocampal electrophysiological plasticity. *Ann NY Acad Sci* 1995; 774: 304-307.
[25] Diamond DM, Fleshner M, Ingersoll N, Rose GM: Psychological stress impairs spatial working memory: relevance to electrophysiological studies of hippocampal function. *Behav Neurosci* 1996; 110: 661-72.
[26] Bremner JD, Randall PR, Scott TM, Bronen RA, Delaney RC, Seibyl JP, Southwick SM, McCarthy G, Charney DS, Innis RB: MRI-based measurement of hippocampal volume in patients with combat-related posttraumatic stress disorder. *Am J Psychiatry* 1995; 152: 973-981.
[27] Bremner JD, Randall P, Vermetten E, Staib L, Bronen RA, Mazure C, Capelli S, McCarthy G, Innis RB, Charney DS: Magnetic resonance imaging-based measurement of hippocampal volume in posttraumatic stress disorder related to childhood physical and sexual abuse—a preliminary report. *Biol Psychiatry* 1997; 41: 23-32.

[28] Bremner JD, Vythilingam M, Vermetten E, Southwick SM, McGlashan T, Nazeer A, Kahn S, Vaccarino LV, Soufer R, Garg PK, Chin KN, Staib LH, Duncan JS, Charney DS: MRI and PET study of deficits in hippocampal structure and function in women with childhood sexual abuse and posttraumatic stress disorder. *Am J Psychiatry* 2003; 160: 924-932.

[29] Emdad R, Bonekamp D, Sondergaard HP, Bjorklund T, Agartz I, Ingvar M, Theorell T: Morphometric and psychometric comparisons between non-substance-abusing patients with posttraumatic stress disorder and normal controls. *Psychother Psychosom* 2006; 75: 122-132.

[30] Freeman TW, Cardwell D, Karson CN, Komoroski RA: In vivo proton magnetic resonance spectroscopy of the medial temporal lobes of subjects with combat-related posttraumatic stress disorder. *Magn Reson Med* 1998; 40: 66-71.

[31] Gilbertson MW, Shenton ME, Ciszewski A, Kasai K, Lasko NB, Orr SP, Pitman RK: Smaller hippocampal volume predicts pathologic vulnerability to psychological trauma. *Nature Neuroscience* 2002; 5: 1242-1247.

[32] Gurvits TV, Shenton ME, Hokama H, Ohta H, Lasko NB, Gilbertson MW, Orr SP, Kikinis R, Jolesz FA, McCarley RW, Pitman RK: Magnetic resonance imaging study of hippocampal volume in chronic, combat-related posttraumatic stress disorder. *Biol Psychiatry* 1996; 40: 1091-1099.

[33] Hedges DW, Allen S, Tate DF, Thatcher GW, Miller MJ, Rice SA, Cleavinger HB, Sood S, Bigler ED: Reduced hippocampal volume in alcohol and substnce naive Vietnam combat veterans with posttraumatic stress disorder. *Cogn and Behav Neurol* 2003; 16: 219-224.

[34] Irle E, Lange C, Sachsse U: Reduced size and abnormal asymmetry of parietal cortex in women with borderline personality disorder. *Biol Psychiatry* 2005; 57: 173-182.

[35] Li L, Chen S, Liu J, Zhang J, He Z, Lin X: Magnetic resonance imaging and magnetic resonance spectroscopy study of deficits in hippocampal structure in fire victims with recent-onset posttraumatic stress disorder. *Can J Psychiatry* 2006; 51: 431-437.

[36] Lindauer RJ, Olff M, van Meijel EP, Carlier IV, Gersons BP: Cortisol, learning, memory and attention in relation to smaller hippocampal volume in police officers with posttraumatic stress disorder. *Biol Psychiatry* 2006; 59: 171-177.

[37] Lindauer RJ, Vlieger EJ, Jalink M, Olff M, Carlier IV, Majoie CB, den Heeten GJ, Gersons BP: Smaller hippocampal volume in Dutch police officers with posttraumatic stress disorder. *Biol Psychiatry* 2004; 56: 356-363.

[38] Lindauer RJ, Vlieger EJ, Jalink M, Olff M, Carlier IV, Majoie CB, Den Heeten GJ, Gersons BP: Effects of psychotherapy on hippocampal volume in out-patients with post-traumatic stress disorder: a MRI investigation. P*sychol Med* 2005; 35: 1421-1431.

[39] Brunson KL, Eghbal-Ahmadi M, Bender R, Chen Y and Baram: Long-term, progressive hippocampal cell loss and dysfunction induced by early-life administration of corticotropin-releasing hormone reproduce the effects of early-life stress. *Proc. Natl. Acad. Sci. U.S.A.* 2001; 98: 8856–8861.

[40] Mahmutyazicioğlu K, Konuk N, Ozdemir H, Atasoy N, Atik L, Gündoğdu S: Evaluation of the hippocampus and the anterior cingulated gyrus by proton MR spectroscopy in patients with post-traumatic stress disorder. *Diagn Interv Radiol* 2005; 11: 125-129.

[41] Narayan M, Bremner JD, Kumar A: Neuroanatomical substrates of late-life mental disorders. *J Geriatr Psychiatry Neurol* 1999; 12, 95-106.

[42] Schuff N, Neylan TC, Lenoci MA, Du AT, Weiss DS, Marmar CR, Weiner MW: Decreased hippocampal N-acetylaspartate in the absence of atrophy in posttraumatic stress disorder. *Biol Psychiatry* 2001; 50: 952-959.

[43] Shin LM, Shin PS, Heckers S, Krangel TS, Macklin ML, Orr SP, Lasko N, Segal E, Makris N, Richert K, Levering J, Schacter DL, Alpert NM, Fischman AJ, Pitman RK, Rauch SL: Hippocampal function in posttraumatic stress disorder. *Hippocampus* 2004; 14: 292-300.

[44] Stein MB, Koverola C, Hanna C, Torchia MG, McClarty B: Hippocampal volume in women victimized by childhood sexual abuse. *Psychol Med* 1997; 27: 951-959.

[45] Villarreal G, Hamilton DA, Petropoulos H, Driscoll I, Rowland LM, Griego JA, Kodituwakku PW, Hart BL, Escalona R, Brooks WM: Reduced hippocampal volume and total white matter volume in posttraumatic stress disorder. *Biol Psychiatry* 2002; 52: 119-125.

[46] Notestine CF, Stein MB, Kennedy CM, Archibald SL, Jernigan TL: Brain morphometry in female victims of intimate partner violence with and without posttraumatic stress disorder. *Biol Psychiatry* 2002; 51: 1089-1101.

[47] Mohanakrishnan Menon P, Nasrallah HA, Lyons JA, Scott MF, Liberto V: Single-voxel proton MR spectroscopy of right versus left hippocampi in PTSD. *Psychiatry Res* 2003; 123: 101-108.

[48] Schuff N, Neylan TC, Fox-Bosetti S, Lenoci M, Samuelson KW, Studholme C, Kornak J, Marmar CR, Weiner MW: Abnormal N-acetylaspartate in hippocampus and anterior cingulate in posttraumatic stress disorder. *Psychiatry Res* 2008; 162: 147-157.

[49] Villarreal G, King CY: Brain imaging in posttraumatic stress disorder. *Semin Clin Neuropsychiatry* 2001; 6: 131-145.

[50] Radley JJ, Rocher AB, Miller M, Janssen WG, Liston C, Hof PR, McEwen BS, Morrison JH: Repeated stress induces dendritic spine loss in the rat medial prefrontal cortex. *Cereb Cortex* 2006; 16: 313-320.

[51] Cerqueira JJ, Catania C, Sotiropoulos I, Schubert M, Kalisch R, Almeida OF, Auer DP, Sousa N: Corticosteroid status influences the volume of the rat cingulate cortex - a magnetic resonance imaging study. *J Psychiatr Res* 2005a; 39: 451-60.

[52] Cerqueira JJ, Pego JM, Taipa R, Bessa JM, Almeida OF, Sousa N: Morphological correlates of corticosteroid-induced changes in prefrontal cortex-dependent behaviors. *J Neurosci* 2005b; 25: 7792-800.

[53] Bremner JD, Vermetten E, Vythilingam M, Afzal N, Schmahl C, Elzinga B, Charney DS: Neural correlates of the classic color and emotional stroop in women with abuse-related posttraumatic stress disorder. *Biol Psychiatry* 2004; 55: 612-620.

[54] Bremner JD, Narayan M, Staib LH, Southwick SM, McGlashan T, Charney DS: Neural correlates of memories of childhood sexual abuse in women with and without posttraumatic stress disorder. *Am J Psychiatry* 1999; 156: 1787-1795.

[55] Yang P, Wu MT, Hsu CC, Ker JH: Evidence of early neurobiological alternations in adolescents with posttraumatic stress disorder: a functional MRI study. *Neurosci Lett* 2004; 370: 13-18.

[56] Lanius RA, Williamson PC, Hopper J, Densmore M, Boksman K, Gupta MA, Neufeld RWJ, Gati JS, Menon RS: Recall of emotional states in posttraumatic stress disorder: an fMRI investigation. *Biol Psychiatry* 2003; 53: 204-210.

[57] Shin LM, Whalen PJ, Pitman RK, Bush G, Macklin ML, Lasko NB, Orr SP, McInerney SC, Rauch SL: An fMRI study of anterior cingulate function in posttraumatic stress disorder. *Biol Psychiatry* 2001; 50: 932-942.

[58] Yamasue H, Kasai K, Iwanami A, Ohtani T, Yamada H, Abe O, Kuroki N, Fukuda R, Tochigi M, Furukawa S, Sadamatsu M, Sasaki T, Aoki S, Ohtomo K, Asukai N, Kato N: Voxel-based analysis of MRI reveals anterior cingulate gray-matter volume reduction in posttraumatic stress disorder due to terrorism. *PNAS* 2003; 100: 9039-9043.

[59] Corbo V, Clement MH, Armony JL, Pruessner JC, Brunet A: Size versus shape differences: contrasting voxel-based and volumetric analyses of the anterior cingulate cortex in individuals with acute posttraumatic stress disorder. *Biol Psychiatry* 2005; 58: 119-124.

[60] Paus T, Tomaiuolo F, Otaky N, MacDonald D, Petrides M, Atlas J, Morris R, Evans AC: Human cingulate and paracingulate sulci: pattern, variability, asymmetry, and probabilistic map. *Cereb Cortex* 1996; 6: 207-214.

[61] De Bellis MD, Keshavan MS, Shifflett H, Iyengar S, Beers SR, Hall J, Moritz G: Brain structures in pediatric maltreatment-related posttraumatic stress disorder: a sociodemographically matched study. *Biol Psychiatry* 2002; 52: 1066-1078.

[62] Innocenti GM, Manzoni T, Spidalieri G: Patterns of the somesthetic messages transferred through the corpus callosum. *Exp Brain Res* 1974; 19: 447-466.

[63] Witelson SF: Hand and sex differences in the isthmus and genu of the human corpus callosum: a postmortem morphological study. *Brain* 1989; 112: 799-835.

[64] Villarreal G, Hamilton DA, Graham DP, Driscoll I, Qualls C, Petropoulos H, Brooks WM: Reduced area of the corpus callosum in posttraumatic stress disorder. *Psychiatry Res* 2004; 131: 227-235.

[65] Suganthy J, Raghuram L, Antonisamy B, Vettivel S, Madhavi C, Koshi R: Gender- and age-related differences in the morphology of the corpus callosum. *Clin Anat* 2003; 16: 396-403.

[66] Luders E, Narr KL, Zaidel E, Thompson PM, Toga AW: Gender effects on callosal thickness in scaled and unscaled space. *NeuroReport* 2006; 17: 1103-1106.

In: Neurobiology of Post-Traumatic Stress Disorder
Editors: L. Sher and A. Vilens, pp. 31-37
ISBN: 978-1-61470-593-2

Chapter III

Voxel Based Morphometric Study of Brain Structure in Patients with Posttraumatic Stress Disorder

***Yang Chunlan*[1], *Wu Shuicai*[1], *Bai Yanping*[1], *Hou Cailan*[2] *and Gao Hongjian*[1]**

[1]College of Life Science and Bioengineering, Beijing University of Technology, Beijing, China

[2]Institute of Mental Health, Central South University, Changsha, China

Abstract

Voxel based morphometry (VBM) methods are used to detect the difference in brain structures between the posttraumatic stress disorder (PTSD) patients and the normal controls. Standard VBM method can detect the difference of the gray matter or white matter densities while the optimized VBM method can detect the difference of gray matter or white matter volumes in the whole brain. The experimental results show that for the patient group, gray matter density or volumes significantly increased in right frontal lobe, middle frontal gyrus, vermis, left caudate and parietal lobe compared with the normal controls. However, in the left frontal lobe and middle frontal gyrus, gray matter density significantly decreased. There is no significant difference in white matter between the two groups. These results are consistent with those studies based on the fMRI. It not only provides the evidence for further study of the pathogeny in PTSD but also validates the efficiency of the VBM methods for detecting the difference in the whole brain structure.

1. Introduction

Neuroimaging is a very emerging research field which is dedicated to finding the relationship between neuroanatomical structure and human diseases. With the development of imaging technique, magnetic resonance imaging (MRI) is popularly used in the anatomical studies of the brain. Accordingly, more and more approaches focusing on the shape and neuroanatomical structure emerged. They are commonly used in healthy and diseased subjects to detect the difference between them.

The approaches can be generally classified into methods which are based on the region of interest (ROI) and based on the voxel in the whole brain. The ROI method needs much more human effort and enough accurate prior knowledge to define the regions which may exhibit distinct difference. On the contrary, the voxel-based method can detect any local difference in the whole brain. So it is more unbiased and objective.

Posttraumatic stress disorder (PTSD) is an anxiety disorder which can develop after exposure to one or more traumatic events that threatened or caused great physical harm [1]. Diagnostic symptoms include reexperience, such as flashbacks and nightmares; avoidance of stimuli associated with the trauma; and increased arousal, such as difficulty falling or staying asleep, anger, and hypervigilance. Currently, PTSD is regarded as one of the most typical psychiatric diseases. The imaging research found that there is atrophy in hippocampus of the PTSD patients [2, 3]. In this paper, the differences of the brain structures between the PTSD patients and normal controls were studied using voxel-based morphometric (VBM) method. It can provide the evidence for further study of the pathogeny in PTSD.

On Jun 8th 2005, mine gas explosion occurred in Loudi, China. There were 21 victims in this accident. 12 survivals were diagnosed as the PTSD. In this study, we analyzed the MRI data of these patients and the matched normal controls using VBM method.

2. VBM Methods

Standard VBM and optimized VBM methods were respectively used in the study. Standard VBM is a commonly used technique for characterizing regional difference which is based on the whole-brain [4, 5]. The main objective of standard VBM is to detect and analyze the differences in the whole brain including the gray matter (GM) and white matter (WM) between groups of people. The main procedure includes four steps. First, normalize all the subject images into a template image. Second, segment the normalized images into GM, WM and CSF images. Third, the segmented images should be smoothed by the isotropic Gaussian kernel. Finally, the smoothed images are performed by a statistical analysis which the result is a statistical parametric map showing regions where GM or WM differs significantly between the groups. We'll give the details of each step as follows.

2.1. Standard VBM

2.1.1. Spatial Normalization

This is a procedure that transforms the subject images into the template image in the stereotactic space. This step is not to make the cortical structures accurately corresponded, but to normalize all the subject images to be analyzed into the approximate global shape and the same stereostatic space. At first an affine transformation which can align the images in global is performed. Then the nonlinear transformation is applied to the affine transformed images. After the normalization, all the subject images are in the same space and the shape of each subject are almost same. The VBM method is based on the "imprecise" registration for if each structure was precisely registered to the template, all the normalized images are totally the same and no difference will be detected from the statistical analysis.

There are two regulations we should pay attention to in normalization. First, we should choose an appropriate template. A proper template can give more accurate spatial normalization. The template can be any template image provided in SPM software (www.fil.ion.ucl.ac.uk/spm/) or the customized one. Second, the voxel size of the template should be close to the subject images for too variant resolution will lead to incorrect registration.

2.1.2. Segmentation

Next the normalized images are segmented into GM, WM and CSF images. The segmentation is based on Bayesian priors, which has incorporated the prior anatomical knowledge in normal subjects. Another character of the segmentation in VBM with SPM is that the image intensity non-uniformity correction is applied to smooth intensity variations. Since the mixture Gaussian model cluster analysis is used to get the segmentation, the three kind of segmented images are the membership probability maps. The probability stands for the degree of certain voxel belonging to GM, WM or CSF. So the probability value of each corresponding voxel in the three images sums up to one. In this sense, the segmentation result also shows the proportion of GM or WM to all the tissues within that local region.

2.1.3. Smoothing

The segmented images are then convolved with an isotropic Gaussian kernel. The kernel size is also an important factor which influences the following result. Kernel size should be comparable to the size of the region which may exist distinct difference. Most researches take 8mm to 12mm as full-width at half-maximum (FWHM) isotropic Gaussian kernel. After smoothing, the probability can be more close to the average density of GM or WM. More important point is that, smoothing by the isotropic Gaussian kernel can make the data more normally distributed which satisfies the assumption for the following statistical analysis.

2.1.4. Statistical Analysis

After we have achieved the smoothed GM or WM images, we can do the voxel-wise statistical analysis. Assuming that the statistical model is accurate, the residuals are approximately normally distributed after the segmented images have been smoothed. Often we use t-test or F-test model to do the statistical analysis. Large size clusters will most possibly occur at the smoothed regions and small size at the rough regions.

So, standard VBM can detect the variation in density of the GM or WM in the whole brain between the groups of analyzed data. Here the density refers to the proportion of GM or WM to the whole tissue structure.

2.2. Optimized VBM

Since VBM performed in SPM doesn't do the "scalp removing", it seems that misclassified tissue which often occurred near the ventricle may have affect on the final GM or WM images analysis. A method which has overcome this limitation is the "optimized VBM" [6]. It uses the segmented GM image to normalize to the GM template and segmented WM image to the WM template. Thus the misclassification induced by non-gray matter or non-white matter during the normalization could be eliminated.

Figure 1 is the illustration of VBM method.

According to the analysis above, VBM only provides the result which reflects the relative GM or WM density. It can't provide the difference of the actual volume for GM or WM. However, "optimized VBM" can provide the certain tissue's volume by the modulation. There are two distinct differences compared to standard VBM. First, it uses segmented GM/WM image to normalize to GM/WM template referred above. Then this normalization parameter is reapplied to the original image. Second, modulation for preservation of the absolute tissue volume in each voxel is used. The Jacobian determinants of the deformation field resulted from normalization are multiplied to the smoothed images. Thus optimized VBM can detect the absolute volume of GM or WM in local brain region.

3. Materials

In this study, we analyzed MRI image data of 12 PTSD patients and 15 normal control subjects.

They are all Chinese men and right-handed. The average age of the two groups are generally matched (the patients are 38.1¸ 5.5 years old and normal subjects are 35.5¸ 5.6 years old). All the analyzed data were achieved by 1.5 T (Tesla) Philips MRI scanner with a T1-weighted RF-FAST sequence. Imaging parameters are as follows: TR 7100ms, TE 31ms, flip angle 15°, matrix size 512 512, FOV 240mm 240mm, 248 slices, slice thickness 1.2mm.

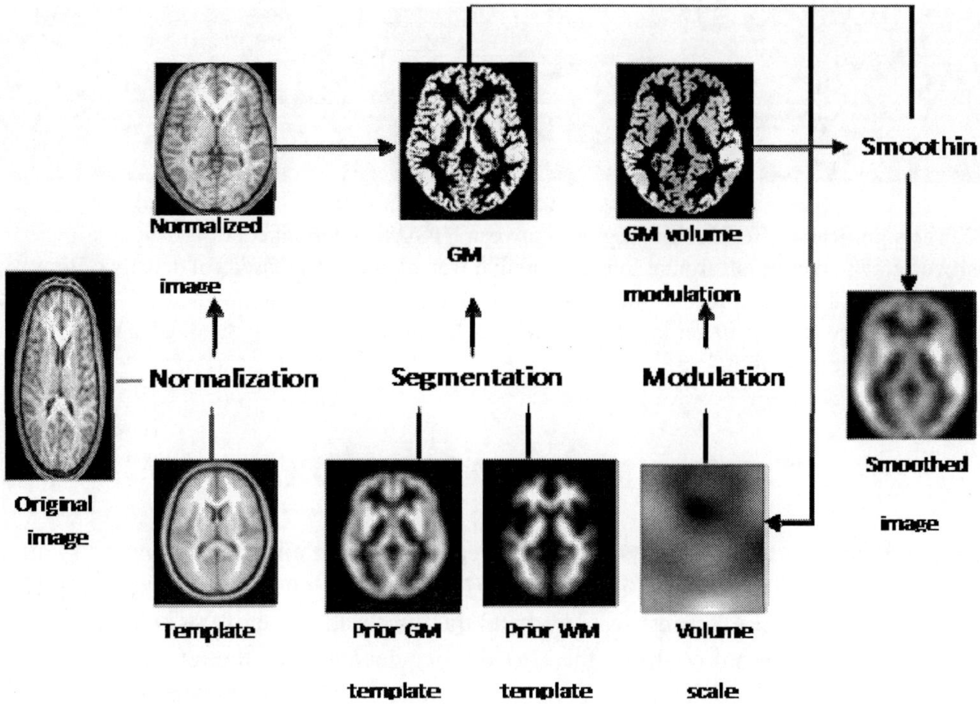

Figure 1. Illustration of VBM method.

Table 1. Different brain regions of the patients and normal controls

Figure ID	Peak's Talairach coordinate (x,y,z/mm)	Brain region	Difference detected (GM/WM)	Increase or decrease (relative to normal controls)
Figure 2.a	-30,27,34	Left frontal lobe, left medial frontal gyrus	GM density	decrease
Figure 2.b	29, 44, 13	Right frontal lobe, left medial frontal gyrus	GM density	increase
Figure 2.c	29,-62,-24	Right vermis	GM density	increase
Figure 2.d	-11,10,12	Left caudate nucleus	GM density	increase
Figure 2.e	-38,-60,50	Left superior parietal lobe	GM volume	increase

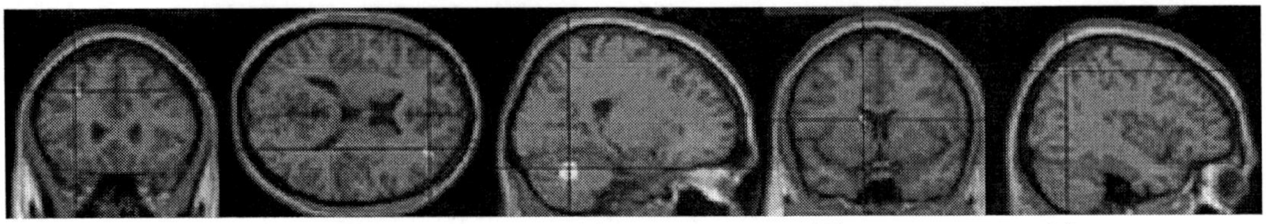

(a) (b) (c) (d) (e)

Figure 2. The significant different brain regions between PTSD and normal controls.
(a) density of gray matter in left frontal lobe and medial frontal gyrus decreased;(b) density of gray matter in right frontal lobe and medial frontal gyrus increased; (c) density of gray matter in right vermis increased; (d) density of gray matter in left caudate nucleus increased; (e) volume of gray matter in left superior parietal lobe increased.

4. Experimental Results

In the experiment, all the data were registered into the Talairach coordinate. T1 image template provided in SPM was used in standard VBM analysis. The Gaussian kernal is 12mm in the smoothing. Two sample t-test was used and the threshold value of P is 0.001.

Compared to the normal controls, the GM density decreases in the left frontal lobe and the medial frontal gyrus (see Figure 2.a) .While it increases at the contralateral positions (see Figure 2.b). The GM density also increases in the right vermis (see Figure 2.c) and the left caudate nucleus (see Figure 2.d). The GM volume increases in the left superior parietal lobe (see Figure 2.e). The cross point shows the position where the significant difference of the brain structure reaches the peak value. The result shows that no significant difference of the WM volume and the WM density was detected.

5. Discussions

Many studies of PTSD focused on its epidemiology, neurobiology, aetiology and the treatment. The researches on neuroimaging are relative few and most of them are focused on the functional mechanism. CT and MRI are the primary analyzed data used in the anatomical research. Most of the anatomical studies are based on ROI method [7]. In this study, there isn't significant difference in hippocampus between the patients and normal controls. It may be related to the difference of sample amount, treatment history, sex (all the subjects are male in this study), etc.

The experimental results show that the GM density decreases in the left frontal lobe and the medial frontal gyrus. There are similar reports of reference [8, 9]. This is consistent with the functional character of the frontal lobe. The GM density increases in the right frontal lobe and the right medial frontal gyrus, left caudate nucleus. This is consistent with reports of reference [10, 11].

The vermis is related to the depressive reaction. Studies based on fMRI found that the decrease of blood stream in young PTSD patients is related to the abuse in their childhood. Our study found the GM density increases in the right vermis of PTSD patients.

We conclude that there are differences in the brain structures between PTSD patients and normal controls. It is consistent with many other functional imaging researches. This study may help to discover the relationship between the brain structural anatomy and the brain functional data of PTSD at a macroscopic level from neuroimaging perspective.

References

[1] American Psychiatric Association, *Diagnostic and statistical manual of mental disorders:* DSM-IV. Washington, DC, 1994.

[2] Bremner J D, Randall P, Scott J M, et al, MRI based measured of hippocampal volume in patients with combat-related post-traumatic stress disorder. *American Journal of Psychiatry,* 1995; 152: 973.

[3] Wang Y L, Xie W, Yang Z, et al., The development of survey for PTSD in China. *Chinese Journal of Applied Psychology*, 2005; 11 (2): 176.

[4] Mechelli A, Price C J, Friston K J, et al., Voxel-Based morphometry of the Human Brain: Methods and Applications. *Current medical Imaging reviews,* 2005; 1: 105.

[5] Ashburner J and Friston K J, Voxel-Based Morphometry-The methods. *NeuroImage,* 2000; 11: 805.

[6] Good C D, Johnsrude I S, Ashburner J et al., A voxel-based morphometric study of aging in 465 Normal Adult Human Brains. *NeuroImage,* 2001;14: 21.

[7] Alastair M. Hull, Neuroimaging findings in post-traumaitc stress disorder. British *Journal of Psychiartry.* 2002;181: 102.

[8] Gil T, Calev A, Greenberg D, et al., Cognitive functioning in post-traumatic stress disorder. *Journal of Traumatic Stress,* 1990; 3: 29.

[9] Uddo M, Vasterling J, Brailey K, et al., Memory and attention in combat-related post-traumatic stress disorder. *Journal of Psychopathology and Behavioural Assessment,* 1993; 15: 43.

[10] Lanius R A, Williamson P C, Densmore M, et al., The Nature of Traumatic Memories: A 4-T fMRI Functional Connectivity Analysis. *American Journal of Psychiatry*, 2004; 161 (1): 36.

[11] Kemp A H, Williams L M, Bryant R A, et al., Implicit perception of fear signals: An fMRI investigation of PTSD. *The world Journal of Biological Psychiatry,* 2004; 5: 135.

[12] Anderson C M, Teicher M H, Polcari A, Renshaw P F, Abnormal T2 relaxation time in the cerebellar vermis of adults sexually abused in childhood: potential role of the vermis in stress-enhanced risk for drug abuse. *Psychoneuroendocrinology,* 2002; 27: 231.

In: Neurobiology of Post-Traumatic Stress Disorder
Editors: L. Sher and A. Vilens, pp. 39-52
ISBN: 978-1-61470-593-2

Chapter IV

Reward and Motivational Systems in Post-traumatic Stress Disorder

Igor Elman, Evelyne Tschibelu, Steven Lowen and David Borsook
McLean Hospital, Harvard Medical School, Belmont, Massachusetts, USA

Abstract

Post-traumatic stress disorder (PTSD) is classified as an anxiety disorder. However, clinical impressions and preclinical work suggest that PTSD may be also associated with dysfunctional reward processing. Here we highlight three lines of research evidence from our laboratory to further support this association. First, PTSD patients' self-reports on a wheel of fortune-type task suggest lower subjective experience of expectancy and satisfaction with receipt of monetary reward. Second, patients with PTSD displayed less motivational effort to extend the viewing time of attractive female faces on a computer key pressing procedure. Third, the abovementioned wheel of fortune-type game administered during functional magnetic resonance imaging was associated with smaller activations of the brain reward/motivational regions in the PTSD group. Further research is needed to unravel the primary versus secondary nature of the observed deficits as well as the place of the reward-enhancing behavioral and pharmacotherapeutic interventions within the PTSD treatment armamentarium.

Introduction

The terrorist attacks on the World Trade Center and the Pentagon along with gruesome episodes in the course of the Iraqi War are just two (albeit highly publicized) examples of stressful events that are capable of exerting an enormous emotional impact on the affected persons. The neural correlates of the emotional sequelae of stress have been extensively investigated in laboratory animals [1, 2]. One of the most consistent findings is that acute stress activates dopamine transmission in the brain's reward circuitry [3-9], whereas chronic stress has the opposite effect [10-23], which is accompanied by reduced motivation towards normally pleasurable stimuli [24-26].

Mild intensity stress may be perceived as pleasant (rewarding), and it is reinforcing across various species, including a variety of stress-seeking behaviors in humans, e.g., skydiving and horror movies [27-30]. Moreover, human neuroimaging studies have reported activation of brain reward circuitry in response to brief stressful experiences including noxious sensory stimuli [31, 32], exposure to facial expressions of fear and disgust [33], mild electric shock [34] and grief induction [35].

Relatively few human data are available regarding the effects of chronic stress on brain reward centers and their emotional/behavioral manifestations. Due to its essential diagnostic criteria (i.e., experience of an extreme traumatic stressor, responding with intense fear, helplessness, or horror, and persistent reliving of the traumatic event) post-traumatic stress disorder (PTSD) may arguably [36] serve as a naturalistic chronic stress paradigm [37, 38].

Although PTSD is classified among anxiety disorders, it is a complex condition with features beyond anxiety. For instance, the "emotional numbing" symptoms, including anhedonia, i.e., loss of interest and pleasure [39], may suggest deficits in the brain reward/reinforcement circuits in PTSD, as such symptoms have been linked with reward function deficits in clinical [40] and mentally healthy [41] samples.

Furthermore, multiple epidemiological surveys firmly link PTSD with the prototype conditions associated with impairment of reward and motivational circuits, namely substance use disorders (SUDs; 42-61]. Indeed, similarly to PTSD, addictive drugs and alcohol are also associated with erratic dopamine surges in the striatum [62], rendering it by their chronic nature hypofunctional and less responsive to natural reinforcers [63-81]. Post-traumatic stress disorder usually heralds the appearance of SUDs and the latter's course tends to parallel the severity of the PTSD symptoms [46, 49, 52], suggesting shared pathophysiological aspects. In view of that, a recent survey of 6099 Vietnam era twin registry members implicated genetic substrate in 36% of comorbid alcohol dependence in PTSD [82].

The significance of the aforesaid epidemiological association is that it may point to molecular and cellular alterations in the reward circuitry [83-85], independently set in motion by either stress (84, 86, 87] or drug/alcohol exposure [84, 88, 83]. These include (among other factors) cAMP response element-binding proteins [84, 88, 89], corticotropin-releasing factor [72, 90, 91] and the brain-derived neurotrophic factor [91, 92]. All are converging on the prodynorphin genes [83, 93] coding for an endogenous κ agonist, dynorphin, which functions as an "antireward" agent that begets motivational deficits and resultant negative affective states [72, 74, 88, 94, 91] by dampening striatal dopaminergic firing [85, 95].

Recent research work has also supported the idea of dysfunction reward system in PTSD. Thus PTSD patients showed less arousal [96] and smaller activations in the emotional [97] and reward areas [98] when exposed to pleasant images [96, 97]. Another study [98) employed a cognitive choice task of memorizing a sequence of two numbers, each repeated three times; correct responses were awarded points but not material rewards. Comparison of females with and without PTSD on that task yielded smaller signal in the classical reward region (viz., ventral striatum) in the former group. Trauma-resilient people conversely demonstrated hyperresponsivity of reward circuitry to rewarding stimuli [99, 100].

During the last few years we have attempted to directly evaluate reward function in PTSD. In the subsequent sections three separate studies will be described [101, 102, 103]. In one study, we reported PTSD-related deficits in anticipatory and consummatory aspects of reward processing as assessed by subjective responses during a wheel of fortune-type game [101]. In the second study, using more objective behavioral measures, we found that male heterosexual combat veterans with current PTSD exerted less motivational effort to extend the viewing time of naturally rewarding stimuli in the form of attractive female faces [102]. The third study employed blood-oxygen-level dependent (BOLD) functional magnetic resonance imaging (fMRI), in conjunction with the same [101] monetary reward probe, to investigate the neural basis of the putative altered reward responsivity in PTSD [103]. While our findings this far suggest reward/motivational deficits in PTSD, the primary vs. secondary nature of these deficits as well as their pathophysiology will have to be assessed in the future studies.

Subjective Reports Study

Being an easily quantifiable reward, valued by most people, money provides a rich framework for clinical and behavioral research. This study utilized a validated [104] monetary reward task that segregates reward expectancy and outcome satisfaction from decision-making without requiring any choices, reward-seeking behavior or instrumental responding. Our wheel of fortune-type gambling procedure assessed subjective responses during 1) an "expectancy phase," when a promising (good), unpromising (bad), or intermediate roulette-like spinner was presented, and 2) an "outcome phase," when the arrow landed on one sector of that spinner and indicated the participant's monetary gain or loss.

Participants comprised 26 male Vietnam combat veterans, 15 with and 11 without current, combat-related PTSD diagnosed according to the Structured Clinical Interview for DSM-IV Axis I Disorders (SCID; 105) and Clinician-Administered PTSD Scale (CAPS; 106). The subjects viewed stimuli and self-ratings questions on a computer monitor. The trial sequence was subdivided into 9 blocks of 17 trials that included a total of 115 spinner/outcome trials. The expectancy phase had three sub-phases: 1) the spinner for each trial was presented, with the following question underneath: "What do you expect the outcome of this trial to be?" 2) participants used the computer mouse to select a point along a 100 mm visual analog scale 3) beneath the question, with anchors "very negative" and "very positive" at the far left and right, respectively. The question and VAS scale then disappeared, and an arrow was superimposed over the spinner, which then commenced spinning. In the outcome phase, the arrow landed on one sector, which then flashed several times to highlight the monetary outcome of that trial. Then the question "How happy are you with the outcome?" appeared below the spinner, and participants rated their satisfaction with the outcome on a 100 mm VAS with anchors of "not happy at all" and "very happy" at the far left and right, respectively.

Patients with PTSD reported both lower expectancy of reward and lower satisfaction with reward when it was received as evidenced by a significant group effect ($p = .01$ for both expectancy and satisfaction). In addition, the latter result was derived from failure to experience the extra satisfaction that is normally incumbent upon obtaining a reward when reward expectancy is low. These results suggest that PTSD patients have diminished subjective perception of reward that also seemed to be relatively insensitive to the normative modulation by the expectancy context.

Many aspects of reward may, however, be quite implicit, that is to say not readily amenable to subjective self-reports [107]. Hence, there is a complimentary value in objective behavioral measures e.g., a key-press task, discussed in the next section.

Behavioral Study

Given that viewing attractive faces was found to be a naturally rewarding stimulus [108], in the present study, we probed reward function in PTSD using the same attractive facial stimuli [102]. Subjects included 23 male heterosexual Vietnam combat veterans, 12 with and 11 without current, combat-related PTSD, diagnosed via SCID and CAPS. Subjects were presented with 80 pictures of nonfamous human faces in the following four categories of 20 each: average female, beautiful female, average male and beautiful male. There were two tasks administered in separate runs: a key-press task, followed by an image attractiveness rating task. For each run, faces were presented in blocks by gender. The order of gender was counterbalanced across subjects. Within each gender block, the average and beautiful faces were presented intermingled, in random order.

For the key-press task, it was explained to the subjects that they could keep the viewing time at 8 sec for a face by not pressing any key, or either increase or decrease this time by up to 4 s (depending upon the frequency of the key presses) by alternately pressing a keyboard's "n" and "m," or "z" and "x" keys, respectively. The former key presses were scored as positive and the latter as negative. The average of these values for the 20 pictures in each of the four facial categories yielded a subject's "net" key presses for each category. Notably, the key press findings were paralleled by activation of major brain-reward regions on a prior neuroimaging study [108].

The groups did not differ in their attractiveness ratings for any of the four facial categories or for the key presses for attractive and unattractive males along with unattractive females. However, PTSD patients expended substantially less effort to extend the viewing time of the beautiful female faces [PTSD: 5.6 ± SD=10.9, non-PTSD: 23.0 ± 15.0; $t(21) = -3.4$, $p < .01$]. Similar subjective valuation of beauty's esthetics by both trauma-exposed groups in conjunction with lesser effort (in units of computer key press) exerted by the PTSD subjects in order to lengthen the viewing time for attractive females were interpreted to represent PTSD-related state deficit in quantified measures of reward valuation.

Neuroimaging Study

Building on our pilot behavioral findings [101, 102], in the present experiment [103] we used 3 Tesla BOLD fMRI, in conjunction with the aforementioned validated wheel of fortune-type game monetary reward task [101], to investigate the neural basis of decreased reward responsivity in PTSD. Subjects with civilian PTSD and healthy individuals were informed that they would passively view a game of chance where they can actually increase, decrease or preserve the endowment of $50 that was given to them at the start of the game. Each spinner was divided into three equal differentially colored sectors. To provide a controlled expectancy condition, the spinners were defined in terms of "good" (generating gains or nothing), "bad" (generating losses or nothing) and "intermediate" (generating a gain, a loss or nothing). Before starting the experiments subjects were trained to correctly identify the spinners, by means of a computerized quiz. The task assessed BOLD fMRI signal changes during a) an "expectancy phase," when a promising (i.e., good), unpromising (i.e., bad), or intermediate spinner was presented, and b) an "outcome phase," when the arrow landed on one sector of that spinner and indicated the participant's monetary gain or loss [101, 104]. A previous study in healthy subjects reported significant activations in the in dopamine terminal fields (striatum, amygdala and prefrontal cortex) that were induced by processing of expectancy and experience of monetary gains and losses on this task [104].

Neuroimaging data from 20 PTSD and 26 healthy subjects withstood quality control procedures and were included in the analyses. In voxelwise and anatomically defined region of interest analyses, when gains were contrasted to losses, between-group comparison revealed smaller bilateral striatal activations in the PTSD subjects. In the PTSD group, less striatal activation to gains vs. losses was associated with more self-reported motivational and social deficits.

Conclusion

The neuroimaging results extended our earlier work with the self-reports and behavioral probes in which we found PTSD-related decrements in reward function and also suggested that this alterations may be related to some aspects of emotional numbing.

An unanswered question of the cross-sectional design is the origin of hyporesponsive reward in PTSD. One possibility is that a hyporesponsive reward system is a preexisting risk factor (c.f., 109). A second possibility is that it is acquired as a result of the traumatic exposure along with the PTSD. A growing body of neuroimaging work indicates that a "reward deficiency" state of depressed striatal dopaminergic activity (perhaps due to a paucity or aberration of the D2 dopaminergic receptors) is a risk factor for the development of SUDs [110-112] because of the heightened reinforcing effects of addictive drugs [113, 112] in this category of people. A reward deficiency state has also been hypothesized to underlie other mental disorders, including PTSD [114]. Thus, individuals afflicted with reward deficiency may perceive their life as bland and unfulfilling and possess a character trait of novelty seeking [115, 116, 28, 117], which could drive their engagement in stressogenic situations with heightened potential for trauma exposure and subsequent PTSD [118, 119].

People with 'reward deficiency' may be additionally prone for the development of PTSD owing to their poor resilience to stress [119]. Intact reward and motivational function are integral to almost every facet of personality through sociability [120], enthusiasm, altruism, self-fulfillment and optimism [121]. Consequently, low motivational/reward capacity may make it difficult, if not impossible, to cope with extremely stressful situations, to the point of feeling helpless, which is an essential precursors for the development of PTSD [39]. Hence, dysfunction in reward and motivational mechanisms may predispose for PTSD.

The causality could also run in the opposite direction. Research in laboratory animals evaluating the effects of chronic stress on the brain reward circuitry consistently show decreased function [11, 14, 15, 18, 19, 21, 122-125]. Hans Selye was the first to suggest that chronic stress causes illnesses in humans. He surmised that illnesses occur because various neurotransmitters and hormones become depleted [30], so the concept of dopaminergic depletion and consequent decrease in the brain reward function noted in substance used disorders [75] may be applied to the PTSD syndrome, where motivational deficits were attributed to emotional depletion caused by chronic stress exposure [126-128]. Nonetheless, conclusive resolution of the primary vs. secondary source of reward deficits in PTSD as well as potential type of interactions between innate and acquired factors (e.g., competition, independence or synergism) may require prospective and/or twin studies. Finally, further research is also needed to evaluate the potential heuristic value of reward-enhancing behavioral and/or pharmacological therapeutic intervention within the PTSD treatment armamentarium.

References

[1] Lu, L., J.D. Shepard, F.S. Hall, and Y. Shaham, Effect of environmental stressors on opiate and psychostimulant reinforcement, reinstatement and discrimination in rats: a review. *Neuroscience and biobehavioral reviews,* 2003. 27(5): p. 457-91.

[2] Zacharko, R.M. and H. Anisman, Stressor-induced anhedonia in the mesocorticolimbic system. *Neuroscience and biobehavioral reviews,* 1991. 15(3): p. 391-405.

[3] Imperato, A., S. Puglisi-Allegra, P. Casolini, A. Zocchi, and L. Angelucci, Stress-induced enhancement of dopamine and acetylcholine release in limbic structures: role of corticosterone. *Eur J Pharmacol,* 1989. 165(2-3): p. 337-8.

[4] Kalivas, P.W. and P. Duffy, Selective activation of dopamine transmission in the shell of the nucleus accumbens by stress. *Brain Res,* 1995. 675(1-2): p. 325-8.

[5] Kaneyuki, H., H. Yokoo, A. Tsuda, M. Yoshida, Y. Mizuki, M. Yamada, and M. Tanaka, Psychological stress increases dopamine turnover selectively in mesoprefrontal dopamine neurons of rats: reversal by diazepam. *Brain Res,* 1991. 557(1-2): p. 154-61.

[6] Piazza, P.V. and M. Le Moal, The role of stress in drug self-administration. *Trends Pharmacol Sci,* 1998. 19(2): p. 67-74.

[7] Rouge-Pont, F., P.V. Piazza, M. Kharouby, M. Le Moal, and H. Simon, Higher and longer stress-induced increase in dopamine concentrations in the nucleus accumbens of animals predisposed to amphetamine self-administration. A microdialysis study. *Brain Res,* 1993. 602(1): p. 169-74.

[8] Thierry, A.M., J.P. Tassin, G. Blanc, and J. Glowinski, Selective activation of mesocortical DA system by stress. *Nature,* 1976. 263(5574): p. 242-4.

[9] Tidey, J.W. and K.A. Miczek, Social defeat stress selectively alters mesocorticolimbic dopamine release: an in vivo microdialysis study. *Brain Res,* 1996. 721(1-2): p. 140-9.

[10] Di Chiara, G. and G. Tanda, Blunting of reactivity of dopamine transmission to palatable food: a biochemical marker of anhedonia in the CMS model? *Psychopharmacology* (Berl), 1997. 134(4): p. 351-3; discussion 371-7.

[11] Gambarana, C., F. Masi, A. Tagliamonte, S. Scheggi, O. Ghiglieri, and M.G. De Montis, A chronic stress that impairs reactivity in rats also decreases dopaminergic transmission in the nucleus accumbens: a microdialysis study. *Journal of neurochemistry,* 1999. 72(5): p. 2039-46.

[12] Gambarana, C., F. Masi, A. Tagliamonte, S. Scheggi, O. Ghiglieri, and M.G. De Montis, A chronic stress that impairs reactivity in rats also decreases dopaminergic transmission in the nucleus accumbens: a microdialysis study. *J Neurochem,* 1999. 72(5): p. 2039-46.

[13] Gambarana, C., S. Scheggi, A. Tagliamonte, P. Tolu, and M.G. De Montis, Animal models for the study of antidepressant activity. *Brain Res Brain Res Protoc,* 2001. 7(1): p. 11-20.

[14] Gambarana, C., P.L. Tolu, F. Masi, M. Rinaldi, D. Giachetti, P. Morazzoni, and M.G. De Montis, A study of the antidepressant activity of Hypericum perforatum on animal models. *Pharmacopsychiatry,* 2001. 34 Suppl 1: p. S42-4.

[15] Imperato, A., S. Cabib, and S. Puglisi-Allegra, Repeated stressful experiences differently affect the time-dependent responses of the mesolimbic dopamine system to the stressor. *Brain research,* 1993. 601(1-2): p. 333-6.

[16] Imperato, A., S. Cabib, and S. Puglisi-Allegra, Repeated stressful experiences differently affect the time-dependent responses of the mesolimbic dopamine system to the stressor. *Brain Res,* 1993. 601(1-2): p. 333-6.

[17] Mangiavacchi, S., F. Masi, S. Scheggi, B. Leggio, M.G. De Montis, and C. Gambarana, Long-term behavioral and neurochemical effects of chronic stress exposure in rats. *J Neurochem,* 2001. 79(6): p. 1113-21.

[18] Masi, F., B. Leggio, G. Nanni, S. Scheggi, M.G. De Montis, A. Tagliamonte, S. Grappi, and C. Gambarana, Effects of long-term acetyl-L-carnitine administration in rats--II: Protection against the disrupting effect of stress on the acquisition of appetitive behavior. *Neuropsychopharmacology,* 2003. 28(4): p. 683-93.

[19] Masi, F., S. Scheggi, S. Mangiavacchi, P. Tolu, A. Tagliamonte, M.G. De Montis, and C. Gambarana, Dopamine output in the nucleus accumbens shell is related to the acquisition and the retention of a motivated appetitive behavior in rats. *Brain research,* 2001. 903(1-2): p. 102-9.

[20] Masi, F., S. Scheggi, S. Mangiavacchi, P. Tolu, A. Tagliamonte, M.G. De Montis, and C. Gambarana, Dopamine output in the nucleus accumbens shell is related to the acquisition and the retention of a motivated appetitive behavior in rats. *Brain Res,* 2001. 903(1-2): p. 102-9.

[21] Puglisi-Allegra, S., A. Imperato, L. Angelucci, and S. Cabib, Acute stress induces time-dependent responses in dopamine mesolimbic system. *Brain research,* 1991. 554(1-2): p. 217-22.

[22] Puglisi-Allegra, S., A. Imperato, L. Angelucci, and S. Cabib, Acute stress induces time-dependent responses in dopamine mesolimbic system. *Brain Res,* 1991. 554(1-2): p. 217-22.

[23] Scheggi, S., B. Leggio, F. Masi, S. Grappi, C. Gambarana, G. Nanni, R. Rauggi, and M.G. De Montis, Selective modifications in the nucleus accumbens of dopamine synaptic transmission in rats exposed to chronic stress. *J Neurochem,* 2002. 83(4): p. 895-903.

[24] Ghiglieri O., G.C., Scheggi S., Tagliamonte A., Willner P., De Montis MG., Palatable food induces an appetitive behaviour in satiated rats which can be inhibited by chronic stress. *Behav Pharmacol,* 1997. 8(6-7): p. 619-28.

[25] Willner P, M.R., Papp M., Chronic mild stress-induced anhedonia: a realistic animal model of depression. *Neuroscience and biobehavioral reviews,* 1992. 16(4): p. 525-34.

[26] Willner P., T.A., Sampson D., Sophokleous S., Muscat R., Reduction of sucrose preference by chronic unpredictable mild stress, and its restoration by tricyclic. *Psychopharmacology (Berl),* 1987. 93(3): p. 358-64.

[27] Gardner, E., The neurobiology and genetics of addiction: implications of the 'Reward Deficiency Syndrome' for theraputic strategies in chemical dependency, in Addictions: entries and exits. *Elster J ed.* 1999, New York: Russell Sage Foundation.

[28] Goeders, N.E., Stress and cocaine addiction. *The Journal of pharmacology and experimental therapeutics,* 2002. 301(3): p. 785-9.

[29] Sapolsky, R.M., *Why Zebras Don't Get Ulcers: An updated Guide to Stress-Related Diseases, and Coping.* 1998: W H Freeman & Co. 434.

[30] Selye, H., *The stress of life.* 1976, New York: McGraw-Hill.

[31] Becerra, L., H.C. Breiter, R. Wise, R.G. Gonzalez, and D. Borsook, Reward circuitry activation by noxious thermal stimuli. *Neuron,* 2001. 32(5): p. 927-46.

[32] Scott DJ., H.M., Koeppe RA, Stohler CS, Zubieta JK., Variations in the human pain stress experience mediated by ventral and dorsal basal ganglia dopamine activity. *J Neurosci.,* 2006. 26(42): p. 10789-95.

[33] Phillips, M.L., et al., Neural responses to facial and vocal expressions of fear and disgust. *Proc R Soc Lond B Biol Sci.,* 1998. 265(1408): p. 1809-17.

[34] Fischer, H., J.L. Andersson, T. Furmark, and M. Fredrikson, Fear conditioning and brain activity: a positron emission tomography study in humans. *Behavioral neuroscience,* 2000. 114(4): p. 671-80.

[35] O'Connor, M.F., D.K. Wellisch, A.L. Stanton, N.I. Eisenberger, M.R. Irwin, and M.D. Lieberman, Craving love? Enduring grief activates brain's reward center. *NeuroImage,* 2008. 42(2): p. 969-72.

[36] Shalev, A.Y., E.J. Videlock, T. Peleg, R. Segman, R.K. Pitman, and R. Yehuda, Stress hormones and post-traumatic stress disorder in civilian trauma victims: a longitudinal study. Part I: HPA axis responses. *The international journal of neuropsychopharmacology,* 2008. 11(3): p. 365-72.

[37] Baum, A., L. Cohen, and M. Hall, Control and intrusive memories as possible determinants of chronic stress. *Psychosomatic medicine,* 1993. 55(3): p. 274-86.

[38] Yehuda, R. and S.M. Antelman, Criteria for rationally evaluating animal models of posttraumatic stress disorder. *Biological psychiatry,* 1993. 33(7): p. 479-86.

[39] American Psychiatric Association (2000): *Diagnostic and Statistical Manual of Mental Disorders,* Fourth Edition, Text Revision (DSM-IV-TR). Washington, DC

[40] Harvey, P.O., J. Pruessner, Y. Czechowska, and M. Lepage, Individual differences in trait anhedonia: a structural and functional magnetic resonance imaging study in non-clinical subjects. *Molecular psychiatry,* 2007. 12(8): p. 703, 767-75.

[41] Pizzagalli, D.A., D. Iosifescu, L.A. Hallett, K.G. Ratner, and M. Fava, Reduced hedonic capacity in major depressive disorder: evidence from a probabilistic reward task. *Journal of psychiatric research,* 2008. 43(1): p. 76-87.

[42] Acierno, R., D.G. Kilpatrick, H. Resnick, B. Saunders, M. De Arellano, and C. Best, Assault, PTSD, family substance use, and depression as risk factors for cigarette use in youth: findings from the National Survey of Adolescents. *Journal of traumatic stress,* 2000. 13(3): p. 381-96.

[43] Acierno, R.A., D.G. Kilpatrick, H.S. Resnick, B.E. Saunders, and C.L. Best, Violent assault, posttraumatic stress disorder, and depression. Risk factors for cigarette use among adult women. *Behavior modification,* 1996. 20(4): p. 363-84.

[44] Beckham, J.C., Smoking and anxiety in combat veterans with chronic posttraumatic stress disorder: a review. *Journal of psychoactive drugs,* 1999. 31(2): p. 103-10.

[45] Beckham, J.C., A.C. Kirby, M.E. Feldman, M.A. Hertzberg, S.D. Moore, A.L. Crawford, J.R. Davidson, and J.A. Fairbank, Prevalence and correlates of heavy smoking in Vietnam veterans with chronic posttraumatic stress disorder. *Addictive behaviors,* 1997. 22(5): p. 637-47.

[46] Bremner, J.D., S.M. Southwick, A. Darnell, and D.S. Charney, Chronic PTSD in Vietnam combat veterans: course of illness and substance abuse. *The American journal of psychiatry,* 1996. 153(3): p. 369-75.

[47] Breslau, N., G.C. Davis, P. Andreski, and E. Peterson, Traumatic events and posttraumatic stress disorder in an urban population of young adults. *Archives of general psychiatry,* 1991. 48(3): p. 216-22.

[48] Breslau, N., G.C. Davis, and L.R. Schultz, Posttraumatic stress disorder and the incidence of nicotine, alcohol, and other drug disorders in persons who have experienced trauma. *Archives of general psychiatry,* 2003. 60(3): p. 289-94.

[49] Chilcoat, H.D. and N. Breslau, Posttraumatic stress disorder and drug disorders: testing causal pathways. *Archives of general psychiatry,* 1998. 55(10): p. 913-7.

[50] Deykin, E.Y. and S.L. Buka, Prevalence and risk factors for posttraumatic stress disorder among chemically dependent adolescents. T*he American journal of psychiatry,* 1997. 154(6): p. 752-7.

[51] Giaconia, R.M., H.Z. Reinherz, A.B. Silverman, B. Pakiz, A.K. Frost, and E. Cohen, Traumas and posttraumatic stress disorder in a community population of older adolescents. *Journal of the American Academy of Child and Adolescent Psychiatry,* 1995. 34(10): p. 1369-80.

[52] Jacobsen, L.K., S.M. Southwick, and T.R. Kosten, Substance use disorders in patients with posttraumatic stress disorder: a review of the literature. *The American journal of psychiatry,* 2001. 158(8): p. 1184-90.

[53] Keane, T., Caddell JM, Martin BW, Zimering RT, Fairbank JA, *Substance abuse among Vietnam veterans with posttraumatic stress disorders Bull Soc Psychol Addict Behav,* 1983. 2: p. 117-22.

[54] Kessler, R.C., A. Sonnega, E. Bromet, M. Hughes, and C.B. Nelson, Posttraumatic stress disorder in the National Comorbidity Survey. *Archives of general psychiatry,* 1995. 52(12): p. 1048-60.

[55] Kulka, R.A., Trauma and the Vietnam War Generation: *Report of Findings From the National Vietnam Veterans Readjustment Study.* 1990, New York, NY: Brunner/Mazel.

[56] McFall, M.E., P.W. Mackay, and D.M. Donovan, Combat-related posttraumatic stress disorder and severity of substance abuse in Vietnam veterans. *Journal of studies on alcohol, 1*992. 53(4): p. 357-63.

[57] NIDA, Stress and Drug Abuse: Epidemiology, Etiology, *Prevention and Treatment.* 2003, NIDA.

[58] Saxon, A.J., T.M. Davis, K.L. Sloan, K.M. McKnight, M.E. McFall, and D.R. Kivlahan, Trauma, symptoms of posttraumatic stress disorder, and associated problems

among incarcerated veterans. *Psychiatric services* (Washington, D.C, 2001. 52(7): p. 959-64.
[59] Stewart, S.H., Alcohol abuse in individuals exposed to trauma: a critical review. *Psychological bulletin,* 1996. 120(1): p. 83-112.
[60] Volkow, N.D., J.S. Fowler, G.J. Wang, J.M. Swanson, and F. Telang, Dopamine in drug abuse and addiction: results of imaging studies and treatment implications. *Archives of neurology,* 2007. 64(11): p. 1575-9.
[61] Volkow, N.D., G.J. Wang, F. Telang, J.S. Fowler, J. Logan, M. Jayne, Y. Ma, K. Pradhan, and C. Wong, Profound decreases in dopamine release in striatum in detoxified alcoholics: possible orbitofrontal involvement. *The Journal of neuroscience,* 2007. 27(46): p. 12700-6.
[62] Dalley, J.W. and B.J. Everitt, Dopamine receptors in the learning, memory and drug reward circuitry. *Seminars in cell & developmental biology,* 2009. 20(4): p. 403-10.
[63] Ahmed, S.H., P.J. Kenny, G.F. Koob, and A. Markou, Neurobiological evidence for hedonic allostasis associated with escalating cocaine use. *Nature neuroscience,* 2002. 5(7): p. 625-6.
[64] Baler, R.D. and N.D. Volkow, Drug addiction: the neurobiology of disrupted self-control. *Trends in molecular medicine,* 2006. 12(12): p. 559-66.
[65] Barr, A.M., D.F. Fiorino, and A.G. Phillips, Effects of withdrawal from an escalating dose schedule of d-amphetamine on sexual behavior in the male rat. *Pharmacology, biochemistry, and behavior,* 1999. 64(3): p. 597-604.
[66] Barr, A.M. and A.G. Phillips, Withdrawal following repeated exposure to d-amphetamine decreases responding for a sucrose solution as measured by a progressive ratio schedule of reinforcement. *Psychopharmacology,* 1999. 141(1): p. 99-106.
[67] Bolla, K.I., J.L. Cadet, and E.D. London, The neuropsychiatry of chronic cocaine abuse. *The Journal of neuropsychiatry and clinical neurosciences,* 1998. 10(3): p. 280-9.
[68] Cassens, G., C. Actor, M. Kling, and J.J. Schildkraut, Amphetamine withdrawal: effects on threshold of intracranial reinforcement. *Psychopharmacology,* 1981. 73(4): p. 318-22.
[69] Garavan, H., J. Pankiewicz, A. Bloom, J.K. Cho, L. Sperry, T.J. Ross, B.J. Salmeron, R. Risinger, D. Kelley, and E.A. Stein, Cue-induced cocaine craving: neuroanatomical specificity for drug users and drug stimuli. *The American journal of psychiatry,* 2000. 157(11): p. 1789-98.
[70] Goldstein, R.Z. and N.D. Volkow, Drug addiction and its underlying neurobiological basis: neuroimaging evidence for the involvement of the frontal cortex. *The American journal of psychiatry,* 2002. 159(10): p. 1642-52.
[71] Harris, G.C. and G. Aston-Jones, Altered motivation and learning following opiate withdrawal: evidence for prolonged dysregulation of reward processing. *Neuropsychopharmacology,* 2003. 28(5): p. 865-71.
[72] Koob, G.F., A role for brain stress systems in addiction. *Neuron,* 2008. 59(1): p. 11-34.
[73] Koob, G.F. and M. Le Moal, Drug abuse: hedonic homeostatic dysregulation. *Science* (New York, N.Y, 1997. 278(5335): p. 52-8.
[74] Koob, G.F. and M. Le Moal, Plasticity of reward neurocircuitry and the 'dark side' of drug addiction. *Nature neuroscience,* 2005. 8(11): p. 1442-4.

[75] Koob, G.F. and M. Le Moal, Addiction and the brain antireward system. *Annual review of psychology,* 2008. 59: p. 29-53.
[76] Martin-Soelch, C., A.F. Chevalley, G. Kunig, J. Missimer, S. Magyar, A. Mino, W. Schultz, and K.L. Leenders, Changes in reward-induced brain activation in opiate addicts. *The European journal of neuroscience,* 2001. 14(8): p. 1360-8.
[77] Martin-Solch, C., S. Magyar, G. Kunig, J. Missimer, W. Schultz, and K.L. Leenders, Changes in brain activation associated with reward processing in smokers and nonsmokers. A positron emission tomography study. Experimental brain research. *Experimentelle Hirnforschung,* 2001. 139(3): p. 278-86.
[78] Self, D.W. and E.J. Nestler, Relapse to drug-seeking: neural and molecular mechanisms. *Drug and alcohol dependence,* 1998. 51(1-2): p. 49-60.
[79] Solomon, R.L. and J.D. Corbit, An opponent-process theory of motivation. I. Temporal dynamics of affect. *Psychological review,* 1974. 81(2): p. 119-45.
[80] Volkow, N.D., J.S. Fowler, and G.J. Wang, The addicted human brain: insights from imaging studies. *The Journal of clinical investigation,* 2003. 111(10): p. 1444-51.
[81] Volkow, N.D., J.S. Fowler, and G.J. Wang, Positron emission tomography and single-photon emission computed tomography in substance abuse research. *Seminars in nuclear medicine,* 2003. 33(2): p. 114-28.
[82] Scherrer, J.F., H. Xian, M.J. Lyons, J. Goldberg, S.A. Eisen, W.R. True, M. Tsuang, K.K. Bucholz, and K.C. Koenen, Posttraumatic stress disorder; combat exposure; and nicotine dependence, alcohol dependence, and major depression in male twins. *Comprehensive psychiatry,* 2008. 49(3): p. 297-304.
[83] Cole, R.L., C. Konradi, J. Douglass, and S.E. Hyman, Neuronal adaptation to amphetamine and dopamine: molecular mechanisms of prodynorphin gene regulation in rat striatum. *Neuron,* 1995. 14(4): p. 813-23.
[84] Pliakas, A.M., R.R. Carlson, R.L. Neve, C. Konradi, E.J. Nestler, and W.A. Carlezon, Jr., Altered responsiveness to cocaine and increased immobility in the forced swim test associated with elevated cAMP response element-binding protein expression in nucleus accumbens. *The Journal of neuroscience,* 2001. 21(18): p. 7397-403.
[85] Steiner, H. and C.R. Gerfen, Dynorphin regulates D1 dopamine receptor-mediated responses in the striatum: relative contributions of pre- and postsynaptic mechanisms in dorsal and ventral striatum demonstrated by altered immediate-early gene induction. *The Journal of comparative neurology,* 1996. 376(4): p. 530-41.
[86] Barrot, M., J.D. Olivier, L.I. Perrotti, R.J. DiLeone, O. Berton, A.J. Eisch, S. Impey, D.R. Storm, R.L. Neve, J.C. Yin, V. Zachariou, and E.J. Nestler, CREB activity in the nucleus accumbens shell controls gating of behavioral responses to emotional stimuli. *Proceedings of the National Academy of Sciences of the United States of America,* 2002. 99(17): p. 11435-40.
[87] Boer, U., I. Cierny, D. Krause, A. Heinrich, H. Lin, G. Mayr, C. Hiemke, and W. Knepel, Chronic lithium salt treatment reduces CRE/CREB-directed gene transcription and reverses its upregulation by chronic psychosocial stress in transgenic reporter gene mice. *Neuropsychopharmacology,* 2008. 33(10): p. 2407-15.
[88] Carlezon, W.A., Jr., J. Thome, V.G. Olson, S.B. Lane-Ladd, E.S. Brodkin, N. Hiroi, R.S. Duman, R.L. Neve, and E.J. Nestler, Regulation of cocaine reward by CREB. *Science* (New York, N.Y, 1998. 282(5397): p. 2272-5.

[89] Carlezon, W.A., Jr., R.S. Duman, and E.J. Nestler, The many faces of CREB. T*rends in neurosciences,* 2005. 28(8): p. 436-45.

[90] Land, B.B., M.R. Bruchas, J.C. Lemos, M. Xu, E.J. Melief, and C. Chavkin, The dysphoric component of stress is encoded by activation of the dynorphin kappa-opioid system. *The Journal of neuroscience,* 2008. 28(2): p. 407-14.

[91] Walker, B.M. and G.F. Koob, Regarding "Dynorphin is a downstream effector of striatal BDNF regulation of ethanol intake". *The FASEB journal,* 2008. 22(7): p. 2113; author reply 2113-4.

[92] Logrip, M.L., P.H. Janak, and D. Ron, Dynorphin is a downstream effector of striatal BDNF regulation of ethanol intake. *The FASEB journal,* 2008. 22(7): p. 2393-404.

[93] Hyman, S.E., Addiction: a disease of learning and memory. *The American journal of psychiatry,* 2005. 162(8): p. 1414-22.

[94] Nestler, E.J., M. Barrot, R.J. DiLeone, A.J. Eisch, S.J. Gold, and L.M. Monteggia, Neurobiology of depression. *Neuron,* 2002. 34(1): p. 13-25.

[95] Spanagel, R., A. Herz, and T.S. Shippenberg, Opposing tonically active endogenous opioid systems modulate the mesolimbic dopaminergic pathway. *Proceedings of the National Academy of Sciences of the United States of America,* 1992. 89(6): p. 2046-50.

[96] Spahic-Mihajlovic, A., J.W. Crayton, and E.J. Neafsey, Selective numbing and hyperarousal in male and female Bosnian refugees with PTSD. *Journal of anxiety disorders,* 2005. 19(4): p. 383-402.

[97] Jatzko, A., A. Schmitt, T. Demirakca, E. Weimer, and D.F. Braus, Disturbance in the neural circuitry underlying positive emotional processing in post-traumatic stress disorder (PTSD). An fMRI study. *European archives of psychiatry and clinical neuroscience,* 2006. 256(2): p. 112-4.

[98] Sailer, U., S. Robinson, F.P. Fischmeister, D. Konig, C. Oppenauer, B. Lueger-Schuster, E. Moser, I. Kryspin-Exner, and H. Bauer, Altered reward processing in the nucleus accumbens and mesial prefrontal cortex of patients with posttraumatic stress disorder. *Neuropsychologia,* 2008. 46(11): p. 2836-44.

[99] Feder, A., E.J. Nestler, and D.S. Charney, Psychobiology and molecular genetics of resilience. *Nature reviews,* 2009. 10(6): p. 446-57.

[100] Vythilingam, M., E.E. Nelson, M. Scaramozza, T. Waldeck, G. Hazlett, S.M. Southwick, D.S. Pine, W. Drevets, D.S. Charney, and M. Ernst, Reward circuitry in resilience to severe trauma: an fMRI investigation of resilient special forces soldiers. *Psychiatry research,* 2009. 172(1): p. 75-7.

[101] Hopper, J.W., R.K. Pitman, Z. Su, G.M. Heyman, N.B. Lasko, M.L. Macklin, S.P. Orr, S.E. Lukas, and I. Elman, Probing reward function in posttraumatic stress disorder: expectancy and satisfaction with monetary gains and losses. *Journal of psychiatric research,* 2008. 42(10): p. 802-7.

[102] Elman, I., D. Ariely, N. Mazar, I. Aharon, N.B. Lasko, M.L. Macklin, S.P. Orr, S.E. Lukas, and R.K. Pitman, Probing reward function in post-traumatic stress disorder with beautiful facial images. *Psychiatry research,* 2005. 135(3): p. 179-83.

[103] Elman I, L.S., Frederick BB, Chi W, Becerra L, Pitman RK, Functional neuroimaging of reward circuitry responsivity to monetary gains and losses in posttraumatic stress disorder. *Biological Psychiatry,* in press, 2009.

[104] Breiter, H.C., I. Aharon, D. Kahneman, A. Dale, and P. Shizgal, Functional imaging of neural responses to expectancy and experience of monetary gains and losses. *Neuron,* 2001. 30(2): p. 619-39.

[105] First, M.B., et al., *Structured Structural Interview for DSM-IV Axis I disorder* (SCID-I), Clinical Version. 1997, Washington D.C.: American Psychiatric Press.

[106] Weathers, F.W., T.M. Keane, and J.R. Davidson, Clinician-administered PTSD scale: a review of the first ten years of research. *Depression and anxiety,* 2001. 13(3): p. 132-56.

[107] Berridge, K.C., T.E. Robinson, and J.W. Aldridge, Dissecting components of reward: 'liking', 'wanting', and learning. *Current opinion in pharmacology,* 2009. 9(1): p. 65-73.

[108] Aharon, I., N. Etcoff, D. Ariely, C.F. Chabris, E. O'Connor, and H.C. Breiter, Beautiful faces have variable reward value: fMRI and behavioral evidence. *Neuron,* 2001. 32(3): p. 537-51.

[109] Gilbertson, M.W., M.E. Shenton, A. Ciszewski, K. Kasai, N.B. Lasko, S.P. Orr, and R.K. Pitman, Smaller hippocampal volume predicts pathologic vulnerability to psychological trauma. *Nature neuroscience,* 2002. 5(11): p. 1242-7.

[110] Bowirrat, A. and M. Oscar-Berman, Relationship between dopaminergic neurotransmission, alcoholism, and Reward Deficiency syndrome. American journal of medical genetics. *Part B, Neuropsychiatric genetics,* 2005. 132B(1): p. 29-37.

[111] Volkow, N.D., G.J. Wang, H. Begleiter, B. Porjesz, J.S. Fowler, F. Telang, C. Wong, Y. Ma, J. Logan, R. Goldstein, D. Alexoff, and P.K. Thanos, High levels of dopamine D2 receptors in unaffected members of alcoholic families: possible protective factors. *Archives of general psychiatry,* 2006. 63(9): p. 999-1008.

[112] Wang, G.J., N.D. Volkow, J. Logan, N.R. Pappas, C.T. Wong, W. Zhu, N. Netusil, and J.S. Fowler, Brain dopamine and obesity. *Lancet,* 2001. 357(9253): p. 354-7.

[113] Volkow, N.D., G.J. Wang, J.S. Fowler, J. Logan, S.J. Gatley, A. Gifford, R. Hitzemann, Y.S. Ding, and N. Pappas, Prediction of reinforcing responses to psychostimulants in humans by brain dopamine D2 receptor levels. *The American journal of psychiatry,* 1999. 156(9): p. 1440-3.

[114] Blum, K., E.R. Braverman, J.M. Holder, J.F. Lubar, V.J. Monastra, D. Miller, J.O. Lubar, T.J. Chen, and D.E. Comings, Reward deficiency syndrome: a biogenetic model for the diagnosis and treatment of impulsive, addictive, and compulsive behaviors. *Journal of psychoactive drugs,* 2000. 32 Suppl: p. i-iv, 1-112.

[115] Bardo, M.T., R.L. Donohew, and N.G. Harrington, Psychobiology of novelty seeking and drug seeking behavior. *Behavioural brain research,* 1996. 77(1-2): p. 23-43.

[116] Dellu, F., P.V. Piazza, W. Mayo, M. Le Moal, and H. Simon, Novelty-seeking in rats--biobehavioral characteristics and possible relationship with the sensation-seeking trait in man. *Neuropsychobiology,* 1996. 34(3): p. 136-45.

[117] Scourfield, J., D.E. Stevens, and K.R. Merikangas, Substance abuse, comorbidity, and sensation seeking: gender differences. *Comprehensive psychiatry,* 1996. 37(6): p. 384-92.

[118] Richman, H. and B.C. Frueh, Personality and PTSD II: personality assessment of PTSD-diagnosed Vietnam veterans using the cloninger tridimensional personality questionnaire (TPQ). *Depression and anxiety,* 1997. 6(2): p. 70-7.

[119] Shumake, J., D. Barrett, and F. Gonzalez-Lima, Behavioral characteristics of rats predisposed to learned helplessness: reduced reward sensitivity, increased novelty

seeking, and persistent fear memories. *Behavioural brain research,* 2005. 164(2): p. 222-30.

[120] Breier, A., L. Kestler, C. Adler, I. Elman, N. Wiesenfeld, A. Malhotra, and D. Pickar, Dopamine D2 receptor density and personal detachment in healthy subjects. *The American journal of psychiatry,* 1998. 155(10): p. 1440-2.

[121] Charney, D.S., Psychobiological mechanisms of resilience and vulnerability: implications for successful adaptation to extreme stress. *The American journal of psychiatry,* 2004. 161(2): p. 195-216.

[122] Di Chiara, G.a.G.T., Blunting of reactivity of dopamine transmission to palatable food: a biochemical marker of anhedonia in the CMS model? *Psychopharmacology,* 1997. 134(4): p. 351-3.

[123] Gambarana, C., S. Scheggi, A. Tagliamonte, P. Tolu, and M.G. De Montis, Animal models for the study of antidepressant activity. *Brain research,* 2001. 7(1): p. 11-20.

[124] Mangiavacchi, S., F. Masi, S. Scheggi, B. Leggio, M.G. De Montis, and C. Gambarana, Long-term behavioral and neurochemical effects of chronic stress exposure in rats. *Journal of neurochemistry,* 2001. 79(6): p. 1113-21.

[125] Scheggi, S., B. Leggio, F. Masi, S. Grappi, C. Gambarana, G. Nanni, R. Rauggi, and M.G. De Montis, Selective modifications in the nucleus accumbens of dopamine synaptic transmission in rats exposed to chronic stress. *Journal of neurochemistry,* 2002. 83(4): p. 895-903.

[126] Flack, W.F., Jr., B.T. Litz, F.Y. Hsieh, D.G. Kaloupek, and T.M. Keane, Predictors of emotional numbing, revisited: a replication and extension. *Journal of traumatic stress,* 2000. 13(4): p. 611-8.

[127] Hoffman, L., P. Burges Watson, G. Wilson, and J. Montgomery, Low plasma beta-endorphin in post-traumatic stress disorder. *The Australian and New Zealand journal of psychiatry,* 1989. 23(2): p. 269-73.

[128] Van Der Kolk, B., et al., Inescapable shock, neurotransmitters, and addiction to trauma: toward a psychobiology of post traumatic stress. *Biological psychiatry,* 1985. 3(20): p. 314-25.

In: Neurobiology of Post-Traumatic Stress Disorder
Editors: L. Sher and A. Vilens, pp. 53-63
ISBN: 978-1-61470-593-2

Chapter V

Representation of Reward in PTSD

Uta Sailer
University of Vienna, Vienna, Austria

Abstract

This chapter summarises current findings on the processing and representation of rewards in PTSD. Behavioural studies using money and attractive faces as rewards point to changes particularly in the evaluation and the motivational impact of rewards. In accordance with these findings, neuroimaging results showed decreased responsiveness in the nucleus accumbens and mesial prefrontal cortex – central structures in the reward pathway – during the processing of gain feedback. Related to these results, reward function has also been demonstrated to be affected by depression, a disorder that is often co-morbid with PTSD, and stress exposure. The common ground of these findings and their implications are discussed to inspire further research on this new topic.

Introduction

Adaptive behaviour is crucially dependent on the appropriate processing and representation of stimuli that signal the level of goal attainment. Such stimuli may come in the form of rather abstract positive and negative feedback, or as monetary or primary (food and liquid) rewards. Whether it is the sight of the bus one was trying to reach disappearing behind the next corner, or the taste of chocolate melting in one's mouth, what is common to these various forms of feedback is that they can drive behaviour modification, learning, and finally, the prediction of upcoming rewards.

There is reason to assume that the processing of rewards and feedback in general may be altered in PTSD. One reason arises from the fact that emotional numbing is one of the defining symptoms of the disorder according to DSM-IV [1]. Emotional numbing in PTSD has been described as anhedonia of low arousal that may alternate with periods of high arousal [2]. It seems plausible to assume that emotional numbing is manifest in decreased hedonic experience and/or altered representation of reward.

A further reason is that PTSD is often co-morbid with major depression (MDD), which has led some authors to subsume both conditions under the term "trauma-spectrum disorders" [3]. Not only are both PTSD and MDD related to stress, they also show common brain mechanisms [4]. Dysfunction of the brain reward system may underlie at least part of the symptomatology of MDD [e.g., 5]. Therefore, similar mechanisms may be involved in the pathophysiology of PTSD.

Finally, individuals with PTSD have been found to learn more slowly than traumatised controls without PTSD [6, 7]. It may be worth investigating whether inefficient processing of feedback contributes to this delay in learning.

The field is only just beginning to be explored, but a few studies already corroborated the assumption of altered reward function in PTSD.

Neuroimaging

In an fMRI study of our own group, patients with PTSD were tested on their ability to learn from feedback in a neutral gambling task [8]. In this task, two independent types of feedback were presented: performance feedback (correct / wrong) and utilitarian (gain / loss) feedback. The subjects were supposed to learn a sequence of correct responses by using the performance feedback. This way, subjects could learn to make 100% correct responses, but still encountered losses.

To investigate whether learning and an accompanying change in brain activation occurred, and whether this was different for patient and controls, performance and brain activation in the first (the so-called "early phase" of learning) and the second half of the experiment (the "late phase") were compared to each other. Prior to the experiment, all subjects underwent three computer tests to determine whether there were general differences in attention, memory and motivation. In all but one of these tests, the patients scored similarly to controls. However, differences were observed in the motivation test (OLMT, objective performance motivation test, version 21.0) [9]. This test measures motivation via perseverance and effort in a motor task with low cognitive requirement. Patients with PTSD scored lower in a subtest assessing the motivation to put effort into the task at hand. They also showed a tendency for a lower individual aspiration level.

The patients were also slower in learning the correct response sequence in the behavioural task, as they reached the learning criterion (three consecutive blocks with 100% correct answers) later than controls did. Patients also tended to obtain a smaller number of correct responses in the late phase of the experiment than controls. Thus, whereas there were no differences in initial learning, controls showed a more pronounced performance increase as learning progressed than patients did.

These learning-dependent behavioural differences were paralleled by differences in brain activation. Following losses, brain activation was similar in patients and controls during both phases of the experiment. Following gains, however, patients showed less brain activation than controls in the late phase. The areas with smaller activation were the nucleus accumbens

(NAcc), thalamus and mesial prefrontal cortex (medial frontal gyrus, Brodman area 10) (see Figure 1).

These differences occurred similarly in patients who learned and who did not learn the correct response sequence, as a separate analysis for learners and non-learners within each group showed. Although no significant differences in thalamus activation occurred, this may be attributable to the reduced power of this analysis in which only a subgroup of subjects was used. Thus, these results suggest that the differences in mesial prefrontal and accumbal activation – and presumaby, also those in thalamic activation – between the subject groups were not due to differences in learning rate, but rather to the PTSD itself.

Patients also showed larger brain activation than patients following gains in the early phase in the dorsomedial prefrontal cortex and the supramarginal area. We will, however, not discuss these early differences any further. Since we did not have any hypotheses on inter-group differences in these regions, only speculations would be possible.

The NAcc, thalamus and prefrontal cortex are part of the ventral frontostriatal circuitry (see Figure 2) that has been shown to play an important role in reward-related behaviour. Central to this circuit are dopamine projections originating in the substantia nigra and ventral tegmental area which project to the striatum and the prefrontal cortex [10, 11]. Studies in rodents and primates have shown that these dopaminergic neurons respond with phasic activity to novel, unexpected stimuli such as rewards or reward-predicting cues [12]. Thus, their burst firing seems to transmit information about the occurrence of positively-valenced salient stimuli. From the striatum, information reaches the cingulate, orbitofrontal and prefrontal cortices via the thalamus [13, 14]. These structures are part of the so-called "limbic loop", which represents one of several parallel circuits transmitting information from the striatum to the cortex. These circuits are often referred to as the basal ganglia thalamocortical circuits [15]. The prefrontal cortex, in turn, is closely interconnected with the striatum, with the NAcc seeming to receive input predominantly from the mesial PFC [16, 17].

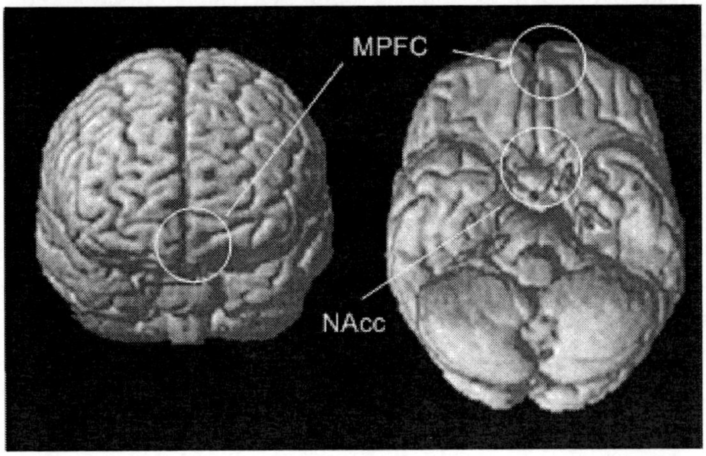

Figure 1. Reduced brain activation following late gains in the mesial prefrontal cortex (MPFC) and nucleus accumbens (NAcc) of patients with PTSD as compared to controls.

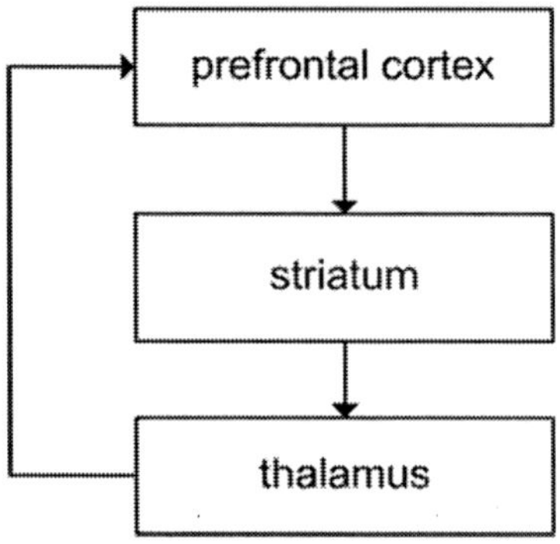

Figure 2. Simplified schematic representation of the cortico-striatal-thalamic-cortical loop.

In humans, the NAcc [18], the mesial PFC [19] and both in combination [20, 21, 22, 23] are activated upon expectation or delivery of primary and secondary rewards. The NAcc receives information from the amygdala, hippocampal formation and limbic association cortex. It responds in a graded way to cues with different predictive value, so that more reliable reward signals elicit larger activation [24, 25, 26]. Together with the ventral striatal area from which it receives dopaminergic inputs, it is one of the most important anatomical substrates for rewards such as food, sex, social interactions, and drugs [27, 28, 29]. Dopamine release in the NAcc has not only been suggested to underly approach responses and motivation [30, 31], but also the experience of pleasure. There is indeed evidence that "neurotransmission in the NAcc can be a sufficient cause for positive affective reactions" [32, p.114] [33].

Neighbouring the orbitofrontal cortex, the mesial PFC has also been implicated in reward [34, 21, 35]. Since the mesial PFC receives input from the rest of the PFC and sends its output to the supplementary motor area, hypothalamus and periaqueductal gray of the brain stem, it has been suggested to facilitate approach behaviour as response to potential reward [36]. The mesial PFC also shares connections with the thalamus [37], an essential structure for the regulation and integration of attentional mechanisms. More specifically, the thalamus has been proposed to synchronise activation in regions which process signals that are in the focus of attention [38]. Possibly related to its function in attentional regulation, the thalamus has also been repeatedly reported to be active in the context of rewards. For example, rewards occurring during a winning streak - i.e. when rewards were increasing rapidly as opposed to stable or decreasing reward levels - led to enhanced activation in the thalamus [20]. In one of the first studies using monetary reward, Thut et al. [39] observed elevated rCBF in the thalamus and OFC, among other regions, following monetary reward in contrast to abstract feedback. Thus, the thalamus acts not only onto the feedback's informational value, but on its rewarding value. As the thalamus receives projections from striatal and limbic regions which are involved in reward-related processes, and in turn projects to mesial prefrontal, orbitofrontal and premotor areas [37], it may act as a link between reward signals, higher cognitive processes and behaviour [20]. Consistent with our findings, reduced activation in the thalamus of PTSD patients has been reported by Lanius et al. [40, 41, 42]. It has been suggested that disrupted thalamus activation could lead to the misinterpretation of external feedback in PTSD [43]. Alternatively, feedback may be correctly interpreted but then fail to induce the appropriate approach behaviour.

The fact that the PTSD patients in our study showed reduced activation in this circuit – encompassing the mesial PFC, nucleus accumbens, and thalamus - suggests that their representation of reward is altered. However, this alteration only occurred when the patients experienced gains in the second part of the experiment (so-called "late gains"). This suggests that the saliency of rewards decreased as the experiment progressed. This was the case independent of whether the patients had learned the correct response sequence or not. Therefore, the differences in brain activation were not related to a difference in the ability to predict upcoming positive feedback.

Given the reduced score of patients in the motivation test, a decrease in motivation is the most straightforward explanation for our findings. The test measured the motivation to put effort into the current task. In relation to this task-oriented motivation, the reduced activation in the nucleus accumbens following late gains may indicate reduced reward-seeking motivation. Thus, our findings suggest an underactive, or, more precisely, a rapidly exhausting appetitive motivational system in PTSD.

Related to this interpretation, the results of electrophysiological studies point to other aspects of motivational deficits in PTSD. In one such study, the P300 was found to be reduced in PTSD [44]. The P300 is an event-related potential that is particularly pronounced following infrequent, unpredictable target stimuli. This decrease in P300 amplitude can be interpreted as reduced motivational significance of the target.

The obvious question given these results is whether the patients also experienced the positive feedback as less rewarding. In this case, the reduced activation in reward-related areas could be related to the emotional numbing in PTSD. Unfortunately, this question cannot be answered with the present data since we did not collect any subjective reports. There are, however, a few behavioural studies that more closely looked at the subjective experience of reward in PTSD.

Behavioural Studies

One of the first studies tackling this question investigated performance in a stop-signal task with either abstract feedback, i.e. by presenting a "+" or an "x" on the screen, or monetary reward [45]. In contrast to the controls, PTSD subjects showed larger stop-signal reaction times with reward than with abstract feedback. Thus, inhibition was reduced in the presence of rewards. Although this finding may suggest increased reward sensitivity, questionnaire data from the same study showed that subjects reported decreased reward dependence on the Tri-dimensional Personality Questionnaire (TPQ) [46]. The authors interpret this discrepance as the subjective ratings obviously not being related to the behavioural effects. Further investigations may hopefully help to clarify this discrepancy.

Using a different type of reward, Elman and colleagues investigated reward function in PTSD with attractive vs average facial images [47]. In their experiment, the subjects could choose to view each picture for either a predefined time by not pressing any key, or to extend the viewing time by performing a combination of key presses. Compared to the veterans without PTSD, those with PTSD exerted less effort to increase the viewing time of attractive images, although they rated the pictures as similarly attractive. Since it was not measured whether viewing the attractive faces was also experienced as rewarding, the authors performed a further study in which they directly measured the satisfaction with experienced monetary rewards [48]. Being presented the question "What do you expect the outcome of this trial to be?" to be answered on a visual analogue scale with the anchors "very negative" and "very positive", the subjects with PTSD generally gave more negative expectancy ratings than the control subjects did, although they were as capable as controls to distinguish between different probabilities of reward. This indicates that whereas the subjects with PTSD were able to correctly predict upcoming outcomes, they labeled them in a less positive way than controls. Thus, the problem does not seem to lie in the interpretation of feedback, but in its effect on motivation.

In addition, when subjects with PTSD expected a negative outcome and then experienced a gain or a loss, they reported to be less satisfied with this outcome than controls did. When they expected a positive outcome, the satisfaction with the actual outcome was similar in both subject groups. Unfortunately, the authors did not analyse the satisfaction with gains and losses separately. Therefore, it is difficult to determine whether subjects with PTSD showed less satisfaction than controls with experienced rewards (i.e., a positive outcome when a negative one was expected) or losses (a negative outcome when a negative one was expected). Nevertheless, the study encourages further research on reward processing in PTSD. It also shows that it may be useful to explore this topic first by purely behavioural studies before turning to the use of brain imaging.

The role of reward representation is only beginning to be analysed in PTSD. However, studies about the effect of stress and depression on reward functions revealed interesting findings that are likely to be related to the effects of PTSD. In fact, it has been suggested that stress induces anhedonia. For example, US Army cadets and college students reported experiencing less pleasure while viewing amusing film clips in the end of a stressful day – with challenging field trainings in the former group, and final examinations in the latter – than of a control day [49]. A different study showed that reward responsiveness in healthy female subjects was reduced when exposed to stress via threat-of shock than in a control condition [50]. Reward responsiveness was defined as the preference for the response paired with the more frequent reward. Whereas this preference was less pronounced when under stress, global performance was not affected, a finding pointing to a specific hedonic impairment. Moreover, the authors also found a negative correlation between their measure of reward responsiveness and self-reported anhedonia in the stress condition. These findings suggest that stress indeed leads to blunted hedonic experience. Since extreme stress is the trigger for PTSD, similar mechanisms are likely to occur in these patients.

Reward Function in Depression

Anhedonia and loss of motivation are not only core symptoms of PTSD, but also of MDD which is frequently comorbid. In MDD, these symptoms have been proposed to be due to dysfunction in the mesolimbic dopamine pathway [51, 29, 52].

There is growing evidence about both reward sensitivity [53, 54, 55], the response to negative feedback, and the use of negative feedback for the adjustment of behaviour during learning being disturbed in MDD [56]. For example, using a progressive reinforcement schedule for financial reward, it was found that depressed subjects worked less for reward

[57]. The behaviour observed in this pioneering study seems reminiscent of PTSD subjects putting less effort into the current task such as in the motivation test used by Sailer et al. [8] or the key press task that allowed to extend the viewing time of attractive faces used by Elman et al. [47]. Furthermore, similarly to the more negative expectancy ratings of spinner outcomes in subjects with PTSD [48], depressed subjects gave a lower estimate of their success rate in a word association task [58].

The notion of the mesolimbic pathway mediating depression and depression-like behaviour emerged from studies with dopamine receptor agonists [59]. Although existing depression research mostly focused on serotonergic and noradrenergic mechanisms, the role of dopaminergic mechanisms in the brain reward system is beginning to be more and more acknowledged [52]. Since PTSD and depression commonly co-occur, similar brain mechanisms may be involved – possibly via a common genetic basis. However, the question of how the two disorders are linked remains to be solved.

Conclusion

Although research on reward in PTSD is still in its early stages, the existing results hint at the processing and/or representation of reward to be altered in these patients. The common finding in these studies seems to be that the patients are unimpaired in predicting upcoming rewards, but show deficits in the motivation to obtain further rewards, and possibly also, the hedonic experience of rewards.

One problem with the existing studies is the variation in ways to measure and define reward. Reward prediction, activation in brain areas processing positive feedback, and hedonic experience are obviously different aspects of reward function. Future studies should profit from simultaneously investigating these aspects in the same experiment. A greater understanding of the changes in reward representation associated with PTSD is likely to also provide clues for a more targeted therapy.

Acknowledgments

This work was supported by the German Research Foundation (DFG), the Austrian National Bank (OeNB), and the Austrian Science Fund (FWF).

References

[1] *Diagnostic and statistical manual of mental disorders* (DSM-IV). Washington, DC: American Psychiatric Association, 1994.

[2] Frewen PA, Lanius RA: Toward a psychobiology of posttraumatic self-dysregulation: reexperiencing, hyperarousal, dissociation, and emotional numbing. *Annals of the New York Academy of Sciences 2006*; 1071: 110-124.

[3] Bremner JD: Does stress damage the brain? *Understanding trauma-related disorders from a mind-body perspective.* New York: Norton, 2002.

[4] Bremner JD, Elzinga B, Schmahl C, Vermetten E: Structural and functional plasticity of the human brain in posttraumatic stress disorder. In: E.R. de Kloet, M.S. Oitzl & E. Vermetten (Eds.*), Progress in Brain Research* 2008; 167: 157-172.

[5] Naranjo CA, Tremblay LK, Busto UA: The role of the brain reward system in depression. *Progress in Neuro-Psychopharmacology & Biological Psychiatry* 2001; 25: 781-823.

[6] Vasterling JJ, Brailey K, Constans JI, Sutker PB: Attention and memory dysfunction in posttraumatic stress disorder. *Neuropsychology* 1998; 12: 125-133.

[7] Vasterling JJ, Duke LM, Brailey K, Constans JI, Allain AN Jr., Sutker PB: Attention, learning, and memory performances and intellectual resources in Vietnam veterans, PTSD and no disorder comparisons. *Neuropsychology* 2002; 16: 5-14.

[8] Sailer U, Robinson S, Fischmeister FP, Konig D, Oppenauer C, Lueger-Schuster B, Moser E, Kryspin-Exner I, Bauer H: Altered reward processing in the nucleus accumbens and mesial prefrontal cortex of patients with posttraumatic stress disorder. *Neuropsychologia* 2008; 46: 2836-2844.

[9] Schmidt-Atzert L: *Objektiver Leistungsmotivations Test OLMT [Objective Achievement Motivation Test (OLMT)].* Moedling, Austria: Schuhfried, 2004.

[10] Haber SN: The primate basal ganglia: parallel and integrative networks. *Journal of Chemical Neuroanatomy* 2003; 26: 317-30.

[11] Nieuwenhuys R: *Chemoarchitecture of the brain.* Springer: New York, 1985.

[12] Schultz W: Getting formal with dopamine and reward. *Neuron* 2002; 36: 241-263.

[13] Martin, JH: *Neuroanatomy: Text and Atlas.* New York, McGraw-Hill Medical, 2003.

[14] Ray JP, Price JL: The organization of projections from the mediodorsal nucleus of the thalamus to the orbital and medial prefrontal cortex in macaque monkeys. *Journal of Comparative Neurology* 1992; 337: 1-31.

[15] Alexander GE, DeLong MR, Strick PL: Parallel organization of functionally segregated circuits linking basal ganglia and cortex. *Annual Review of Neuroscience* 1986; 9: 357-81.

[16] Ferry AT, Öngür D, An X, Price JL: Prefrontal cortical projections to the striatum in macaque monkeys: evidence for an organization related to prefrontal networks. *Journal of Comparative Neurology* 2000; 425: 447–470.

[17] Roberts AC, Tomic DL, Parkinson CH, Roeling TA, Cutter DJ, Robbins TW, Everitt BJ: Forebrain connectivity of the prefrontal cortex in the marmoset monkey (Callithrix jacchus): an anterograde and retrograde tract-tracing study. *Journal of Comparative Neurology* 2007; 502: 86–112.

[18] Delgado MR, Nystrom LE, Fissell C, Noll DD, Fiez JA: Tracking the hemodynamic responses to reward and punishment in the striatum. *Journal of Neurophysiology* 2000; 84: 3072-3077.

[19] Bush G, Vogt BA, Holmes J, Dale AM, Greve D, Jenike MA, Rosen BR: Dorsal anterior cingulate cortex: a role in reward-based decision making. *Proceedings of the National Academy of Sciences of the United States of America* 2002; 99: 523-8.

[20] Elliott R, Friston KJ, Dolan RJ: Dissociable Neural Responses in Human Reward Systems. *Journal of Neuroscience* 2000; 20: 6150-6165.

[21] Knutson B, Fong GW, Bennett SM, Adams CM, Hommer D: A region of mesial prefrontal cortex tracks monetarily rewarding outcomes: characterization with rapid event-related fMRI. *Neuroimage* 2003; 18: 263-72.

[22] McClure SM, Li J, Tomlin D, Cypert KS, Montague LM, Montague PR: Neural correlates of behavioral preference for culturally familiar drinks. *Neuron* 2004; 44: 379-387.

[23] Bjork JM, Hommer DW: Anticipating instrumentally obtained and passively-received rewards: A factorial fMRI investigation. *Behavioural Brain Research* 2007; 177: 165-170.

[24] Abler B, Walter H, Erk S, Kammerer H, Spitzer M: Prediction error as a linear function of reward probability is coded in human nucleus accumbens. *Neuroimage* 2006; 31: 790-5.

[25] Heekeren HR, Wartenburger I, Marschner A, Mell T, Villringer A, Reischies FM: Role of ventral striatum in reward-based decision making. *Neuroreport* 2007; 18: 951-5.

[26] Spicer J, Galvan A, Hare TA, Voss H, Glover G, Casey B: Sensitivity of the nucleus accumbens to violations in expectation of reward. *Neuroimage* 2007; 34:455-61.

[27] Smith GP: Dopamine and food reward. *Progress in Psychobiology and Physiolical Psychology* 1995; 16: 83–144.

[28] Wise RA: Drug-activation of brain reward pathways. *Drug Alcohol Dependence* 1998; 51: 13-22.

[29] Nestler EJ, Carlezon WA Jr: The mesolimbic dopamine reward circuit in depression. *Biological Psychiatry* 2006; 59: 1151-9.

[30] Koob GF: Drug addiction: the yin and yang of hedonic homeostasis. *Neuron* 1996, 16: 893-896.

[31] Salamone JD, Cousins MS, Snyder BJ: Behavioral functions of nucleus accumbens dopamine: empirical and conceptual problems with the anhedonia hypothesis. *Neuroscience and Biobehavioral* Reviews 1997, 21: 341-359.

[32] Berridge KC. Pleasures of the brain. *Brain and Cognition* 2003; 52: 106-28.

[33] Peciña S, Smith KS, Berridge KC: Hedonic hot spots in the brain. *Neuroscientist* 2006; 12: 500-11.

[34] Knutson B, Adams CM, Fong GW, Hommer D: Anticipation of increasing monetary reward selectively recruits nucleus accumbens. *Journal of Neuroscience* 2001; 21: RC159.

[35] O'Doherty J, Kringelbach ML, Rolls ET, Hornak J, Andrews C: Abstract reward and punishment representations in the human orbitofrontal cortex. *Nature Neuroscience* 2001; 4: 95-102.

[36] Knutson B, Westdorp A, Kaiser E, Hommer D: FMRI visualization of brain activity during a monetary incentive delay task. *Neuroimage* 2000; 12: 20-7.

[37] Price JL: Prefrontal cortical networks related to visceral function and mood. *Annals of the New York Academy of Sciences* 1999; 877: 383-396.

[38] Frith CD, Friston KJ. The role of the thalamus in "top down" modulation of attention to sound. *Neuroimage* 1996; 4: 210-5.

[39] Thut G, Schultz W, Roelcke U, Nienhusmeier M, Missimer J, Maguire RP, Leenders KL: Activation of the human brain by monetary reward. *Neuroreport* 1997; 8: 1225-8.

[40] Lanius RA, Williamson PC, Densmore M, Boksman K, Gupta MA, Neufeld RW, Gati JS, Menon RS: Neural correlates of traumatic memories in posttraumatic stress disorder: a functional MRI investigation. *American Journal of Psychiatry* 2001; 158: 1920-2.

[41] Lanius RA, Williamson PC, Hopper J, Densmore M, Boksman K, Gupta MA, Neufeld RW, Gati JS, Menon RS: Recall of emotional states in posttraumatic stress disorder: an fMRI investigation. *Biological Psychiatry* 2003; 53: 204-10.

[42] Lanius RA, Williamson PC, Densmore M, Boksman K, Neufeld RW, Gati JS, Menon RS: The nature of traumatic memories: a 4-T FMRI functional connectivity analysis. *American Journal of Psychiatry* 2004; 161: 36-44.

[43] Francati V, Vermetten E, Bremner JD: Functional neuroimaging studies in posttraumatic stress disorder: review of current methods and findings. *Depression and Anxiety* 2007, 24: 202-218.

[44] Felmingham KL, Bryant RA, Kendall C, Gordon E: Event-related potential dysfunction in posttraumatic stress disorder: the role of numbing. *Psychiatry Research* 2002; 171-179.

[45] Casada JH, Roache JD: Behavioral inhibition and activation in posttraumatic stress disorder. *Journal of Nervous and Mental disease* 2005; 193: 102-109.

[46] Cloninger CR, Przybeck TR, Svrakic DM: The Tridimensional Personality Questionnaire: U.S. normative data. *Psychological Reports* 1991; 69: 1047-57.

[47] Elman I, Ariely D, Mazar N, Aharon I, Lasko NB, Macklin ML, Orr SP, Lukas SE, Pitman RK: Probing reward function in post-traumatic stress disorder with beautiful facial images. *Psychiatry Research* 2005; 135: 179-183.

[48] Hopper JW, Pitman RK, Su Z, Heyman GM, Lasko NB, Macklin ML, Orr SP, Lukas SE, Elman I: Probing reward function in posttraumatic stress disorder: expectancy and satisfaction with monetary gains and losses. *Journal of Psychiatric Research* 2008; 42: 802-807.

[49] Berenbaum H, Connelly J: The effect of stress on hedonic capacity, *Journal of Abnormal Psychology* 1993; 102: 474–481.

[50] Bogdan R, Pizzagalli DA: Acute stress reduces reward responsiveness: implications for depression. *Biologica Psychiatry* 2006; 60: 1147-54.

[51] Willner P: Dopaminergic mechanisms in depression and mania. In: *Psychopharmacology: the Fourth Generation of Progress*. Edited by Bloom FE, Kupfer DJ. New Work: Raven Press, 1995, pp 921-931.

[52] Nestler EJ, Barrot M, DiLeone RJ, Eisch AJ, Gold SJ, Monteggia LM: Neurobiology of depression. *Neuron* 2002; 34:13-25.

[53] Henriques JB, Davidson RJ: Decreased responsiveness to reward in depression. *Cognition and Emotion* 2000; 14: 711-724.

[54] Henriques JB, Glowacki JM, Davidson RJ: Reward fails to alter response bias in depression. *Journal of Abnormal Psychology* 1994, 103: 460-466.

[55] Shankman SA, Klein DN, Tenke CD, Bruder GE: Reward sensitivity in depression: a biobehavioral study. *Journal of Abnormal Psychology* 2007, 116: 95-104.

[56] Martin-Soelch C: Is depression associated with dysfunction of the central reward system? *Biochem Soc Trans* 2009, 37: 313-317.

[57] Hughes JR, Pleasants CN, Pickens RW. Measurement of reinforcement in depression: A pilot study. *Journal of Behavior Therapy and Experimental Psychiatry* 1985; 16: 231–236.

[58] Wener AE, Rehm L: Depressive affect: a test of behavioral hypotheses. *Journal of Abnomral Psychology* 1975, 84: 221-227.

[59] Wise RA: Neuroleptics and operant behavior: The anhedonia hypothesis. *Behav Brain Sci* 1982, 5: 39-87.

In: Neurobiology of Post-Traumatic Stress Disorder
Editors: L. Sher and A. Vilens, pp. 65-89
ISBN: 978-1-61470-593-2

Chapter VI

Genetic Substrates of Posttraumatic Stress Disorder: Review of Extant Research and Neurobiologically Informed Future Directions

Ananda B. Amstadter[1], *Kristyn Zajac*[1], *Nicole R. Nugent*[2] *and Karestan C. Koenen*[3]
[1]Medical University of South Carolina, Charleston, South Carolina, USA
[2]Bradley Hasbro Children's Research Center, Providence, Rhode Island, USA
[3]Harvard School of Public Health, Boston, Massachusetts, USA

Abstract

Extant research suggests genetic and biological factors shape individuals' responses to trauma and play a role in determining who develops posttraumatic stress disorder (PTSD). However, the role of genetic and biological risk factors in the etiology of PTSD has been largely neglected by many trauma researchers. The aim of this chapter is to review current evidence for the role of genetic and biological risk factors in PTSD etiology in a way that is broadly accessible to the field of traumatic stress. In order to accomplish this aim, we first operationalize our definition of risk factors and discuss the specific challenges to identifying biological risk factors for PTSD. Next, we review the evidence for genetic risk factors for PTSD. Then, we review the evidence for other biological risk factors, with a specific emphasis on neurocognitive factors. Finally, we conclude by reflecting on future research needed to better understand the role of genetic and biological risk factors in PTSD etiology.

Introduction

Although an estimated 50-70% of individuals in the United States experience at least one traumatic event during their lifetime [1-3], only a minority of trauma-exposed individuals develop posttraumatic stress disorder (PTSD) [1]. Conditional risk of PTSD is estimated to range from 8-32%, depending on the type of trauma exposure [1, 4]. PTSD is highly comorbid with other mental disorders, frequently co-occurring with substance abuse/dependence, major depression, other anxiety disorder, and antisocial personality disorder [1, 3, 5-12]. Despite gains in the psychosocial literature to identify variables related to risk and resilience in the aftermath of exposure to a traumatic event [13], a large amount of variance remains unexplained, calling for additional variables to be included in models of risk and resilience. The present chapter aims to review the literature on biologic and genetic risk factors for PTSD. Genetic and biologic research has the potential to greatly inform our understanding of PTSD with regards to (a) why some individuals are vulnerable and others resilient to the effect of trauma, (b) genetic underpinnings co-occuring psychiatric concerns, and (c) putative functional genetic and biologic vulnerabilities.

Challenges to Identifying Biologic Risk Factors for PTSD

Very few biological correlates shown to be associated with PTSD actually qualify as risk factors for the disorder. Our understanding of what constitutes a risk factor has been succinctly described by Kraemer et al. (p.337), "A risk factor is a measurable characterization of each subject in a specified population that precedes the outcome of interest and which can be used to divide the population into two groups (the high-risk and the low-risk groups that comprise the total population"[14]. Temporality is a key aspect of this definition; a biological factor associated with PTSD only qualifies as a risk factor if it has been shown to precede the disorder. In most biologically-oriented PTSD research, traditional case-control designs are used to assess whether a given biological marker is more prevalent in individuals with PTSD compared to trauma-exposed controls who did not develop PTSD. Such studies show whether a biological factor is a correlate of PTSD. However, traditional case-control designs can not determine whether the biological factor is a part of the disease process, also referred to as a 'PTSD sign,' or a risk factor for the disorder. This limitation is present even in prospective studies that test whether biological factors assessed in the acute aftermath of trauma exposure predict the development of PTSD. Biological factors assessed in the acute aftermath of trauma exposure that predict the development of PTSD may be early indicators of the emerging disorder rather than true risk factors. Thus, to be considered in this review, there must be evidence that the biological factor was: (1) present prior to trauma exposure, and (2) associated with risk of PTSD.

Genetic Study Designs

An individual's genotype or DNA sequence is determined at conception. Thus, genetic variants associated with risk for PTSD qualify as risk factors according to our definition above, with the exception of epigenetic modifications of the genome that occur as a result of environmental input. Genetic investigations of PTSD have implemented family, twin and molecular genetic study designs. *Family studies* explore whether risk of having PTSD is higher among family members of individuals with PTSD (called *probands* in genetic studies) than non-relatives. If relatives of probands have a higher risk of PTSD, then PTSD is presumed to "run in families" or be "familial." However, as family members share more environmental experiences than unrelated individuals, it is difficult to determine whether observed differences are attributable to genetic or environmental reasons. Family studies are often the first step in genetic research.

Twin studies are needed to disentangle the role of genetic and environmental influences. The twin design is used to calculate the "heritability" of PTSD, or the proportion of the variance in an anxiety disorder that can be explained by genetic factors. Basically, the twin method compares similarity between identical or *monozygotic (MZ)* pairs relative to similarity between fraternal or *dizygotic (DZ)* pairs. MZ twins share 100% of their genes as well as 100% of the shared environment; DZ twins share roughly 50% of their genes and 100% of their environment. If MZ twins are significantly more similar on a characteristic than DZ twins, then this *phenotype* (observed characteristic) is genetically influenced. The *heritability* estimate is derived by 2(rMZ – rDZ), where r = the intraclass twin correlation [15]. For categorical phenotypes, such as PTSD, the tetrachoric correlation is used to calculate heritability.

Molecular genetic studies have sought to understand the role of specific genes in the etiology of PTSD. Genetic association studies identify whether variation in DNA sequence in specific genes is associated with the PTSD diagnosis. These studies focus on the tiny fraction (1%) of the DNA sequences that differs among individuals. Studies of PTSD have made use of two types of DNA sequence variation. The first are single nucleotide polymorphisms (SNPs, pronounced "snips") which occur when a single nucleotide (A, T, C or G) in the DNA sequence is altered. An example of a SNP is a change the DNA sequence from CGTTGG to CGATGG. The second type of polymorphisms is variable number tandem repeat polymorphisms (VNTR). Tandem repeats occur where a pattern of two or more nucleotides is repeated and the repetitions are directly adjacent to each other. VNTR's occur where the number of these repeats varies between individuals. *Candidate gene association studies* use extant research to guide selection of genes which are then tested for potential associations with the disorder. In contrast to genetic association studies which focus on DNA sequence variation, *gene expression studies* have examined whether patterns of gene activity distinguish individuals with PTSD from trauma-exposed controls who do not have PTSD.

Family Studies of PTSD

A diagnosis of PTSD requires exposure to at least one traumatic event, making comparisons among family members who may not have experienced trauma difficult.

Accordingly, few family studies have examined PTSD. One recent study examined PTSD symptoms in 12 multigenerational families (N = 200) exposed to the Spitak earthquake [16]. After adjusting for sex, age, and environmental risk factors (i.e., other traumatic experiences, earthquake-related injury, death of a family member, home destruction) PTSD symptoms were reported to be 41% heritable (unadjusted heritability = 57%). Although other published family studies have not been able to test formal heritability, given the constraints of family research in PTSD, their findings remain noteworthy. Early family studies of PTSD generally focused on genetically-related dyads who have known (often shared) trauma exposure. Most often these studies have examined the association between parent and child PTSD in trauma-exposed samples. For example, Cambodian refugee children whose mother and father both had PTSD were five times more likely to receive the diagnosis than refugee children whose parents did not have PTSD [17]. Similarly, adult children of Holocaust survivors with PTSD had a higher risk of PTSD following trauma compared to adult children of Holocaust survivors without PTSD [18].

Family studies have also explored the association between child and parent post-traumatic stress symptoms following child traumatic exposure. Parent-child PTSD symptom correlations following child traumatic exposure have been mixed, with some studies reporting significant associations [19-22] whereas other studies have found no relationship [23-25]. Attempts to resolve this apparent discrepancy have highlighted the importance of time lapsed between traumatic exposure and assessment [26]. Studies examining acute distress levels soon after the trauma typically find no association between parent and child PTSD symptoms [27, 28], while longitudinal studies show increases in parent-child PTSD symptom associations over time [29-31]. Initial symptoms in the child have been found to impact subsequent symptoms in the parent [29] and initial parent symptoms have been found to predict subsequent child symptoms [32-34]. However, since family members frequently share important environmental experiences, family studies cannot tell us whether a disorder runs in families for genetic or environmental reasons. Twin studies have been used to distinguish between genetic and environmental contributions to risk for PTSD.

Family studies have also looked more broadly at whether there is a familial association between PTSD and other mental disorders. Family history studies of combat veterans have reported greater prevalence of anxiety disorders in families of individuals with PTSD [35, 36]. Similarly, a civilian epidemiological study found familial anxiety to be a risk factor in the development of PTSD [7] and parental mental disorders are associated with risk for PTSD in male civilians [37]. A family history of depression has been associated with proband PTSD in trauma-exposed veterans, albeit inconsistently [35, 36, 38]. Female rape victims with PTSD and comorbid lifetime depression also evidenced an increased prevalence of familial depression [39]. Finally, Goenjian and colleagues' [16] investigation of PTSD symptoms in 200 family members following the Spitak earthquake additionally examined post-earthquake symptoms of depression and anxiety, revealing high genetic correlations between PTSD and anxiety (*[rho]*g = .75) and depression (*[rho]*g = .71). These findings suggest PTSD has a familial association with anxiety disorders and major depression.

Twin Studies of PTSD

Twin studies have made three important contributions to our understanding of the genetic etiology of PTSD. First, and perhaps most notably, they indicate that genetic factors influence exposure to PTEs. This is referred to as gene-environment correlation, whereby selection of environment, and subsequently potential for exposure to trauma, is partly determined by genetic factors [40]. For example, twin studies have demonstrated that genetic factors influence exposure to PTEs. Lyons et al. [41] included members of the Vietnam Era Twin Registry (VET) and studied variables indicative of combat-related trauma (e.g., volunteering for service in Southest Asia, service in Southeast Asia, combat exposure, being awarded a combat medal). Heritability estimates ranged from 35% for Southeast Asia service to 54% for being awarded a combat medal. Additionally, a civilian study found evidence for a gene-environment correlation for assaultive violence [42].

Second, twin studies suggest genetic influences explain a substantial proportion of vulnerability to PTSD even after accounting for genetic influences on PTE exposure. An early examination of the Vietnam Era Twin (VET) Registry reported that 30% of the variance in PTSD was accounted for by genetic factors, even after controlling for combat exposure [43]. Similarly, a twin study of male and female civilian volunteers identified similar heritability of PTSD, with further variance accounted for by non-shared environmental factors [42]. Taken together with family research designs using disaster-exposed samples (i.e., [16], which are less vulnerable to the self-selection bias that may exist in samples exposed to military or assault trauma, these findings suggest that genetic factors play an important role in vulnerability to developing PTSD.

Third, data from twin studies on PTSD indicate some degree of specificity of genetic influences on PTSD as well as some degree of overlap in genetic contributions with other mental disorders. For example, genetic influences on major depression account for the majority of the genetic variance in PTSD [44, 45]. Genetic influences common to generalized anxiety disorder and panic disorder symptoms account for approximately 60% of the genetic variance in PTSD [46] and those common to alcohol and drug dependence [47] and nicotine dependence [48] account for over 40% of the variance associated with PTSD. Thus, the majority of genes that affect risk for PTSD also influence risk for other psychiatric disorders and vice versa.

Molecular Genetic Studies of PTSD

Although twin and family studies support the notion that PTSD is moderately heritable, one key limitation of these forms of investigations is that they cannot tell us which genes are important in PTSD etiology. The literature on molecular genetic investigations, which have the potential to identify genetic markers of vulnerability or resilience, is rapidly growing. For example, when our group first reviewed this literature just two years ago there were 11 molecular genetic studies of PTSD [49]. This literature has nearly doubled in two years, as currently there are 21 published molecular genetic investigations of PTSD. Our group [49-51] and other groups [52] have reviewed the molecular genetic PTSD literature in detail, and the interested reader is referred to those sources for a more in-depth description of these studies.

For the purposes of the present chapter, a brief overview of the 21 candidate gene studies of PTSD that have been conducted is provided.

Dopamine System

As shown in Table 1, most of the molecular genetic studies of PTSD conducted to date have focused on the dopaminergic system (i.e., *DRD2, DRD4, DAT*), as both animal and human studies have implicated the DA system in the etiology of PTSD [53, 54]. Five out of the seven investigations of dopamine system genes studied the relation between marker alleles at the D2 dopamine receptor gene (*DRD2*) and PTSD. Initial investigations reported a positive association with the *DRD2A1* allele [55, 56]. However, a subsequent investigation, that did not assess for trauma exposure in the control group, found no association with the *DRD2A1* allele or with any combination of alleles for the *DRD2* locus [57]. Of note, the Comings and colleagues [56] investigation consisted of a relatively small sample of substance abusers with PTSD (N = 37) compared with substance abusers without PTSD (N = 19), limiting generalizability to a substance abusing population. Comorbid PTSD and substance abuse was also addressed in an investigation of combat veterans with and without PTSD, with analyses revealing a positive association between *DRD2A1* and PTSD only in the subset of PTSD cases who engaged in harmful drinking [58]. Another investigation of the *DRD2* gene in a case-control study of war veterans with PTSD and general population controls has been conducted [59]. This study examined three SNPs in the *DRD2* gene, and found a significant association between the 957C>T SNP and PTSD, with those with PTSD having a higher likelihood of having the C allele than were controls. No significant associations were found for the *TaqlA or -141delC* polymorphisms; however, it is not clear whether control participants had a traumatic event history. In an investigation of yet another dopamine receptor gene (*DRD4),* one recent study of flood survivors reported that *DRD4* long allele (seven or eight repetitions) carriers had more intense PTSD symptoms [60]. The final study examined a slightly different facet of dopamine transmission in patients with chronic PTSD and trauma-exposed healthy controls, reporting a positive association between of the DA transporter *SLC6A3* (DAT1) 3'polymorphism and chronic PTSD [61].

Serotonin System

The serotonergic system (i.e., *SLC6A4, 5-HTR2A*) has also been studied in case-control studies of PTSD. There have been four investigations of a VNTR in the promoter region of the serotonin transporter (*SLC6A4,* locus 5-HTTLPR), a functional polymorphism, with the short, s, allele being less transcriptionally efficient than the long, l, allele [62]. The first investigation of this gene was conducted by Lee and colleagues who found an excess of s/s genotypes in Korean PTSD patients compared with normal controls [63]. Another investigation in a sample of Rwandan refugees indicated a main effect for genotype, with the resiliency associated with the presence of an l allele diminishing across exposure to increasing numbers of traumatic events [64]. More specifically, whereas the fitted probability of lifetime PTSD for participants homozygous for the low expression allele (s/s) was 100% regardless of the number of traumatic experiences, carriers of the l alleles evidenced lower levels of PTSD following exposure to few trauma experiences but showed significantly increased risk associated with increasing numbers of traumatic experiences.

Table 1. Review of published case-control candidate gene associations studies of PTSD by neurobiological system

First author	Year	Trauma Exposed Controls?	Trauma Type	Gene Name (Symbol)	Finding
Dopamine System					
Comings	1991	No	Combat	Dopamine Receptor D2 (DRD2)	Excess D2A1 allele in PTSD cases p=.007
Comings	1996	Yes	Combat	Dopamine Receptor D2 (DRD2)	Excess D2A1 allele in PTSD cases p=.041
	1996	Yes	Combat	Dopamine Receptor D2 (DRD2)	Excess D2A1 allele in PTSD cases p=.002
Gelernter	1999	No	Combat	Dopamine Receptor D2 (DRD2)	No significant association between D2A1 allele/DRD2 haplotypes and PTSD
Young	2002	No	Combat	Dopamine Receptor D2 (DRD2)	Excess D2A1 allele only in PTSD cases with harmful drinking p<.001
Voisey	2008	Yes	Combat	Dopamine Receptor D2 (DRD2)	Excess C allele in 957C>T in PTSD; no differences in Taq1A or -141delC
Dragan	2009	Yes	Flood	Dopamine Receptor D4 (DRD4)	DRD4 long allele predicted more intense symptoms of PTSD p=.048
Segman	2002	Yes	Various	Dopamine Transporter (DAT1)	Excess 9-repeat allele in PTSD cases p=.012
Serotonin System					
Lee	2005	No	Various	Serotonin Transporter (SLC6A4)	Excess s allele in PTSD cases p = .04

Table 1. Continued

First author	Year	Trauma Exposed Controls?	Trauma Type	Gene Name (Symbol)	Finding
Kolassa*	In press	Yes	War	Serotonin Transporter (SLC6A4)	The fitted probability of PTSD in 's' homozygotes, regardless of number of traumas reported. However, participants with the s'/l' and l'/l' genotypes showed an S-shaped dose-response, with increasing rates of PTSD reported as a function of number of traumatic experiences. Little difference for genotype was observed in participants reporting 15+ traumas.
Kilpatrick*	2007	Yes	Hurricane	Serotonin Transporter (SLC6A4)	Significant association between s/s genotype and PTSD in adults with high hurricane exposure and low social support
Koenen*	2009	Yes	Hurricane; Various	Serotonin Transporter (SLC6A4)	Significant interaction between genotype and both county-level environmental variables. 's' allele was associated with decreased risk of PTSD in low-risk environments (low crime/unemployment), and increased risk of PTSD in high risk environments
Lee	2007	Yes	Not specified	Serotonin 2A receptor (5-HTR2A)	Excess GG genotype in female PTSD p = .04
Alternative Neurobiological Systems					
Lee	2006	Yes	Not specified	Brain derived Neurotrophic Factor (BDNF)	No significant association between BDNF Val66Met and PTSD
Zhang	2006	Not specified	Not specified	Brain Derived Neurotrophic Factor (BDNF)	No significant association between three BDNF variants and PTSD
Lappalainen	2002	No	Combat	Neuropeptide Y (NPY)	No significant association between Leu7Pro polymorphism and PTSD
Bachman	2005	Yes	Combat	Glucocorticoid Receptor (GCCR)	No significant association between GCCR polymorphisms and PTSD
Mustapic	2007	Yes	Combat	Dopamine Beta-Hydroxylase	No association of -1021C/T variant with PTSD; PTSD CC

				(DBH)	genotype cases had lower DBH activity than non-PTSD CC cases p<.001
Lu	2008	No	Not specified	Cannabinoid Receptor Gene (CNR1)	Excess C-A and C-G variants in Caucasian parents of youth with ADHD who report PTSD
Binder*	2008	Yes	Various	FKBP5	4 SNPs in *FKBP5* significantly interacted with severity of child abuse in prediction of adult PTSD symptoms ps<.0004
Amstadter*	2009	Yes	Hurricane; Various	Regulator of G-Protein Signaling 2 (RGS2)	Significant association between CC genotype and post-hurricane PTSD in adults with high hurricane exposure and low social support; significant association between CC genotype and lifetime PTSD in adults with prior trauma exposure and low social support
Nelson*	2009	Not Specified	Various	γ-aminobutyric acid receptor subunit A2 (GABRA2)	Significant interaction between composite lifetime history of trauma exposure and three out of four risk alleles for adult PTSD (the third SNP was of marginal significance).

* = GxE study; PTSD = posttraumatic stress disorder; SNPs = single nucleotide polymorphisms; D2DA1 = Al one allele of DRD2 gene; s allele = short version (versus long) of the serotonin transporter promoter polymorphism.

The 5-HTTLPR polymorphism has been found to be triallelic, in that a third functional allele, 'L_G', has been identified [65]; L_G is characterized by an A>G substitution at nucleotide 6 of the first of two extra 22-bp repeats in the l allele, resulting in transcriptional capacity comparable to the s allele. Therefore, two investigations [66, 67] have categorized the L_G as s. The first of these triallelic investigations of this polymorphism was conducted by Kilpatrick and colleagues [68], who also documented a significant association between the 5-HTTLPR genotype and PTSD in a sample of hurricane-exposed adults. However, 5-HTTLPR genotype was only associated with increased risk of PTSD among adults with high stress exposure (low social support, high hurricane exposure). This was the first report of a significant genotype-environment interaction in PTSD. A second gene-environment interaction was documented for this genotype with the same disaster-exposed sample [67], however, the environmental variables were *county level* variables (county level crime rate, county level unemployment rate). Results indicated that county-level crime rate and percent unemployment modified the association between the 5-HTTLPR genotype and PTSD; s allele carriers were at increased risk for PTSD in high environmental stress conditions (high unemployment; high crime), and were at decreased risk for PTSD in low-risk environments (low unemployment, low crime). The final investigation of the serotonergic system has examined a different polymorphism, the serotonin 2A receptor (*5-HTR2A*) [69], reporting an excess of GG genotypes in females with PTSD.

Alternative Neurobiologic Systems

Other neurobiologic systems have also been studied, including markers of the hypothalamic pituitary adrenal axis (i.e., *GCCR, FKBP5, CNR1*), components of the locus coeruleus/noradrenergic system (i.e., *NPY, DBH, GABRA2*), and neurotrophins (i.e., *BDNF*). Out of these remaining nine studies, five have been negative reports. Two investigations found no association between polymorphisms in the brain derived neurotrophic factor (*BDNF*) gene and chronic PTSD [70, 71]. Similarly, no significant associations were found between PTSD and either the *Leu7Pro* polymorphism in the neuropeptide Y (*NPY*) gene [72], two glucocorticoid receptor polymorphisms (*N363S and BclI*) [73], or the -1021C/T variant of dopamine beta-hydroxylase (*DBH*) [74].

Of the remaining four positive reports, only one study is a main effect and the other three are GxE effects. A positive main effect association was reported between variants in a haplotype in the cannabinoid receptor gene (*CNR1*) and PTSD, with adults having PTSD being more likely to have C-A and C-G variants compared to adults without PTSD [75]. Binder and colleagues examined variants in *FKBP5* in a sample of lower-income, predominately African American participants recruited from an inner city clinic [76]. The results show that the variants in *FKBP5* were associated with higher levels of PTSD symptoms, but only among individuals exposed to two or more types of child abuse. There was no effect of genotype for individuals who experienced only one type of child abuse or no child abuse. Using data from the hurricane sample, Amstadter and colleagues [77] reported a significant effect of regulator of G-protein signaling 2 (*RGS2*) and PTSD under high stress conditions (high trauma exposure and low social support). The final molecular genetics study to date has been recently published by Nelson and colleagues [78], who examined SNPs in *GABRA2*, a gene that has been predominately studied in relation to alcoholism. A composite lifetime trauma exposure score was used in GxE analyses in a sample of 259 adults, 46 of whom met criteria for PTSD; analyses revealed no significant main effect of risk alleles, but significant GxE interactions for three SNPs in this gene (with marginal significance found for the fourth SNP).

Endophenotypes of PTSD

Another promising approach to identification of candidate genes is investigation of the association between genetic polymorphisms and a range of phenotypes evidenced by individuals diagnosed with PTSD. Moreover, as neurobiological mechanisms are unlikely to show effects on "diagnostic status" as a construct, careful consideration of the PTSD-related phenotype is also warranted. The first example of a successful endophenotype investigation for PTSD was conducted by Koenen and colleagues [79] who found that polymorphisms in *FKBP5* are associated with peritraumatic dissociation pediatric injury patients. Another key example of PTSD-phenotype investigations was borne from a series of studies that first noted that a deletion variant of *ADRA2B* was related to enhanced memory for emotional information in a sample of young adults [80]. A subsequent study examined *ADRA2B* in trauma-exposed Rwandan refugees, with higher re-experiencing symptoms found in deletion carriers than in non-carriers, though no differences were found in hyperarousal, avoidance, or diagnostic status. In this case, given the mechanism whereby *ADRA2B* influences posttraumatic phenotypes, the role for *ADRA2B* would have been missed if PTSD, instead of the refined phenotype of reexperiencing symptoms, had been measured.

Another example of a line of research that underscores the importance of examining different endophenotypes or symptoms instead of just diagnostic data is provided by investigations of apolipoprotein E (APOE). Given evidence supporting both an association between APOE and cognitive disturbances [81] and PTSD and neurocognitive deficits [82], Freeman and colleagues [83] reported that patients with the APOE 2 allele reported increased symptoms of re-experiencing and demonstrated poorer performance on facets of memory function relative to patients without the APOE allele.

Gene Expression and PTSD

A relatively new literature is that of gene expression studies and PTSD; notably, this category of genetic investigations does not fit the "risk factor" definition, as this methodology examines epigenetic modification of the genome as a result of environmental input. Nonetheless, a brief warrant of this growing research area is warranted. To date, three gene expression studies have been conducted on this phenotype [84-86]. In the first expression study conducted, Segman and colleagues measured peripheral blood mononuclear cell gene expression in individuals seeking treatment following a trauma at the emergency room, and at four month follow-up [84]; findings supported differential gene expression in those with and without PTSD. Of the several genes that were found to differentiate between PTSD positive and negative participants, *FKBP5* was one that has received increased attention due to its role in the HPA axis regulation. Recently, Yehuda and colleagues conducted a gene expression study in 15 individuals with PTSD and 20 matched control participants who were exposed to the 9/11 attacks on the World Trade Center [85], and among the 17 genes that were differentially expressed between cases and controls was *FKBP5*, which showed reduced expression in PTSD cases. A third expression study has been conducted in a sample of eight cases and eight controls who were exposed to the Ramstein air show catastrophe, which occurred in 1989 [86]. Despite the long time elapsed since the potentially traumatic event exposure, four up-regulated and 14 down-regulated genes were identified as significantly different between those with and without PTSD, many of which are related to immune function. This developing literature is promising, and there are many areas in need of future study in this trajectory. For example, a key question is whether difference in observed gene expression are due to DNA sequence variation and/or epigenetic changes resulting from environmental exposures.

Biological Risk Factors for PTSD

Numerous studies have documented biological correlates of PTSD. That is, factors such as cognitive ability, reduced hippocampal volume, and heart rate response to loud tones, have been associated with PTSD in case-control studies. However, such studies are unable to determine whether such correlates of PTSD are actually risk factors (e.g., were present prior to trauma exposure and increased risk for developing the disorder) or an acquired PTSD sign (e.g., developed as part of a pathological post-trauma response) [87]. A thorough review of all biologic correlates is outside of the scope for this chapter, and therefore we focus on biologic risk factors.

Twin Studies of Neurocognitive Risk Factors for PTSD

The discordant MZ twin research design offers an alternative strategy to case-control designs in determining whether a biological correlate of PTSD is a risk factor for the disorder or a sign of PTSD and related post-trauma sequelae (see Figure 1 adapted from Gilbertson et al. [88]). The MZ-discordant design is similar to that used in other case-controls studies when individuals with and without PTSD are compared *across* twin pairs on a specific biological correlate to determine whether that marker is associated with the PTSD diagnosis. However, by allowing comparisons *within* MZ twin pairs discordant for trauma exposure, the MZ-discordant design offers the opportunity to examine whether the biological marker is a risk factor for PTSD or develops along with PTSD (i.e., a PTSD sign). MZ twin pairs share 100% of their genes and family environment in youth and, therefore, are matched on a number of factors that would be impossible to control for in a non-twin sample. The design includes four participant groups: (1) trauma-exposed index twins who developed PTSD, (2) their "high-risk" trauma-unexposed co-twins who did not have PTSD (they are considered high risk because their identical twin developed PTSD when exposed to trauma), (3) trauma-exposed index twins with no PTSD, and (4) their "low-risk" trauma unexposed co-twins who did not have PTSD (they are considered low risk because their identical twin did not develop PTSD when exposed to trauma).

The logic behind the MZ-discordant design is that if a biological marker associated with PTSD is an underlying familial risk or vulnerability factor for the disorder then it should meet several criteria. First, it should be associated with PTSD across twin pairs; that is, it should be more prevalent in MZ twins with PTSD than in trauma-exposed MZ twins who did not develop PTSD. In Figure 1, contrast A is similar to the traditional case-control design and tests whether a biological marker is correlated with PTSD diagnosis (group 1 > group 3). Second, it should be similarly prevalent in both MZ twins with PTSD and their high-risk co-twins. In essence, the high-risk co-twin is serving as a proxy for what his brother would have been like if he had never been exposed to trauma and developed PTSD. If the biological marker is equally prevalent in the high-risk co-twins and the MZ twins with PTSD, then this suggests the marker existed prior to the development of PTSD. Third, the biological marker should be more prevalent in the "high-risk co-twins" than in the MZ twins exposed to trauma who did not develop PTSD or their "low-risk co-twins." Contrast C tests whether the biological correlate is a risk factor for PTSD (group 2 > group 3). In contrast, if a biological marker is a PTSD sign, it should be more prevalent in the twins with PTSD than in their high-risk co-twins (Contrast A, group 1 > group 3), trauma-exposed MZ twins without PTSD (Contrast B, group 1 > group 2) or their low-risk co-twins (group 1 > group 4).

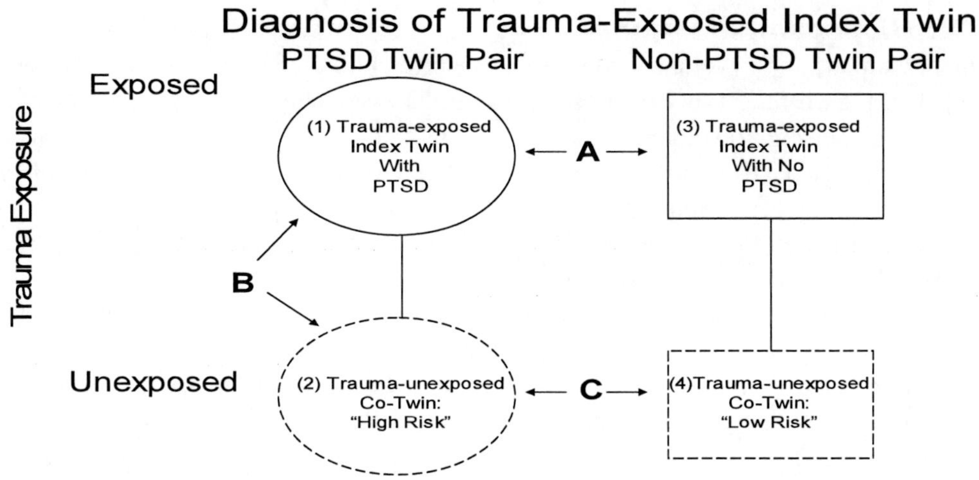

Figure 1. Discordant monozygotic twin design for testing whether biological correlates of posttraumatic stress disorder (PTSD) are actually risk factors for the disorder. The design includes four participant groups: (1) trauma-exposed index twins who developed PTSD; (2) their "high-risk" trauma-unexposed co-twins who did not have PTSD. They are considered high risk because their identical twin developed PTSD when exposed to trauma; (3) trauma-exposed index twins with no PTSD; and (4) their trauma unexposed co-twins who did not have PTSD. They are considered low risk because their identical twin did not develop PTSD when exposed to trauma. Contrast A is similar to the traditional case-control design and tests whether a biological marker is corrected with PTSD diagnoses (group 1 > group 3). If a biological marker is a correlate of PTSD, then contrast B tests whether the biological correlate is a PTSD sign (group 1 > group 2). Contrast C tests whether the biological correlate is a risk factor for PTSD (group 2 > group 3). If the biological correlate is a risk factor group 1 should also ~= group 2 for contrast B.
Adapted from Gilbertson et al. (2006). p. 486.

Pitman and colleagues have used the MZ discordant design to test whether commonly found biological correlates of PTSD are actually risk factors for developing the disorder [87]. Results from these investigations provide some of the only available evidence that four specific biological correlates of PTSD may in fact serve as risk factors for developing the disorder. These include an abnormally large cavum septum pellucidum (CSP), a smaller hippocampus, more neurological soft signs (NSS), and deficits on standard neuropsychological measures.

Abnormally Large Cavum Septum Pellucidum

The CSP exists when the medial walls of the lateral ventricles in the brain fail to fuse leaving a small cavity. The CSP is present at birth but is usually absent by 3 to 6 months of age. The prevalence of an abnormally large CSP is thought to be approximately 20% in adults and associated with abnormal limbic system development [89]. In the MZ twin discordant design, an abnormally large CSP was found more often in those with PTSD and their high-risk non-combat exposed co-twins than in combat-exposed veterans without PTSD and their low-risk co-twins [90]. These results suggest an abnormal CSP serves as a risk-factor for chronic PTSD.

Smaller Hippocampal Volume

Smaller hippocampal volume has been correlated with PTSD in several case-control studies [91-95]. Two recent meta-analyses confirm the association between PTSD diagnosis and smaller hippocampal volume [96, 97], although a recent investigation suggests the association between PTSD and smaller hippocampal volume is due to comorbid alcohol dependence [98]. The association between PTSD diagnosis and smaller hippocampal volume has also been attributed to the adverse effects of stress hormones on the brain [99], as animal models show hippocampal damage in response to extreme stress [100, 101]. However, a study of hippocampal volume and PTSD using the MZ discordant design found smaller hippocampal volume was observed in both combat veterans with chronic PTSD and in their non-combat exposed co-twins, as compared to combat veterans who did not develop the disorder. These results suggest that small hippocampal volume is a risk factor for chronic PTSD [102].

Neurological Soft Signs

NSS are assessed by neurological exam and represent subtle indices of neurological functioning that cannot be linked to a specific brain region. NSS have been correlated with PTSD in several studies [103, 104]. Greater NSS are thought to represent "subtle cortical dysfunction" that confers vulnerability the developing chronic PTSD possibly via "failure of inhibitory control over conditioned emotional responses" [87] p. 249. In the MZ discordant design, combat exposed index twins with PTSD had more NSS than those without, and moreover, the high-risk co-twins of the combat veterans with PTSD had greater NSS than the low-risk co-twin of the combat veterans who did not develop PTSD [105].

Neurocognitive Performance Deficits

An inverse association between pre-trauma IQ and risk of PTSD has been well-documented in both veteran [106-108] and civilian samples [109, 110]. Findings for IQ from the MZ discordant design were consistent with these previous studies; IQ was also found to be a risk factor for chronic PTSD. Moreover, a recent study from the VET Registry suggests the association between IQ and PTSD is genetically mediated. Findings from the MZ discordant design also suggest other specific domains of neurocognitive performance, including attention and verbal declarative memory deficits are familial risk factors for PTSD [82].

Prospective Studies of Biological Risk Factors for PTSD

Evidence for whether a biological correlate of PTSD is a actually a pre-trauma risk factor could be gathered via a prospective cohort design whereby a large random sample of respondents from the general population are assessed on key biological risk factors prior to trauma exposure and then followed to see who develops PTSD. Until recently, this type of design was not viewed as feasible except possibly in military samples due to the expense and long follow-up periods required for such studies. However, in the past few years, data on risk factors for PTSD have emerged from two prospective cohort studies begun in childhood. These studies offer stronger evidence than was previously available on a wide-range of risk factors for trauma exposure and PTSD. Since this review focuses on genetic and biological risk factors, our discussion here will focus on only two aspects of their findings: the inverse association between childhood IQ and risk of developing PTSD when exposed to trauma, and increased startle response.

Whereas some researchers have argued that low IQ and other cognitive deficits may be PTSD-sequelae [111], prospective studies of military samples have used measures of cognitive ability from military records (e.g., Armed Forces Qualifying Test) to show that pre-military IQ is inversely associated with risk for developing combat-related PTSD [106-108]. Breslau and colleagues have published data from a prospective study of children first assessed at 6 years of age and followed through age 17. A key finding from this cohort was that high IQ (>115) was associated with decreased risk of developing PTSD following trauma exposure [109]. Similarly, Koenen and colleagues examined the association between early childhood neurocognitive factors and risk for developing PTSD in participants from the Dunedin Multidisciplinary Health and Development Study, a birth cohort followed through age 32. The authors also found that childhood IQ assessed as early as age 5 was inversely associated with risk for developing PTSD by age 32. No associations were found between PTSD and other neurodevelopmental factors assessed in this cohort, including number of perinatal insults or gross motor skills in childhood. This suggests the IQ-PTSD association was not a marker of broader neurodevelopmental deficits [110]. Parslow and Jorm [112] were able to provide additional evidence that neurocognitive deficits represent a risk factor for PTSD when the community sample in which they were collecting epidemiological data was struck by widespread fires. Using the subsample of their original study that had experienced fire-related trauma, they found that pre-trauma neurocognitive weaknesses in several areas, including verbal intelligence, working memory, processing speed, and word recall, were risk factors for the development of PTSD symptoms; however, it should be noted that prior history of trauma was not controlled for. In large-scale prospective study using a military sample, Gale and colleagues replicated the link between low IQ and PTSD, but found a stronger link between PTSD with comorbidities, including Generalized Anxiety Disorder and substance abuse or dependence [113]. These results suggest that severity of the psychiatric disorder may play a role in the IQ-PTSD link.

The mechanism for the IQ-PTSD association is not well understood. It is possible that individuals with greater cognitive resources are better able to translate their traumatic event into a narrative [114] and make meaning out of it [115]. In addition, low IQ is a risk factor for other forms of psychopathology [116], including externalizing disorders, which are in turn associated with increased risk of trauma exposure and PTSD. However, in both the Breslau et al. (2006) and Koenen et al. studies (2007), the effect of low IQ on risk for PTSD remained significant after controlling for externalizing characteristics in multivariate analyses. This suggests that IQ may have a direct effect on risk for the disorder. Further research is needed to identify the mechanism by which cognitive ability confers risk for or protection against developing PTSD.

Another biological risk factor for PTSD that has been studied prospectively in a recent study is startle response. Even before PTSD was recognized as a psychiatric diagnosis, combat veterans complained of physiological hypersensitivity to loud noises years after returning from active duty [117]. However, many researchers attributed exaggerated startle responses to an overgeneralized conditioned response to threat stimuli (e.g., [118], without considering the possibility that this hypersenstivity may represent a preexisting risk factor for

PTSD. In a recent prospective study, Pole and colleagues examined whether physiological response to startling stimuli among police academy cadets prior to entering police duty predicted symptoms of PTSD one year later [119]. They found that increased skin conductance and slow skin conductance habituation in response to startling sounds at baseline predicted greater symptom severity after exposure to police-related trauma, even when controlling for prior trauma history and baseline psychiatric distress. These results suggest that physiological hypersensitivity is a risk factor for PTSD rather than a response to traumatic experiences.

Considerations for Future Research

In the final section of this review, we highlight two areas for consideration in research seeking to identify genetic and biologic research in PTSD. The investigation of new genetic risk factors for PTSD is largely informed by the observation of biological risk factors, as they may indicate potential underlying genetic influences. Integration of the literature on biological and genetic risk factors is an ongoing endeavor and is complicated by certain methodological issues that are areas to be addressed in future studies. Two such considerations include the distinction between onset and course of PTSD and developmental considerations in the study of risk factors, and are discussed below. The interested reader is referred to two recent in-depth reviews [50, 120].

Onset versus Course of PTSD

Studies of genetic and biological risk factors for PTSD have largely included only PTSD cases with current PTSD, where current PTSD often involves a chronic disorder extending over many years or even decades. When considering disorder etiology, it is useful to distinguish between risk factors for onset or development of the disorder and risk factors for course or chronicity of the disorder. Factors that influence who develops the disorder in the first place may differ from those factors that influence who recovers from the disorder once it develops. For example, members of disadvantaged ethnic groups are not at higher risk for the development of psychiatric disorders; however, once they develop a psychiatric disorder, their disorders are more chronic compared to the course for non-Hispanic white populations [121].

Twin studies have relied almost exclusively on diagnoses of lifetime PTSD and, therefore, heritability estimates from such studies explain the proportion of variation in risk for developing PTSD explained by genetic factors. It is not known whether the same genes that influence risk for developing PTSD affect PTSD chronicity. Studies are needed that distinguish between genetic influences on risk for developing PTSD versus persistence of the disorder.

The same can be said for studies of other biological risk factors for PTSD. Studies by Pitman and colleagues using the MZ discordant design have only included cases with chronic PTSD. Thus, it is unclear whether the biological risk factors they have identified would be associated broadly with increased risk of developing PTSD or are specific to the chronic form of the disorder. The exceptions are studies of IQ and risk of PTSD. Prospective data suggest pre-trauma IQ is inversely related both to risk of developing PTSD [107, 109, 110] and to PTSD chronicity [82]. Prospective cohort designs that assess biological risk factors prior to trauma exposure and follow individuals over an adequate time period are needed to distinguish between biological risk factors for onset versus course of the disorder.

Developmental Perspective

The effect of genetic and other biological risk factors in the etiology of PTSD may depend on developmental timing of trauma exposure. Childhood trauma has been associated with a variety of long-term deleterious consequences across social, emotional, and cognitive/academic functioning (e.g., [122]). Research has suggested that children may be more vulnerable to developing PTSD than adults, with one meta-analysis concluding that traumatized children were 1.5 times more likely to develop PTSD following a traumatic event than were similarly traumatized adults [123]. Developmental stage at time of trauma exposure may be an important factor in determining level, or possibly even type, of risk conferred.

Sensitivity to developmental stage may be particularly important in neurobiological aspects of PTSD, as animal research has suggested that early life events may be particularly relevant for subsequent cortisol reactivity [124, 125]. This is consistent with evidence that women with PTSD who experienced childhood sexual abuse excreted higher levels of cortisol than both non-PTSD adults with childhood sexual abuse and nonabused controls [126]. Moreover, different patterns of acute and chronic cortisol response in children with PTSD relative to adults with PTSD [30, 93, 127, 128]. It is possible that the lower cortisol levels seen in many adults with PTSD, as De Bellis and colleagues [129] have argued, may be a long-term developmental consequence of trauma exposure during childhood.

In addition to developmental stage of trauma exposure, studies of genetic and biological risk factors for PTSD might benefit from an examination of the effects of repeated trauma exposure. A recent meta-analysis found prior trauma history to be one of the most consistent predictors of PTSD [130]. Resnick and colleagues [131] found that female rape victims with a history of assault were 6.7 times more likely to develop PTSD and than victims with no trauma history. Furthermore, prior trauma has been shown to be associated with altered HPA axis responses to subsequent stressors [128, 131-133]. More specifically, women with trauma histories exhibited lower plasma cortisol levels following the rape than victims without such history [131]. Delahanty and colleagues [132] found that acute urinary cortisol levels mediated the relationship between prior trauma history and PTSD symptoms. Thus, repeated trauma exposure may modify the association between genetic and other biological risk factors and PTSD.

Conclusion

The genetic literature on PTSD is growing rapidly. As noted above, this literature has doubled in the past two years, adding vital information to the PTSD field. As the hypothesis-driven candidate gene association approach is reliant on extant literature regarding the biological endophenotype of PTSD, candidate gene association strategies are significantly limited by the relative paucity of information regarding the biological underpinnings of PTSD, therefore underscoring the necessity for continuation of this important research.

References

[1] Kessler, R.C., et al., Posttraumatic stress disorder in the National Comorbidity Survey. *Archives of General Psychiatry,* 1995. 52(12): p. 1048-1060.

[2] Kilpatrick, D.G., et al., *Drug-facilitated, incapacitated, and forcible rape: A national study.* Final report submitted to the National Institute of Justice. 2007.

[3] Kilpatrick, D.G., et al., Violence and risk of PTSD, major depression, substance abuse/dependence, and comorbidity: results from the National Survey of Adolescents. *J Consult Clin Psychol,* 2003. 71(4): p. 692-700.

[4] Breslau, N., et al., Trauma and posttraumatic stress disorder in the community: The 1996 Detroit area survey of trauma. *Archives of General Psychiatry,* 1998. 55: p. 626-632.

[5] Breslau, N., Epidemiologic studies of trauma, posttraumatic stress disorder, and other psychiatric disorders. *Canadian Journal of Psychiatry,* 2002. 47(10): p. 923-9.

[6] Breslau, N., et al., Epidemiological findings on posttraumatic stress disorder and co-morbid disorders in the general population., in *Adversity, stress, and psychopathology.* 1998, University Press: London, Oxford. p. 319-328.

[7] Breslau, N., et al., Traumatic events and posttraumatic stress disorder in an urban population of young adults. *Archives of General Psychiatry,* 1991. 48: p. 216-222.

[8] Davidson, J.R., et al., Post-traumatic stress disorder in the community: An epidemiological study. *Psychological Medicine,* 1991. 21: p. 713-721.

[9] Kessler, R.C., Posttraumatic stress disorder: the burden to the individual and to society. *Journal of Clinical Psychiatry,* 2000. 61: p. 4-12.

[10] Kulka, R.A., et al., Trauma and the Vietnam war generation: *Report of findings from the National Vietnam Veterans Readjustment study.* 1990, New York: Bruner/Mazel.

[11] Perkonigg, A., et al., Traumatic events and post-traumatic stress disorder in the community: Prevalence, risk factors, and comorbidity. *Acta Psychiatrica Scandinavica,* 2000. 101: p. 46-59.

[12] Resnick, H.S., et al., Prevalence of civilian trauma and posttraumatic stress disorder in a representative national sample of women. *Journal of Consulting and Clinical Psychology,* 1993. 61: p. 984-991.

[13] Brewin, C.R., B. Andrews, and J.D. Valentine, Meta-analysis of risk factors for posttraumatic stress disorder in trauma-exposed adults. *Journal of Consulting and Clinical Psychology,* 2000. 68(5): p. 748-766.

[14] Kraemer, H.C., et al., *Coming to terms with the terms of risk.* Arch Gen Psychiatry, 1997. 54(4): p. 337-43.

[15] Plomin, R., et al., *Behavioral Genetics.* 2001, New York: Worth Publishers.

[16] Goenjian, A.K., et al., Heritabilities of symptoms of posttraumatic stress disorder, anxiety and depression in earthquake exposed Armenian families. *Psychiatric Genetics,* 2008. 18(6): p. 261-266.

[17] Sack, W.H., G.N. Clarke, and J. Seeley, Posttraumatic stress disorder across two generations of Cambodian refugees. *Journal of the American Academy of Child and Adolescent Psychiatry,* 1995. 34(9): p. 1160-1166.

[18] Yehuda, R., S.L. Halligan, and L.M. Bierer, Relationship of parental trauma exposure and PTSD to PTSD, depressive and anxiety disorders in offspring. *Journal of Psychiatric Research,* 2001. 35(5): p. 261-270.

[19] Barakat, L.P., et al., Families surviving childhood cancer: a comparison of posttraumatic stress symptoms with families of healthy children. *Journal of Pediatric Psychology,* 1997. 22: p. 843-859.

[20] de Vries, A.P.J., et al., Looking beyond the physical injury: posttraumatic stress disorder in children and parents after pediatric traffic injury. *Pediatrics,* 1999. 104(6): p. 1293-1299.

[21] Hall, E., et al., Posttraumatic Stress Symptoms in Parents of Children with Acute Burns. *Journal of Pediatric Psychology,* 2006. 31: p. 403-412.

[22] Pelcovitz, D., et al., Posttraumatic stress disorder and family functioning in adolescent cancer. Journal of Traumatic Stress, 1998. 11: p. 205-221.

[23] Landolt, M.A., et al., Incidence and associations of parental and child posttraumatic stress symptoms in pediatric patients. *Journal of Child Psychology & Psychiatry,* 2003. 44: p. 1199-1207.

[24] McDermott, B.M. and A. Cvitanovich, Posttraumatic stress disorder and emotional problems in children following motor vehicle accidents: An extended case series. *Australian and New Zealand Journal of Psychiatry,* 2000. 34(3): p. 446-452.

[25] Stuber, M.L., et al., Posttrauma symptoms in childhood leukemia survivors and their parents. *Psychosomatics,* 1996. 37: p. 254-261.

[26] Pfefferbaum, B. and R.L. Pfefferbaum, Contagion in stress: An infectious disease model for posttraumatic stress in children. *Child and Adolescent Psychiatric Clinics of North America,* 1998. 7: p. 183-194.

[27] Bryant, R.A., M. Moulds, and R.M. Guthrie, Cognitive strategies and the resolution of acute stress disorder. *Journal of Traumatic Stress,* 2004. 14(1): p. 213-219.

[28] Winston, F.K., et al., Acute stress disorder symptoms in children and their parents after pediatric traffic injury. *Pediatrics,* 2002. 109: p. 90-99.

[29] Koplewicz, H.S., et al., Child and parental response to the 1994 World Trade Center bombing. *Journal of Traumatic Stress,* 2002(15): p. 77-85.

[30] Ostrowski, S., A., N.C. Christopher, and D.L. Delahanty, Brief report: The impact of maternal posttraumatic stress disorder symptoms and child gender on risk for persistent posttraumatic stress disorder symptoms in child trauma victims. *Journal of Pediatric Psychology,* 2007. 32: p. 338-342.

[31] Smith, P., et al., War exposure and maternal reactions in the psychological adjustment of children from Bosnia–Hercegovina. *Journal of Child Psychology and Psychiatry,* 2001. 42: p. 395-404.

[32] Daviss, W.B., et al., Predicting posttraumatic stress after hospitalization for pediatric injury. *Journal of the American Academy of Child and Adolescent Psychiatry,* 2000. 39: p. 573–583.

[33] McFarlane, A.C., S.K. Policansky, and C. Irwin, A longitudinal study of the psychological morbidity in children due to natural disaster. *Psychological Medicine.,* 1987. 17: p. 727-738.

[34] Nugent, N.R., et al., Parental posttraumatic stress symptoms as a moderator of child's acute biological response and subsequent posttraumatic stress symptoms in pediatric injury patients *Journal of Pediatric Psychology,* 2007. 32: p. 309-318.

[35] Davidson, J.R., et al., A diagnostic and family study of posttraumatic stress disorder. *American Journal of Psychiatry,* 1985. 142: p. 90-93.

[36] Davidson, J.R., R. Smith, and H. Kudler, Familial psychiatric illness in chronic posttraumatic stress disorder. *Comparative Psychiatry,* 1989. 30: p. 339-345.

[37] Bromet, E., A. Sonnega, and R.C. Kessler, Risk factors for DSM-III-R posttraumatic stress disorder: Findings from the National Comorbidity Survey. *American Journal of Epidemiology,* 1998. 147: p. 353-361.

[38] Koenen, K.C., et al., A twin registry study of familial and individual risk factors for trauma exposure and posttraumatic stress disorder. *Journal of Nervous and Mental Disease,* 2002. 190: p. 209-218.

[39] Davidson, J.R., et al., A family study of chronic post-traumatic stress disorder following rape trauma. *Journal of Psychiatric Research,* 1998. 32: p. 301-309.

[40] Kendler, K.S. and L.J. Eaves, Models for the joint effects of genotype and environment on liability to psychiatric illness. *American Journal of Psychiatry,* 1986. 143(33): p. 279-289.

[41] Lyons, M.J., et al., Do genes influence exposure to trauma? A twin study of combat. *American Journal of Medical Genetics*, 1993. 48(1): p. 22-27.

[42] Stein, M.B., et al., Genetic and environmental influences on trauma exposure and posttraumatic stress disorder: A twin study. *American Journal of Psychiatry,* 2002. 159(10): p. 1675-1681.

[43] True, W.J., et al., A twin study of genetic and environmental contributions to liability for posttraumatic stress symptoms. *Archives of General Psychiatry,* 1993. 50(4): p. 257-264.

[44] Fu, Q., et al., Differential etiology of posttraumatic stress disorder with conduct disorder and major depression in male veterans. *Biological Psychiatry,* 2007. 62(10): p. 1088-1094.

[45] Koenen, K.C., et al., Common genetic liability to major depression and posttraumatic stress disorder in men. *Journal of Affective Disorders,* 2008. 105(1-3): p. 109-115.

[46] Chantarujikapong, S.I., et al., A twin study of generalized anxiety disorder symptoms, panic disorder symptoms and post-traumatic stress disorder in men. *Psychiatry Research,* 2001. 103(2-3): p. 133-145.

[47] Xian, H., et al., Genetic and environmental influences on posttraumatic stress disorder, alcohol, and drug dependence in twin pairs. *Drug and Alcohol Dependence,* 2000. 61(1): p. 95 - 102.

[48] Koenen, K.C., et al., A twin registry study of the relationship between posttraumatic stress disorder and nicotine dependence in men. *Arch Gen Psychiatry,* 2005. 62(11): p. 1258-65.

[49] Koenen, K.C., N.R. Nugent, and A.B. Amstadter, Posttraumatic stress disorder: A review and agenda for gene-environment interaction research in trauma. *European Archives of Psychiatry and Clinical Neuroscience,* 2008. 258(2): p. 82-96.

[50] Amstadter, A.B., N.R. Nugent, and K.C. Koenen, Genetics of PTSD: Fear conditioning as a model for future research. *Psychiatric Annals,* 2009. 39(6): p. 358-367.

[51] Nugent, N.R., A.B. Amstadter, and K.C. Koenen, Genetics of post-traumatic stress disorder: informing clinical conceptualizations and promoting future research. *American Journal of Medical Genetics Part C: Seminars in Medical Genetics,* 2008. 148: p. 127-132

[52] Broekman, B.F.P., M. Olff, and F. Boer, The genetic background to PTSD. *Neuroscience & Biobehavioral Reviews,* 2007. 31: p. 348-362.

[53] Inglis, F.M. and B. Moghaddam, Dopaminergic innervation of the amygdala is highly responsive to stress. *J Neurochem,* 1999. 72(3): p. 1088-94.

[54] Yehuda, R., et al., Urinary catecholamine excretion and severity of PTSD symptoms in Vietnam combat veterans. *J Nerv Ment Dis,* 1992. 180(5): p. 321-5.

[55] Comings, D.E., et al., The dopamine D2 receptor locus as a modifying gene in neuropsychiatric disorder. *JAMA,* 1991. 266(13): p. 1793-1800.

[56] Comings, D.E., D. Muhleman, and R. Gysin, Dopamine D2 receptor (DRD2) gene and susceptibility to posttraumatic stress disorder: A study and replication. *Biological Psychiatry,* 1996. 40(5): p. 368-372.

[57] Gelernter, J., et al., No association between D2 dopamine receptor (DRD2) 'A' system alleles, or DRD2 haplotypes, and posttraumatic stress disorder. *Biological Psychiatry,* 1999. 45(5): p. 620-625.

[58] Young, B.R., et al., Harmful drinking in military veterans with posttraumatic stress disorder: Association with the D2 dopamine receptor A1 allele. *Alcohol & Alcoholism,* 2002. 37(5): p. 451-456.

[59] Voisey, J., et al., The DRD2 gene 957C>T polymorphism is associated with posttraumatic stress disorder in war veterans. *Depression and Anxiety,* 2009. 26(1): p. 28-33.

[60] Dragan, W.L. and W.Ç. Oniszczenko, The association between dopamine D4 receptor exon III polymorphism and intensity of PTSD symptoms among flood survivors. Anxiety, *Stress & Coping: An International Journal,* 2009.

[61] Segman, R.H., et al., Association between the dopamine transporter gene and posttraumatic stress disorder. *Molecular Psychiatry,* 2002. 7(8): p. 903-907.

[62] Lesch, K.P., et al., Association of anxiety-related traits with a polymorphism in the serotonin transporter gene regulatory region. *Science,* 1996. 274(5292): p. 1527-31.

[63] Lee, H.J., et al., Influence of the serotonin transporter promoter gene polymorphism on susceptibility to posttraumatic stress disorder. *Depress Anxiety,* 2005. 21(3): p. 135-9.

[64] Kolassa, I., et al., Association study of trauma load and SLC6A4 promoter polymorphism in PTSD: evidence from survivors of the Rwandan genocide. *Journal of Clinical Psychiatry,* in press.

[65] Nakamura, M., et al., The human serotonin transporter gene linked polymorphism (5-HTTLPR) shows ten novel allelic variants. *Molecular Psychiatry,* 2000. 5(1): p. 32-38.

[66] Kilpatrick, D.G., et al., Serotonin transporter genotype and social support and moderation of posttraumatic stress disorder and depression in hurricane-exposed adults. *American Journal of Psychiatry,* 2007. 164(11): p. 1-7.

[67] Koenen, K.C., et al., County-level social environment modifies the association between serotonin transporter genotype and risk of post-traumatic stress disorder in adults. *American Journal of Epidemiology,* 2009. 169(6): p. 704-711.

[68] Kilpatrick, D.G., et al., The serotonin transporter genotype and social support and moderation of posttraumatic stress disorder and depression in hurricane-exposed adults. *Am J Psychiatry,* 2007. 164(11): p. 1693-9.
[69] Lee, H., et al., Association between serotonin 2A receptor gene polymorphism and posttraumatic stress disorder. *Psychiatry Investigations,* 2007. 4(2): p. 104-108.
[70] Lee, H., et al., No association between the brain-derived nurotrophic factor gene Val66Met polymorphism and post-traumatic stress disorder. *Stress and Health,* 2006. 22(2): p. 115-119.
[71] Zhang, H., et al., Brain derived neurotrophic factor (BDNF) gene variants and Alzheimer's disease, affective disorders, posttraumatic stress disorder, schizophrenia, and substance dependence. *Am J Med Genet B Neuropsychiatr Genet,* 2006. 141(4): p. 387-93.
[72] Lappalainen, J., et al., A functional neuropeptide Y Leu7Pro polymorphism associated with alcohol dependence in a large population sample from the United States. *Arch. Gen. Psychiatry,* 2002. 59(9): p. 825-831.
[73] Bachmann, A.W., et al., Glucocorticoid receptor polymorphisms and post-traumatic stress disorder. *Psychoneuroendocrinology,* 2005. 30(3): p. 297-306.
[74] Mustapic, M., et al., Dopamine beta-hydroxylase (DBH) activity and -1021C/T polymorphism of DBH gene in combat-related posttraumatic stress disorder. American Journal of Medical Genetics Part B: *Neuropsychiatric Genetics,* 2007. 144B(1087-1089).
[75] Lu, A.T., et al., Association of the cannabinoid receptor gene (CNR1) with ADHD and post-traumatic stress disorder. *American Journal of Medical Genetics Part B: Neuropsychiatric Genetics,* 2008. 147B(8): p. 1488-1494.
[76] Binder, E.B., et al., Association of FKBP5 polymorphisms and childhood abuse with risk of posttraumatic stress disorder symptoms in adults. *Journal of the American Medical Association,* 2008. 299(11): p. 1291-1305.
[77] Amstadter, A.B., et al., Variant in RGS2 moderates posttraumatic stress symptoms following potentially traumatic event exposure. *Journal of Anxiety Disorders,* 2009. 23(3): p. 369-373.
[78] Nelson, E.C., et al., Association of childhood trauma exposure and GABRA2 polymorphisms with risk of posttraumatic stress disorder in adults. *Molecular Psychiatry,* 2009. 14: p. 234-238.
[79] Koenen, K.C., et al., Polymorphisms in FKBP5 are associated with peritraumatic dissociation in medically injured children. *Molecular Psychiatry,* 2005. 10(12): p. 1058-9.
[80] de Quervain, D.J.-F., et al., A deletion variant of the 2b-adrenoceptor is related to emotional memory in Europeans and Africans. *Nature Neuroscience,* 2007. 10: p. 1137-1139.
[81] Gallagher-Thompson, D., et al., Apolipoprotein E epsilon4 allele affects the relationship between stress and depression in caregivers of patients with Alzheimer's disease. *J Geriatr Psychiatry Neurol,* 2001. 14(3): p. 115-9.
[82] Gilbertson, M.W., et al., Neurocognitive function in monozygotic twins discordant for combat exposure: relationship to posttraumatic stress disorder. *J Abnorm Psychol,* 2006. 115(3): p. 484-95.

[83] Freeman, T., et al., Neuropsychiatric associations of apolipoprotein E alleles in subjects with combat-related posttraumatic stress disorder. *J Neuropsychiatry Clin Neurosci,* 2005. 17(4): p. 541-3.

[84] Segman, R.H., et al., Peripheral blood mononuclear cell gene expression profiles identify emergent post-traumatic stress disorder among trauma survivors. *Molecular Psychiatry,* 2005. 10: p. 500-513.

[85] Yehuda, R., et al., Gene expression patters associated with posttraumatic stress disorder following exposure to the World Trade Center attacks. *Biological Psychiatry,* in press.

[86] Zieker, J., et al., Differential gene expression in peripheral blood of patients suffering from post-traumatic stress disorder. *Molecular Psychiatry,* 2007. 12(116-118).

[87] Pitman, R.K., et al., Clarifying the origin of biological abnormalities in PTSD through the study of identical twins discordant for combat exposure. *Ann N Y Acad Sci,* 2006. 1071: p. 242-54.

[88] Gilbertson, M.W., et al., Neurocognitive function in monozygotic twins discordant for combat exposure: relationship to posttraumatic stress disorder. *Journal of Abnormal Psychology,* 2006. 115(3): p. 484-495.

[89] Sarwar, M., The septum pellucidum: normal and abnormal. *AJNR Am J Neuroradiol,* 1989. 10(5): p. 989-1005.

[90] May, F.S., et al., Cavum septum pellucidum in monozygotic twins discordant for combat exposure: relationship to posttraumatic stress disorder. *Biol Psychiatry,* 2004. 55(6): p. 656-8.

[91] Hedges, D.W., et al., Reduced hippocampal volume in alcohol and substance naive Vietnam combat veterans with posttraumatic stress disorder. *Cogn Behav Neurol,* 2003. 16(4): p. 219-24.

[92] Villarreal, G., et al., Reduced hippocampal volume and total white matter volume in posttraumatic stress disorder. *Biol Psychiatry,* 2002. 52(2): p. 119-25.

[93] De Bellis, M.D., et al., A pilot longitudinal study of hippocampal volumes in pediatric maltreatment-related posttraumatic stress disorder. *Biological Psychiatry,* 2001. 50(4): p. 305-309.

[94] Gurvits, T.V., et al., Magnetic resonance imaging study of hippocampal colume in chronic combat-related posttraumatic stress disorder. *Blogical Psychiatry,* 1996. 40: p. 1091-1099.

[95] Bremner, J.D., et al., MRI-based measurement of hippocampal volume in patients with combat-related posttraumatic stress disorder. *American Journal of Psychiatry,* 1995. 152: p. 973-981.

[96] Kitayama, N., et al., Magnetic resonance imaging (MRI) measurement of hippocampal volume in posttraumatic stress disorder: a meta-analysis. *J Affect Disord,* 2005. 88(1): p. 79-86.

[97] Smith, M.E., Bilateral hippocampal volume reduction in adults with post-traumatic stress disorder: a meta-analysis of structural MRI studies. *Hippocampus,* 2005. 15(6): p. 798-807.

[98] Woodward, S.H., et al., Hippocampal volume, PTSD, and alcoholism in combat veterans. *Am J Psychiatry,* 2006. 163(4): p. 674-81.

[99] Sapolsky, R.M., Glucocorticoids and hippocampal atrophy in neuropsychiatric disorders. *Archives of General Psychiatry,* 2000. 57: p. 925-935.

[100] Sapolsky, R.M., Why stress is bad for your brain. *Science,* 1996. 273: p. 749-750.
[101] Sapolsky, R.M., et al., Hippocampal damage associated with prolonged glucocorticoid exposure in primates. *Journal of Neuroscience,* 1990. 10: p. 2897-2902.
[102] Gilbertson, M.W., et al., Smaller hippocampal volume predicts pathologic vulnerability to psychological trauma. *Nature Neuroscience,* 2002. 5(11): p. 1242-1247.
[103] Gurvits, T.V., et al., Neurological status of combat veterans and adult survivors of sexual abuse PTSD. *Ann N Y Acad Sci,* 1997. 821: p. 468-71.
[104] Gurvits, T.V., et al., Neurological status of Vietnam veterans with chronic posttraumatic stress disorder. *J Neuropsychiatry Clin Neurosci,* 1993. 5(2): p. 183-8.
[105] Gurvits, T.V., et al., Subtle neurologic compromise as a vulnerability factor for combat-related posttraumatic stress disorder: results of a twin study. *Arch Gen Psychiatry,* 2006. 63(5): p. 571-6.
[106] Pitman, R.K., et al., Pre-Vietnam contents of posttraumatic stress disorder veterans' service medical and personnel records. *Comprehensive Psychiatry,* 1991. 32(5): p. 416-22.
[107] Kremen, W.S., et al., Pretrauma cognitive ability and risk for posttraumatic stress disorder: a twin study. *Archives of General Psychiatry,* 2007. 64(3): p. 361-8.
[108] Macklin, M.L., et al., Lower precombat intelligence is a risk factor for posttraumatic stress disorder. *Journal of Consulting and Clinical Psychology,* 1998. 66(2): p. 323-6.
[109] Breslau, N., V.C. Lucia, and G.F. Alvarado, Intelligence and other predisposing factors in exposure to trauma and posttraumatic stress disorder: a follow-up study at age 17 years. *Arch Gen Psychiatry,* 2006. 63(11): p. 1238-45.
[110] Koenen, K.C., et al., Early childhood factors associated with the development of post-traumatic stress disorder: results from a longitudinal birth cohort. *Psychol Med,* 2007. 37(2): p. 181-92.
[111] De Bellis, M.D., Developmental traumatology: The psychobiological development of maltreated children and its implications for research, treatment, and policy. *Development & Psychopathology,* 2001. 13: p. 539-564.
[112] Parslow, R.A. and A.F. Jorn, Pretrauma and posttrauma neurocognitive functioning and PTSD symptoms in a community sample of young adults. *American Journal of Psychiatry*, 2007. 164: p. 509 – 515.
[113] Gale, C.R., et al., Cognitive ability in early adulthood and risk of 5 specific psychiatric disorders in middle age: the Vietnam experience study. *Archives of General Psychiatry,* 2008. 65(12): p. 1410-1418.
[114] Pennebaker, J.W., The effects of traumatic disclosure on physical and mental health: the values of writing and talking about upsetting events. *International Journal of Emerging Mental Health,* 1999. 1(1): p. 9-18.
[115] Janoff-Bulman, R., *Shattered assumptions: Towards a New Psychology of Trauma.* 1992, New York: The Free Press.
[116] Fergusson, D.M., L.J. Horwood, and E.M. Ridder, Show me the child at seven II: Childhood intelligence and later outcomes in adolescence and young adulthood. *Journal of Child Psychology and Psychiatry and Allied Disciplines,* 2005. 46(8): p. 850-8.
[117] Meakins, J.C. and R.M. Wilson, The effect of certain sensory stimulations of respiratory and heart rate in cases of so-called "irritable heart". *Heart,* 1918. 7: p. 17 – 22.

[118] Butler, R.W., et al., Physiologic evidence of exaggerated startle response in a subgroup of Vietnam veterans with combat-related PTSD. *American Journal of Psychiatry,* 1990. 147: p. 1308-1312.

[119] Pole, N., et al., Prospective prediction of Posttraumatic Stress Disorder symptoms using fear potentiated auditory startled responses. *Biological Psychiatry,* 2009. 65: p. 235 – 240.

[120] Koenen, K.C., A.B. Amstadter, and N.R. Nugent, Gene-environment Interaction in Posttraumatic Stress Disorder: An Update. *Journal of Traumatic Stress,* in press.

[121] Breslau, J., et al., Lifetime risk and persistence of psychiatric disorders across ethnic groups in the United States. *Psychol Med,* 2005. 35(3): p. 317-27.

[122] Lansford, J.E., et al., A 12-year prospective study of the long-term effects of early child physical maltreatment on psychological, behavioral, and academic problems in adolescence *Archives of Pediatric Adolescent Medicine,* 2002. 156: p. 824-830.

[123] Fletcher, K.E., Childhood posttraumatic stress disorder. In: Mash EJ, Barkley RA, editors. *Child Psychopathology.* New York: Guilford Publications, Inc., 1996: p. 242-276.

[124] Anisman, H., et al., Do early life events permanently alter behavioural and emotional response to stressors? *International Journal of Developmental Neuroscience,* 1998. 16: p. 149-164.

[125] Vasquez, D.M., Stress and the developing limbic-hypothalamic-pituitary-adrenal axis *Psychoneuroendocrinology,* 1998. 23: p. 663-700.

[126] Lemieux, A.M. and C.L. Coe, Abuse-related posttraumatic stress disorder: Evidence for chronic neuroendoctrine activation in women. *Psychosomatic Medicine,* 1995. 57(2): p. 105-115.

[127] De Bellis, M.D., et al., A.E. Bennett Research Award. Developmental traumatology. Part I: Biological stress symptoms. *Biological Psychiatry,* 1999. 45: p. 1259-1270.

[128] Delahanty, D.L., et al., Initial urinary epinephrine and cortisol levels predict acute PTSD symptoms in child trauma victims. *Psychoneuroendocrinology,* 2005. 30(2): p. 121-128.

[129] De Bellis, M.D., G.P. Chrousos, and L.D. Dorn, Hypothalamic-pituitary-adrenal axis dysregulation in sexually abused girls. *Journal of Clinical Neuroendocrinology and Metabolism,* 1994. 78: p. 249-255.

[130] Ozer, E.J., et al., Predictors of posttraumatic stress disorder and symptoms in adults: A meta-analysis. *Psychological Bulletin,* 2003. 129(1): p. 52-73.

[131] Resnick, H.S., et al., Effects of previoius trauma on acute plasma cortisol level following rape. *American Journal of Psychiatry,* 1995. 152: p. 1675-1677.

[132] Delahanty, D.L., et al., Injury severity, prior trauma history, urinary cortisol levels and acute PTSD in motor vehicle accident victims. *Journal of Anxiety Disorders,* 2003. 17: p. 149-164.

[133] Heim, C., et al., Pituitary-adrenal and autonomic responses to stress in women after sexual and physical abuse in childhood. *JAMA,* 2000. 284: p. 592-597.

In: Neurobiology of Post-Traumatic Stress Disorder
Editors: L. Sher and A. Vilens, pp. 91-106
ISBN: 978-1-61470-593-2

Chapter VII

Early Postnatal Stress and Emotional Vulnerability: Focus on Neural Circuits Underlying Fear Memory in Rats

Machiko Matsumoto and Hiroko Togashi
School of Pharmaceutical Science, Health Sciences University of Hokkaido, Ishikari-Tobetsu, Hokkaido, Japan

Abstract

Traumatic events during early life may affect the neural systems associated with emotional regulation, depending on exposure period, and lead to altered sensitivity to stress later in life. We elucidated whether early postnatal stress alters the neural circuits associated with fear conditioning and fear extinction during the postadolescent period in rats using electrophysiological approaches combined with behavioral analysis. Aversive stimuli during the second postnatal week (during postnatal day (PND) 14 to PND 18; 2W-FS) caused a marked reduction in fear-related freezing behavior during exposure to contextual fear conditioning, presumably because of disturbance of the retention of fear memory, an effect that was attributable to changes in synaptic transmission in the hippocampal CA1 field. Aversive stress exposure during the third postnatal week (during PND 21 to PND 25; 3W-FS group) impaired the extinction of context-dependent fear memory evaluated by the expression of freezing behavior, an effect that was associated with dysfunction of the hippocampal-medial prefrontal cortex neural pathway. Thus, the altered behavior observed in adulthood may likely be the result of neurodevelopmental perturbations elicited by early life stress, suggesting that a "critical period" may exist for neural circuits involved in emotional expression. Furthermore, we found that extinction deficits, i.e., behavioral and synaptic dysfunction observed in the 3W-FS group was abolished by repeated treatment with the partial *N*-methyl-D-aspartate receptor agonist D-cycloserine. Considering the similarities between exposure psychotherapy in humans and fear extinction training in animals, both pharmacological interventions and psychotherapy targeting the neural circuits underlying fear memory may have significant clinical value in the treatment of anxiety-related disorders, such as posttraumatic stress disorder

(PTSD). A better understanding of the neural mechanisms of extinction deficits induced by early postnatal stress may help clarify the pathophysiology of PTSD in which fear extinction is compromised.

Introduction

Stressful life events increase the susceptibility to emotional stress, termed emotional vulnerability, which contributes to the development of depression and anxiety disorders [1,2]. Animal studies have shown that early postnatal stress such as maternal separation promote long-lasting and sometimes permanent changes in brain function including structural remodeling [3,4], synaptic plasticity [4,5,6] and receptor expression [7,8], and neuroendocrine function [9,10], which are attributable to emotional vulnerability later in life. Stressful events also affect memory function. Traumatic stress especially results in functional changes in neural circuits involved in fear memory [11,12]. For instance, chronic restraint stress impaired the recall of extinction based on fear memory, an effect in which dysfunction of the medial prefrontal cortex is involved [11]. Neuroimaging studies have shown that the pathophysiology of posttraumatic stress disorder (PTSD), in which extinction deficits of fear memory are implicated as a critical risk factor, is associated with impairments in the functional network of the amygdala, anterior cingulate, and hippocampus [13,14]. Recent studies focused on the influence of pre-existing vulnerability on the neurobiology of PTSD. Thus, environmental factors such as developmental exposure to stress, rather than the consequences of the trauma exposure itself, induce profound and long-lasting neurobiological effects. In other words, the "programming" factors may increase subsequent stress reactivity and result in vulnerability to develop PTSD [15,16,17]. These findings lead us to hypothesize that early life stress precipitates long-lasting alterations in the neural circuits underlying emotional expression and alters the sensitivity to stress later in life, which may be implicated in the pathophysiology of stress-related disorders such as PTSD.

Based on this hypothesis, we investigated whether early postnatal stress alters the neural circuits and/or neural mechanisms underlying fear memory during postadolescence in rats. In this review, we first provide evidence showing the hippocampal synaptic changes that occur during fear conditioning and then present results from our recent study, which focused on hippocampal-medial prefrontal cortex (mPFC) circuits underlying extinction of fear memory. Finally, we discuss the possible relationship between emotional vulnerability caused by early postnatal stress exposure and the pathophysiology of PTSD with regard to translational research.

Early Postnatal Stress and Emotional Vulnerability

I. Neural Circuits Undelying Fear Conditioning

Pavlovian fear conditioning is a form of learning in which pairings of a neutral conditioned stimulus (e.g., tone or light) with a fear-inducing unconditioned stimulus (e.g., footshock) comes to elicit a fear-related response (e.g., freezing behavior). A context-cued stimulus (i.e., contextual fear conditioning [CFC]) is a widely used experimental paradigm for investigating the neural circuitry involved in fear memory. In this section, we summarize our data using the CFC paradigm, with a focus on synaptic function in the hippocampus, a structure that is critically involved in learning and memory processes.

Rat pups received an aversive footshock as an early postnatal stressor during either the second postnatal week (i.e., during postnatal day [PND] 14-18; 2W-FS group) or the third postnatal week (during PND 21-25; 3W-FS group). Non-footshock (Non-FS) controls remained in the FS box for the same time period without FS stimuli. During the postadolescent period (10-14 weeks old), electrophysiological approaches combined with behavioral analysis were performed to elucidate the neural circuits underlying context-dependent fear conditioning and fear extinction (Figure 1). Freezing behavior determined immediately after footshock conditioning (i.e., acquisition session) was similar among FS- and Non-FS groups, however the behavior during exposure to CFC (i.e., retention session) was markedly reduced in the 2W-FS group (Figure 2B). This behavioral change was not attributable to motor impairment because locomotor activity measured by unconditioned open field stress did not differ from Non-FS controls [18]. Thus, the 2W-FS group acquired the learning process however exhibited low levels of fear-related behavior, presumably because of disturbance in the retention and/or consolidation of fear memory.

The hippocampus is known to be a key structure responsible for spatial learning or context-dependent fear memory [19,20]. For example, contextual environmental information related to exploratory behavior was encoded by cell activity [21,22], and the firing activity of CA1 pyramidal neurons reduced during exposure to CFC [23]. Consistent with these findings, the evoked potential in the CA1 field by stimulation of Shaffer collaterals was decreased during a CFC retention session in Non-FS controls and the 3W-FS group. The electrophysiological profile of CA1 pyramidal neurons, assessed by input-output curves in the footshock groups, did not differ from Non-FS controls, however CFC-induced synaptic inhibition was not observed in the 2W-FS group (Figure 2A). Hippocampal synaptic plasticity, long-term potentiation (LTP) is considered to be the electrophysiological basis of learning and memory [24,25]. Hippocampal LTP induction is known to be impaired by severe stressor exposure, including CFC [26,27,28]. Interestingly, LTP induction is suppressed not only by stress paradigms but also by low frequency stimulation (LFS) prior to LTP-inducing high-frequency (tetanus) stimulation. This phenomenon, termed homosynaptic metaplasticity [29,30] was mimicked by CFC (i.e., CFC inhibited synaptic transmission and subsequent LTP induction). Given the similarities between the electrophysiological and neurochemical profiles [31], we suggest that homosynaptic metaplasticity in the CA1 is related to the neural basis of stress experience-dependent fear memory. In other words, the historical synaptic event may influence the later efficacy of synaptic transmission and synaptic plasticity. As shown in Figure 2C, LFS-induced LTP suppression was observed in the Non-FS controls and the 3W-FS group. In the 2W-FS group, LFS did not suppress subsequent LTP; rather LTP was induced, although the magnitude of LTP itself was not different among the groups [32]. Thus, aversive stress exposure during the second postnatal week precipitates long-lasting alterations in the hippocampal synaptic function, which may be associated with abnormal fear expression (i.e., low anxiety-like behavior) during adulthood.

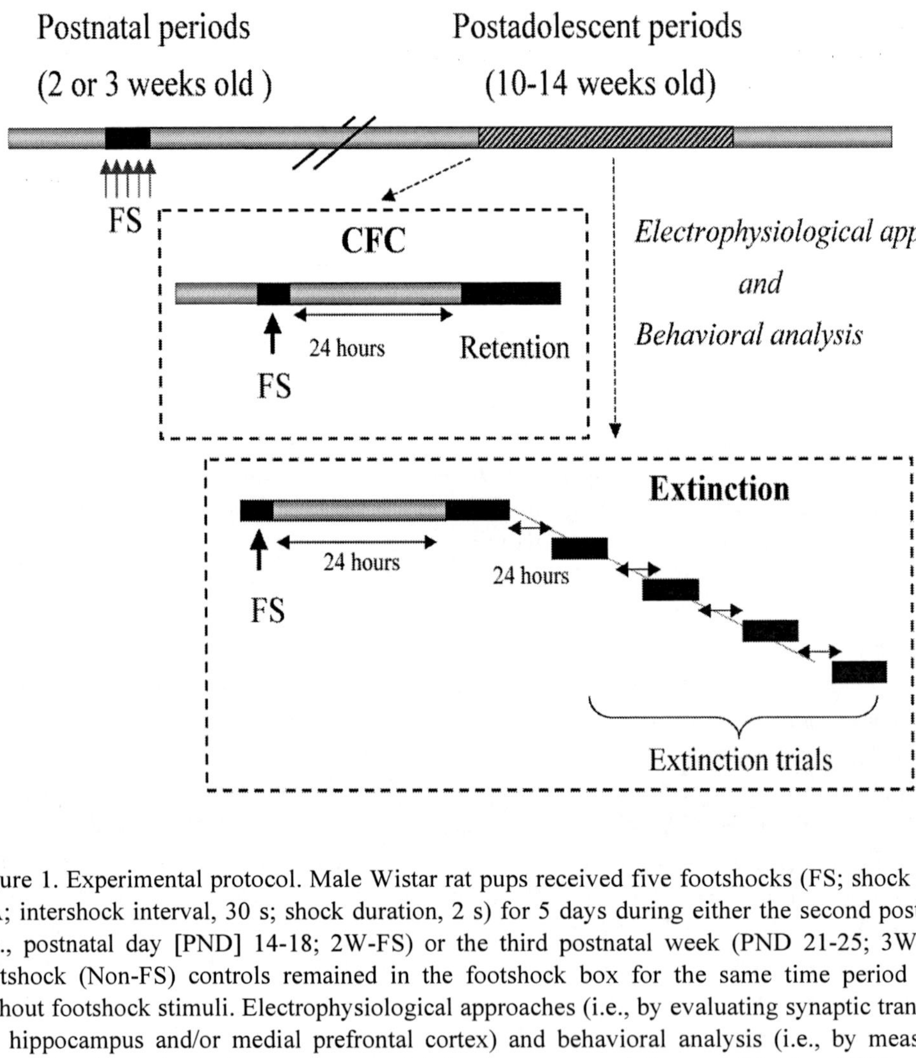

Figure 1. Experimental protocol. Male Wistar rat pups received five footshocks (FS; shock intensity, 1 mA; intershock interval, 30 s; shock duration, 2 s) for 5 days during either the second postnatal week (i.e., postnatal day [PND] 14-18; 2W-FS) or the third postnatal week (PND 21-25; 3W-FS). Non-footshock (Non-FS) controls remained in the footshock box for the same time period (12.5 min) without footshock stimuli. Electrophysiological approaches (i.e., by evaluating synaptic transmission in the hippocampus and/or medial prefrontal cortex) and behavioral analysis (i.e., by measuring fear-related freezing behavior) were performed following contextual fear conditioning (CFC) paradigm or extinction processes during the postadolescent period (10-14 weeks old). CFC paradigm: Rats were subjected to five footshocks (shock intensity, 1 mA; intershock interval, 30 sec; shock duration, 2 sec) and exposed to the conditioning chamber in the absence of footshock stimuli (retention session). The presence or absence of freezing behavior was estimated every 5 sec for 30 min. Extinction processes: Rats subjected to footshock (see above) were exposed to the conditioning chamber in the absence of footshock stimuli for 5 min, and extinction trials were repeated for consecutive days. Extinction was estimated by measuring freezing behavior for 5 min.

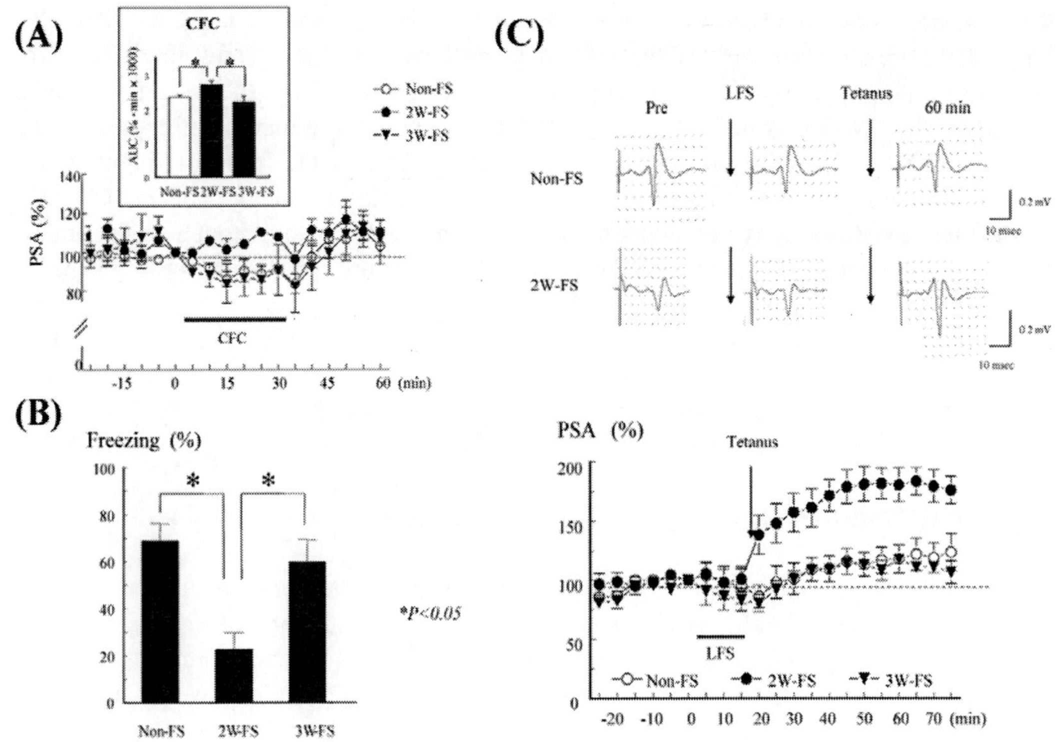

Figure 2. Time-course response and area under the curve (AUC) (inset) of the population spike amplitude (PSA) in the hippocampal CA1 field evoked by Schaffer collaterals stimulation (A) and freezing behavior (B) during the retention session of contextual fear conditioning (CFC). Hippocampal synaptic transmission did not decrease but rather increased in the 2W-FS group, which exhibited low levels of freezing expression (Modified from ref. 18). (C) Specimen recordings and time-course response of hippocampal synaptic transmission, i.e., low frequency stimulation (LFS, 900 pulses at 1 Hz) prior to long-term potentiation (LTP)-inducing high frequency stimulation (HFS; tetanus, 10 trains of stimuli, 10 pulses at 100 Hz, 10 s intertrain interval) in the hippocampal CA1 field. LFS-induced LTP suppression was observed in the Non-FS control and the 3W-FS group. In the 2W-FS group, LFS did not suppress subsequent LTP; rather, LTP was induced.

Early postnatal period is a critical time for the functional development of the brain. In rodents, tremendous synaptogenesis and neuronal differentiation occur 2 to 3 weeks after birth. The density of synapses in the cortex increases five-fold between PND 10 and PND 15 and gradually approaches adult levels [33]. Aversive stress exposure during the second postnatal week, therefore, may affect the developmental processes of the hippocampus, presumably due to elevated corticosterone levels. The hypothalamic pituitary adrenal (HPA) axis stress-hyporesponsive period is well known to protect the developing juvenile brain against the negative effects of stress [34]. Thus, during the first 2 weeks, pups show a markedly diminished adrenocortical response to stress. Aversive stress exposure during the second postnatal week, therefore, may overcome this protective mechanism and may consequently enhance corticosterone levels for a prolonged period of time. Several studies, on the other hand, have indicated that the amygdala, a key structure involved in the conditioned fear response [35,36], can mediate the effects of stress on hippocampal function [37,38,39] via its projection to the hippocampus. Kim et al. [37,38], using rat brain slices, reported that lesions to the amygdala prevented stress-induced LTP suppression in the CA1 field. We recently found that inactivation of the amygdala by local infusion of muscimol prevented LFS-induced suppression of LTP *in vivo*. Furthermore, low levels of freezing behavior observed in the 2W-FS groups were increased by electrical stimulation of the amygdala (unpublished data). These results lead us to hypothesize that aversive stress might induce persistent changes in the maturation or naive characteristics of the hippocampus and/or amygdala that are particularly vulnerable period in brain development. Altogether, the present data support the hypothesis that stressful events during early life induce long-lasting synaptic changes in the hippocampus and influence later sensitivity to emotional stress. In turn, these findings suggest the existence of a critical period for developing "proper" behavioral responses to emotional stress. The second postnatal week may be a critical time for establishing lifelong emotional expression including fear conditioning.

II. Neural Circuits Underlying Fear Extinction

Extinction of fear memory is defined as a reduction of the conditioned fear response when the conditioned stimulus is repeatedly presented in the absence of an unconditioned stimulus. Extinction does not result from loss of the fear memory itself, but rather from a learning process based on the formation of new inhibitory associations between conditioned and unconditioned stimuli [40,41,42]. Similar to other forms of learning, extinction is characterized by acquisition, consolidation, and retrieval [43,44]. In this section, we first evaluate the expression of freezing behavior during repeated extinction trials on consecutive days (Figure 1). We then provide electrophysiological data combined with behavioral analysis during extinction trials within one day. Based on these results, we discuss the relevance of synaptic plasticity in the hippocampal-mPFC pathway as key neural circuits of extinction processes related to early postnatal stress.

As shown in Figure 3A, the 3W-FS group exhibited an attenuated decline in freezing behavior during extinction trials on consecutive days. The 2W-FS group exhibited low levels of freezing behavior during the initial post-training session (Day 1), however the behavior decreased depending on extinction trials [45]. Figure 3B shows representative specimen recordings and the time-course response of the evoked potential in the mPFC during extinction trials within one day. Synaptic transmission was reduced during the first extinction trial, an effect reversed by repeated extinction trials, accompanied by decreases in fear-related freezing behavior in Non-FS controls and the 2W-FS group. In the 3W-FS group, however, synaptic inhibition continued through the third extinction trial [46]. Thus, the reversal of synaptic inhibition in the hippocampal-mPFC pathway appears to be associated with reducing freezing behavior during extinction trials. In other words, sustained synaptic inhibition observed in the 3W-FS group may reflect the impairment of extinction processes, which is presumably attributable to dysfunction of the hippocampal-mPFC neural pathway.

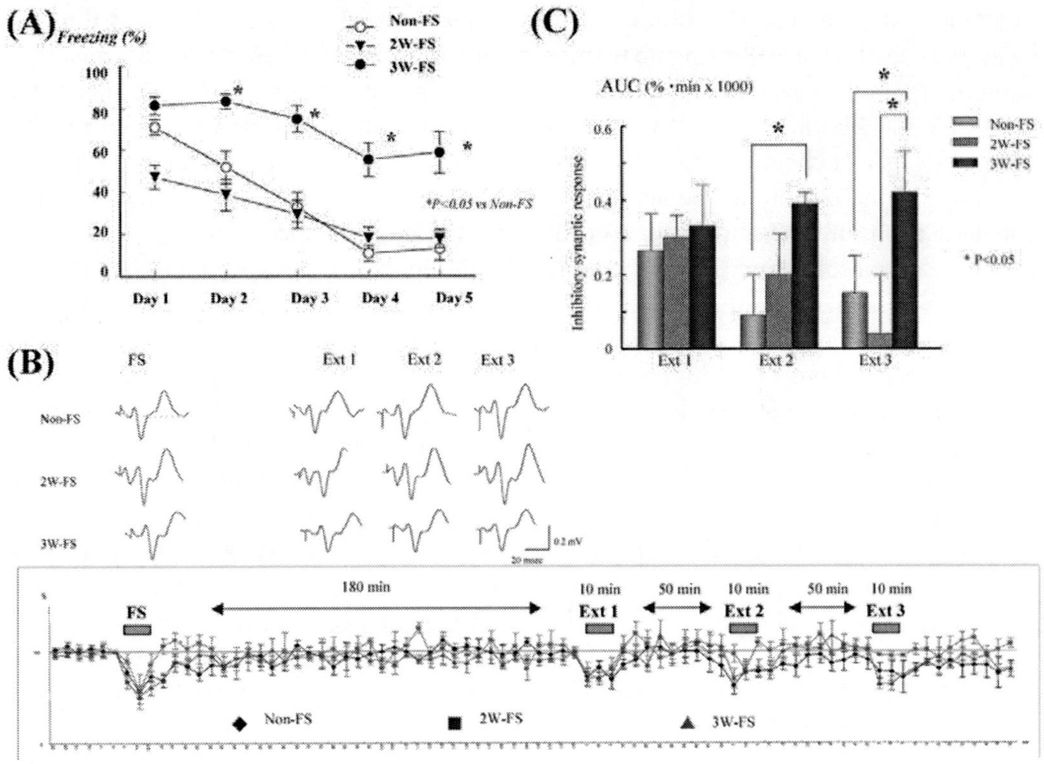

Figure 3. (A) Behavioral response during the extinction trial of context-dependent fear memory following the experimental protocol (see Figure 1). Extinction was estimated by measuring freezing behavior during 5 min exposure to the conditioning chamber in the absence of footshock (FS) stimuli. The 3W-FS group exhibited an attenuated decline in freezing behavior during extinction trials on consecutive days (Modified from ref. [45]). (B) Representative specimen recordings and time-course response of the population spike amplitude (PSA) in the medial prefrontal cortex evoked by stimulation of the hippocampal CA1 field (subicular region) during extinction trials within one day. Extinction trials were repeated three times (i.e., Ext 1, Ext 2, and Ext 3; 10 min per extinction trial), 180 min after footshock (FS) conditioning. (C) Areas under the curve (AUC) of the PSA calculated for 20 min after the start of each extinction trial. Synaptic inhibition observed during the first extinction trial (Ext 1) was reversed by repeated trials in the Non-FS control and the 2W-FS group. In the 3W-FS group, synaptic inhibition continued throughout the third extinction trial (Ext. 3) (Modified from ref. [46]).

Substantial evidence has demonstrated a role of the infralimbic cortex (i.e., a subregion of the ventral mPFC) in the neural and molecular mechanisms of extinction [42,47], however several studies have provided evidence for the functional significance of the prelimbic area of the mPFC in mediating extinction retrieval [48,49,50]. For example, Herry et al. [48,49] found a correlation between extinction of tone-elicited fear and synaptic plasticity in mediodorsal thalamus-mPFC projections. Extinction of fear conditioning was accompanied by delayed synaptic potentiation, and this synaptic facilitation was associated with the maintenance of fear extinction. Recent studies further explored synaptic changes in the hippocampal-mPFC pathway [50,51,52]. Farinelli et al. [52] showed that LTP-like responses in the mPFC after extinction training were suppressed by LFS in the hippocampal CA1 field, leading to impaired retrieval of tone-dependent fear extinction. Furthermore, retrieval of extinction was facilitated by tetanic stimulation applied several hours later, but LTP failed to be reestablished. We also found that synaptic transmission in the hippocampal-mPFC pathway was facilitated following extinction retrieval of context-dependent fear memory. Synaptic inhibition, rather than facilitation, occurred without extinction training, suggesting that the inputs to the mPFC from the hippocampus are potentiated by extinction training. Thus, the effects of extinction trials may depend on a transition from synaptic depression to long-lasting normalization or potentiation of synaptic efficacy, indicating that synaptic potentiation in the mPFC may be a crucial neurophysiological mechanism underlying the maintenance or consolidation of fear extinction. Interestingly, we did not find any changes in synaptic transmission induced by extinction retrieval in the 3W-FS group (unpublished data). These findings indicate the possibility that the 3W-FS group underwent long-lasting neurocircuitry alterations in the hippocampal-mPFC pathway, which would be consistent with our previous findings showing that early stress exposure during the third postnatal week impaired prefrontal synaptic function associated with extinction trials [46].

Although synaptic density dramatically increased between PND 10 and PND 15, the maturation of the cortex and hippocampus are not yet fully developed until the third postnatal week [53]. Aversive stress exposure during the third postnatal week, therefore, might permanently influence the growth and maturation of the mPFC and/or hippocampus, an effect possibly attributable to elevated levels of corticosterone, and consequently precipitate long-lasting alterations in their functional properties underlying extinction. We have previously reported that the behavioral and neuronal dysfunction observed during extinction trials in the 3W-FS group were prevented by pretreatment with the serotonin$_{1A}$ (5-hydroxytryptamine-1A, 5-HT$_{1A}$) receptor agonist tandospirone [45,46]. Thus, activation of 5-HT$_{1A}$ receptors may trigger the onset of fear extinction and may consequently result in long-term storage of fear extinction memory. This possibility is partially consistent with a study showing that another 5-HT$_{1A}$ receptor agonist, buspirone, facilitated the responses observed during extinction of conditioning [54]. 5-HT is well known to regulate not only the maturation of its own neurons but also developmental processes such as neurogenesis, synaptogenesis, and cortical organization [55,56]. Given the association between the development of the serotonergic system and its prominent role in maturation, early stress exposure may lead to persistent changes in serotonergic function mediated by 5-HT$_{1A}$ receptors. Indeed, a reduction in the number of 5-HT-immunoactive cells was found in the median raphe nuclei, where 5-HT$_{1A}$ autoreceptors are located [57], and dysfunction of postsynaptic 5-HT$_{1A}$ receptors was found in the mPFC [58] in the 3W-FS group. Several reports have demonstrated that elevated corticosterone levels decreased 5-HT$_{1A}$ receptor expression in the hippocampus [59,60]. These findings led to the hypothesis that early stress exposure during the third postnatal week affects serotonergic function mediated by 5-HT$_{1A}$ receptors during development and leads to deficits in extinction processes. We cannot determine, however, whether early postnatal stress exposure changes either 5-HT$_{1A}$ receptor expression itself or the developmental transition in signal transduction [61]. Further detailed research will be required.

Converging evidence from lesion studies [20,62] suggests that the hippocampus is a pivotal region responsible for not only context-dependent fear conditioning but also context-specific fear extinction [20,63,64]. The hippocampus has strong reciprocal connections with the mPFC and amygdala. These brain regions appear to interactively coordinate the encoding and expression of fear extinction. The hippocampal-cortical circuit is especially considered to undergo major reorganization between the recent and remote stages of contextual fear memory [65]. Considering the involvement of the hippocampus in contextual encoding processes, the time-dependent reorganization of neural networks may mediate hippocampus-dependent memory processes. Hypothetically, fear conditioning and fear extinction may yield excitatory and inhibitory memory, respectively, and these memories may compete with each other for behavioral expression gated by neural systems that mediated context-appropriate recall. The present findings indicate that the extinction of fear memory correlates with LTP-like synaptic changes in the hippocampal projections to the mPFC. Thus, the mPFC is a critical structure involved in the retrieval and storage of fear extinction, and the interaction with the hippocampus appears to play an important role in mediating extinction processes. Synaptic plasticity in the hippocampal-mPFC neural pathway may be a good index of the maintenance of extinction.

Conclusion

Extinction of fear memory is an important component of exposure psychotherapy in PTSD [66,67]. Fear conditioning to contextual cues is a good model of intrusive memories activated by trauma-related stimuli in PTSD [68]. Accordingly, understanding the neural mechanisms of extinction of context-dependent fear memory can provide useful information of not only the pathophysiology of PTSD but also more effective therapeutic strategies (e.g., facilitation of extinction or reduction in the number of exposure sessions). Animal studies have focused on the clinical relevance of the partial *N*-methyl-D-aspartate (NMDA) receptor agonist D-cycloserine [69,70,71], which has therapeutic effects in anxiety disorders [72,73,74]. Indeed, we observed that repeated treatment with DCS ameliorated extinction impairment in the 3W-FS group [45], and interestingly, that the dysfunction of synaptic response during extinction retrieval in the 3W-FS group was reversed by D-cycloserine administration immediately after extinction training (unpublished data). These data suggest that D-cycloserine treatment induced a transition from synaptic inhibition to potentiation in the PFC during extinction trials, presumably accompanied by protein synthesis. The formation of long-term memory has generally been linked to a molecular cascade involving NMDA-mediated Ca^{2+} entry, protein synthesis, and gene expression. DCS activates NMDA receptors and consequently contributes to the enhancement of new inhibitory learning [75]. Therefore, extinction training can trigger NMDA receptors through activation of molecular cascades, such as protein kinase [50,76], and may lead to extinction-induced gene expression. This hypothesis is supported by previous studies in which extinction of conditioned fear enhanced immediate early gene c-*fos* expression in the mPFC [77,78].

The most interesting finding is that the behavioral alterations and synaptic changes underlying fear memory depend on the postnatal period of aversive stress exposure. Thus, early life stress alters the emotional response at different stages of emotional expression (i.e., fear conditioning and fear extinction), depending on neurocircuitry development (Figure 4). Aversive stress exposure during periods of hippocampus and/or mPFC development may permanently influence the growth and maturation of these systems which are particularly vulnerable period, and alter sensitivity to stress later in life. A wealth of clinical and preclinical data has provided evidence that the amygdala, together with the hippocampus and mPFC, plays a critical role in fear extinction [36,38,66,79]. Several studies suggested a prominent role for mPFC-amygdala circuits in mediating fear extinction [12,14,80]. PTSD patients, in whom fear extinction is compromised, exhibit amygdala hyperactivity [12,66,81]. Amygdala damage, in turn, confers resistance to PTSD through diminished expression of fear-related responses [82]. Hypothetically, the extinction impairments associated with dysfunction of hippocampal-mPFC neurocircuitry in the 3W-FS group may be involved in amygdala perturbation (e.g., exaggerated amygdala function). In contrast, low levels of anxiety-like behavior associated with hippocampal dysfunction in the 2W-FS group may be attributable to amygdala dysregulation. Altogether, not only pharmacological interventions but also psychotherapy targeting the mPFC and/or amygdala may have therapeutic potential in the treatment of PTSD. Further understanding of the neurocircuitry underlying fear memory in juvenile stress models may lead to well-characterized translational research to elucidate the pathophysiology of PTSD.

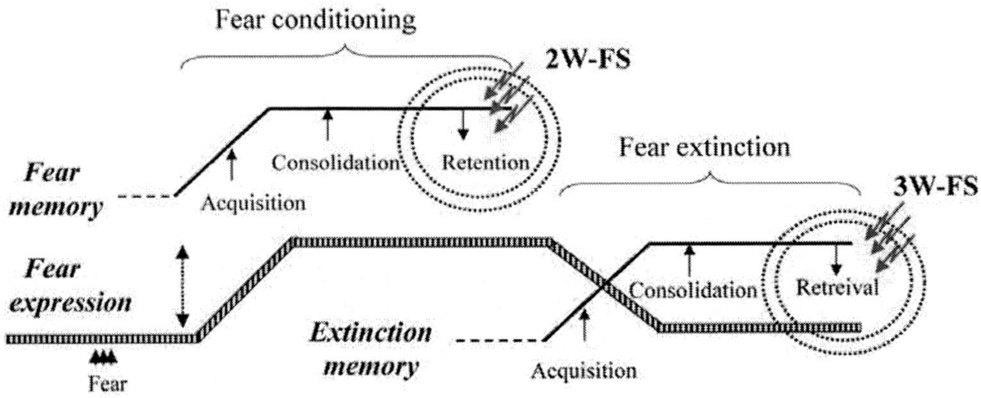

Figure 4. Hypothesized processes of fear conditioning and fear extinction associated with developmental dysfunction caused by early postnatal stress. Fear expression increases in parallel with acquisition and consolidation of fear conditioning, whereas it decreases depending on the extinction learning of fear memory. Extinction processes do not result from loss of the fear memory itself (i.e., excitatory associations between conditioned and unconditioned stimuli), but rather from a learning process based on the formation of new inhibitory associations between conditioned and non-unconditioned stimuli. Aversive stress exposure during the second postnatal week (2W-FS) attenuates fear expression, freezing behavior during the retention and/or consolidation of contextual fear conditioning (CFC), presumably due to alterations in hippocampal synaptic function during brain development. Aversive stress exposure during the third postnatal week (3W-FS) attenuates the decline in freezing behavior induced by extinction trials, which may result from dysfunction of the hippocampal-medial prefrontal cortex (mPFC) pathway underlying regulation of extinction storage.

Acknowledgments

The authors thank Dr. H. Koseki, Dr. R. Hirata, Dr. K. Konno and Dr. K. Higuchi for their excellent techniques. The authors also thank Prof. M. Yoshioka, Dr. T. Yamaguchi, and Dr. T. Izumi for their generous support. This study was supported by a Grant-in-Aid for Scientific Research from the Ministry of Education and Science, Japan, a Grant-in-Aid for Scientific Research from the High Technology Research Program, and the Academic Frontier Research Program organized by the Education and Science Department of Japan, and was supported in part by Grant-in-aid for the 2009-2010 Research Project of the Research Institute of Personalized Health Sciences, Health Sciences University of Hokkaido, Japan.

References

[1] de Kloet ER, Sibug RM, Helmerhorst FM, Schimidt M: Stress, genes and the mechanisms of programming the brain for later life. *Neurosci Biobehav Rev* 2005; 29:271-281.

[2] Pryce CR, Ruedi-Bettschen D, Dettling AC, Weston A, Russig H, Ferger B, Feldon J: Long-term effects of early-life environmental manipulations in rodents and primates: Potential animal models in depression research. *Neurosci Biobehav Rev* 2005;29:649-374.

[3] Huot RL, Plotsky PM, Lenox RH, McNamara RK: Neonatal maternal separation reduces hippocampal mossy fiber density in adult Long Evans rats. *Brain Res* 2002;950:52-63.

[4] Brunson KL, Kramár E, Lin B, Chen Y, Colgin LL, Yanagihara TK, Lynch G, Baram TZ: Mechanisms of late-onset cognitive decline after early-life stress. *J Neurosci* 2005;25:9328-9338.

[5] Akers KG, Nakazawa M, Romeo RD, Connor JA, McEwen BS, Tang AC: Early life modulators and predictors of adult synaptic plasticity. *Eur J Neurosci* 2006;24:547-554.

[6] Blaise JH, Koranda JL, Chow U, Haines KE, Dorward EC:Neonatal isolation stress alters bidirectional long-term synaptic plasticity in amygdalo-hippocampal synapses in freely behaving adult rats. *Brain Res* 2008;1193:25-33.

[7] Vázquez DM, López JF, Van Hoers H, Watson SJ, Levine S: Maternal deprivation regulates serotonin 1A and 2A receptors in the infant rat. *Brain Res* 2000;855:76-82.

[8] Lukkes JL, Summers CH, Scholl JL, Penner KJ, Forster GL: Early life social isolation alteres corticotrophin-releasing factor responses in adult rats. *Neurosci* 2009; 158:845-855.

[9] Ladd CO, Huot RL, Thrivikraman KV, Nemeroff CB, Plotsky PM: Long-term adaptations in glucocorticoid receptor and mineralocorticoid receptor mRNA and negative feedback on the hypothalamo-pituitary-adrenal axis following neonatal maternal separation. *Biol Psychiatry* 2004;55:367-375.

[10] De la Fuente M, Llorente R, Baeza I, De Castro NM, Arranz L, Cruces J, Viveros MP: Early maternal deprivation in rats: a proposed animal model for the study of developmental neuroimmunoendocrine interactions. *Ann N Y Acad Sci* 2009;1153:176-183.

[11] Miracle AD, Brace MF, Huyck KD, Singler SA, Wellman CL: Chronic stress impairs recall of extinction of conditioned fear. *Neurobiol Learn Mem* 2006;85:213-218.

[12] Akirav I, Maroun M: The role of the medial prefrontal cortex-amygdala circuit in stress effects on the extinction of fear. *Neural Plast* 2007;2007 ID30873:1-11.

[13] Gilboa A, Shalev AY, Laor L, Lester H, Louzoun Y, Chisin R, Bonne O: Functional connectivity of the prefrontal cortex and the amygdala in posttraumatic stress disorder. *Biol Psychiatry* 2004;55:263-272.

[14] Shin LM, Rauch SL, Pitman RK: Amygdala, medial prefrontal cortex, and hippocampal function in PTSD. *Ann N Y Acad Sci* 2006;1071:67-79.

[15] Nemeroff CB: Neurobiological consequence of childhood trauma. *J Clin Psychiatry* 2004;65:18-28.

[16] Secke JR, Meaney MJ: Glucocorticoid “programming” and PTSD risk. *Am N Y Acad Sci* 2006;1071:351-378.

[17] Heim C, Nemeroff CB: Neurobiology of posttraumatic stress disorder. *CNS Spectr* 2009;14(1 Suppl 1):13-24.

[18] Koseki H, Matsumoto M, Togashi H, Yamaguchi T, Izemi T, Yoshioka M, Fukushima K: Effects of aversive stress during brain development on hippocampal synaptic and behavioral responses to emotional stress at postadolescence. *Nihon Shinkei Seishin Yakurigaku Zasshi* 2007;27:19-27.

[19] Eldridge LL, Knowlton BJ, Furmanski CS, Bookheimer SY, Engel SA: Remembering episodes: a selective role for the hippocampus during retrieval. *Nat Neurosci* 2000;3:1149-1152.

[20] Corcoran KA, Maren S: Hippocampal inactivation disrupts contextual retrieval of fear memory after extinction. *J Neurosci* 2001; 21:1720-1726.

[21] Thompson LT, Best PJ: Long-term stability of the place-field activity of single units recorded from the dorsal hippocampus of freely behaving rats. *Brain Res* 1990;509:299-308.

[22] Hollup SA, Molden S, Donnett JG, Moser MB, Moser EI: Place fields of rat hippocampal pyramidal cells and spatial learning in the watermaze. *Eur J Neurosci* 2001;13:1197-1208.

[23] Tada K, Kasamo K, Suzuki T, Matsuzaki Y, Kojima T: Endogenous 5-HT inhibits firing activity of hippocampal CA1 pyramidal neurons during contextual fear stress-induced freezing behavior through stimulating 5-HT_{1A} receptors. *Hippocampus* 2004;14:143-147.

[24] Malenka RC, Nicoll RA: Long-term potentiation: a decade of progress? *Science* 1999;285:1870-1874.

[25] Martin SJ, Grimwood PD, Morris RG: Synaptic plasticity and memory: an evaluation of the hypothesis. *Annu Rev Neurosci* 2000;23:649-711.

[26] Shakesby AC, Anwyl R, Rowan MJ: Overcoming the effects of stress on synaptic plasticity in the intact hippocampus: rapid actions of serotonergic and antidepressant agents. *J Neurosci* 2002;22:3638-3644.

[27] Matsumoto M, Tachibana K, Togashi H, Tahara K, Kojima T, Yamaguchi T, Yoshioka M: Chronic treatment with milnacipran reverses the impairment of synaptic plasticity induced by conditioned fear stress. *Psychopharmacol* 2005;179:606-612.

[28] Hirata R, Togashi H, Matsumoto M, Yamaguchi T, Izumi T, Yoshioka M: Characterization of stress-induced suppression of long-term potentiation in the hippocampal CA1 field of freely moving rats. *Brain Res* 2008;1226:27-32.

[29] Abraham WC, Bear MF: Metaplasticity: the plasticity of synaptic plasticity. *Trends Neurosci* 1996; 19:126-130.

[30] Zelcer I, Cohen H, Richter-Levin G, Lebiosn T, Grossberger T, Barkai E: A cellular correlate of learning-induced metaplasticity in the hippocampus. *Cereb Cortex* 2006;16:460-468.

[31] Hirata R, Matsumoto M, Judo C, Yamaguchi T, Izumi T, Yoshioka M, Togashi H: Possible relationship between the stress-induced synaptic response and metaplasticity in the hippocampal CA1 field of freely moving rats. *Synapse* 2009;63:549-556.
[32] Matsumoto M, Higuchi K, Togashi H, Koseki H, Yamaguchi T, Kanno M, Yoshioka M: Early postnatal stress alters the 5-HTergic modulation to emotional stress at postadolescent periods of rats. *Hippocampus* 2005;15:775-781.
[33] Micheva KD, Beaulieu C: Quantitative aspects of synaptogenesis in the rat barrel field cortex with special reference to GABA circuitry. *J Comp Neurol* 1996;373:340-354.
[34] Levine S: Primary social relationships influence the development of the hypothalamic-pituitary-adrenal axis in the rat. *Physiol Behav* 2001;73:255-260.
[35] Davis M, Shi C: The amygdala. *Curr Biol* 2000;10:R131.
[36] LeDoux JE: Emotion circuits in the brain. *Annu Rev Neurosci* 2000;23:155-184.
[37] Kim JJ, Lee HJ, Han JS, Packard MG: Amygdala is critical for stress-induced modulation of hippocampal long-term potentiation and learning. *J Neurosci* 2001;21:5222-5228.
[38] Kim JJ, Koo JW, Lee HJ, Han JS: Amygdalar inactivation blocks stress-induced impairments in hippocampal long-term potentiation and spatial memory. *J Neurosci* 2005;25:1532-1539.
[39] Roozendaal B, Griffith QK, Buranday J, De Quervain DJ, McGaugh JL: The hippocampus mediates glucocorticoid-induced impairment of spatial memory retrieval: dependence on the basolateral amygdala. *Proc Natl Acad Sci U S A* 2003;100:1328-1333.
[40] Myers KM, Davis M: Behavioral and neural analysis of extinction. *Neuron* 2002;36:567-584.
[41] Kim JJ, Jung MW: Neural circuits and mechanisms involved in Pavlovian fear conditioning: a critical review. *Neurosci Biobehav Rev* 2006;30:188-202.
[42] Quirk GJ, Mueller D: Neural mechanisms of extinction learning and retrieval. *Neuropsychopharmacol* 2008;33:56-72.
[43] Abel T, Lattal KM: Molecular mechanisms of memory acquisition, consolidation and retrieval. *Curr Opin Neurobiol* 2001;11:180-187.
[44] Rodrigues SM, Schafe GE, LeDoux JE: Molecular mechanisms underlying emotional learning and memory in the lateral amygdala. *Neuron* 2004;44:75-91.
[45] Matsumoto M, Togashi H, Konno K, Koseki H, Hirata R, Izumi T, Yamaguchi T, Yoshioka M: Early postnatal stress alters the extinction of context-dependent conditioned fear in adult rats. *Pharmacol Biochem Behav* 2008;89:247-252.
[46] Koseki H, Matsumoto M., Togashi H, Miura Y, Fukushima K, Yoshioka M: Alteration of synaptic transmission in the hippocampal-mPFC pathway during extinction trials of context-dependent fear memory in juvenile rat stress models. *Synapse* 2009;63:805-813.
[47] Corcoran KA, Quirk GJ: Recalling safety: cooperative functions of the ventromedial prefrontal cortex and the hippocampus in extinction. CNS Spectr 2007;12:200-206.
[48] Herry C, Vouimba RM, Garcia R: Plasticity in the mediodorsal thalamo-prefrontal cortical transmission in behaving mice. *J Neurophysiol* 1999;82:2827-2832.
[49] Herry C, Garcia R: Prefrontal cortex long-term potentiation, but not long-term depression, is associated with the maintenance of extinction of learned fear in mice. *J Neurosci* 2002;22:577-583.

[50] Hugues S, Chessel A, Lena I, Marsault R, Garcia R: Prefrontal infusion of PD098059 immediately after fear extinction training blocks extinction-associated prefrontal synaptic plasticity and decreases prefrontal ERK2 phosphorylation. *Synapse* 2006; 60: 280-287.

[51] Hugues S, Garcia R: Reorganization of learning-associated prefrontal synaptic plasticity between the recall of recent and remote fear extinction memory. *Learn Mem* 2008;14:520-524.

[52] Farinelli M, Deschaux O, Hugues S, Thevenet A, Garcia R: Hippocampal train stimulation modulates recall of fear extinction independently of prefrontal cortex synaptic plasticity and lesions. *Learn Mem* 2008;13:329-334.

[53] Van Eden CG, Uylings HB:Cytoarchitectonic development of the prefrontal cortex in the rat. *J Comp Neurol* 1985;241:253-267.

[54] Bond AJ, Wingrove J, Baylis M, Dalton J: Buspirone decreases physiological reactivity to unconditioned and conditioned aversive stimuli. *Psychopharmacol* 2003;165:291-295.

[55] Janusonis S, Gluncic V, Rakic P: Early serotonergic projections to Cajal-Retzius cells: relevance for cortical development. *J Neurosci* 2004;24: 1652-1659.

[56] Sodhi, MS, Sanders-Bush E: Serotonin and brain development. *Int Rev Neurobiol* 2004; 59: 111-174.

[57] Konno K, Matsumoto M, Togashi H, Yamaguchi T, Izumi T, Watanabe M, Iwanaga T, Yoshioka M: Early postnatal stress affects the serotonergic function in the median raphe nuclei of adult rats. *Brain Res* 2007;1172:60-66.

[58] Matsuzaki M, Izumi T, Matsumoto M, Togashi H, Yamaguchi T, Yoshida T, Watanabe M, Yoshioka M: Early postnatal stress affect $5\text{-}HT_{1A}$ receptor function in the medial prefrontal cortex in adult rats. *Eur J Pharmacol* 2009 *(in press)*.

[59] Korte SM, Buwalda B, Meijer O, De Kloet ER, Bohus B: Socially defeated male rats display a blunted adrenocortical response to a low dose of 8-OH-DPAT. *Eur J Pharmacol* 1995; 272: 45-50.

[60] López JF, Chalmers DT, Little KY, Watson SJ: Regulation of serotonin1A, glucocorticoid, and mineralocorticoid receptor in rat and human hippocampus: implications for the neurobiology of depression. *Biological Psychiatry* 1998; 43: 547-573.

[61] Béïque J-C, Campbell B, Perring P, Hamblin MW, Walker P, Mladenovi L, Andrade R: 5-HTergic regulation of membrane potential in developing rat prefrontal cortex: coordinated expression of 5-hydroxytryptamine $(5\text{-}HT)_{1A}$, $5\text{-}HT_{2A}$, and $5\text{-}HT_7$ receptors. *J Neurosci* 2004;24:4807-4817.

[62] Corcoran KA, Desmond TJ, Frey KA, Maren S: Hippocampal inactivation disrupts the acquisition and contextual encoding of fear extinction. *J Neurosci* 2005;25:8978-8987.

[63] Fischer A, Sananbenesi F, Schrick C, Spiess J, Radulovic J: Distinct roles of hippocampal *de novo* protein synthesis and actin rearrangement in extinction of contextual fear. *J Neurosci* 2004;24:1962-1966.

[64] Heldt SA, Stanek L, Chhatwal JP, Ressler KJ: Hippocampus-specific deletion of BDNF in adult mice impairs spatial memory and extinction of aversive memories. *Mol Psychiatry* 2007;12:656-670.

[65] Frankland PW, Josselyn SA, Anagnostaras SG, Kogan JH, Takahashi E, Silva AJ: Consolidation of CS and US representations in associative fear conditioning. *Hippocampus* 2004;14:557-569.
[66] Rauch SL, Shin LM, Phelps EA: Neurocircuitry models of posttraumatic stress disorder and extinction: human neuroimaging research-past, present, and future. *Biol Psychiatry* 2006;60:376-382.
[67] Dirikx T, Hermans D, Vansteenwegen D, Baeyens F, Eelen P: Reinstatement of conditioned responses in human differential fear conditioning. *J Behav Ther Exp Psychiatry* 2007;38:237-251.
[68] Grillon C, Southwich SM, Charney DS: The psychobiological basis of posttraumatic stress disorder. *Mol Psychiatry* 1996;1:278-297.
[69] Walker DL, Ressler KJ, Lu KT, Davis M: Facilitation of conditioned fear extinction by systemic administration or intra-amygdala infusions of D-cycloserine as assessed with fear-potentiated startle in rats. *J Neurosci* 2002;22:2343-2351.
[70] Ledgerwood L, Richardson R, Cranney J: D-cycloserine and the facilitation of extinction of conditioned fear: consequences for reinstatement. Behav *Neurosci* 2004;118:505-513.
[71] Parnas AS, Weber M, Richardson R: Effects of multiple exposures to D-cycloserine on extinction of conditioned fear in rats. *Neurobiol Learn Mem* 2005;83:224-231.
[72] Davis M, Ressler K, Rothbaum BO, Richardson R: Effects of D-cycloserine on extinction: translation from preclinical to clinical work. *Biol Psychiatry* 2006;60:369-375.
[73] Otto MW, Basden SL, Leyro TM, McHugh RK, Hofmann SG: Clinical perspectives on the combination of D-cycloserine and cognitive-behavioral therapy for the treatment of anxiety disorders. *CNS Spectr* 2007;12:51-66.
[74] Norberg MM, Krystal JH, Tolin DF: A meta-analysis of D-cycloserine and the facilitation of fear extinction and exposure therapy. *Biol Psychiatry* 2008;63:1118-1126.
[75] Rouaud E, Billard JM: D-cycloserine facilitates synaptic plasticity but impairs glutamatergic neurotransmission in rat hippocampal slices. *Br J Pharmacol* 2003;140:1051-1056.
[76] Szapiro G, Vianna MR, McGaugh JL, Medina JH, Izquierdo I: The role of NMDA glutamate receptors, PKA, MAPK, and CAMKII in the hippocampus in extinction of conditioned fear. *Hippocampus* 2003;13:53-58.
[77] Santini E, Ge H, Ren K, Peña de Ortiz S, Quirk GJ: Consolidation of fear extinction requires protein synthesis in the medial prefrontal cortex. *J Neurosci* 2004;24:5704-5710.
[78] Mickley GA, Kenmuir CL, Yocom AM, Wellman JA, Biada JM: A role for prefrontal cortex in the extinction of a conditioned taste aversion. Brain Res 2005;1051:176-182.
[79] Maren S, Quirk GJ: Neuronal signalling of fear memory. *Nat Rev Neurosci* 2004;5:844-852.
[80] Myers KM, Davis M: Mechanisms of fear extinction. *Mol Psychiatry* 2007;12:120-150.
[81] Etkin A, Wager TD: Functional neuroimaging of anxiety: a meta-analysis of emotional processing in PTSD, social anxiety disorder, and specific phobia. *Am J Psychiatry* 2007;164:1476-1488.

[82] Koenigs M, Huey ED, Raymont V, Cheon B, Solomon J, Wassermann EM, Grafman J: Focal brain damage protects against post-traumatic stress disorder in combat veterans. *Nat Neurosci* 2008;11:232-237.

In: Neurobiology of Post-Traumatic Stress Disorder
Editors: L. Sher and A. Vilens, pp. 107-118
ISBN: 978-1-61470-593-2

Chapter VIII

Neurobiology of PTSD: Effects on the Spinal Cord

Hirotaka Sakamoto[1,2], S. Marc Breedlove[3] and Mitsuhiro Kawata[2]
[1]Okayama University, Kashino, Ushimado, Setouchi, Okayama, Japan
[2]Kyoto Prefectural University of Medicine, Kawaramachi-Hirokoji, Kamigyo-ku, Kyoto, Japan
[3]Michigan State University, East Lansing, Michigan, USA

Abstract

Many men suffering from stress, including post-traumatic stress disorder (PTSD), report sexual dysfunction, which is traditionally treated *via* psychological counseling. Recently, we identified a gastrin-releasing peptide (GRP) system in the lumbar spinal cord that is a primary mediator for male reproductive functions. To ask whether acute severe stress could alter the male-specific GRP system, we used a single-prolonged stress (SPS), a putative rat model for PTSD. Exposure of male rats to SPS decreases both the local content and axonal distribution of GRP in the lower lumbar spinal cord and results in an attenuation of penile reflexes *in vivo*. Remarkably, pharmacological stimulation of GRP receptors restores penile reflexes in SPS-exposed males, and induces spontaneous ejaculation in a dose-dependent manner. Furthermore, although the level of plasma testosterone is normal one week after SPS exposure, there is a significant decrease in the expression of androgen receptor protein in this spinal center, which may make the spinal center less responsive to circulating androgens. We conclude that the spinal GRP system for male reproductive function is vulnerable to stress, which may provide new insights into clinical treatment of erectile dysfunction triggered by stress and psychiatric disorders.

Introduction

Sexual behaviors differ in males and females, suggesting that the mechanisms underlying these behaviors also differ between the sexes. Male sexual behavior is complex,depending on intrinsic and extrinsic factors, including olfactory, somatosensory and visceral cues [1, 2]. The lumbar spinal cord is of particular interest because it is sexually dimorphic and contains several neural circuits that are important in regulating male sexual behaviors [3, 4]. Recently, we identified a gastrin-releasing peptide (GRP) system in the lumbar spinal cord that is a primary mediator for male reproductive functions [5, 6]. However, the mechanisms involved in the stress-related male sexual dysfunction within spinal reproductive centers have never been investigated. Here, we review data indicating that acute severe stress can alter the male-specific GRP system, such that exposure to a single-prolonged stress (SPS), a putative rat model for post-traumatic stress disorder (PTSD) [7], decreases both the local content and axonal distribution of GRP in the lumbar spinal cord, as well as the capacity of the system to respond to androgen, resulting in an attenuation of penile reflexes *in vivo* more than a week later [8].

PTSD Patients and Erectile Dysfunction

It is widely accepted that PTSD is a psychiatric disorder involving long-lasting symptoms that may occur after exposure to a life-threatening traumatic event, and is characterized by intrusive memories (flashbacks), a hyperarousal state and avoidance of stimuli associated with the trauma [9, 10]. Clinical data have indicated increased rates of sexual dysfunction, including erection and ejaculation difficulties in men with PTSD [11-13]. Most combat veterans with PTSD experience clinically relevant sexual difficulties and 69% have erectile dysfunction (ED) [14]. Although erections are clearly androgen-dependent, as evidenced by a marked reduction in the frequency, amplitude, and rigidity of erections in men with hypogonadism [15], little is known about the role of androgen-dependent neuropathy within the central nervous system in the development of psychogenic ED. Previous studies of stress and the hypothalamic-pituitary-gonadal axis have indicated that circulating testosterone (T) fluctuates in response to physical and psychological stress [16-18]. In contrast to other stress-related diseases, there is evidence that plasma or serum T levels do not change in combat-related PTSD patients [19] or in refugees suffering from PTSD [20], so any sexual difficulties accompanying the disorder appear to be unrelated to circulating levels of androgen.

Bombesin-like Peptides and their Receptors

Bombesin was first discovered in the skin of the European frog *Bombina bombina* [21, 22] with potent bioactivity in the mammalian nervous system. Subsequently, the mammalian bombesin-like peptides, GRP [22] and neuromedin B (NMB) [23], were isolated, and considerable evidence indicates that they play a role in many physiological processes, including food intake [24-26], circadian rhythms [27-29], itch [30, 31], and fear memory consolidation [32-34]. In mammals, bombesin-like peptides act through a family of at least three G protein-coupled receptors, namely the GRP-preferring receptor (GRP-R) [35], the NMB-preferring receptor [36] and bombesin receptor subtype-3 which is considered an orphan receptor [37]. Thus, we focused on the function of GRP/GRP-R in the reproductive system since earlier studies have shown that the bombesin-like peptides play a significant role in the central regulation of the neuroendocrinological and/or autonomic systems [38, 39] that are generally considered to be relevant to sexual behaviors in mammals.

Local Circuits in the Lumbar Spinal Cord Controlling Male Sexual Behavior

Onuf's nucleus, located in the ventral horn of the sacral spinal cord of many mammals, including humans, is a sexually dimorphic nucleus that innervates the perineal muscles involved in copulatory behavior. The number of neurons in Onuf's nucleus is greater in males than in females [40-43], and the distribution of serotonergic fibers and terminals in this nucleus also differs between the sexes [44-46]. The spinal nucleus of the bulbocavernosus (SNB) of rats in the fifth and sixth lumbar segments is homologous to Onuf's nucleus in that it innervates the striated perineal muscles attached to the base of the penis [47, 43]. Like Onuf's nucleus, male rats have more and larger SNB motoneurons than do females, a dimorphism that results from differences in perinatal androgen signaling *via* an androgen receptor (AR)-mediated mechanism [47]. Furthermore, a number of studies have revealed another group of neurons exhibiting sexual dimorphism in the lumbar spinal cord: a population of galanin [48]-, cholecystokinin [49]- and enkephalin [50]-expressing neurons. These neurons are dorsolateral to the central canal in lamina X within the third and fourth lumbar spinal cord and project to the thalamus. These so-called lumbar spinothalamic (LSt) neurons [51, 52] are also sexually differentiated, with males possessing a greater number of neurons than females. Fos induction in LSt neurons is specifically associated with ejaculation, but not with mounting or intromission, indicating that a specific subpopulation of LSt neurons conveys information associated with ejaculation [52]. Truitt and Coolen [53] reported that a population of neurons in the upper lumbar spinal cord acts as a 'spinal ejaculation generator' because toxin treatments that selectively lesion the galanin-expressing neurons in this region eliminate ejaculation in rats. While it is clear that these galaninergic neurons project to the thalamus [51, 52], it is unclear whether there are also direct connections between this center and the lower spinal cord regions that directly trigger ejaculation.

Identification and Characterization of GRP Neurons in the Lumbar Spinal Cord

Via investigation by immunocytochemistry directed at GRP, we found a separate group of neurons in the spinal region (L3–4) that appears to play a role in ejaculation [54]. These GRP-expressing neurons project axons to the more caudal spinal cord and are much more prominent in males than in females [54, 5] (Figure 1). To test whether ARs direct sexual dimorphism of these neurons, we examined genetically male (XY) Long-Evans rats carrying the testicular feminization mutation (Tfm) of the *Ar* gene [55, 56]. These rats develop testes embryologically and secrete testosterone prenatally, but because their AR protein is dysfunctional, they develop a wholly feminine exterior phenotype [55, 56]. The spinal cord of genetic male rats carrying the *Tfm* allele for *Ar* was hyperfeminine in terms of GRP system, having even fewer GRP-immunoreactive neurons in this region than wild-type females [5]. GRP-expressing neurons in males also express AR (Figure 1), but do not express estrogen receptors.

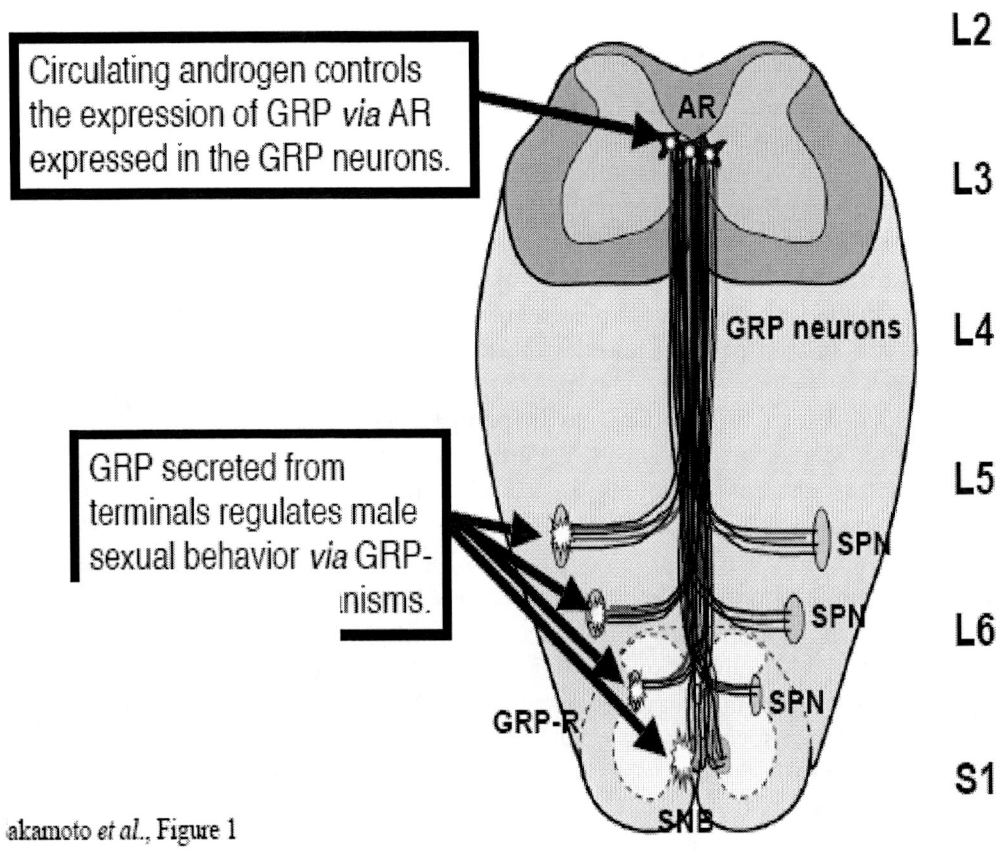

Figure 1. Schematic drawing summarizing the gastrin-releasing peptide (GRP)/GRP receptor (GRP-R) system in the lumbar spinal cord that controls male reproductive functions. A sexually dimorphic spinal cord system of GRP-containing neurons in the upper lumbar spinal cord "ejaculation center" projects axons to both the autonomic (sacral parasympathetic nucleus; SPN) and somatic (spinal nucleus of the bulbocavernosus; SNB) regions of the lower lumbar spinal cord that mediate penile reflexes and trigger ejaculation. The figure was reproduced from Sakamoto and Kawata [6] with permission.

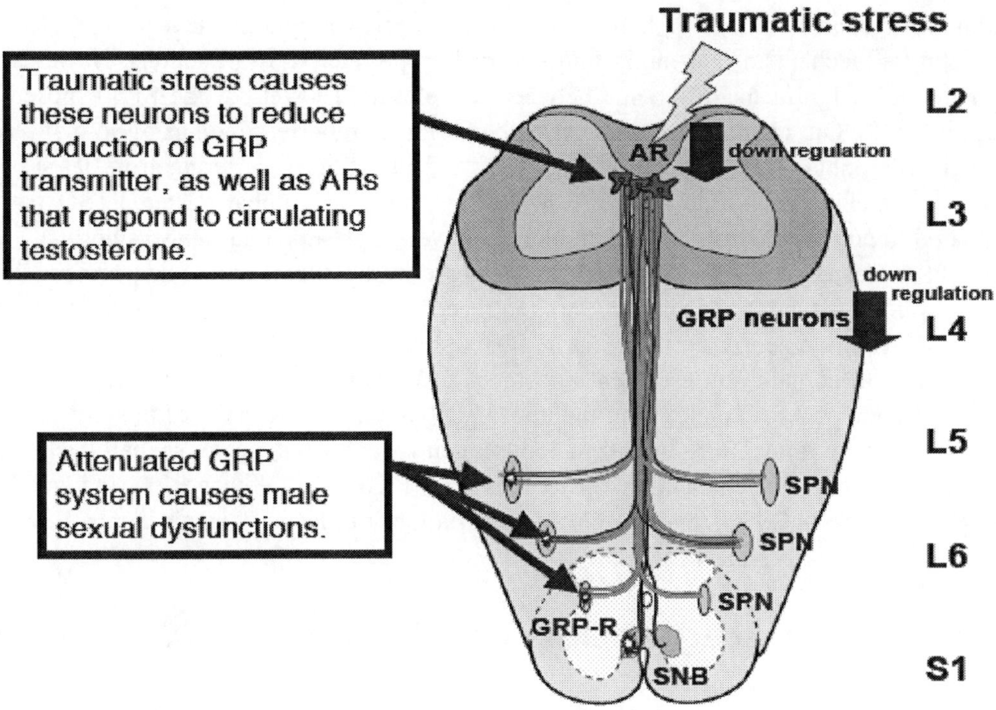

Figure 2. Possible mechanisms for the attenuation of the spinal GRP system in the post-traumatic stress disorder (PTSD). A traumatic stress (single-prolonged stress; SPS) significantly affects the spinal GRP system regulating male reproductive function, reducing expression of both GRP and the androgen receptor-mediated testosterone effects on the system, interfering with male reproductive function *in vivo*.

Androgen Regulates GRP Expression in the Lumbar Spinal Cord of Adult Males but not of Females: A Perspective in the Stress Vulnerability of ED at the Spinal Level

Orchidectomy of adult male rats significantly reduces the intensity of GRP-immunoreactive fibers in the lumbar spinal cord 28 days later and this reduction is averted by androgen replacement [5, 57]. Manipulating androgens in adulthood also has marked effects on the SNB system [58]. Castration of adult male rats leads to the shrinkage of SNB neuronal somata and dendrites, which can be prevented by treating castrates with testosterone [58-61]. Treatment of castrated female rats with testosterone in adulthood also increases SNB neuronal cell size, although not to the level seen in males [58]. Consequently, these results indicate that both T and AR, as an androgen signaling system, are necessary to maintain both the GRP and the SNB systems in the lumbar spinal cord [5, 6]. On the other hand, lower levels of T in major depressed men and a negative correlation between T levels and the severity of depression has been observed [62, 17]. It has also been reported that exposing rats to chronic restraint stress, as a model of human depression, decreases plasma levels of T [18, 63], suggesting a mechanism that might underlie male reproductive dysfunctions. However, we did not find a significant difference between the plasma T levels of the SPS-exposed and control rats [8]. Our results are consistent with previously reported findings of normal plasma T levels in combat-related PTSD patients [19] and in refugees suffering from PTSD [20]. However, cerebrospinal fluid (CSF) T levels are lower in combat-related PTSD patients compared to normal controls [19]. Although the level of plasma T in rats was normal 7 days after SPS exposure, we found a significant decrease in the expression of AR proteins in the lumbar spinal cord, but no change in estrogen receptorproteins [8]. We also found that nearly every GRP-positive neuron in the upper lumbar spinal cord normally contains AR, but these neurons do not contain any estrogen receptors [5, 6]. Taken together, these results suggest that the decline of AR expression in the upper lumbar spinal cord induces the attenuation of the GRP system after SPS exposure, consequently contributing to sexual dysfunction, including erection and ejaculation difficulties (see Figure 2). However, we do not know whether SPS affects CSF T levels or circadian variations of plasma T levels in rats.

The Central Regulation of Libido in PTSD Patients

PTSD has been found to affect emotional and social functioning, and to reduce aggression and libido [9, 64, 65]. Reduced libido is a higher brain dysfunction that is also a concern in both men and women [66]. Clinically, several findings have focused on the limbic system. Neuroimaging studies of patients with PTSD have suggested that pathology in the hippocampus [67, 68], amygdala [69] and anterior cingulate cortex [70, 71] plays an important role in the disorder. Thus, severe psychological stress may influence these brain regions, and may undermine the higher brain functions that regulate libido. The decreased libido, combined with effects on the sexual centers in the lumbar spinal cord as observed previously [8] could act synergistically to impair sexual function. Further studies of brain-spinal cord interactions, focusing on the pathogenesis of loss of libido in the brain and sexual dysfunctions in the spinal cord, including GRP system, should clarify the neural basis of different symptom clusters in PTSD patients.

Usefulness of the PTSD Animal Model

A number studies using the SPS paradigm in rats have indicated that the procedure successfully reproduces neuroendocrine characteristics of PTSD, including enhanced hypothalamo-pituitary-adrenal (HPA) negative feedback and exaggerated acoustic startle response, and also alters the expression of glucocorticoid receptors (GR) and mineralcorticoid receptors (MR), both of which serve as corticosterone (CORT) receptors, in the hippocampus [7, 72, 73]. In addition, a striking increase of dendritic arborization in pyramidal neurons in the amygdala was observed 7 days after SPS [74]. These findings indicate that SPS induces time-dependent sensitization, which resembles the developmental course of PTSD and may represent a better model for PTSD in the central nervous system [75, 7]. SPS consists of three different stress paradigms (restraint for 2 hours, forced swimming for 20 minutes, and exposure to ether vapor until loss of consciousness) [7]. Each of these individual stressors is known to stimulate the HPA axis markedly; however, administration of the three different stressors consecutively, followed by a 7-day undisturbed period, was required to reproduce the enhanced inhibition of the HPA axis, which is a major neuroendocrine abnormality in PTSD [7, 72, 76, 77].

Conclusion

In summary, we have found that a severe stress significantly attenuates the spinal GRP system regulating male reproductive functions, reducing expression of both GRP and the ARs mediating T effects on the system, interfering with male sexual function. These findings suggest that the use of selective GRP-R agents may provide new avenues for the treatment of stress-related ED in men.

Acknowledgments

The authors are grateful to Drs. Ken-Ichi Matsuda, Mayumi Nishi, Miwako Masugi-Tokita, Yukie Hirahara-Wada, Keiko Takanami, Hisayuki Hongu, Nobuko Nishiura, Hiroaki Mohri, Etsuko Wada, Keiji Wada, Damian G. Zuloaga, and Cynthia L. Jordan for their valuable discussions and collaborations. This work was supported by Grants-in-Aid for Scientific Research from the Ministry of Education, Science, Sports, Culture and Technology, Japan (to H.S. and M.K.) and from the National Institutes of Health (to S.M.B.).

References

[1] Rosen RC, Sachs BD: Central mechanisms in the control of penile erection: current theory and research. *Neurosci Biobehav Rev* 2000; 24: 503-505.
[2] Coolen LM: Neural control of ejaculation. *J Comp Neurol* 2005; 493: 39-45.
[3] Breedlove SM: Hormonal control of the anatomical specificity of motoneuron-to-muscle innervation in rats. *Science* 1985; 227: 1357-1359.
[4] Morris JA, Jordan CL, Breedlove SM: Sexual differentiation of the vertebrate nervous system. *Nat Neurosci* 2004; 7: 1034-1039.
[5] Sakamoto H, Matsuda K-I, Zuloaga DG, Hongu H, Wada E, Wada K, Jordan CL, Breedlove SM, Kawata M: Sexually dimorphic gastrin releasing peptide system in the spinal cord controls male reproductive functions. *Nat Neurosci* 2008; 11: 634-636.
[6] Sakamoto H, Kawata M: Gastrin-releasing peptide system in the spinal cord controls male sexual behaviour. *J Neuroendocrinol* 2009; 21: 432-435.
[7] Liberzon I, Krstov M, Young EA: Stress-restress: effects on ACTH and fast feedback. *Psychoneuroendocrinology* 1997; 22: 443-453.

[8] Sakamoto H, Matsuda K-I, Zuloaga DG, Nishiura N, Takanami K, Jordan CL, Breedlove SM, Kawata M: Stress affects a gastrin-releasing peptide system in the spinal cord that mediates sexual function: Implications for psychogenic erectile dysfunction. *PLoS ONE* 2009; 4: e4276.

[9] Pitman RK: Overview of biological themes in PTSD. *Ann N Y Acad Sci* 1997; 821: 1-9.

[10] Yehuda R: Post-traumatic stress disorder. *N Engl J Med* 2002; 346: 108-114.

[11] Kaplan HS: Anxiety and sexual dysfunction. *J Clin Psychiatry* 1988; 49 Suppl: 21-25.

[12] Kaplan HS: Post-traumatic stress syndrome and sexual dysfunction. *J Sex Marital Ther* 1989; 15: 74-77.

[13] Cosgrove DJ, Gordon Z, Bernie JE, Hami S, Montoya D, Stein MB, Monga M: Sexual dysfunction in combat veterans with post-traumatic stress disorder. *Urology* 2002; 60: 881-884.

[14] Letourneau EJ, Schewe PA, Frueh BC: Preliminary evaluation of sexual problems in combat veterans with PTSD. *J Trauma Stress* 1997; 10: 125-132.

[15] Rajfer J: Relationship between testosterone and erectile dysfunction. *Rev Urol* 2000; 2: 122-128.

[16] Kreuz LE, Rose RM, Jennings JR: Suppression of plasma testosterone levels and psychological stress. A longitudinal study of young men in Officer Candidate School. *Arch Gen Psychiatry* 1972; 26: 479-482.

[17] Mason JW, Giller EL, Kosten TR: Serum testosterone differences between patients with schizophrenia and those with affective disorder. *Biol Psychiatry* 1988; 23: 357-366.

[18] Retana-Marquez S, Bonilla-Jaime H, Vazquez-Palacios G, Martinez-Garcia R, Velazquez-Moctezuma J: Changes in masculine sexual behavior, corticosterone and testosterone in response to acute and chronic stress in male rats. *Horm Behav* 2003; 44: 327-337.

[19] Mulchahey JJ, Ekhator NN, Zhang H, Kasckow JW, Baker DG, Geracioti TD, Jr.: Cerebrospinal fluid and plasma testosterone levels in post-traumatic stress disorder and tobacco dependence. *Psychoneuroendocrinology* 2001; 26: 273-285.

[20] Bauer M, Priebe S, Graf KJ, Kurten I, Baumgartner A: Psychological and endocrine abnormalities in refugees from East Germany: Part II. Serum levels of cortisol, prolactin, luteinizing hormone, follicle stimulating hormone, and testosterone. *Psychiatry Res* 1994; 51: 75-85.

[21] Anastasi A, Erspamer V, Bucci M: Isolation and structure of bombesin and alytesin, 2 analogous active peptides from the skin of the European amphibians Bombina and Alytes. *Experientia* 1971; 27: 166-167.

[22] McDonald TJ, Jornvall H, Nilsson G, Vagne M, Ghatei M, Bloom SR, Mutt V: Characterization of a gastrin releasing peptide from porcine non-antral gastric tissue. *Biochem Biophys Res Commun* 1979; 90: 227-233.

[23] Minamino N, Kangawa K, Matsuo H: Neuromedin B: a novel bombesin-like peptide identified in porcine spinal cord. *Biochem Biophys Res Commun* 1983; 114: 541-548.

[24] Ladenheim EE, Taylor JE, Coy DH, Moore KA, Moran TH: Hindbrain GRP receptor blockade antagonizes feeding suppression by peripherally administered GRP. *Am J Physiol* 1996; 271: R180-184.

[25] Yamada K, Wada E, Wada K: Bombesin-like peptides: studies on food intake and social behaviour with receptor knock-out mice. *Ann Med* 2000; 32: 519-529.

[26] Ladenheim EE, Behles RR, Bi S, Moran TH: Gastrin-releasing peptide messenger ribonucleic acid expression in the hypothalamic paraventricular nucleus is altered by melanocortin receptor stimulation and food deprivation. *Endocrinology* 2009; 150: 672-678.

[27] Shinohara K, Tominaga K, Isobe Y, Inouye ST: Photic regulation of peptides located in the ventrolateral subdivision of the suprachiasmatic nucleus of the rat: daily variations of vasoactive intestinal polypeptide, gastrin-releasing peptide, and neuropeptide Y. *J Neurosci* 1993; 13: 793-800.

[28] Okamura H, Ibata Y: GRP immunoreactivity shows a day-night difference in the suprachiasmatic nuclear soma and efferent fibers: comparison to VIP immunoreactivity. *Neurosci Lett* 1994; 181: 165-168.

[29] Karatsoreos IN, Romeo RD, McEwen BS, Silver R: Diurnal regulation of the gastrin-releasing peptide receptor in the mouse circadian clock. *Eur J Neurosci* 2006; 23: 1047-1053.

[30] Sun YG, Chen ZF: A gastrin-releasing peptide receptor mediates the itch sensation in the spinal cord. *Nature* 2007; 448: 700-703.

[31] Swain MG: Gastrin-releasing peptide and pruritus: more than just scratching the surface. *J Hepatol* 2008; 48: 681-683.

[32] Shumyatsky GP, Tsvetkov E, Malleret G, Vronskaya S, Hatton M, Hampton L, Battey JF, Dulac C, Kandel ER, Bolshakov VY: Identification of a signaling network in lateral nucleus of amygdala important for inhibiting memory specifically related to learned fear. *Cell* 2002; 111: 905-918.

[33] Roesler R, Lessa D, Venturella R, Vianna MR, Luft T, Henriques JA, Izquierdo I, Schwartsmann G: Bombesin/gastrin-releasing peptide receptors in the basolateral amygdala regulate memory consolidation. *Eur J Neurosci* 2004; 19: 1041-1045.

[34] Merali Z, Bedard T, Andrews N, Davis B, McKnight AT, Gonzalez MI, Pritchard M, Kent P, Anisman H: Bombesin receptors as a novel anti-anxiety therapeutic target: BB1 receptor actions on anxiety through alterations of serotonin activity. *J Neurosci* 2006; 26: 10387-10396.

[35] Battey JF, Way JM, Corjay MH, Shapira H, Kusano K, Harkins R, Wu JM, Slattery T, Mann E, Feldman RI: Molecular cloning of the bombesin/gastrin-releasing peptide receptor from Swiss 3T3 cells. *Proc Natl Acad Sci U S A* 1991; 88: 395-399.

[36] Wada E, Way J, Shapira H, Kusano K, Lebacq-Verheyden AM, Coy D, Jensen R, Battey J: cDNA cloning, characterization and brain region-specific expression of neuromedin B-preferring bombesin receptor. *Neuron* 1991; 6: 421-430.

[37] Fathi Z, Corjay MH, Shapira H, Wada E, Benya R, Jensen R, Viallet J, Sausville EA, Battey JF: BRS-3: a novel bombesin receptor subtype selectively expressed in testis and lung carcinoma cells. *J Biol Chem* 1993; 268: 5979-5984.

[38] Panula P, Nieminen O, Falkenberg M, Auvinen S: Localization and development of bombesin/GRP-like immunoreactivity in the rat central nervous system. *Ann N Y Acad Sci* 1988; 547: 54-69.

[39] Battey J, Wada E: Two distinct receptor subtypes for mammalian bombesin-like peptides. *Trends Neurosci* 1991; 14: 524-528.

[40] Onufrowicz B: Notes on the arrangement and function of the cell groups in the sacral region of the spinal cord. *J Nerv Mental Dis* 1899; 26: 498-504.

[41] Sato M, Mizuno N, Konishi A: Localization of motoneurons innervating perineal muscles: a HRP study in cat. *Brain Res* 1978; 140: 149-154.

[42] Nakagawa S: Onuf's nucleus of the sacral cord in a South American monkey (Saimiri): its location and bilateral cortical input from area 4. *Brain Res* 1980; 191: 337-344.

[43] Forger NG, Breedlove SM: Sexual dimorphism in human and canine spinal cord: role of early androgen. *Proc Natl Acad Sci U S A* 1986; 83: 7527-7531.

[44] Kojima M, Takeuchi Y, Goto M, Sano Y: Immunohistochemical study on the localization of serotonin fibers and terminals in the spinal cord of the monkey (Macaca fuscata). *Cell Tissue Res* 1983; 229: 23-36.

[45] Kojima M, Matsuura T, Kimura H, Nojyo Y, Sano Y: Fluorescence histochemical study on the noradrenergic control to the anterior column of the spinal lumbosacral segments of the rat and dog, with special reference to motoneurons innervating the perineal striated muscles (Onuf's nucleus). *Histochemistry* 1984; 81: 237-241.

[46] Kojima M, Sano Y: Sexual differences in the topographical distribution of serotonergic fibers in the anterior column of rat lumbar spinal cord. *Anat Embryol (Berl)* 1984; 170: 117-121.

[47] Breedlove SM, Arnold AP: Hormone accumulation in a sexually dimorphic motor nucleus of the rat spinal cord. *Science* 1980; 210: 564-566.

[48] Newton BW: A sexually dimorphic population of galanin-like neurons in the rat lumbar spinal cord: functional implications. *Neurosci Lett* 1992; 137: 119-122.

[49] Phan DC, Newton BW: Cholecystokinin-8-like immunoreactivity is sexually dimorphic in a midline population of rat lumbar neurons. *Neurosci Lett* 1999; 276: 165-168.

[50] Nicholas AP, Zhang X, Hokfelt T: An immunohistochemical investigation of the opioid cell column in lamina X of the male rat lumbosacral spinal cord. *Neurosci Lett* 1999; 270: 9-12.

[51] Ju G, Melander T, Ceccatelli S, Hokfelt T, Frey P: Immunohistochemical evidence for a spinothalamic pathway co-containing cholecystokinin- and galanin-like immunoreactivities in the rat. *Neuroscience* 1987; 20: 439-456.

[52] Truitt WA, Shipley MT, Veening JG, Coolen LM: Activation of a subset of lumbar spinothalamic neurons after copulatory behavior in male but not female rats. *J Neurosci* 2003; 23: 325-331.

[53] Truitt WA, Coolen LM: Identification of a potential ejaculation generator in the spinal cord. *Science* 2002; 297: 1566-1569.

[54] Sakamoto H, Kawata M: Distribution of gastrin releasing peptide in the rat lumbar spinal cord is sexually dimorphic and regulated by androgen. *Front Neuroendocrinol* 2006: 46.

[55] Bardin CW, Bullock L, Blackburn WR, Sherins RJ, Vanha-Perttula T: Testosterone metabolism in the androgen-insensitive rat: a model for testicular feminization. *Birth Defects Orig Artic Ser* 1971; 7: 185-192.

[56] Yarbrough WG, Quarmby VE, Simental JA, Joseph DR, Sar M, Lubahn DB, Olsen KL, French FS, Wilson EM: A single base mutation in the androgen receptor gene causes androgen insensitivity in the testicular feminized rat. *J Biol Chem* 1990; 265: 8893-8900.

[57] Sakamoto H, Takanami K, Zuloaga DG, Matsuda KI, Jordan CL, Breedlove SM, Kawata M: Androgen regulates the sexually dimorphic gastrin-releasing peptide system in the lumbar spinal cord that mediates male sexual function. *Endocrinology* 2009; 150: 3672-3679.

[58] Breedlove SM, Arnold AP: Sexually dimorphic motor nucleus in the rat lumbar spinal cord: response to adult hormone manipulation, absence in androgen-insensitive rats. *Brain Res* 1981; 225: 297-307.

[59] Kurz EM, Sengelaub DR, Arnold AP: Androgens regulate the dendritic length of mammalian motoneurons in adulthood. *Science* 1986; 232: 395-398.

[60] Goldstein LA, Kurz EM, Sengelaub DR: Androgen regulation of dendritic growth and retraction in the development of a sexually dimorphic spinal nucleus. *J Neurosci* 1990; 10: 935-946.

[61] Yang LY, Verhovshek T, Sengelaub DR: Brain-derived neurotrophic factor and androgen interact in the maintenance of dendritic morphology in a sexually dimorphic rat spinal nucleus. *Endocrinology* 2004; 145: 161-168.

[62] Yesavage JA, Davidson J, Widrow L, Berger PA: Plasma testosterone levels, depression, sexuality, and age. *Biol Psychiatry* 1985; 20: 222-225.

[63] Matsumoto A: Testosterone prevents synaptic loss in the perineal motoneuron pool in the spinal cord in male rats exposed to chronic stress. *Stress* 2005; 8: 133-140.

[64] Gruden V, Gruden V, Jr.: Libido and PTSD. Coll Antropol 2000; 24: 253-256.

[65] Newport DJ, Nemeroff CB: Neurobiology of posttraumatic stress disorder. *Curr Opin Neurobiol* 2000; 10: 211-218.

[66] Green B: Post-traumatic stress disorder: symptom profiles in men and women. *Curr Med Res Opin* 2003; 19: 200-204.

[67] Gilbertson MW, Shenton ME, Ciszewski A, Kasai K, Lasko NB, Orr SP, Pitman RK: Smaller hippocampal volume predicts pathologic vulnerability to psychological trauma. *Nat Neurosci* 2002; 5: 1242-1247.

[68] Villarreal G, Hamilton DA, Petropoulos H, Driscoll I, Rowland LM, Griego JA, Kodituwakku PW, Hart BL, Escalona R, Brooks WM: Reduced hippocampal volume and total white matter volume in posttraumatic stress disorder. *Biol Psychiatry* 2002; 52: 119-125.

[69] Matsuoka Y, Yamawaki S, Inagaki M, Akechi T, Uchitomi Y: A volumetric study of amygdala in cancer survivors with intrusive recollections. *Biol Psychiatry* 2003; 54: 736-743.

[70] Rauch SL, Shin LM, Segal E, Pitman RK, Carson MA, McMullin K, Whalen PJ, Makris N: Selectively reduced regional cortical volumes in post-traumatic stress disorder. *Neuroreport* 2003; 14: 913-916.

[71] Yamasue H, Kasai K, Iwanami A, Ohtani T, Yamada H, Abe O, Kuroki N, Fukuda R, Tochigi M, Furukawa S, Sadamatsu M, Sasaki T, Aoki S, Ohtomo K, Asukai N, Kato N: Voxel-based analysis of MRI reveals anterior cingulate gray-matter volume reduction in posttraumatic stress disorder due to terrorism. *Proc Natl Acad Sci U S A* 2003; 100: 9039-9043.

[72] Liberzon I, Lopez JF, Flagel SB, Vazquez DM, Young EA: Differential regulation of hippocampal glucocorticoid receptors mRNA and fast feedback: relevance to post-traumatic stress disorder. *J Neuroendocrinol* 1999; 11: 11-17.

[73] Kohda K, Harada K, Kato K, Hoshino A, Motohashi J, Yamaji T, Morinobu S, Matsuoka N, Kato N: Glucocorticoid receptor activation is involved in producing abnormal phenotypes of single-prolonged stress rats: a putative post-traumatic stress disorder model. *Neuroscience* 2007; 148: 22-33.

[74] Cui H, Sakamoto H, Higashi S, Kawata M: Effects of single-prolonged stress on neurons and their afferent inputs in the amygdala. *Neuroscience* 2008; 152: 703-712.

[75] Yehuda R, Antelman SM: Criteria for rationally evaluating animal models of posttraumatic stress disorder. *Biol Psychiatry* 1993; 33: 479-486.

[76] Khan S, Liberzon I: Topiramate attenuates exaggerated acoustic startle in an animal model of PTSD. *Psychopharmacology (Berl)* 2004; 172: 225-229.

[77] Takahashi T, Morinobu S, Iwamoto Y, Yamawaki S: Effect of paroxetine on enhanced contextual fear induced by single prolonged stress in rats. *Psychopharmacology (Berl)* 2006; 189: 165-173.

In: Neurobiology of Post-Traumatic Stress Disorder
Editors: L. Sher and A. Vilens, pp. 119-135
ISBN: 978-1-61470-593-2

Chapter IX

Sleep Disturbances in Posttraumatic Stress Disorder

Mitsunari Habukawa, Masaharu Maeda and Naohisa Uchimura
Kurume University School of Medicine, Kurume, Fukuoka, Japan

Abstract

Sleep disturbances are common complaints among patients with posttraumatic stress disorder (PTSD), as well as intrusive recollection, avoidance of disturbing stimuli, difficulty in concentration, and depressive symptoms. Patients with PTSD frequently report difficulty falling or staying asleep and recurrent nightmares related to the traumatic events. These sleep disturbances are part of the diagnostic criteria for PTSD and are included in both the re-experiencing (Criterion B) and increased arousal (Criterion D) symptom cluster in the DSM-IV-TR. Despite these sleep complaints, polysomnographic studies have not provided consistent evidence of sleep disturbances, especially regarding non-rapid eye movement (REM) sleep. However, the characteristic sleep disturbances, in particular the repetitive, stereotypical nightmares may be relatively specific to PTSD. The presence of nightmares in PTSD has been hypothesized as a dysfunction of REM sleep mechanisms and investigators have reported various types of REM sleep abnormalities. Treatment studies also have shown that antidepressants, such as nefazodone, trazodone, fluvoxamine, and image rehearsal therapy are beneficial in the treatment of PTSD-related sleep disturbances as well as core symptoms of this anxiety disorder; however, these sleep disturbances may persist and impair the quality of life in PTSD patients for several decades. It is important, therefore, to understand the characteristics of sleep disturbances in PTSD and these treatments. In this review, we discuss the clinical features and objective findings of sleep disturbances in PTSD and these treatments based on the published literature, including our findings.

1. Introduction

Posttraumatic stress disorder (PTSD) is an anxiety disorder that develops in individuals who experience a traumatic event capable of eliciting intense feelings of fear, helplessness, or horror, and frequently becomes chronic, recurrent, or progressive. PTSD first appeared in the Diagnostic and Statistical Manual of Mental Disorders (DSM-III) in 1980 [1]. Although the majority of people will have only transient psychological and behavioral symptoms that represent the normal reaction to traumatic events, if symptoms persist, the diagnostic criteria for PTSD may be met. The presence of symptoms is determined on the basis of subjective reports, but not any objective verification.

Sleep complaints are very common, in both the immediate and long-term aftermath of exposure to traumatic events [2]. In fact, some authors have labeled sleep disturbance as the hallmark of PTSD [3]. PTSD patients frequently report difficulty falling or staying asleep and recurrent nightmares related to the traumatic events [4-6]. These sleep disturbances are part of the diagnostic criteria for PTSD and are included in both the re-experiencing (Criterion B) and increased arousal (Criterion D) symptom cluster in the DSM-IV-TR [7].

Objective sleep studies of PTSD using a polysomnograph have been conducted mainly for combat-related patients with chronic PTSD typically decades after the traumatic events. However, laboratory sleep studies have not provided consistent evidence of sleep disturbances, although subjective sleep complaints are very common among patients with PTSD. Generally, objective studies of PTSD have demonstrated far fewer disturbances of sleep than would be expected [2]; however, the characteristic sleep disturbance of PTSD, in particular the repetitive, stereotypical nightmares related to the traumatic events, may be relatively specific to PTSD. The presence of repeated nightmares in PTSD has been hypothesized as a dysfunction of REM sleep mechanisms [3] and several studies have reported various types of REM sleep abnormalities. In this review, we discuss the clinical features and objective findings of sleep disturbances in PTSD and these treatments based on the published literature including our findings.

Table 1. Sleep Disturbances in PTSD (citation from reference 8)

Insomnia	
	Difficulty falling asleep
	Sleep maintenance insomnia
	Nonrestrative/unrefreshing sleep
Nightmares	
	Trauma-related nightmares
	Nontrauma related nightmares
Others	
	Startle/fear awakenings (without dream)
	Panic-like awakenings
	Thrashing movements during sleep
	Rapid eye movement (REM) sleep behavior disorder
	Sleep-disordered breathing

2. Clinical Features

Patients with PTSD clinically present with various types of sleep disturbances, although they most frequently report insomnia, such as difficulty falling and staying asleep and nightmares related to the traumatic events [8]. Table 1 shows various types of sleep disturbances in PTSD [8].

2.1. Insomnia

Insomnia in PTSD patients includes difficulty initiating or maintaining sleep or nonrestorative sleep. Mellman et al. compared the sleep of combat veterans with and without PTSD and reported that combat veterans with PTSD complained of significantly poorer sleep [4]. In this study, 59 and 73% of PTSD subjects reported insomnia and nonrestorative sleep, respectively, while 19 and 38% of combat veterans without PTSD reported the same complaints. Among the insomnia patterns, middle insomnia (awakening after sleep onset) was associated most specifically with PTSD.

In the National Vietnam Veterans Readjustment Study, Neylan et al. compared the frequency of sleep disturbances in Vietnam combat veterans to that in Vietnam-era veteran and civilian comparison subjects using questionnaire items that addressed complaints about sleep [6]. The sample consisted of 1,200 male combat veterans, 412 male Vietnam-era veterans, and 450 male civilian comparison subjects. Combat veterans with PTSD reported significantly poorer sleep than combat veterans without PTSD, era veterans, and civilian comparison subjects. Difficulty falling asleep occurred in 44% combat veterans with PTSD, 5.5% combat veterans without PTSD, 9.4% Vietnam-era veterans, and 5.0% civilian comparison subjects. Difficulty staying asleep occurred in 90.7% combat veterans with PTSD, 62.5% combat veterans without PTSD, 63.1% Vietnam-era veterans, and 52.9% civilian subjects. This study remains the most comprehensive survey to date that examines the relationships among war zone exposure, PTSD, and sleep complaints. Insomnia and poor subjective sleep seems to be significantly higher in PTSD patients across most studies and these sleep disturbances may persist and impair the quality of life in PTSD patients for several decades.

2.2. Nightmares

Repetitive and stereotypical nightmares related to traumatic events are a characteristic sleep complaint of PTSD patients. Horowitz et al. found that 54% PTSD patients reported nightmares [9] and van der Kolk et al. reported that more than 60% Vietnam combat veterans had one or more nightmare per month [10]. In contrast, studies of healthy students suggest that 20–40% normal young adult population have nightmares at a frequency of less than one per year [11, 12]. From these studies, it is apparent that patients with PTSD more frequently report nightmares than normal control subjects.

In the National Vietnam Veterans Readjustment Study, frequent nightmares appeared to be specific to PTSD. Nightmares occurred in 52.4% combat veterans with PTSD, 4.8%

combat veterans without PTSD, 5.7% Vietnam-era veterans, and 3.4% civilian subjects [6]. Combat exposure strongly correlated with the frequency of nightmares. Alcohol abuse, chronic medical illness, panic disorder, major depression, and mania did not predict the frequency of nightmares after control for nonsleep PTSD symptoms.

The occurrence of nightmares in the sleep laboratory has been found to be rare in most studies that have tested chronic PTSD patients typically 20–30 years after the traumatic events [13-17]. On the other hand, many earlier Israeli studies with more common nightmares tested civilians and younger veterans in the more immediate aftermath of exposure to traumatic events [18-20]. The frequency of nightmares appears to be related to the amount of time since the trauma and the acuity of traumatic events.

2.3. Other Sleep Disturbances

PTSD patients have reported sleep complaints other than insomnia and nightmares, including startle awakening without dreaming, panic-like awakenings, thrashing movements during sleep, rapid eye movement (REM) sleep behavior disorder, and sleep-disordered breathing (Table 1) [8]. Mellman et al. found that hospitalized veterans with chronic PTSD reported startle or fear awakenings without dream recall, panic-like awakenings (with physiological symptoms), and thrashing movements in sleep diary recordings [4]. Ross et al. compared the activity of anterior tibial muscles between 12 patients with Vietnam-related PTSD and 10 normal controls, and reported a high frequency of twitch burst during REM sleep and periodic limb movements during non-REM sleep in PTSD patients [21]. In general, the motor system during sleep appears to show elevated activities in PTSD patients compared with normal controls.

In the earlier Israeli studies, Hefez et al. reported that two sea disaster survivors had REM-related nightmares with violent body movements, vocalizations and falling out of bed [20]. Whenever these patients were awakened from these "stormy" REM periods, they reported having vivid nightmares related to the sea disaster. REM sleep behavior disorder (RBD) is a parasomnia characterized by intermittent loss of muscle atonia in REM sleep accompanied by motor activity associated with dream recall [22]. The first description of patients with RBD came from Schenck et al. who noted frequent comorbidity with degenerative neurologic disorders but not with psychiatric disorders [23-25]. The aforementioned two survivors in sea disaster reported by Hefez et al. are speculated to have comorbid RBD. Husain et al. subsequently reported a unique series of veterans with RBD and PTSD [25]. Of the 27 patients with RBD, 15 (56%) had concurrent PTSD. They suggested that similar neuropathologic processes might be responsible for both disorders.

3. Objective Findings

3.1. Polysomnographic Findings

Polysomnography is very useful to elucidate sleep architecture, the depth of sleep, and abnormalities of REM and non-REM sleep. Polysomnographic studies of PTSD have been conducted mainly for combat-related patients with chronic PTSD. Most studies have tested middle-aged chronic PTSD subjects typically decades after the traumatic events, except for a handful of studies that were conducted during the immediate aftermath of trauma [2].

Table 2 shows 11 polysomnographic results, including our findings, that have been compared between PTSD patients and normal controls [13-17, 19, 20, 26-29]. In earlier Israel studies before 1990 that tested various PTSD patients, including combat neurosis, Holocaust survivors and sea disaster survivors, total sleep time (TST) and sleep efficiency (SE) consistently have been found to be reduced in PTSD patients. On the other hand, many studies since the 1990s mainly of Vietnam veterans who had been suffering from PTSD for more than 20–30 years have not demonstrated reduced total sleep time (TST) and sleep efficiency (SE).

Sleep latency (SL) and the percentage of slow-wave sleep (%SWS) are typically normal, with some exceptions. Regarding non-REM sleep in PTSD, polysomnographic studies generally have not provided consistent evidence of sleep disturbances, although some earlier studies found reduced total sleep time (TST) and sleep efficiency (SE). There are likely a number of factors affecting sleep when PTSD persists for many years. Possible factors considered to explain the discrepancies of polysomnographic findings may include differences of 1) the amount of time since the trauma, 2) the age of subjects, 3) the source of the trauma, 4) the acuity of the trauma, 5) many comorbid psychiatric disorders, 6) comorbidities of substance dependence, and 7) the administration of psychotropic medications.

We investigated young adult and drug-naïve or drug-free PTSD patients without comorbidities of substance dependence and/or personality disorders and tested polysomnographic recordings within 1 to 3.5 years after the trauma [27]. Table 3 shows the comparison of polysomnographic data between our PTSD patients and age- and sex-matched normal controls. PTSD patients showed significantly reduced sleep efficiency caused by increased wake time after sleep onset. Our PTSD patients demonstrated unequivocal sleep maintenance impairments, along with decreased slow-wave sleep (Stage 3 and 4 sleep). These results diverge from most previous studies [13-17]. These different findings may be due to the high sleep efficiency (94.9%) of the normal controls in this study; however, in fact, we found significant improvement of these sleep measures, along with the amelioration of PTSD symptoms after psychotropic medication. Our patients were young adult and drug-naïve or drug-free PTSD patients without a history of substance dependence; therefore, they may be vulnerable to PTSD-related sleep disturbances.

The characteristic sleep disturbances of PTSD, in particular repetitive, stereotypical nightmares, may be relatively specific to PTSD. Many aspects of PTSD-related nightmares would predict their emergence from rapid eye movement (REM) sleep [14]; therefore, some investigators have hypothesized that PTSD involves a dysfunction of REM sleep mechanisms. Ross and Mellman et al. have found elevated REM sleep phasic events, such as greater REM density and motor activity during REM sleep, in chronic combat-related PTSD patients [14, 16, 21].

Table 2. Comparison of polysomnographic data between PTSD patients and normal controls

Authors	No. of subjects	Subjects	Time since the trauma	TST	SE	SL	%SWS	RL	%REM	REM interruption
Lavie (1979)	11	Combat neurosis	2-2.5 Y	-	-	ns	ns	+	-	0
Hefez (1987)	5	Holocaust survivors	40-45 Y	-	-	ns	+	ns	-	0
	6	Combat neurosis Sea disaster survivors	0.5-1.5 Y	-	-	+	ns	+	-	0
Glaubman (1990)	7	Combat neurosis Holocaust survivors	3-35 Y (mean: 12.3 Y)	ns	-	ns	-	+	ns	0
van Kammen (1990)	12	World War II	45-50 Y	-	-	+	ns	ns	+	0
Ross (1994)	11	Vietnam veterans	20-30 Y	ns	ns	ns	ns	ns	+	0
Dow (1996)	14	Vietnam veterans	20-30 Y	-	ns	ns	ns	ns	ns	0
Mellman (1997)	24	Vietnam veterans	20-30 Y	-	-	ns	ns	ns	ns	0
Woodward (2002)	88	Vietnam veterans	20-30 Y	ns	ns	ns	ns	ns	ns	0
Mellman (2002)	21	Motor vehicle accident Industrial accident	within 1 M	ns	ns	ns	ns	ns	ns	+
Breslau (2004)	71	Young-adult community with lifetime PTSD	within 5 Y	0	ns	ns	ns	ns	ns	+
Habukawa (2007)	10	Sea disaster survivors Motor vehicle accident	0.3-3.3Y (mean: 1.3 Y)	ns	-	ns	-	ns	ns	+

Time since the trauma: Y=years, M=months; TST=total sleep time; SE=sleep efficiency; SL=sleep latency; %SWS=slow-wave sleep time as ratio of TST; RL=REM latency; %REM=REM sleep time as ratio of TST.

ns=no significant difference; - significantly lower in PTSD subjects; + significantly greater in PTSD subjects; 0 insufficient data.

More recent studies have tested patients during the more immediate aftermath of trauma and observed the fragmentation and interruption of REM sleep [27-29]. Mellman et al. tested prospective polysomnographic recordings in 21 injured subjects within a month of injury and observed a more fragmented pattern of REM sleep in subjects developing PTSD by calculating the average duration of continuous REM sleep [28]. Breslau et al. compared polysomnographic measures between 71 lifetime PTSD subjects and 212 control subjects in a community sample and reported increased brief arousal from REM sleep in PTSD subjects by calculating the rate per hour of shifts to stage 1 sleep and waking from REM sleep [29]. We also compared polysomnographic data between 10 young adult PTSD patients and 10 age- and sex-matched normal controls and found increased REM interruptions in PTSD patients by determining REM interruptions by summing intrusive wake times during REM periods and adding the subsequent wake time to the last epoch of REM periods before emerging to more than 2 minutes of non-REM sleep for each REM period [27]. Our PTSD patients showed 10.8% REM interruptions, while normal subjects showed 2.2% (Table 3). For example, Figure 1 shows sleep hypnograms of REM interruption in two PTSD patients (Patient 1 and 2). Patient 1 demonstrated REM interruption for about 40 minutes following awakening from REM sleep. The percentage of REM interruption was 24.6%. Patient 2 demonstrated clear evidence of REM interruption insomnia. REM periods were fragmented by multiple arousals. The percentage of REM interruptions was 19.8%. They had often reported trauma-related nightmares, although they reported no nightmares on this polysomnographic night. This REM abnormality of REM interruption has not been seen in previous controlled studies of patients with other psychiatric disorders, including major depressive disorder; therefore, REM interruption may be a sleep finding specific to PTSD.

Table 3. Comparison of polysomnographic data between PTSD patients and normal controls

	PTSD Group (N=10)		Control Group (N=10)		Analysis
Measurement	Mean	SD	Mean	SD	P value
Total sleep time (min)	450.3	76.0	466.7	72.4	n.s.
Sleep efficiency (%)	85.5	2.5	94.9	2.1	.0002
Sleep latency (min)	25.6	21.0	12.9	7.5	n.s.
Waking time after sleep onset (min)	35.9	19.7	11.1	6.6	.0041
Number of arousals	22.7	10.1	12.1	7.3	.0280
Stage 1 sleep (%)	11.6	5.7	6.8	2.8	n.s.
Stage 2 sleep (%)	52.5	10.2	50.6	8.2	n.s.
Stage 3 and 4 sleep (%)	7.9	7.4	18.1	7.9	.0041
REM sleep (%)	20.7	4.6	22.7	4.9	n.s.
REM latency (min)	88.8	37.3	92.5	46.0	n.s.
REM density (%)	30.8	9.0	25.6	7.0	n.s.
REM interruption (min)	12.8	12.1	2.3	1.8	.0354
Percentage of REM interruptions (%)	10.6	8.1	2.2	1.8	.0247

Finally, regarding the correlation between sleep measures and PTSD symptoms, Woodward et al. investigated the relationships between polysomnographic data and nightmare complaints using a sleep history questionnaire in 63 unmedicated Vietnam combat veterans undergoing inpatient treatment for PTSD [17]. They reported that trauma-related nightmare, but not non-trauma-related nightmare, was associated with increased wake-after-sleep-onset in the sleep laboratory. In an aforementioned study [27], we found a strong positive correlation between the severity of trauma-related nightmares, as measured by the Clinician Administered PTSD Scale (CAPS) [30], and the percentage of REM interruptions, as well as wake time after sleep onset (Table 4). These studies indicate that trauma-related nightmares are an important factor resulting in increased REM interruptions and wake time after sleep onset in PTSD.

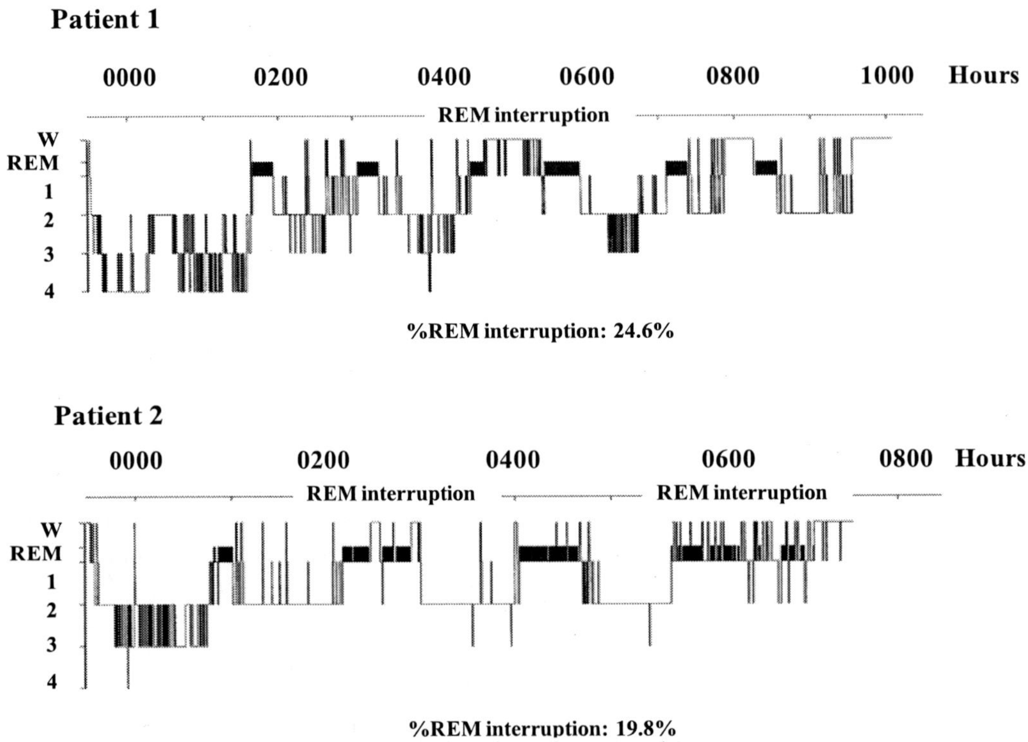

Figure 1. Sleep hypnograms of REM interruptions in patients 1 and 2.

3.2. Actigraphic Findings

The actigraph is a microcomputer attached to the wrist of the subject and differentiates between sleep and waking based on the amount of movement in the limb [31]. The main advantage of this method is in-home study of sleep, for several days, with minimal discomfort and disruption to the patient's routine. The disadvantage of the method is the inability to evaluate stage-specific abnormalities, such as an altered percentage of slow-wave sleep, REM sleep latency or awakening from REM.

Only a few studies have evaluated objective measures of sleep in PTSD patients using actigraphic recordings in their natural environment at home. Data from the few actigraphic studies have not supported the clear effect of PTSD on sleep. Dagan et al. compared objective actigraphic data between 16 PTSD patients with complaints of poor sleep from the Lebanon war and 11 healthy volunteers [32]. Their results indicated that PTSD patients did not suffer from poorer sleep than the control group, based on actigraphic measures. PTSD patient's subjective sleep evaluation was inconsistent with objective sleep measures using actigraphic recordings. Klein et al. prospectively compared objective sleep between 102 motor vehicle collision (VMC) survivors and 19 comparison subjects using actigraphic records, and found no significant differences among the three groups, VMC survivors with and without PTSD 12 months after the collision and comparison subjects, in the actigraphic measures, which were largely normal [33]. They therefore concluded that altered sleep perception, rather than sleep disturbances per se, might be the key problem in PTSD.

Table 4. Correlations between the total score and subscale scores in CAPS and polysomnographic data

	TST (min)	SE (%)	SL (min)	WASO (min)	No. arousal	%SWS (%)	REM Density (%)	Percentage of REM Interruptions (%)
Total score in CAPS	n.s.	n.s.	n.s.	n.s.	n.s.	n.s.	n.s.	n.s.
Re-experiencing (Criterion B)	n.s.	n.s.	n.s.	n.s.	n.s.	n.s.	n.s.	n.s.
Nightmare (Criterion B-2)	n.s.	n.s.	-.78 (P=.014)	.71 (P=.036)	n.s.	n.s.	n.s.	.82 (P=.014)
Avoidance of stimuli (Criterion C)	-.65 (P=.049)	n.s.	n.s.	n.s.	n.s.	n.s.	n.s.	n.s.
Hyperarousal (Criterion D)	n.s.	n.s.	n.s.	n.s.	n.s.	n.s.	n.s.	n.s.
Difficulty initiating and maintaining sleep (Criterion D-1)	n.s.	n.s.	n.s.	n.s.	n.s.	n.s.	n.s.	n.s.

In our study, the relationships between actigraphic findings and nightmare complaints were analyzed (unpublished data). Our PTSD patients wore an actigraph on the wrist of their non-dominant hand for fourteen consecutive days and were asked to record the previous night's dreams in a dream diary in the morning of the same consecutive days along with actigraphic monitoring. They were also asked to record events such as the occurrence of nightmares or panic attacks during the night, and the time of awakening following those events. Combined assessment in two PTSD patients (patients 3 and 4) by actigraphic recording and a self-reported dream diary provided evidence of difficulty maintaining sleep following the occurrence of trauma-related nightmares (Figure 2). For instance, Patient 3 suddenly awakened following the occurrence of trauma-related nightmares and had difficulty maintaining sleep for about 30–40 minutes on two nights out of eight actigraphic records. He reported vivid trauma-related nightmares in which he relived the sea accident and was falling from the sinking ship, and his classmates who had died suddenly appeared. He had experienced this trauma 8 months before the study. The actigraphic record in patient 4 demonstrated increased activity for about 20–60 minutes during the night almost every night. She also suddenly awakened following trauma-related nightmares and had difficulty maintaining sleep on two nights out of seven actigraphic records. She also reported vivid trauma-related nightmares, in which she was being hunted and killed with a knife by her attacker. She had been raped 10 months before the study. Combined assessment by consecutive actigraphic monitoring and a self-reported dream diary may be useful to evaluate sleep in some PTSD patients, but needs to be systematically studied in a large sample of PTSD patients.

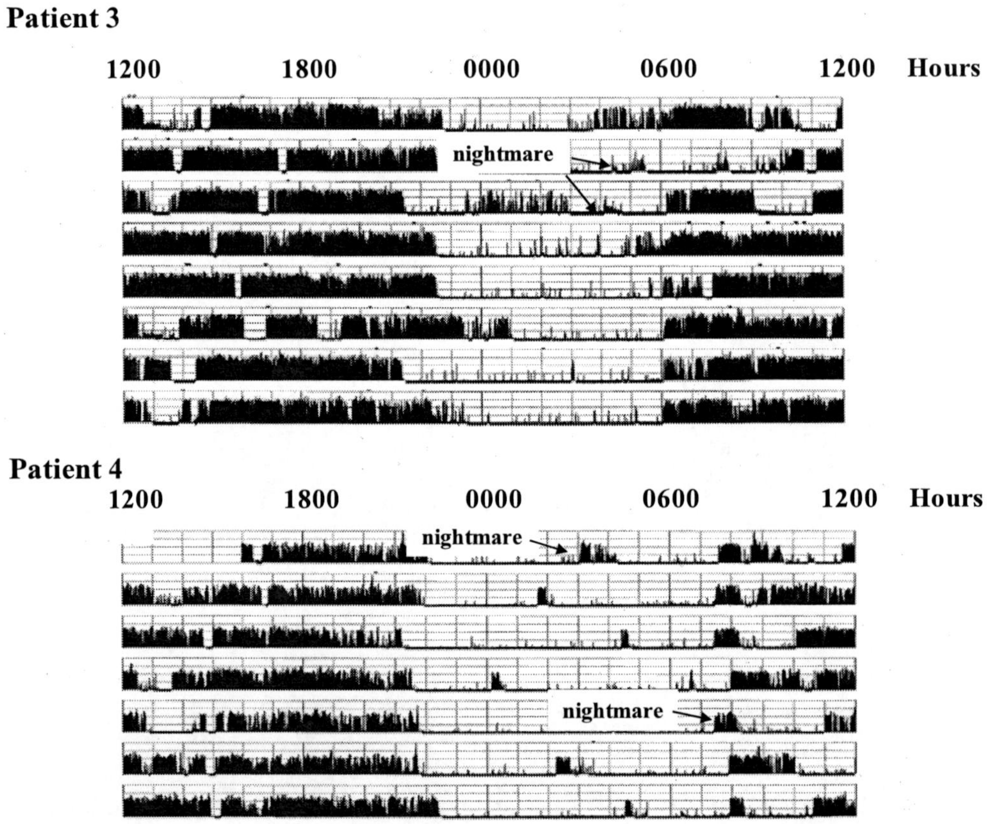

Figure 2. Combined assessments by actigraphic records and dream diary (Patients 3 and 4).

4. Treatment

Sleep disturbances may be the antecedents or precursors of PTSD and exacerbate chronic PTSD. Considering that insomnia, repetitive and stereotypical nightmares, and REM sleep dysfunction are major sleep-related disturbances in PTSD, sleep disturbances in PTSD should be treated as an independent clinical entity rather than as the effect of a psychopathologic condition [2, 8]. There are two types of sleep-focused interventions: pharmacologic and cognitive behavioral therapies.

4.1. Selective Serotonin Reuptake Inhibitors (SSRIs)

Selective serotonin reuptake inhibitors (SSRI), such as fluoxetine, paroxetine, fluvoxamine, and sertraline, are effective in the treatment of PTSD [34-37]. In fact, this group of medications is considered to be the first line of agents for the treatment of patients with PTSD [38]. Several placebo-controlled trials have been performed in PTSD patients; however, only limited studies have evaluated the improvement of sleep complaints as a primary outcome measure [39]. Placebo-controlled trials have shown the efficacy of paroxetine in the treatment of PTSD symptoms, including sleep disturbances [40]. A placebo-controlled trial showed a significant decrease in 'troubled sleeping' after fluoxetine treatment, as measured by a self-administered questionnaire [41]; however, the improvement was not significant, as measured by a structured interview. Fluoxetine did not have an effect on nightmares in this study. In two open-label studies, the effectiveness of fluvoxamine in the treatment of sleep complaints in combat-related PTSD patients was evaluated. The first study showed no statistically significant decrease in insomnia and nightmares in 24 combat-related PTSD patients [42]. A large number of patients terminated this study prematurely, mainly due to gastrointestinal complaints. Another study assessed the efficacy of fluvoxamine treatment on different domains of subjective sleep quality in 21 Vietnam combat veterans with chronic PTSD [43]. The largest effect was on trauma-related nightmares. Sleep maintenance insomnia and "troubled sleep" showed a large treatment response, while sleep onset insomnia improved less substantially.

However, as SSRIs are known to have activating effects on sleep electroencephalogram (EEG) such as an increase in the number of arousals and the percentage of Stage 1 sleep, a decrease in total sleep time, suppression of rapid eye movement (REM) sleep, and increase of phasic rapid eye movements [44, 45], further studies are needed to examine the efficacy of SSRIs on PTSD-related sleep disorders.

4.2. Nefazodone and Trazodone

Antidepressants acting on serotonin-$_2$ (5-HT_2) receptors, such as nefazodone and trazodone, are reported to improve sleep disturbances in PTSD patients. Nefazodone is a 5-HT_2 receptor antagonist and has α_1-adrenoceptor blocking properties. Several open-label studies on the effect of nefazodone in PTSD patients have been published [39]. Pooled analysis of these studies revealed significant improvement in sleep duration and reduction of nightmares [46]. Three of the open-label studies evaluated the efficacy of nefazodone by using both subjective and objective measures. In the study of Mellman et al., trauma-related nightmares and total sleep time did not improve significantly in the 11 PTSD patients, whereas overall PTSD symptoms did [47]. Polysomnographic recordings obtained from four patients demonstrated a trend for a decreased number of awakenings and arousals after treatment. In another study, Gillin et al. examined the effect of nefazodone on polysomnographic sleep measures and subjective reports of sleep quality and nightmares in 12 male combat veterans with chronic PTSD [48]. The patients reported significantly fewer nightmares and sleep problems during 12 weeks of treatment. Nevertheless, nefazodone did not significantly affect polysomnographic sleep measures compared with the baseline. Finally, Neylan et al. assessed the efficacy of nefazodone on both subjective and objective sleep quality in 10 Vietnam combat veterans with chronic PTSD [49]. Nefazodone treatment led to a significant decrease in PTSD and depressive symptoms, an improvement in subjective sleep quality, and a reduction in nightmares. Nefazodone also resulted in a substantial improvement in objective measures of sleep quality, particularly increased total sleep time, sleep maintenance, and delta sleep, as measured by period amplitude analysis.

Another 5-HT_2 antagonist, trazodone, promotes sleep continuity and results in a decrease in REM sleep [45]. Trazodone blocks the 5-HT_{1A}, 5-HT_2 and α_1-adrenoceptor. An open-label study with trazodone reported that sleep was the first symptom to improve in six patients with PTSD [50]. In a survey of 74 PTSD patients by Warner et al., trazodone was effective in decreasing insomnia and nightmares [51]. More than 70% of the 60 PTSD patients who were able to tolerate the medication reported significant improvement in nightmares, sleep onset insomnia, and sleep maintenance; however, most of these patients were on other psychotropic medications (SSRIs, nefazodone, valproic acid, and benzodiazepines) and the questionnaire was not standardized. Of the 74 patients surveyed, 9 (12%) reported priapism as a side effect.

4.3. Antipsychotics

The utility of antipsychotics in patients with PTSD has been reviewed by Ahearn et al. [52]. The use of olanzapine in the treatment of sleep complaints in PTSD patients has been found to be effective in a few case reports and in one placebo-controlled study. The case reports show that the addition of olanzapine to SSRIs is useful in reducing nightmares and insomnia in PTSD patients [53, 54]. A randomized placebo-controlled trial of adjunctive olanzapine confirmed these findings [55]. Nineteen PTSD patients who were minimally responsive to 12 weeks of treatment with SSRI were included in this double-blind, placebo-controlled study in which olanzapine was added to SSRI treatment. Olanzapine augmentation was associated with a statistically significant reduction of nightmares and improvement of insomnia, although clinician-rated global response rates did not significantly differ between groups. In an open-label study, quetiapine improved nightmares and insomnia complaints in 20 combat-related PTSD patients [56]. Levomepromazine was studied in a 4-week open-label study in 21 PTSD patients, and robust improvement was seen in nightmares and insomnia [57].

4.4. Cognitive Behavioral Therapy

Among the psychotherapeutic modalities, imagery rehearsal therapy was initially found to be useful in case reports [58, 59]. Imagery rehearsal therapy is a form of cognitive behavioral therapy by which the patients practice envisioning pleasant images and learn cognitive tools for dealing with unpleasant images of nightmares [60]. An open-label study of 62 crime victims with PTSD found significant improvement in nightmares, insomnia, and PTSD symptoms with imagery rehearsal therapy, sleep hygiene, stimulus control (eliminating learned sleep preventing associations and training to create a new association between the bed and bedroom and the rapid onset of sleep) and sleep restriction (curtailing the amount of time in bed in order to increase sleep quality and efficiency) [61]. Furthermore, a randomized controlled trial of female sexual assault survivors with PTSD demonstrated that imagery rehearsal therapy significantly improved nightmares, sleep quality, and PTSD symptoms [60].

This study, however, was limited by the lack of a placebo control and large drop-out rates at follow up.

5. Conclusion

Sleep complaints such as insomnia and nightmares are common among PTSD patients. Despite these complaints, polysomnographic studies mainly of combat veterans with chronic PTSD generally have not provided consistent evidence of sleep disturbances, especially regarding non-REM sleep; however, the characteristic sleep disturbances, in particular repetitive, stereotypical nightmares, may be relatively specific to PTSD. The presence of nightmares in PTSD has been hypothesized as a dysfunction of REM sleep mechanisms. Some investigators have reported elevated REM sleep phasic events, such as greater REM density and motor activity during REM sleep, in chronic combat-related PTSD patients. Furthermore, recent studies have found REM sleep abnormalities, such as fragmented REM sleep or REM interruption, in PTSD patients in the more immediate aftermath of the trauma. This REM sleep abnormality has not been seen in patients with other psychiatric disorders, including major depressive disorder; therefore, REM interruption may be a sleep measure specific to PTSD, particularly in the immediate aftermath of trauma. Some studies have also indicated that trauma-related nightmares are an important factor resulting in increased wake time after sleep onset and REM interruption. Attempts to test these findings under controlled conditions in large samples may further advance the understanding of the relationship between REM sleep dysfunction and PTSD.

Some treatment studies have shown that antidepressants, such as nefazodone, trazodone, and fluvoxamine, and image rehearsal therapy are beneficial in the treatment of PTSD-related sleep disturbances as well as core symptoms of this anxiety disorder; however, only a handful of studies have tested the efficacy of these treatments on sleep using objective measures such as polysomnography. Future studies need to focus on the role of REM sleep abnormalities and polysomnographic changes after treatments such as antidepressants, antipsychotics, and cognitive behavioral therapies to further understand the pathophysiology of PTSD-related sleep disturbances.

References

[1] American Psychiatric Association: Diagnostic and Statistical Mannual of Mental Disorders, Third Edition, Revised. Washington, DC: *American Psychiatric Press* 1987, pp 247-251.

[2] Lavie P: Sleep disturbances in the wake of traumatic events. *N Engl J Med* 2001; 345: 1825-1832.

[3] Ross RJ, Ball WA, Sullivan KA, Caroff SN: Sleep disturbance as the hallmark of posttraumatic stress disorder. *Am J Psychiatry* 1989; 146: 697-707.

[4] Mellman TA, Kulick-Bell R, Ashlock LE, Nolan B: Sleep events among veterans with combat-related posttraumatic stress disorder. *Am J Psychiatry* 1995; 152: 110-115.

[5] Mellman TA, Davis GC: Combat-related flashbacks in posttraumatic stress disorder: Phenomenology and similarity to panic attack. *J Clin Psychiatry* 1985; 46: 379-382.

[6] Neylan TC, Marmar CR, Metzler TJ, Weiss DS, Zatzick DF, Delucchi KL, Wu RM, Schoenfeld FB: Sleep disturbance in the Vietnam generation: Findings from a nationally representative sample of male Vietnam veterans. *Am J Psychiatry* 1998; 155: 929-933.

[7] American Psychiatric Association: *Diagnostic and Statistical Mannual of Mental Disorders*, Fourth Edition, Revised. Washington, DC: American Psychiatric Association 2000.

[8] Singareddy RK, Balon R: Sleep in posttraumatic stress disorder. *Ann Clin Psychiatry* 2002; 14: 183-190.

[9] Horowitz MJ, Wilner N, Kaltreider N, Alvarez W: Signs and symptoms of posttraumatic stress disorder. *Arch Gen Psychiatry* 1980; 37: 85-92.

[10] van der Kolk BA, Hartmann E, Burr A, Blitz R: A survey of nightmare frequency in a veterans outpatient clinic. *Sleep Res* 1980; 9: 229

[11] Belicky D, Belicky K: Nightmares in a university population. *Sleep Res* 1982; 11: 116

[12] Feldman MJ, Hersen M: Attitudes toward death in nightmare subjects. *J Abnorm Psychol* 1967; 72: 421-425.

[13] van Kammen WB, Christiansen C, van Kammen DP, Reynolds CF III: Sleep and the prisoner-of-war experience-40 years later. In Giller EL Jr (ed), *Biological Assessment and Treatment of Posttraumatic Stress Disorder.* Washington DC: American Psychiatric Press, pp 159-172.

[14] Ross RJ, Ball WA, Dinges DF, Kribbs NB, Morrison AR, Silver SM, Mulvaney FD: Rapid eye movement sleep disturbance in posttraumatic stress disorder. *Biol Psychiatry* 1994; 35: 195-202.

[15] Dow BM, Kelsoe JR, Gillin JC: Sleep and dreams in Vietnam PTSD and depression. *Biol Psychiatry* 1996; 39: 42-50.

[16] Mellman TA, Nolan B, Hebding J, Kulick-Bell R, Dominguez R: A polysomnographic comparison of veterans with combat-related PTSD, depressed men, and non-ill controls. *Sleep 1997*; 20: 46-51.

[17] Woodward SH, Arsenault NJ, Murray C, Bliwise DL: Laboratory sleep correlates of nightmare complaint in PTSD inpatients. *Biol Psychiatry* 2000; 48: 1081-1087.

[18] Schlosberg A, Benjamin M: Sleep patterns in three acute combat fatigue cases. *J Clin Psychiatry* 1978; 39: 546-549.

[19] Lavie P, Hefez A, Halperin G, Enoch D: Long-term effects of traumatic war-related events on sleep. *Am J Psychiatry* 1979; 136: 175-178.

[20] Hefez A, Metz L, Lavie P: Long-term effects of extreme situational stress on sleep and dreaming. *Am J Psychiatry* 1987; 144: 344-347.

[21] Ross RJ, Ball WA, Dinges DF, Kribbs NB, Morrison AR, Silver SM, Mulvaney FD: Motor dysfunction during sleep in posttraumatic stress disorder. *Sleep* 1994; 17: 723-732.

[22] American Sleep Disorders Association. REM sleep behavior disorder. In: *International classification of sleep disorders revised.* Rochester, MN: Davies Printing, 1997: 177-180.

[23] Schenck CH, Bundlie SR, Ettinger MC, Mahowald MW: Chronic behavioral disorders of human REM sleep; A new category of parasomnia. *Sleep* 1986; 9: 293-308.

[24] Schenck CH, Bundlie SR, Patterson AL, Mahowald MW: Rapid eye movement sleep behavior disorder: a treatable parasomnia affecting older adults. *JAMA* 1987; 257: 1786-1789.

[25] Husain AF, Miller PP, Carwile ST: REM sleep behavior disorder: potential relationship to post-traumatic stress disorder. *J Clin Neurophysiol* 2001; 18: 148-157.

[26] Glaubman H, Mikulincer M, Porat A: Sleep of chronic posttraumatic patients. *J Traumatic Stress Studies* 1990; 3: 255-263.

[27] Habukawa M, Uchimura N, Maeda M, Kotorii N, Maeda H: Sleep findings in young adult patients with posttraumatic stress disorder. *Biol Psychiatry* 2007; 62: 1179-1182.

[28] Mellman TA, Bustamante V, Fins AI, Pigeon WR, Nolan B: REM sleep and the early development of posttraumatic stress disorder. *Am J Psychiatry* 2002; 159: 1696-1701.

[29] Breslau N, Roth T, Burduvali E, Kapke A, Schultz L, Roehrs T: Sleep in lifetime posttraumatic stress disorder: a community-based polysomnographic study. *Arch Gen Psychiatry* 2004; 61: 508-516.

[30] Blake DD, Weathers FW, Nagy LM, Kaloupek DG, Gusman FD, Charney DS, Keane TM: The development of a clinician-administered PTSD scale. *J Traumatic Stress* 1995; 8: 75-90.

[31] Sadeh A, Aesler DU, Lawre P. Actigraphically-based automatic bedtime sleep-wake recording: validity and clinical applications. *J Ambul Monit* 1989; 2: 209-216.

[32] Dagan Y, Zinger Y, Lavie P. Actigraphic sleep monitoring in posttraumatic stress disorder (PTSD) patients. *J Psychosom Res* 1997; 42: 577-581.

[33] Klein E, Koren D, Arnon I, Lavie P. Sleep complaints are not corroborated by objective sleep measures in post-traumatic stress disorder: a 1-year prospective study in survivors of motor vehicle crashes. *J Sleep Res* 2003; 12: 35-41.

[34] Davidson JRT, Rothbaum BO, van der Kolk BA, Sikes CR, Farfel GM. Multicenter, double-blind comparison of sertraline and placebo in the treatment of posttraumatic stress disorder. *Arch Gen Psychiatry* 2001; 58: 485-492.

[35] van der Kolk BA, Dreyfuss D, Michaels M, Shera D, Berkowitz R, Fisler R, Saxe G. Fluoxetine in posttraumatic stress disorder. *J Clin Psychiatry* 1994; 55: 517-522.

[36] Marshall RD, Beebe KL, Oldham M, Zaninelli R. Efficacy and safety of paroxetine treatment of chronic PTSD: A fixed dose, placebo controlled study. *Am J Psychiatry* 2001; 158: 1982-1988.

[37] Tucker P, Zaninelli R, Yehuda R, Ruggiero L, Dillingham K, Pitts C: Paroxetine in the treatment of chronic posttraumatic stress disorder: Results of placebo-controlled, flexible dosage trial. *J Clin Psychiatry* 2001; 62: 860-868.

[38] Foa EB, Davidson JRT, Frances A. Expert consensus guideline series: Treatment of posttraumatic stress disorder. *J Clin Psychiatry* 1999; 60: 1-76.

[39] van Liempt S, Vermetten E, Geuze E, Westenberg HG. Pharmacotherapy for disordered sleep in post-traumatic stress disorder: a systematic review. *Int Clin Psychopharmacol* 2006; 21: 193-202.

[40] Stein DJ, Davidson J, Seedat S, Beebe K. Paroxetine in the treatment of posttraumatic stress disorder: pooled analysis of placebo-controlled studies. *Expert Opin Pharmacother* 2003; 4: 1829-1838.

[41] Meltzer-Brody S, Connor KM, Churchill E, Davidson JR. Symptom-specific effects of fluoxetine in post-traumatic stress disorder. *Int Clin Psychopharmacol* 2000; 15: 227-231.

[42] De Boer M, Op den Velde W, Falger PJ, Hovens JE, De Groen JH, Van Duijn H. Fluvoxamine treatment for chronic PTSD: a pilot study. *Psychother Psychosom* 1992; 57: 158-163.

[43] Neylan TC, Metzler TJ, Schoenfeld FB, Weiss DS, Lenoci M, Best SR, Lipsey TL, Marmar CR. Fluvoxamine and sleep disturbances in posttraumatic stress disorder. *J Trauma Stress* 2001; 14: 461-467.

[44] Rush AJ, Armitage R, Gillin JC, Yonkers KA, Winokur A, Moldofsky H, Vogel GW, Kaplita SB, Fleming JB, Montplaisir J, Erman MK, Albala BJ, McQuade RD. Comparative effects of nefazodone and fluoxetine on sleep in outpatients with major depressive disorder. *Biol Psychiatry* 1998; 44: 3-14.

[45] Wilson S, Argyropoulos S. Antidepressants and sleep: a qualitative review of the literature. *Drugs* 2005; 65: 927-947.

[46] Hidalgo R, Hertzberg MA, Mellman T, Petty F, Tucker P, Weisler R, Zisook S, Chen S, Churchill E, Davidson J. Nefazodone in post-traumatic stress disorder: results from six open-label trials. *Int Clin Psychopharmacol* 1999; 14: 61-68.

[47] Mellman TA, David D, Barza L. Nefazodone treatment and dream reports in chronic PTSD. *Depress Anxiety* 1999; 9: 146-148.

[48] Gillin JC, Smith-Vaniz A, Schnierow B, Rapaport MH, Kelsoe J, Raimo E et al. An open-label, 12-week clinical and sleep EEG study of nefazodone in chronic combat-related posttraumatic stress disorder. *J Clin Psychiatry;* 62: 789-796.

[49] Neylan TC, Lenoci M, Maglione ML, Rosenlicht NZ, Leykin Y, Metzler TJ, Schoenfeld FB, Marmar CR. The effect of nefazodone on subjective and objective sleep quality in posttraumatic stress disorder. *J Clin Psychiatry* 2003; 64: 445-450.

[50] Hertzberg MA, Feldman ME, Beckham JC, Davidson JR. Trial of trazodone for posttraumatic stress disorder using a multiple baseline group design. *J Clin Psychopharmacol* 1996; 16: 294-298.

[51] Warner MD, Dorn MR, Peabody CA. Survey on the usefulness of trazodone in patients with PTSD with insomnia or nightmares. *Pharmacopsychiatry* 2001; 34: 128-131.

[52] Ahearn EP, Krohn A, Connor KM, Davidson JR. Pharmacologic treatment of posttraumatic stress disorder: a focus on antipsychotic use. *Ann Clin Psychiatry* 2003; 15: 193-201.

[53] Labbate LA, Douglas S. Olanzapine for nightmares and sleep disturbances in posttraumatic stress disorder (PTSD) (letter). *Can J Psychiatry* 2000; 45: 667-668.

[54] Jakovljevic M, Sagud M, Mihaljevic-Peles A. Olanzapine in treatment-resistant, combat-related PTSD – a series of case reports. *Acta Psychiatr Scand* 2003; 107: 394-396.

[55] Stein MB, Kline NA, Matloff JL. Adjunctive olanzapine for SSRI-resistant combat-related PTSD: a double-blind, placebo-controlled study. *Am J Psychiatry* 2002; 159: 1777-1779.

[56] Robert S, Hamner MB, Kose S, Ulmer HG, Deitsch SE, Loberbaum JP. Quetiapine improves sleep disturbances in combat veterans with PTSD: sleep data from a prospective, open-label study. *J Clin Psychopharmacol* 2005; 25: 387-388.

[57] Aukst-Margetic B, Margetic B, Tosic G, Bilic-Prcic A. Levomepromazine helps to reduce sleep problems in patients with PTSD. *Eur Psychiatry* 2004; 19: 235-236.

[58] Halliday G. Direct psychological therapies for nightmares: A review. *Clin Psychol Rev* 1987; 7: 501-523.

[59] Krakow B, Sandoval D, Schrader R, Keuhne B, McBride L, Yau CL, Tandberg AD. Treatment of chronic nightmares in adjudicated adolescent girls in a residential facility. *J Adolesc Health* 2001; 29: 94-100.

[60] Krakow B, Hollifield M, Johnston L, Koss M, Schrader R, Warner TD, Tandberg D, Lauriello J, McBride L, Cutchen L, Cheng D, Emmons S, Germain A, Melendrez D, Sandoval D, Prince H. Imagery rehearsal therapy for chronic nightmares in sexual assault survivors with posttraumatic stress disorder: A randomized controlled trial. *JAMA* 2001; 286: 537-545.

[61] Krakow B, Johnston L, Melendrez D, Hollifield M, Warner TD, Chavez-Kennedy D, Herlan MJ. An open-label trial of evidence-based cognitive behavioral therapy for nightmares and insomnia in crime victims with PTSD. *Am J Psychiatry* 2001; 158: 2043-2047.

In: Neurobiology of Post-Traumatic Stress Disorder
Editors: L. Sher and A. Vilens, pp. 137-152
ISBN: 978-1-61470-593-2

Chapter X

Neurobiology of Posttraumatic Stress Disorder in Childhood and Adolescence

Panagiota Pervanidou[1], Gerasimos Kolaitis[2] and George P. Chrousos[1]

1. Developmental and Behavioral Pediatrics Unit, First Department of Pediatrics and
2. Department of Child Psychiatry, Athens University Medical School, Aghia Sophia Children's Hospital, Athens, Greece

Abstract

Since Posttraumatic Stress Disorder (PTSD) was recognized as an anxiety disorder in childhood, a limited number of studies investigated the neuroendocrine parameters underlying the clinical manifestations of the disease. In contrast to the majority of adult studies, where low or normal peripheral cortisol concentrations are reported in individuals with PTSD, cortisol is frequently found high in pediatric PTSD patients, whereas high catecholaminergic activity has been reported in both age groups. Developmental stage of the individual, and time since the trauma occurred, seem to be crucial variables in explaining discrepancies between pediatric and adult studies.

Longitudinal psychobiological studies in children, after motor vehicle accidents, have shown that initially elevated cortisol concentrations gradually normalize while norepinephrine concentrations become greater, revealing a progressive interrelation between these two main hormones of the stress system that may underlie the maintenance of the disorder. Thus, we hypothesized, that low cortisol levels, together with the high catecholamines in adults with chronic PTSD may reflect a late event in the natural history of the disorder, months or years after the trauma.

1. Introduction

Dysregulation of the stress system, including the hypothalamic-pituitary-adrenal (HPA) axis and the locus caeruleus / norepinephrine-sympathetic nervous system (LC/NE-SNS), is involved in the pathophysiology of Posttraumatic Stress Disorder (PTSD) [1]. Normally, activation of the stress system leads to behavioral and physical adaptive changes that improve an organism's ability to survive. However, excessive and prolonged activation of the two axes of the stress system, might lead to long-term emotional and behavioural symptoms and disorders as well as physical complications. Increased and prolonged production of Corticotropin Reseasing Hormone (CRH), cortisol and catecholamines could explain many of the emotional-behavioral, circulatory, metabolic and immune manifestations of such syndromes. However, chronic stress might also result in hypo-activation of the HPA axis, likely reflecting a compensatory physiologic adaptation [2,3].

Adaptation to stress has been primarily studied in adults, however, the same principles may also apply to children and adolescents [4]. Furthermore, there is evidence that early adverse experiences can cause alterations in the functioning of neuroendocrine structures and networks in the central nervous system [5]. Endocrine systems, which are crucial for growth and puberty, including the gonadal, growth hormone and thyroid axes, are influenced by the stress system. Chronic changes in these networks during critical periods of development could have irreversible effects on growth and development [4,5]. In addition, it has been hypothesized that childhood trauma may alter CNS development: early severe stress results in elevated levels of CRH and glucocorticoids affecting specific, vulnerable brain areas [5,6]. In this view, the impact of childhood stress and trauma, in addition to concurrent symptoms and disorders, may "program" the developing organism to be more susceptible in future stressors and thus, more vulnerable to develop mental and physical disorders.

2. PTSD as a Developmental Disorder

Since posttraumatic stress disorder (PTSD) emerged as a clinical diagnosis in childhood, in 1987, several studies investigated the development and the clinical manifestations of the disease in the youth, as well as the underlying biological mechanisms involved in the pathophysiology of the disorder [7]. Similarly to adults, PTSD in childhood describes the clustering of symptoms that develop after exposure to traumatic life events that involved actual death, injury or a threat to the physical integrity of the individual or others and that evoked intense fear, helplessness, or horror. Traumatic events have been recognized most often in children in cases of natural disasters, war, sexual or physical abuse and accidents [8-10]. Symptoms in school aged kids older than 6 and adolescents include: 1) Re-experience of the initial trauma via intrusive memories and/or dreams about the event, feeling as if the trauma was continuing to occur, and intense distress on exposure to cues that recall the event. In young children, repetitive play or trauma specific re-enactment may occur, in which themes or aspects of the trauma are expressed. 2) Avoidance of stimuli associated with the trauma and numbing of overall responsiveness. 3) Symptoms of excessive arousal, including insomnia, angry outbursts, hyper-vigilance, an exaggerated startle response, and difficulty concentrating [7-9]

Additionally, age-related clinical manifestations and assessment of PTSD in infants and pre-school children are frequently based on behaviors observed, since younger children are not able to describe their internal status or express responses to the environment [11-12]. Infants or toddlers may become very aggressive or distressed at exposure to cues related to the trauma. Posttraumatic play, repeatedly mimicking the trauma, is also frequent in young children. They may also exhibit nightmares and significant sleep disruption. In school age children aggression, withdrawal or avoidance and antisocial behaviour, at home and at school, may prevail. Regression to earlier developmental stages is not rare in this age group. A significant number of preschool and school age children are reported by their parents or teachers as being involved in repetitive play or drawings involving themes related to the trauma. Adolescents are often focused on the social stigma of trauma, particularly sexual trauma [9,11,12].

Similarly to many psychiatric disorders of adulthood, PTSD is often comorbid with other clinical conditions, such as major depressive disorder (MDD) and generalized anxiety disorder (GAD), as well as with a number of psychopathologic manifestations such as somatic complains, depressive symptoms or symptoms of dissociation [13]. A study in adolescents showed that depression was the most common comorbid diagnosis of PTSD, being reported in around one third of the PTSD population, while GAD was prevalent in 27%. Overlap of all three diagnoses may occur in 14% of the patients [13].

3. PTSD Pathophysiology

Neuroendocrine studies in adults with PTSD have shown altered activity of the HPA axis, as evidenced by elevated basal cerebrospinal fluid (CSF) CRH concentrations [14] and differing results from peripheral measurements [15]. In the majority of adult studies, low 24h excretion of urinary free cortisol [1,16,17], low or normal circulating cortisol levels [16,18] or even high plasma cortisol levels [19,20] are reported. The direction of HPA activity, as evidenced by peripheral cortisol measures, differs between studies and may depend on several factors such as a genetic predisposition, epigenetic changes, the age of the individual, type and chronicity of stressor, time since the trauma occurred and co-morbid disorders such as depression, other psychopathology, alcohol or other drug abuse.

In examining the locus caeruleus / norepinephrine-sympathetic nervous system, peripheral biomarkers of sympathetic nervous system activity are more consistent, showing in the majority of studies increased 24h urinary catecholamine excretion in PTSD patients compared to control subjects [21-23]. However, the hypoactive HPA with the hyperactive LC/NE-SNS is the most prevalent finding among adults with chronic PTSD. Indeed, it has been suggested that exaggerated catecholamine secretion without the concurrent regulatory effect of accompanying cortisol elevations during stress, could result in defective consolidation of traumatic memories in the brain and subsequent PTSD symptomatology [24].

Furthermore, the functioning of the immune system has been investigated in adults with PTSD and the majority of the studies have reported increased levels of circulating or CSF pro-inflammatory cytokines [25-27], such as interleukin (IL)-6, as indicators of the inflammatory response to physical and emotional stress and disruption of the HPA axis [28].

4. Studies in Maltreated Children

PTSD is the most common trauma-related disorder, occurred in 40-90% of children exposed in sexual abuse and in 11-50% among physically abused children, however, the majority of the first studies in maltreated children have not clearly examined PTSD *per se*, but children with a history of maltreatment exhibiting several or comorbid symptoms or disorders [29-32].

Childhood maltreatment, defined most frequently as physical or sexual abuse or neglect, represents the most widely studied chronic stressor in childhood. One of the first neuroendocrine studies, in 1994, showed that sexually abused girls with histories of depression and suicidal behavior exhibited significantly greater 24-hour urinary concentrations of catecholamines and their metabolites than matched controls [31]. The same group of abused girls exhibited reduced evening basal and ovine CRH-stimulated plasma ACTH concentrations compared with matched non-abused girls. Plasma cortisol responses to CRH stimulation did not differ between the two groups [32]. Another study, in 2001 [33] investigated cortisol secretion in maltreated children, in the context of a day camp research program. Significant variation in cortisol secretion patterns were found, based on the subtypes of maltreatment the children had experienced: maltreated children who had both been physically and sexually abused (as well as neglected and/or emotionally abused) exhibited substantial elevations in morning cortisol levels. In contrast to the multiple abuse group, the subgroup of only physically abused children showed a trend toward lower morning cortisol relatively to non-maltreated children, and a smaller decrease in cortisol levels from morning to afternoon. In the same groups, maltreated children with internalizing problems demonstrated higher morning, afternoon and average daily cortisol levels, while nonmaltreated boys with externalizing problems were more likely not to show the expected diurnal decrease in cortisol. This very important study, was one of the first to clarify the diversity of neuroendocrine patterns, depending on the type of maltreatment and the concurrent psychopathology in childhood.

More recent studies, have also examined neuroendocrine measures in maltreated youth, and found an asymmetry between salivary alpha amylase (sAA), an indicator of the sympathetic nervous system, and cortisol reactivity to a social stressor: The researchers concluded that the maltreatment status moderated the relation between sAA and cortisol responses, in contrast to the comparison group that showed significant links between the sAA and cortisol responses [34]. A second study in girls with previous history of child maltreatment, showed that during the reactivity phase of the Trier Social Stress Test (TSST), the maltreated group exhibited an attenuated response whereas the control group showed a typical increase and gradual flattening over time [35]. However, children in both studies were not assessed for PTSD diagnosis.

However, child maltreatment represents a chronic and complex stressor. Physical and/or sexual abuse may co-exist with other psychosocial risk factors, such as low socioeconomic status, unemployment, parental psychopathology, drug and alcohol abuse. Terr differentiated single from repeated traumas and suggested that greater dissociation was related to repeated trauma [36].

5. Studies after the Impact of an Acute Stressor

Studies examining the pathophysiology of childhood PTSD after an acute stressor are limited. In a study after the 1988 Armenian earthquake, adolescents who lived close to the epicenter and experienced a direct threat to their lives, had more severe PTSD and lower baseline mean salivary cortisol levels and greater afternoon suppression of cortisol by dexamethasone than adolescents who lived 20 miles from the epicentre [37]. Another study, in 2005, found that high initial urinary cortisol and epinephrine levels, immediately following an accident, were significantly correlated with subsequent PTSD symptoms, especially in boys [38].

Our study [39,40] examined prospectively and longitudinally sixty children and adolescents hospitalised for motor vehicle accidents (MVAs) and the development of PTSD one and six months later in relation to neuroendocrine factors within 24 hours after hospitalization, and 1 and 6 months later. Children's K-SADS [41] semi-structured interview was used for PTSD diagnosis and the PTSD-RI [42] questionnaire to assess symptoms severity. Serum cortisol and interleukin-6 and plasma catecholamine concentrations were measured, while salivary cortisol was measured 5 times a day to examine circadian variation, in all three assessments.

In the cross sectional analysis, 30% of the children were diagnosed with PTSD at month 1 and they exhibited higher plasma noradrenaline concentrations than the non-PTSD and the control groups. Salivary afternoon (6.00 pm) and evening (9.00) cortisol concentrations were also higher in the PTSD than the non-PTSD and the control groups. In month 6, children with PTSD were fewer (15%) and plasma noradrenaline was similarly higher in the PTSD than the non-PTSD and the control group. In the longitudinal analysis, we examined children exhibiting continuously the PTSD diagnosis, in both months 1 and 6. Results from the longitudinal analysis, showed that children with continuous PTSD demonstrated progressively increasing plasma noradrenaline concentrations 1 and 6 months after the accident. Indeed, within the same group, noradrenaline was significantly higher in month 6 than in month 1. In contrast, evening salivary cortisol levels were elevated at month 1, but became normal at month 6, and this difference was statistically significant. These findings revealed a natural history of the disorder after the impact of a single stressor, showing the continuous interrelations of the two axes of the stress system, as evidenced by peripheral measures of cortisol and catecholamine concentrations [39].

In the same study, we also examined biologic factors within 24hours after the accident in relation to later PTSD development. We found that initially high evening salivary cortisol and high morning serum IL-6 concentrations were predictive of PTSD diagnosis at month 6. IL6 was positively correlated with evening salivary cortisol but not with salivary cortisol levels measured at other time-points or morning plasma cortisol [40].

Another study, examined the impact of September 11, 2001 Terror attacks in New York City, in bereaved children compared with children selected from the community. In this study, bereaved children had significantly higher rates of psychiatric disorders, including PTSD. Morning (am) and 4.00 pm baseline cortisol was significantly higher for bereaved, in total, than nonbereaved children, however, bereaved children with PTSD had significantly lower 4.00 pm baseline cortisol and significantly greater 4.00 pm cortisol suppression compared with bereaved children without psychopathology [43].

6. Developmental Psychobiology

HPA Hyper- and Hypo-Activation: Elucidating Discrepancies

In accordance to adult studies, a limited number of neuroendocrine studies on pediatric PTSD have found lower or variable [33, 44] cortisol levels in children with PTSD or a history of maltreatment. Diversity in HPA peripheral measures may be due to methodological differences across studies: indeed, type and duration of the trauma, length of time since trauma, age and developmental stage of the individual, timing of the assessments, chronicity of PTSD and other variables, are factors influencing HPA activity in individuals with PTSD.

The direction of HPA axis responses under acute or chronic stress may depend on our genetic background, previous stress history and the nature of the stressor, as well as its chronicity, severity and predictability. Age and developmental stage are also determinants not only of the clinical manifestations of a syndrome of chronic stress but also of the biological patterns of disease. The extent to which the above mentioned parameters contribute to PTSD biological endophenotypes is unclear.

It is a clinical observation that prior trauma exposure is a risk factor that predisposes for later PTSD development. Similarly, previous trauma may also be the cause of altered HPA axis responses and the low cortisol levels observed in adults with chronic PTSD [45, 46]. This view hypothesizes that after an initial normal increase in cortisol levels, HPA axis functioning may be altered to a state of low basal cortisol levels, as time passes since the traumatic event. Indeed, a recent study in children [47] found that cortisol and PTSD symptoms were significantly positively correlated in children having recently experienced trauma, but negatively in children with distal trauma.

Thus, low cortisol levels found immediately or shortly after the trauma in adult studies, may reflect the impact of a previous stressor, which represents a preexisting vulnerability factor, rather than a biologic factor responsible *per se* for the subsequent development of PTSD [47,48]. Children, because of their age, are generally less likely to have experienced a great majority of prior traumas, compared to adults, and less likely to exhibit comorbid disorders or alcohol or drug use and abuse. This might be the explanation of high cortisol levels reported more frequently in pediatric than in adult studies.

Furthermore, the study of our group following children after motor vehicle accidents, clearly demonstrated that evening salivary cortisol is a better peripheral biomarker in indicating HPA activity and in predicting later PTSD development than cortisol values in other time-points. These initially elevated evening cortisol levels shortly after the trauma suggest a CRH hypersecretion-driven flattening of the circadian rhythm after the accident that might better disclose increased activity of the HPA axis related to later PTSD development, compared to morning values. Finally, the same data showed that serum morning IL-6 after the trauma was also predictive of PTSD development 6 months later and significantly inter-related only with evening cortisol [39, 40].

The Key Role of Norepinephrine in PTSD Development and Maintenance

The most consistent finding in PTSD neuroendocrine studies is increased noradrenergic activity, both centrally [49] and peripherally [21-23] in adults, while children with PTSD and post-traumatic symptomatology have been reported to also have increased peripheral sympathetic nervous system activity [31, 34, 38, 39]. In the clinical practice, increased blood pressure, heart rate, skin conductance and an exaggerated startle response have been reported in adults suffering from PTSD, reflecting increased noradrenergic activity [50]. The majority of clinical studies on adult and pediatric PTSD shows elevated urine or plasma norepinephrine concentrations [21-23, 31, 38, 39].

Indeed, norepinephrine plays major physiologic roles in arousal, attention, and memory and dysregulation of the SNS neurons may be involved in the hyperarousal and re-experience symptoms of PTSD [50, 51]. Stressful stimuli may enhance memory encoding in PTSD individuals, causing defective consolidation for the stressful event, leading to the development of intrusive thoughts, images, flashbacks, and repetitive nightmares [50, 51]. Our longitudinal study in children after motor vehicle accidents, being in accordance with the current literature, demonstrated significantly elevated plasma norepinephrine concentrations in children with PTSD, at both months 1 and 6 after the accident compared to the non-PTSD and control groups [39]. Furthermore, we demonstrated that plasma norapinephrine concentrations increased progressively over time in PTSD children, while these differences within group were statistically significant. Interestingly, the same pattern of progressive norepinephrine elevation, but in a lower setting, was similar in non-PTSD individuals, who were characterized as those experiencing an MVA but not fulfilling the full PTSD diagnostic criteria, and could be also characterized as partial PTSD individuals [39].

Moreover, examining the PTSD group serially, we found that higher plasma norepinephrine concentrations at month 6 than month 1 was not related PTSD severity, as measured by the PTSD-RI Questionnaire, since the PTSD diagnosis was less severe in six months than in month 1 within the same group.

Interactions between Cortisol and Norepinephrine in Pediatric PTSD Maintenance

Though adult and pediatric studies have convincingly reported a dysregulation of the HPA and the SNS axes, the longitudinal interaction between these two axes have not been reported. In our study in children after Motor Vehicle Accidents, as cortisol concentrations and circadian rhythm normalized in the PTSD subjects at month 6, norepinephrine elevations became greater. We hypothesized that this could be the result of lifting a cortisol-mediated noradrenergic system restraint [39]. Clinical symptoms observed in PTSD patients may thus also be attributed to cortisol dysregulation that fails to shut down the catecholaminergic response in limbic structures leading to PTSD development and maintenance through time.

The Role of IL-6

IL-6 is an inflammatory cytokine with a variety of immune, endocrine and metabolic actions. Physical and/or psychologic stress is associated with immune activation and the release of IL-6 [28, 54] and this effect can be partly explained as resulting from SNS activation, as catecholamines stimulate IL-6 secretion by activating β-adrenergic receptors [28, 52, 53]. Several studies in adults with PTSD have investigated the role of this proinflammatory cytokine in the pathophysiology of the disorder. The majority of those studies have demonstrated increased levels of circulating inflammatory cytokines [26, 27], including IL-6, as well as increased levels of CSF IL-6 [25]. Evidence in children [40] after Motor Vehicle accidents, revealed that morning serum IL-6 was involved in the initial biological alterations in the aftermath of the trauma, and predictive of PTSD development 6 months later. In the same study, IL-6 was positively inter-related with evening salivary cortisol which was similarly predictive of later PTSD. Interestingly, morning serum IL-6 did not correlate with morning serum cortisol, neither with morning, noon or afternoon salivary cortisol concentrations, revealing an interrelation only with evening salivary cortisol that seemed to be a better marker of HPA dysregulation. However, circulating IL-6 concentrations were normal and did not differ between the PTSD and the comparison groups in months 1 and 6.

Another interesting observation, related to IL-6 and pediatric PTSD after motor vehicle accidents, was that this cytokine was not related to injury severity. Furthermore, in investigating the impact of the physical stress in this population, we found that IL-6 was almost equally elevated in the physically stressed and the emotionally stressed groups [54]. Indeed, several studies have shown that perceived fear and subjective reports of injury severity better predict PTSD development than objective severity scores [55, 56].

7. Familial Biological Factors Responsible for PTSD Development

Lifetime prevalence of exposure to traumatic events is between 40% and 90% in the general population, however, the overall lifetime prevalence of PTSD is estimated at 7-12% [59, 60]. The complex nature of PTSD, for which exposure to an external traumatic event is a necessary condition, makes it difficult to identify the extent to which each risk factor contributes to the disorder.

The Genetic Component of PTSD Vulnerability

Evidence for a genetic component for PTSD vulnerability is based on transgenerational research and epidemiological studies on twins. However, this genetic vulnerability may develop in interaction with environmental factors. Candidate genes associated with basic biochemical, endocrine, neuroanatomical, cognitive and psychiatric characteristics of the disorder, include the serotonin transporter (5-HTT), the dopamine transporter (DAT), the glucocorticoid receptor (GR), the GABA (A) receptor, the apolipoprotein E (APOE), the brain-derived neurotrophic factor (BDNF) and the neuropeptide Y (NPY) [59]. Recently, the cannabinoid receptor gene (CNR1) was also associated with Attention Deficit Hyperactivity Disorder (ADHD) and PTSD [60]. The results indicate a role for the serotonin transporter gene in the degree of response to traumatic life events, while regarding the dopamine system results may be influenced by comorbid alcohol abuse [61, 62]. Also, recently, the GR was found to be epigenetically affected through promoter methylation in victims of childhood abuse resulting in a dysfunctional HPA axis regulation in later life [63].

Transgenerational Effects of Trauma, Parental PTSD and Fetal Programming of the HPA Axis

Parental PTSD is a well recognized risk factor for child PTSD. Although this relation may underlie a common genetic background, predisposing both parent and child in PTSD development, there in evidence that other biologic parameters contribute as well. Adult offspring of Holocaust survivors with PTSD presented with a higher prevalence of PTSD, without trauma exposure, and with low cortisol levels compared to control individuals [64]. Recent evidence also suggests that stress in utero, through fetal programming of the HPA axis, may also contribute to PTSD in the offspring. Indeed, in mothers exposed to the 11th September attacks during pregnancy, maternal PTSD was related to awakening and bedtime salivary cortisol levels in their babies and lower cortisol values were noted in infants of PTSD mothers compared to the exposed but non-PTSD affected ones [65]. These data suggest that epigenetic mechanisms are also involved in biological risk factors related to PTSD development.

Evidence also supports that stress during pregnancy, through elevations in glucocorticoid levels, influences fetal brain development and program the HPA axis [66]. Reduced activity of placental 11 beta-hydroxycorticosteroid dehydrogonase type 2, the enzyme that catalyses the conversion of maternal cortisol to cortisone, result in an increased exposure of the fetus to glucocorticoids and may contribute to HPA programming [66].

8. Conclusion

In conclusion, depending on the time point of the assessment after the traumatic event, PTSD in children and adolescents is associated with elevated norepinephrine and variable cortisol levels. High cortisol levels, reported in the majority of pediatric studies, and, in contrast to most of adult studies (Figure 1) [67], may represent an initial alteration of the HPA axis in response to the traumatic experience. Data from our longitudinal study on pediatric PTSD development and maintenance, show that the peripheral measures of the two main axes of the stress system diverge, with norepinephrine increasing further and initially elevated cortisol normalizing (Figure 2) [48]. These data reveal a biological natural history of pediatric PTSD, with an underlying interaction of changes of the SNS and the HPA axis, characterizing those that develop the disorder, and may lead to the understanding of some of the discrepancies in the biological background between children and adults with chronic PTSD.

Furthermore, the genetic background, previous stress history, epigenetic programming, and other variables, may also determine the biologic vulnerability to PTSD and the initial hormonal responses to the traumatic experience. Longitudinal biologic studies following children with PTSD in adulthood would be of great importance for understanding and treating PTSD.

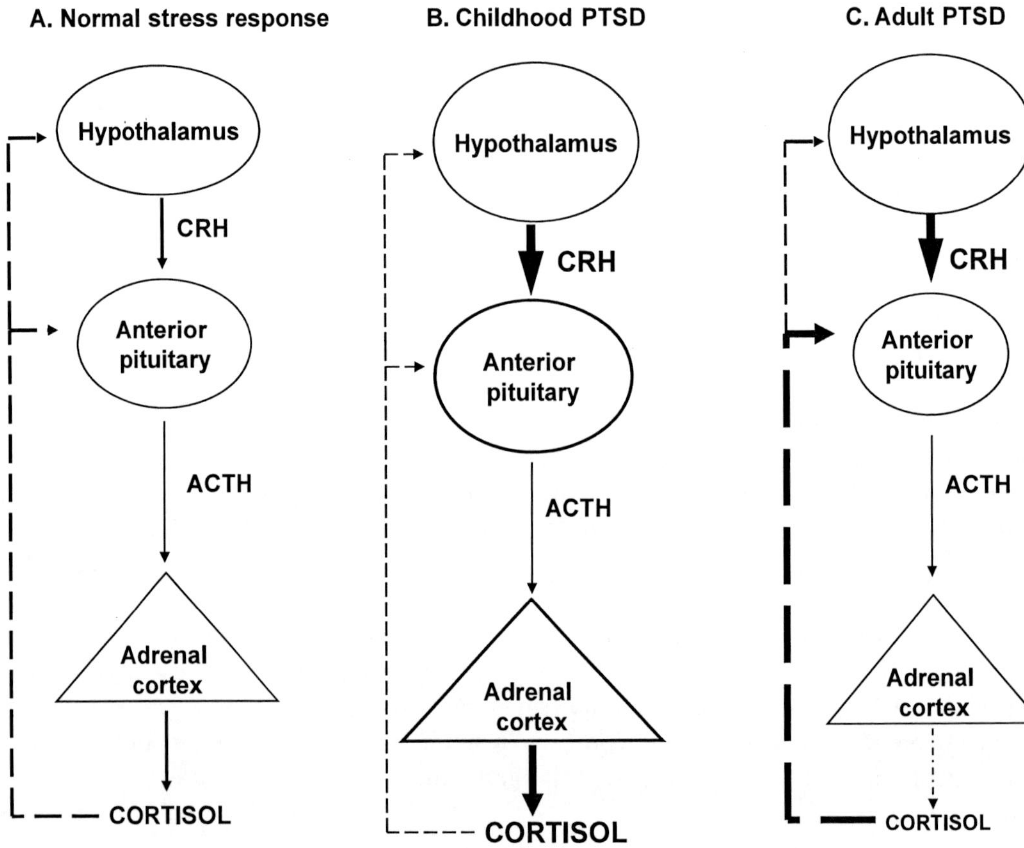

Figure 1. Normal stress response, pediatric patients with PTSD and adult patients with PTSD. In each figure the thickness of the arrows represents the magnitude of the response.
A. Normally, acute stress reaction includes brief elevation of CRH and cortisol.
B. In children with PTSD, elevated CRH and cortisol are reported. Cortisol inhibits the release of CRH from the hypothalamus and ACTH from the pituitary.
C. In adults with PTSD, the majority of studies reports high CRH and low cortisol in the periphery. An enhanced negative inhibition of the hypothalamic-pituitary-adrenal axis is suggested in adults with PTSD.
Reference for Figure 1: 67 in list of references, with permission of the Publisher.

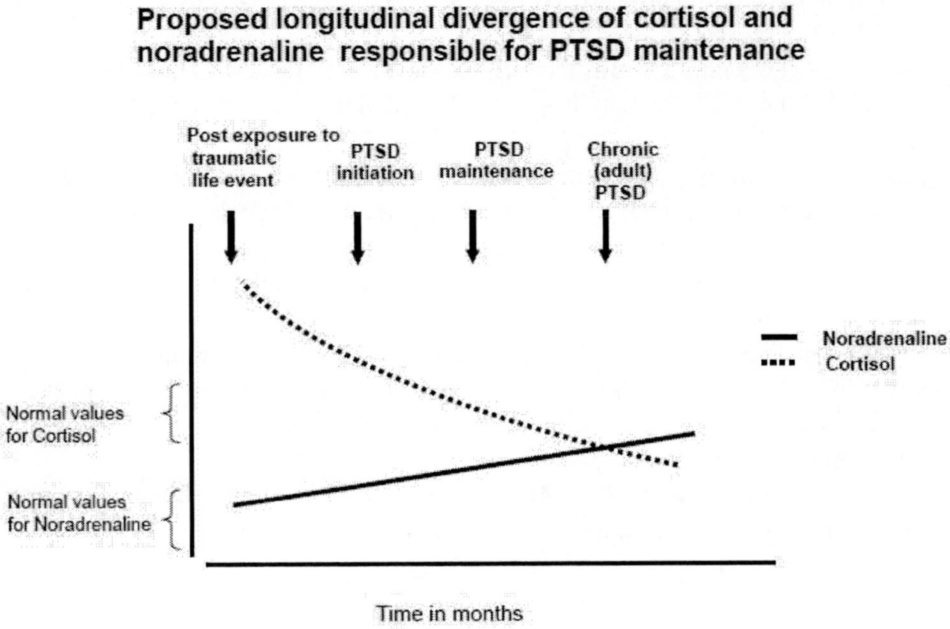

Figure 2. A simplified figure showing the longitudinal neuroendocrine changes in a proportion of pediatric PTSD leading to adult PTSD. Only the peripheral measures of the two main axes (HPA and SNS) are mentioned here: Early extreme stress, in the absence of previous trauma in the majority of typical children, results in immediate elevation of cortisol in the periphery. PTSD by definition begins at least one month later, and cortisol levels are found still high or normal while noradrenaline is gradually elevated. Progressive divergence of cortisol and noradrenaline concentrations over time may be the underlying pathophysiologic mechanism leading to PTSD maintenance. Low cortisol and high nodadrenaline is a common finding in chronic adult PTSD and may represent the natural history of the disorder as well as a biologic risk factor for further PTSD vulnerability.
Reference for Figure 2: 48 in list of references, with permission of the Publisher.

References

[1] Yehuda R, Giller EL, Southwick SM, Lowy MT, Mason JW. Hypothalamic-pituitary-adrenal dysfunction in posttraumatic stress disorder. *Biol Psychiatry* 1991; 30(10):1031-1048.

[2] Chrousos GP, Gold PW. The Concepts of Stress and Stress System Disorders. *JAMA* 1992; 267:1244-1252.

[3] Chrousos GP. Stress and disorders of the stress system. *Nat Rev Endocrinol.* 2009 Jul;5(7):374-81.

[4] Charmandari E, Kino T, Souvatzoglou E, Chrousos GP. Pediatric stress: hormonal mediators and human development. *Horm Res.* 2003;59(4):161-79.

[5] Teicher MH, Andersen SL, Polcari A, Anderson CM, Navalta CP. Developmental neurobiology of childhood stress and trauma. *Psychiatr Clin North Am* 2002;25(2):397-426.

[6] De Bellis MD, Baum AS, Birmaher B, Keshavan MS, Eccard CH, Boring AM, Jenkins FJ, Ryan ND. A.E. Bennett Research Award. Developmental traumatology. Part I: Biological stress systems. *Biol Psychiatry* 1999; 45(10):1259-1270.

[7] *American Psychiatric Association 1994 Diagnostic and statistical manual of mental disorders-DSM-IV,* 4TH ed. Washington DC: American Psychiatric Press.

[8] Yule W. Post-traumatic stress disorder. *Arch Dis Child* 1999; 80(2):107-109.

[9] Yule W. Posttraumatic stress disorder in the general population and in children. *J Clin Psychiatry* 2001; 62 Suppl 17:23-28.

[10] Pervanidou P, Kolaitis G, Chrousos GP. Bio-behavioral consequences of traumatic stress in childhood and adolescence: The effects of war on children' s mental health, growth and development. In: Sher L, Vilens A (editors). *War and Suicide.* Hauppauge, New York: Nova Science Publishers, in press.

[11] Scheeringa MS, Zeanah CH, Myers L, Putnam FW. New findings on alternative criteria for PTSD in preschool children. *J Am Acad Child Adolesc Psychiatry.* 2003 May;42(5):561-70.

[12] Scheeringa MS. Developmental considerations for diagnosing PTSD and acute stress disorder in preschool and school-age children. *Am J Psychiatry.* 2008 Oct;165(10):1237-9.

[13] Kar N, Bastia BK. Post-traumatic stress disorder, depression and generalised anxiety disorder in adolescents after a natural disaster: a study of comorbidity. *Clin Pract Epidemol Ment Health* 2006; 2:17.

[14] Baker DG, West SA, Nicholson WE, Ekhator NN, Kasckow JW, Hill KK, Bruce AB, Orth DN, Geracioti TD Jr. Serial CSF corticotropin-releasing hormone levels and adrenocortical activity in combat veterans with posttraumatic stress disorder. *Am J Psychiatry* 1999; 156(4):585-8. Erratum in: *Am J Psychiatry* 1999; 156(6):986.

[15] Rasmusson AM, Vythilingam M, Morgan CA 3rd. The neuroendocrinology of posttraumatic stress disorder: new directions. *CNS Spectr.* 2003; 8(9):651-656, 665-667.

[16] Yehuda R Biology of posttraumatic stress disorder. *J Clin Psychiatry* 2001; 62 (17):41-46.

[17] Yehuda R, Kahana B, Binder-Brynes K, Southwick SM, Mason JW, Giller EL. Low urinary cortisol excretion in Holocaust survivors with posttraumatic stress disorder. *Am. Journal of Psychiatry* 1995; 152:982-986.

[18] Boscarino JA. Posttraumatic stress disorder, exposure to combat, and lower plasma cortisol among Vietnam veterans-findings and clinical implications. *J Consult Clin Psychol* 1996; 64:191-201.

[19] Lemieux AM and Coe CL. Abuse related posttraumatic stress disorder: evidence for chronic neuroendocrine activation in women. *Psychosom. Med* 1995; 57: 105-115.

[20] Maes M, Lin A, Bonaccorso S, van Hunsel F, Van Gastel A, Delmeire L, Biondi M, Bosmans E, Kenis G, Scharpe S. Increased 24-hour urinary cortisol excretion in patients with post-traumatic stress disorder and patients with major depression, but not in patients with fibromyalgia. *Acta Psychiatr Scand* 1998; 98(4): 328-335.

[21] Yehuda R, Southwick S, Giller EL, Ma X, Mason JW. Urinary catecholamine excretion and severity of PTSD symptoms in Vietnam combat veterans. *J Nerv Ment Dis* 1992; 180(5):321-325.

[22] Southwick SM, Paige S, Morgan CA 3rd, Bremner JD, Krystal JH, Charney DS. Neurotransmitter alterations in PTSD: catecholamines and serotonin. *Semin Clin Neuropsychiatry* 1999; 4(4):242-248.

[23] Pitman R and Orr S Twenty –four urinary cortisol and catecholamine excretion in combat–related posttraumatic stress disorder. *Biol Psychiatry* 1990; 27:245-247.

[24] Pitman RK, Orr SP, Shalev AY. Once bitten, twice shy: beyond the conditioning model of PTSD. *Biol Psychiatry* 33: 145-146.

[25] Baker DG, Ekhator NN, Kasckow JW, Hill KK, Zoumakis E, Dashevsky BA, Chrousos GP, Geracioti TD Jr. Plasma and cerebrospinal fluid interleukin-6 concentrations in posttraumatic stress disorder. *Neuroimmunomodulation* 2001; 9(4):209-217.

[26] Rohleder N, Joksimovic L, Wolf JM, Kirschbaum C Hypocortisolism and Increased glucocorticoid sensitivity of Proinflammatory Cytokine Production in Bosnian War Refugees with Posttraumatic Stress Disorder. *Biological Psychiatry* 2004; 55:745-751.

[27] Maes M, Lin AH, Delmeire L, Van Gastel A, Kenis G, De Jongh R, Bosmans E. 1999 Elevated serum interleukin-6 (IL-6) and IL-6 receptor concentrations in posttraumatic stress disorder following accidental man-made traumatic events. *Biological Psychiatry* 1999; 45:833-839.

[28] Chrousos GP. The hypothalamic-pituitary-adrenal axis and immune-mediated inflammation. *N Engl J Med* 1995; 332(20): 1351-1362.

[29] Children of Trauma: Stressful Life Events and their Effects on Children and Adolescents / edited by Thomas Miller, 1998, International Universities Press, Inc

[30] Kiser LJ, Heston J, Millsap PA, Pruitt DB. Physical and sexual abuse in childhood: relationship with post-traumatic stress *disorder. J Am Acad Child Adolesc Psychiatry. 1991 Sep;30(5):776-8375*

[31] De Bellis MD, Lefter L, Trickett PK, Putnam FW Jr. Urinary catecholamine excretion in sexually abused girls. *J Am Acad Child Adolesc Psychiatry. 1994, Mar-Apr;33(3):320-7.*

[32] De Bellis MD, Chrousos GP, Dorn LD, Burke L, Helmers K, Kling MA, Trickett PK, Putnam FW. Hypothalamic-pituitary-adrenal axis dysregulation in sexually abused girls. *JCEM* 1994; 78:249-255.

[33] Ciccheti D, Rogosch FA. Diverse patterns of neuroendocrine activity in maltreated children. *Dev Psychopathol* 2001;13(3): 677-693.

[34] Gordis EB, Granger DA, Susman EJ, Trickett PK. Salivary alpha amylase-cortisol asymmetry in maltreated youth. *Horm Behav*. 2008 Jan;53(1):96-103.

[35] MacMillan HL, Georgiades K, Duku EK, Shea A, Steiner M, Niec A, Tanaka M, Gensey S, Spree S, Vella E, Walsh CA, De Bellis MD, Van der Meulen J, Boyle MH, Schmidt LA. Cortisol response to stress in female youths exposed to childhood maltreatment: results of the youth mood project. *Biol Psychiatry*. 2009 Jul 1;66(1):62-8. Epub 2009 Feb 12.

[36] Terr LC: Childhood Traumas-an outline and overview. *American Journal of Psychiatry,* 1991; 27:96-104.

[37] Goenjian AK, Yehuda R, Pynoos RS, Steinberg AM, Tashjian M, Yang RK, Najarian LM, Fairbanks LA. Basal cortisol, dexamethasone suppression of cortisol, and MHPG in adolescents after the 1988 earthquake in Armenia. *Am J Psychiatry* 1996 153(7):929-934.

[38] Delahanty DL, Nugent NR, Christopher NC, Walsh M. Initial urinary epinephrine and cortisol levels predict acute PTSD symptoms in child trauma victims. Psychoneuroendocrinology. 2005 Feb; 30(2):121-8.

[39] Pervanidou P, Kolaitis G, Charitaki S, Lazaropoulou Ch, Hindmarsh P, Bakoula Ch, Papassotiriou I, Tsiantis J, Ghrousos GP. The natural history of neuroendocrine changes in pediatric posttraumatic stress disorder after motor vehicle accidents: progressive divergence of noradrenaline and cortisol concentrations over time. *Biol Psychiatry* 2007 15;62(10):1095-1102.

[40] Pervanidou P, Kolaitis G, Charitaki S, Margeli A, Lazaropoulou C, Papassotiriou I, Ferentinos S, Bakoula C, Tsiantis J, Chrousos GP. Elevated morning serum IL-6 or evening salivary cortisol predict PTSD in children and adolescents 6 months after motor vehicle accidents. *Psychoneuroendocrinology 2007 32(8-10):991-992.*

[41] Kaufman J, Birmaher B, Brent D, Rao U, Flynn C, Moreci P, Williamson D, Ryan N. Schedule for Affective Disorders and Schizophrenia for School-Age Children-Present and Lifetime Version (K-SADS-PL): initial reliability and validity data. *J Am Acad Child Adolesc Psychiatry* 1997; 36(7):980-988.

[42] Rodriguez N, Steinberg A, Pynoos RS. *UCLA Post Traumatic Stress Disorder Reaction Index for DSM-IV, Child, Adolescent, and Parent Versions;* 1998 (available from the authors).

[43] Pfeffer CR, Altemus M, Heo M, Jiang H. Salivary cortisol and psychopathology in children bereaved by the september 11, 2001 terror attacks. *Biol Psychiatry*. 2007 Apr 15;61(8):957-65.

[44] De Bellis MD, Thomas LA. Biological findings of posttraumatic stress disorder and child maltreatment. *Curr Psychiatry Rep* 2003 5(2):108-117.

[45] Yehuda R, McFarlane AC and Shalev AY. Predicting the development of post-traumatic stress disorder from the acute response to a traumatic event. *Biol. Psychiatry* 1998; 44: 1305–1313.

[46] Resnick HS, Yehuda R, Pitman RK, Foy DW. Effect of previous trauma on acute plasma cortisol level following rape. *Am J Psychiatry* 1995; 152:1675-1677.

[47] Weems CF, Carrión VG. Brief report: diurnal salivary cortisol in youth--clarifying the nature of posttraumatic stress dysregulation. *J Pediatr Psychol.* 2009 May;34(4):389-95.

[48] Pervanidou P. Biology of Posttraumatic Stress Disorder in childhood and adolescence *J. Neuroendocrinol.* 2008 May;20(5):632-638.

[49] Geracioti TD Jr, Baker DG, Ekhator NN, West SA, Hill KK, Bruce AB, Schmidt D, Rounds-Kugler B, Yehuda R, Keck PE Jr, Kasckow JW. CSF norepinephrine concentrations in posttraumatic stress disorder. *Am J Psychiatry* 2001; 158(8):1227-1230.

[50] O Donnell TO, Hegadoren KM, Coupland NC. Noradrenergic mechanisms in the pathophysiology of post-traumatic stress disorder. *Neuropsychobiology* 2004; 50:273-283.

[51] Southwick SM, Bremner JD, Rasmusson A, Morgan CA 3rd, Arnsten A, Charney DS. Role of norepinephrine in the pathophysiology and treatment of posttraumatic stress disorder. *Biol Psychiatry* 1999; 46(9): 1192-1204.

[52] Papanicolaou DA, Wilder R, Manolagas S, Chrousos GP. The pathophysiologic Roles of Interleukin-6 in Human Disease. *Annals of Internal Medicine* 1998; 128(2): 127-137.

[53] Elenkov IJ, Iezzoni DG, Daly A., Harris, AG, Chrousos GP. Cytokine dysregulation, inflammation and well-being. *Neuroimmunomodulation* 2005;12(5): 255-269.

[54] Pervanidou P, Margeli A, Lazaropoulou Ch, Papassotiriou I, Chrousos GP. The immediate and long-term impact of physical and/or emotional stress from motor vehicle accidents on circulating stress hormones and adipo-cytokines in children and adolescents. *Stress.* 2008;11(6):438-47.

[55] Blanchard EB, Hickling EJ, Mitnick N, Taylor AE, Loos, WR, Buckley TC. The impact of severity of physical injury and perception of life threat in the development of posttraumatic stress disorder in motor vehicle accident victims. *Behav Res Ther* 1995; 33(5): 529-534.

[56] Holbrook TL, Hoyt DB, Stein MB Sieber WJ. Perceived threat to life predicts posttraumatic stress disorder after major trauma: Risk factors and functional outcome. *J. Trauma* 2001; 51:287–293.

[57] Kessler RC, Sonnega A, Bromet E, Hughes M, Nelson CB: Posttraumatic stress disorder in the National Comorbidity Survey. *Arch Gen Psychiatry* 1995; 52(12):1048-1060.

[58] Breslau N: The epidemiology of posttraumatic stress disorder: what is the extent of the problem? *J Clin Psychiatry.* 2001;62 Suppl 17:16-22.

[59] Broekman FFP. Olff M, Boer F: The genetic background of PTSD. *Neuroscience and Behavioral Reviews.* 2007.31:348-362.

[60] Lu AT, Ogdie MN, Järvelin MR, Moilanen IK, Loo SK, McCracken JT, McGough JJ, Yang MH, Peltonen L, Nelson SF, Cantor RM, Smalley SL. Association of the cannabinoid receptor gene (CNR1) with ADHD and post-traumatic stress disorder. *Am J Med Genet B Neuropsychiatr Genet.* 2008 Dec 5;147B(8):1488-94.

[61] Lee HJ, Lee MS, Kang RH, Kim H, Kim SD, Kee BS, Kim YH, Kim YK, Kim JB, Yoon JS, Lee C, Jung HY, Chee IS, Paik ICH. Influence of the serotonin transporter promoter gene polymorphism on susceptibility to posttraumatic stress disorder. *Depression and Anxiety* 2005;21:135–139.

[62] Young RM, Lawford BR, Noble EP, Kann B , Wilkie A, Ritchie T, Arnold L, Shadforth S. Harmful drinking in military veterans with post-traumatic stress disorder: association with the D2 dopamine receptor A1 allele. *Alcohol and Alcoholism* 2002; 37: 451–456.

[63] McGowan PO, Sasaki A, D'Alessio AC, Dymov S, Labonté B, Szyf M, Turecki G, Meaney MJ. Epigenetic regulation of the glucocorticoid receptor in human brain associates with childhood abuse. *Nat Neurosci.* 2009 Mar;12(3):342-8.

[64] Yehuda R, Halligan SL, Bierer LM. Cortisol levels in adult offspring of Holocaust survivors: relation to PTSD symptom severity in the parent and child. *Psychoneuroendocrinology.* 2002; 27(1-2): 171-180.

[65] Yehuda R, Engel SM, Brand SR, Seckl J, Marcus SM, Berkowitz GS. Transgenerational effects of posttraumatic stress disorder in babies of mothers exposed to the World Trade Center attacks during pregnancy. *JCEM* 2005; 90(7): 4115-4118.

[66] Seckl JR Prenatal glucocorticoids and long-term programming. *Eur J Endocrinol* 2004 151(Suppl 3):U49–U62.

[67] Pervanidou P, Chrousos GP. Post-traumatic Stress Disorder in children and adolescents: from Sigmund Freud's "trauma" to psychopathology and the (Dys)metabolic syndrome. *Horm Metab Res.* 2007 Jun;39(6):413-9. © Georg Thieme Verlag Stuttgart, New York.

In: Neurobiology of Post-Traumatic Stress Disorder
Editors: L. Sher and A. Vilens, pp. 153-173
ISBN: 978-1-61470-593-2

Chapter XI

EEG Asymmetries in Survivors of Severe Motor Vehicle Accidents: Association with Posttraumatic Stress Disorder and its Treatment as well as Posttraumatic Growth

Sirko Rabe[1] and Anke Karl[2]
[1]Saxonian Hospital Großschweidnitz, Großschweidnitz, ,Germany
[2]School of Psychology, University of Exeter, Exeter, U.K.

Abstract

Posttraumatic stress disorder (PTSD) has been related to a number of functional and structural brain alterations. This chapter aims to review the current understanding of the role of functional brain asymmetries for negative (PTSD) and positive posttraumatic outcomes (posttraumatic growth) and for the successful treatment of PTSD. Theoretically, this chapter is based on theories of brain asymmetry and emotion which propose that asymmetries of brain activation are related to certain features of human emotion (e.g. valence, approach or withdrawal tendencies, arousal). These theories are supported by asymmetries found in the alpha frequency of the electroencephalogram (EEG). Recent findings of EEG alpha asymmetry in patients with PTSD are reviewed and implications for its theoretical understanding, treatment and further research are discussed.

Introduction

Theories of brain asymmetry and emotion [1-3] propose that asymmetries of brain activation are related to certain features of human emotion (e.g. valence, approach or withdrawal tendencies, arousal). Whereas an enormous increase in the understanding of structural and functional abnormalities in Posttraumatic Stress Disorder (PTSD) has been achieved in the last decades due to neuroimaging research (see [4-7] for review), there are still numerous unanswered questions. Especially, there is only little research explicitly examining activation asymmetries in PTSD. Furthermore, research is sparse investigating alterations of brain function which are associated with successful psychological treatment of PTSD [8]. In recent years, observations of positive changes following trauma called Posttraumatic Growth (PTG), have stimulated research of associated psychological processes and factors. However, there is a lack of research on associated biological correlates of PTG. The following chapter will give a short overview about theories and research of EEG asymmetries. Then, results of research relating EEG asymmetries and PTSD; treatment of PTSD and PTG are presented with a more detailed description of three studies recently conducted by our research group. Furthermore, some generally important issues in this area of research will be considered such as (a) the relation of the results of these studies to the state-trait debate of EEG asymmetry; (b) the association of the results to psychological theories of PTSD and PTG; and (c) the use of brain research for psychotherapy.

Brain Asymmetry and Emotion

A substantial body of research suggests that the left and right cerebral hemispheres are differentially involved in emotion regulation and processing [9]. Theories of brain asymmetries and emotion propose two fundamental systems that underlie approach and withdrawal–related emotion and motivation, or positive and negative affect [1, 10-12]. Davidson [1, 13, 14] proposed a model in which structures within the left prefrontal cortex are part of a circuit that processes emotions associated with approach behaviors and emotions. Conversely, the right anterior hemisphere is a major component within a brain circuit that is specialized for withdrawal behavior and activation of this system is associated with the experience of negative emotions. Evidence for the model comes from research investigating EEG alpha asymmetry in relation to emotional/motivational traits, emotional responding, and psychopathology [10]. In these studies, EEG activation of left and right cortical regions is compared. EEG alpha activity is an indicator of cortical activation. An inverse relationship has been shown between brain activation as measured using PET and EEG alpha [15, for critical review see also 16, 17]. In contrast to this EEG asymmetry research, human neuroimaging literature mostly did not perform systematic statistical comparisons to ascertain true asymmetric effects. This requires a test of the interaction between Condition and Hemisphere (see [18] for an extensive discussion).

Reduced left anterior activation has been shown for depression [19]. For anxiety, Wiedemann and associates [20] found relative right frontal activation during rest and confrontation with anxiety-related pictures in panic patients. Comparable findings of relative right-sided activation in anterior-temporal, lateral-frontal, and parietal regions were reported by Davidson and associates [21] in a sample of patients with social phobia during the anticipation of a public speech. Studies of resting EEG asymmetries in anxiety disorders have produced conflicting results (see [22] for review). A number of studies investigating brain asymmetries in relation to psychopathology showed results inconsistent with Davidson's model [23, 24]. Heller and colleagues [25] proposed an alternative model in which anterior cortical regions are involved in the modulation of valence whereas the right parieto-temporal region is involved in the modulation of emotional arousal. In a refinement of their theory, Heller and associates [26] proposed a distinction between subtypes of anxiety related to different patterns of brain activation. Anxious arousal should be associated with greater right parieto-temporal activation, while anxious apprehension (e.g. worry and rumination) should be associated with greater left anterior activity.

Brain Asymmetry and PTSD

There is some evidence for hemispheric differences in functional imaging studies with a favour for enhanced right hemisphere activation of several brain regions during processing of negative stimuli in PTSD subjects [27-29]. However, there is only one neuroimaging study in PTSD which directly investigated hemispheric asymmetry and which found support for the assumption of enhanced right hemisphere involvement in PTSD [30]. Further support of enhanced right hemisphere involvement in PTSD comes from a study investigating regional lesions in injury survivors [31]. Subjects with PTSD re-experiencing symptoms had fewer lesions in limbic brain structures on the right hemisphere. A study using functional connectivity analysis of brain activity (fMRI) during recall of traumatic memories found greater correlation of brain activity within right hemisphere regions in PTSD patients as compared to traumatized healthy controls who showed an opposite pattern of greater left hemisphere interregional correlations [32]. These differences in brain connectivity have been interpreted as evidence for the nonverbal nature of traumatic memory recall in PTSD subjects, compared to a more verbal pattern of traumatic memory recall in comparison subjects. There are only few studies investigating electroencephalographic (EEG) asymmetries in PTSD. One study reported increased right-sided activation in response to trauma-related olfactory stimuli in a small sample ($n = 5$) of Vietnam veterans with PTSD [33]. Metzger et al. [34] reported that PTSD arousal symptoms were associated with increased relative right parietal baseline activity in a sample of female Vietnam War nurse veterans. However, recently Shankman and colleagues [35] did not find any differences in brain asymmetry during rest in a sample of PTSD subjects with mixed trauma as compared to a control group. In a recent study of our group we examined EEG alpha asymmetry among MVA survivors with PTSD, with subsyndromal PTSD, and without PTSD as well as non-exposed healthy controls during resting baseline and in response to neutral, positive, negative, and trauma-related pictures [36]. Based on research highlighting the role of right anterior regions in withdrawal-related emotions [1] and right posterior regions in anxious arousal [3], we expected that patients with PTSD and subsyndromal PTSD would exhibit increased activation (i.e., EEG alpha reduction) of right hemisphere anterior and posterior regions during exposure to a trauma-related picture. On the basis of previous findings highlighting the role of resting brain activity as a trait marker for psychopathology (for review see [10]), we assumed that PTSD and subsyndromal PTSD patients would show increased relative right anterior and posterior baseline activity. In accordance with our first hypothesis, both PTSD and subsyndromal PTSD patients displayed a pattern of increased right-sided anterior activation during the trauma-related condition (see Figure 1). This activation pattern was similar to that previously reported in patients with PTSD [33] and other anxiety disorders [20, 21] during symptom provocation. Our finding of increased relative right posterior activation in PTSD and subsyndromal PTSD was consistent with recent studies reporting right posterior activation during anxiety-provoking situations in anxious participants [21, 26]. Interestingly, posterior asymmetry in non-traumatized healthy controls varied with gender, with female participants showing a pattern of higher right posterior activation. This finding underscores the importance of gender in EEG asymmetry research [37, 38]. Contrary to our second hypothesis, we observed no group differences in EEG alpha activity during resting baseline which was assumed because of recent findings in a PTSD sample [34] and other anxiety populations [20]. However, this finding is in accordance with a recent study that also failed to show abnormalities in baseline asymmetry in PTSD subjects [35]. In summary, our study showed a context-dependent (trauma-specific) alteration of hemispheric activation in symptomatic MVA-survivors (patients with PTSD and subsyndromal PTSD) which was associated with anxious arousal and symptoms of PTSD. The advantage of this study was the parallel assessment of trait (resting baseline) and state (emotion induction) measures of EEG asymmetry in a relatively large sample of treatment-seeking MVA survivors. Furthermore, it was also the first study assessing brain asymmetry in PTSD which used two control groups (healthy controls with and without previous exposure to MVAs).

Changes in Brain Asymmetry after Cognitive Behavioral Therapy for PTSD

There is some preliminary evidence of alterations of brain activation and structure due to treatment (pharmacological or psychological) of PTSD. Interestingly, pharmacological studies demonstrated an increase in hippocampal volume [39, 40], indicating plastic changes (e.g. neurogenesis, altered efficacy of synaptic connections) in brain networks associated with PTSD. However, psychological treatment did not produce hippocampal volume changes in PTSD subjects [41]. Functional neuroimaging studies showed treatment-related functional changes in diverse subcortical and cortical brain regions due to psychotherapy in PTSD [42-47]. However, these results should be cautiously interpreted since the majority of studies were non-experimental pretest/posttest comparisons without control groups or have other methodological limitations (e.g. patients receiving concomitant pharmacotherapy).

In a recent study [48], we examined whether increased right-sided hemispheric activation during exposure to a trauma-related accident picture in patients with full and subsyndromal PTSD would change as a consequence of cognitive behaviour therapy (CBT). Therefore, we measured EEG activity before and after CBT in comparison to an assessment-only Wait-list condition as part of a randomized controlled treatment trial for MVA-related PTSD [49]. As predicted, there was a change in relative right anterior activation in the CBT group but not the Wait-list group (see Figure 2). Both groups showed a reduction in relative right posterior activity. In addition, reduction of PTSD symptoms was correlated with a decrease in right anterior activation during trauma-related stimulus presentation. The pattern of relative reduced right hemisphere activation in the CBT group was similar to that observed in MVA victims without PTSD in study our previous study [36].

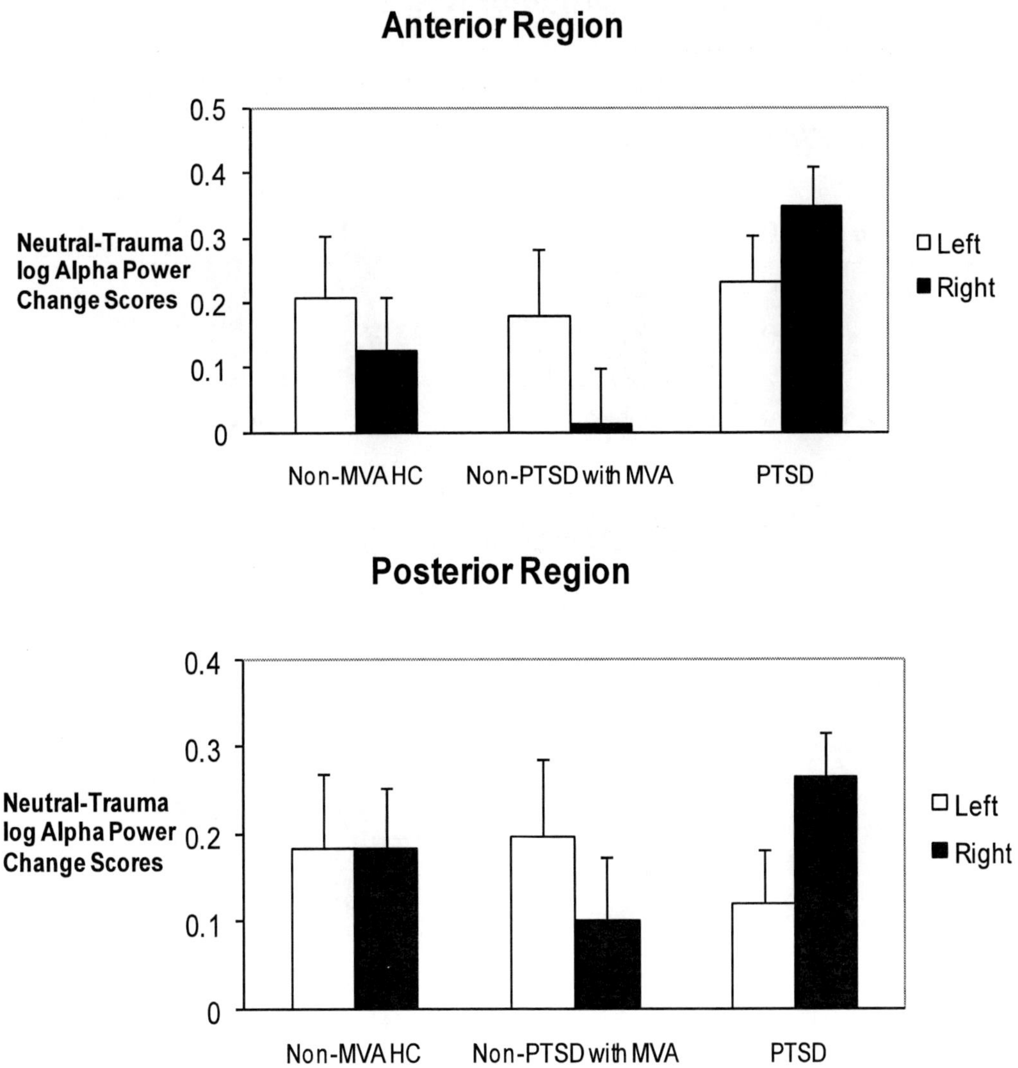

Figure 1. EEG alpha power change scores (neutral minus trauma condition) for the left and right anterior (upper panel) and posterior regions (lower panel) in healthy controls (n = 23), non-PTSD controls with MVA (n = 21), patients with subsyndromal PTSD (n = 21) and full PTSD (n=22). Since EEG alpha power is inversely related to activity positive change score values denote increased activation during exposure to the trauma-related picture. Left = left hemisphere; Right = right hemisphere. (from Rabe et al. [36]).

We suggested that this reduction of right anterior activation might reflect a reduction of avoidance tendencies. Also it might reflect adaptive information processing in terms of relative reduction of predominant visual-emotional processing of trauma-related information which has been associated with PTSD [50, 51]. To the authors' knowledge, this study presents the first study reporting changes in brain activation as assessed by EEG asymmetries in PTSD in a randomized controlled trial of cognitive-behavioral therapy.

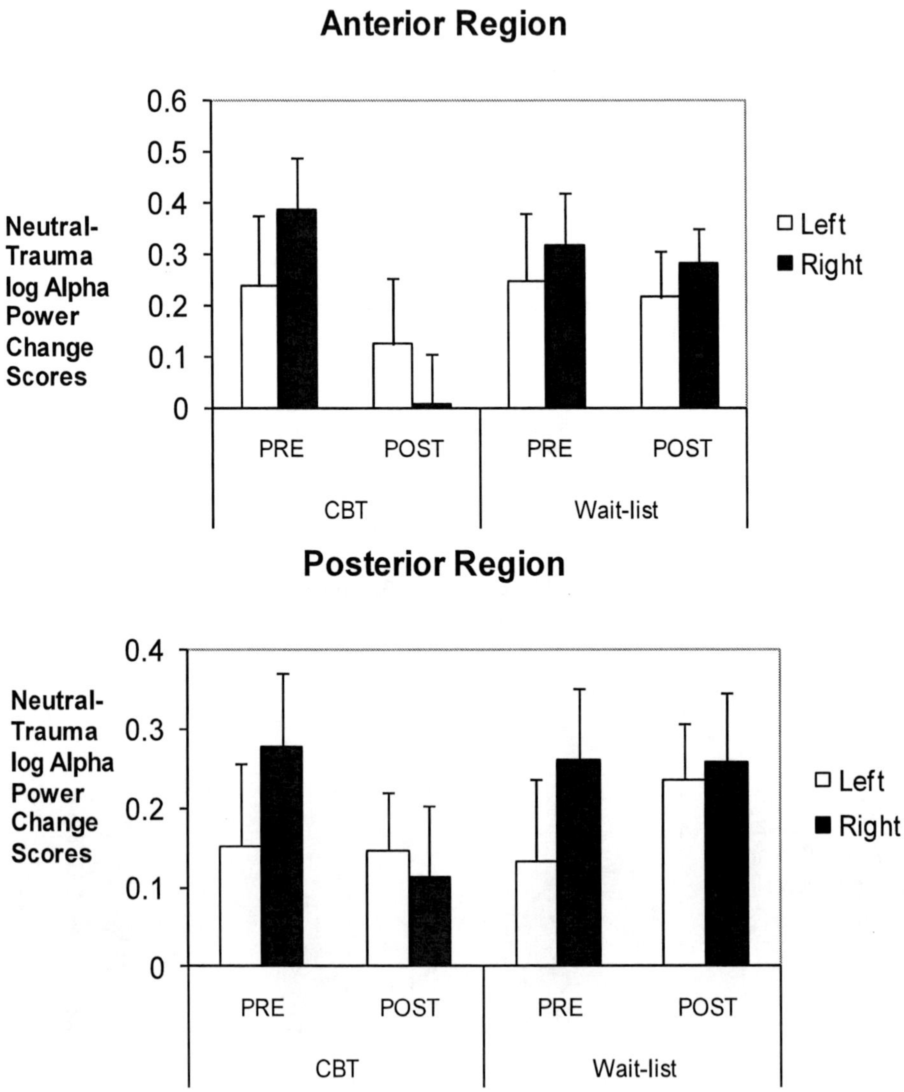

Figure 2. EEG alpha power change scores (neutral minus trauma condition) for the left and right anterior (upper panel) and posterior regions (lower panel) for CBT group (n = 17) and Wait-list controls (n = 18) during pre-therapy assessment (pre) and post-CBT-assessment (post). Since EEG alpha power is inversely related to activity positive change score values denote increased activation during exposure to the accident picture. Left = left hemisphere; Right = right hemisphere. (from Rabe et al. [48]).

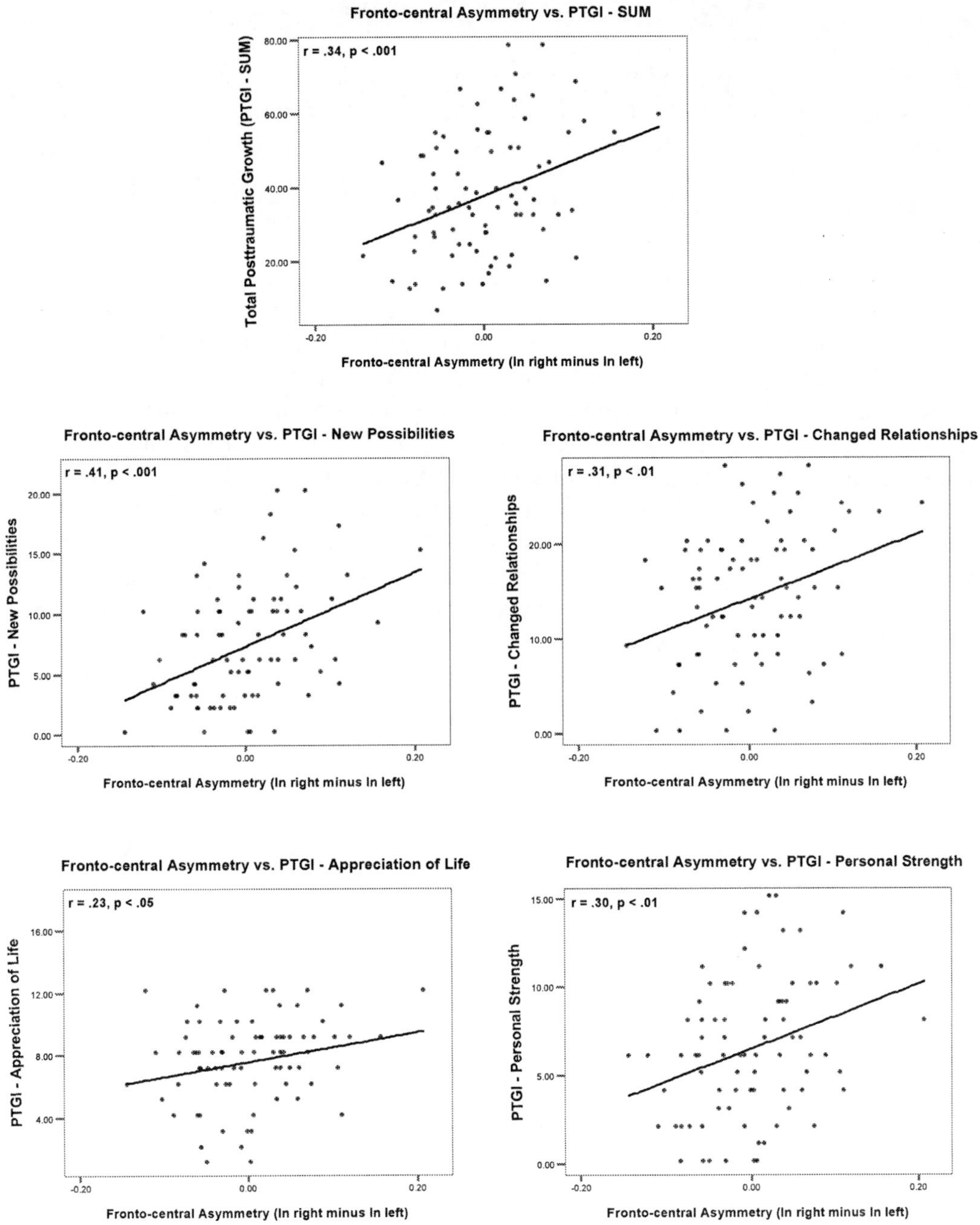

Figure 3. Scatter plots of fronto-central (Fc2-Fc1) asymmetry scores with total Posttraumatic Growth (PTGI-SUM) and the PTGI dimensions: New Possibilities, Changed Relationships, Appreciation of Life, and Personal Strength (n = 82). More positive asymmetry scores indicate greater relative left-sided activity. (from Rabe et al. [55]).

Brain Asymmetry and PTG

In contrast to PTSD, positive changes following trauma and adversity have been studied systematically only in recent years [52]. These positive changes have been labeled posttraumatic growth (PTG) referring to a positive psychological change arising from the struggle with a major life crisis [53]. Furthermore, neural correlates of individual differences in posttraumatic growth are unknown to date. Based on research highlighting the role of left anterior regions of the cortex in positive emotion, approach-related behavioral tendencies and personal well-being [10; 54], we conducted an exploratory analysis on the relation of anterior resting asymmetry and subjective perception of Posttraumatic Growth after MVAs [55]. We expected that higher relative left-sided anterior activity would be associated with higher levels of PTG. The results confirmed our prediction; i.e., left frontal cortical activation was related to the sum score of the Posttraumatic Cognition Inventory (PTGI), a measure of PTG. In addition, our data showed that the different domains of PTG were differentially related to anterior asymmetry (see Figure 3). Significant correlations between anterior asymmetry and the PTG domains were also present after controlling for trait positive affect which in previous studies was associated with asymmetry. This suggests PTG and anterior brain asymmetry are associated over and above the role of positive affect. The findings suggest that trait-like relative left prefrontal baseline activity, which has been associated with goal-directed approach tendencies [12], may promote adaptive emotion-regulation [56, 57] and thus facilitate the achievement of growth after traumatic events [55]. To the authors knowledge this is the first study relating measures of brain function to self perceived Posttraumatic Growth.

The Relation of Trait and State Asymmetry with PTSD and PTG

In the study of Rabe et al. [36], we predicted that patients with full and subsyndromal PTSD would differ from healthy controls in trait EEG asymmetry (resting baseline) and EEG asymmetry as state measure during exposure to a trauma relevant stimulus. However, we only found group differences in asymmetry during an emotional state but not during resting baseline. Furthermore, as shown in the RCT, changes in PTSD severity due to CBT were associated only with changes in state-related but not trait-like baseline EEG asymmetry.

It has been hypothesized that resting EEG asymmetry represents a trait-like vulnerability factor or diathesis that is associated with risk for emotion-related psychopathology [58]. This proposal has been extensively tested for depression showing that relatively reduced left frontal activity characterizes depressed individuals not only when depressed, but also when in remission or when successfully treated [10, 19, 59, 60]. However, our findings suggest that baseline EEG may not represent a trait marker for increased risk of PTSD. In line with research showing that PTSD and depression may exhibit different biological correlates such as HPA function [61], measures of baseline EEG asymmetry may thus be useful differentiating between these two conditions. However, one caveat is that PTSD is associated with a high comorbidity of depression (see also below). In addition, resting EEG asymmetry may be a distinguishing biological marker for different anxiety disorders or predispositions. For example, no effects have been found in patients with social phobia [21] whereas in individuals with panic disorder [20], high trait anxiety [62], or trait levels of anxious arousal [63] baseline asymmetry effects were revealed. Our results of no association of PTSD with trait asymmetry have been recently replicated in a mixed sample of patients with civilian trauma [35]. However, they are in contrast to previous research showing that posterior baseline asymmetry was associated with PTSD hyperarousal in female Vietnam veterans [34]. Reasons for discrepant findings may be differences in type of trauma (civilian vs. combat), chronicity of PTSD, and time since trauma which has been highlighted by research that revealed tonic psychophysiological alterations (e.g. increased baseline heart rate) more consistently in individuals with repeated, interpersonal as compared to single accidental trauma [64-66]. In contrast, psychophysiological reactivity to trauma-related stimuli is a well replicated finding in PTSD [66] and particularly in MVA-related PTSD [67, 68]. Baseline EEG asymmetry may be useful discriminating between different PTSD samples or symptom patterns.

Our findings suggest that individuals with MVA-related PTSD do not exhibit abnormal affective traits (as general withdrawal tendencies or general hyperarousal), but show abnormal trauma-specific affective states, which changes as a consequence to CBT. In contrast, there was a relation of trait-like resting anterior asymmetry and dimensions of posttraumatic growth. Thus, instead being a risk factor for the development of PTSD, resting brain asymmetry might rather promote the development of PTG. In particular, the approach-related and goal-oriented dimensions of PTG (e.g., the PTGI subscale New Possibilities) were associated with anterior trait asymmetry. This is consistent with the model of Davidson [1, 69], which suggests that trait-like left anterior activation represents a disposition to experience rather positive and especially approach-related emotions. However, the assumption that baseline brain asymmetry represents a predisposition to respond with PTG to traumatic stress can only be answered using longitudinal designs. For example, future studies could assess brain asymmetry in persons with high risk for traumatic experiences (e.g. fire fighters) and examine later growth in response to stress. In another design one might assess brain asymmetry shortly after traumatic experience and assess later development of PTG. Finally, it should be noted, that there is possibly no single brain characteristic that predicts development of PTG. Rather, recent research suggests that posttraumatic growth is a dynamic process across time [70] indicating that the development of growth involves several psychological and biological processes which may interact and vary across time. This has important implications for study designs.

The Relation of Findings of Brain Asymmetry to Psychological Models of PTSD and PTG

Recent theories of PTSD based on neuroimaging findings have been criticized for being etiologically reductionistic, suggesting PTSD might be just a manifestation of altered function of some brain regions [6, 71] and for their little convergence of psychological and neurobiological research of PTSD (for exception see Brewin et al. [72, 73]). Therefore, the current chapter tries to discuss the EEG asymmetry findings with respect to psychological theories of PTSD and PTG.

With regard to models of brain asymmetry and emotion; [1, 3, 74], the finding of trauma-specific right anterior activation in PTSD [36] might reflect the activation of a brain system involved in avoidance–withdrawal tendencies and right posterior activation may reflect activation of a system mediating anxious arousal. The finding of treatment-related reduction in right frontal cortical activation [48] might reflect a decrease in anxiety-related withdrawal tendencies in the CBT group. What might be a possible link of EEG asymmetry research and models of PTSD? First, nearly all recent psychological models propose that avoidance of trauma-related stimuli, situations, and thoughts is a key factor for development and maintenance of PTSD [51, 75, 76]. Thus, right anterior activation in PTSD may be the neural correlate of anxiety-related avoidance and the reduction of this right anterior activation in the CBT group in the Rabe et al. study [48] might reflect decreased anxiety and a disruption of avoidance tendencies in response to trauma-relevant stimuli due to successful treatment. A second link might be the association of brain asymmetry and the dual representation theory which postulates a dysfunction in two memory systems in PTSD, a situationally accessible, vivid sensory system and a verbally accessible memory system [50]. Whereas the left hemisphere might be specialized for processing of verbal information, the right hemisphere seems to be specialized for processing of visuo-spatial information. Furthermore, especially the right prefrontal cortex might be involved in cognitive processes such as episodic memory retrieval, sustained attention, and visual vigilance [77, 78]. Altered information processing and memory is suggested in understanding key features of PTSD such as intrusions [51, 75]. These models propose that externally or internally triggered intrusive memories are mostly visual, very vivid and associated with negative emotions. In contrast, ordinary autobiographical memories are structured and the sensory elements are verbally accessible and integrated into a personal narrative. Additionally, they seem to be at least in part dependent on functioning of the left hemisphere [79, 80]. Thus, enhanced right anterior activation during processing of trauma-related information in PTSD subjects could be a correlate of nonverbal-emotional information processing and memory observed in PTSD. Furthermore, reduction in right anterior activation in the CBT group could reflect a normalization of these processes due to therapy [48].

To further test these assumptions, joint assessment of brain activity and psychological processes based on models of PTSD might be useful. Future studies could evaluate brain activity during manipulation of avoidance strategies that are postulated to maintain PTSD such as thought-suppression, safety seeking behaviors, or distraction. It has been suggested that although there are some studies examining brain function (e.g. hippocampal) during tasks that are thought to activate certain brain structures [81-84], studies that directly link psychological models of PTSD with measures of brain function are lacking. Current neurobiological models tend to focus on symptoms, since most of the neuroimaging studies used symptom provocation paradigms. Studies directly examining brain function during cognitive processes (cognitive activation paradigms) proposed to be central for development of PTSD are needed such as data-driven vs. conceptually-driven processing of traumatic information [51] or verbal vs. visuo-spatial encoding [85].

Conceptually, the potential connection between research of PTG and EEG asymmetry is intuitively appealing. There are several theoretical links between PTG and anterior brain asymmetry research. Both, PTG and anterior asymmetry have been associated with concepts of well-being and positive affect. Within the theoretically framework of well-being, PTG has been postulated to represent an element of psychological well-being [86]. In this view, growth and distress may be two independent dimensions of well-being and thus, PTG may co-occur with distress. In fact, findings of a recent meta-analysis showed that perceived growth was unrelated to anxiety and indexes of distress, but positively related to positive affect and less depression but more PTSD symptoms (intrusions and avoidance) [70]. The results of the study by Rabe et al. [55] support the link of PTG with psychological well-being and positive affect. The latter two concepts also have been found to be associated with anterior asymmetry [54; 87]. However, although growth reports were related to positive affect, the correlation of PTG and brain asymmetry remained significant after positive affect was statistically controlled for.

A second link between PTG and anterior brain asymmetry may be coping. On the one hand it is widely acknowledged that coping with trauma represents a central mechanism for achieving growth as an outcome. On the other hand, PTG is viewed as a process and therefore as a coping mechanism itself (for discussion see also [88, 89]). Empirically, self reported growth has been associated with several coping mechanisms such as positive reinterpretation coping [90-92] but also with acceptance coping and denial (see [70, 89] for review). A possible link to EEG asymmetry research may be the finding that trait-like relative left hemisphere activity is associated with a positive affective style which means a tendency to react rather positive to emotion-elicitors. Furthermore, left hemisphere activity has been linked with positive emotion-regulation [56], positive reappraisal [57], but also a self enhancing coping style called repression or defensiveness [93, 94]. This partly self-deceptive coping style has been associated with self-enhancing cognitive and regulatory processes [94], and lower risk for depression [95]. The theoretical association of repressive coping and PTG, which both are associated with anterior asymmetry, might explain the positive association between growth and distress often found. However, research is necessary to more directly assess the relationship between PTG, resting frontal asymmetry and its associated self-regulatory mechanisms. At this point it is noteworthy that EEG asymmetry can not be viewed as a mechanism itself, but rather, as a marker of some underlying neural process [96]. Given the sizable literature, EEG asymmetry research has established high psychological construct validity as a measure of an underlying approach-related or withdrawal-related motivational style and emotions [74]. Thus, the connection of EEG asymmetry and growth may also help to increase construct validity of PTG. There are a number of questions regarding validity, conceptualization, and assessment of PTG and its association with adjustment and well-being that future studies need to address.

The Relation of Measures of Brain Function and Psychotherapy

One important new area concerns the neuroscientific investigation of psychological treatment in general [97, 98] and especially in PTSD [99]. Even before the era of neuroimaging, it was suggested by Freud 1895 [100] that changes in affect, cognition, and behavior due to psychotherapy are mediated by the so called "in between", the brain. Nobel Prize winner Eric Kandel [101] more directly suggested, "that all mental processes, even the most complex psychological processes, derive from operations of the brain“ (p. 460) and changes in the patients mind due to psychotherapy should be related to changes in the patient's brain.

Neuroscience has been suggested to be useful to psychotherapy research in several ways: to help understanding mechanisms underlying the development and maintenance of the psychological disorder; for the validation of existing psychotherapeutic theories and interventions; for the understanding of change mechanisms of efficacious psychological treatments and therefore for the development of more efficient psychological therapies [98, 102, 103].

Several possible directions for further research on the question how psychotherapy is related to brain mechanisms have been suggested. First, one of the most important questions of psychotherapy research is what the mechanisms of change due to treatment are [98]. For CBT of PTSD, there are several mechanisms postulated to underlie successful treatment such as reduction of avoidance, habituation, cognitive reappraisal, elaboration of autobiographical memory or integration of competing memories [51, 73, 104]. One possible contribution of measures of brain activity to disentangling specific effects of various intervention strategies would be to compare changes of brain function as a consequence of different psychological treatment components for PTSD such as cognitive restructuring, different kinds of exposure, relaxation training, stress management, imagery rescripting or combinations of these interventions. Furthermore, it might be interesting to compare the effects of different psychological treatments with medication or combinations of both on measures of brain activity in PTSD. To date, there are no such studies for PTSD available but for depression [105] and social phobia [106], suggesting that clinical recovery due to psychological or pharmacological treatment seems to have some common but also different neural correlates.

A possibility for increasing treatment efficacy might be research predicting treatment response based on neurobiological profiles. For example, in a recent meta-analysis, Seminowicz et al. [107] investigated the differential effects of psychotherapy and medication. They found that categorization of types of depression based on neural responses may differentiate responders and non-responders of medication and psychotherapy and thus may increase treatment efficacy. For PTSD, preliminary evidence suggests that structural [108] and functional [109] neuroimaging measures may provide critical information for predicting response to CBT. Given the association of EEG asymmetry and PTSD and its successful treatment [36, 48] and studies showing the predictive value of EEG asymmetry for antidepressant treatments [110-112], a goal of future studies might be to determine the predictive value of this measure for psychological treatment.

A further field of investigation would be the link of neural mechanisms and general and specific mechanisms of different psychotherapy approaches for PTSD. There is an ongoing debate that different kinds of psychotherapeutic approaches are associated with common or general mechanisms of change [113-118]. For PTSD there have been similar patterns of brain alterations following different treatments (CBT, EMR, brief eclectic psychotherapy). For example, the results of the study of Rabe et al. [48] (using CBT), the study of Lindauer et al. [46] (using brief eclectic psychotherapy), and the study of Lansing et al. [44] found a reduction of activation in right frontal regions after therapy. Furthermore, some studies [44; 46, 119] reported a common pattern of increase in left hemisphere cortical regions in response to different psychological treatments for PTSD (brief eclectic psychotherapy, EMDR, CBT). This common pattern of alterations in brain activity due to diverse psychological treatments suggests some common mechanisms of change, despite different theoretical approaches of these treatments. This might in part explain the fact, that for PTSD several psychological treatments have been shown to be effective to reduce symptoms [120, 121]. Finding the most cost-effective approach that allows access to psychological treatment for a higher percentage of patients might be of increasing interest for health care systems.

From a more theoretical perspective, the link between EEG asymmetry research and research of general psychotherapy might be an interesting field for further research. For example, Grawe [113, 114] proposed a framework for general psychotherapy based on its "Konsistenztheorie" [theory of consistency] in which motivational goals (such as approach goals and avoidance goals) and their conflicts with each other and with the reality are central for the development and treatment of psychological disorders. Given the common interest in emotional-motivational aspects between Grawe's assumption and the approach-withdrawal model of Davidson [58], studies linking these two theoretical approaches are of unique interest. This theory would predict that treatments intending a reduction of avoidance goals or an increase of approach goals due to therapeutic interventions are associated with changes in frontal asymmetry. Results of our study [48] may signify a first step in this line of research.

An important question for neurobiological research of psychotherapy is: Do we really measure mechanisms of change when we compare pre and post assessments of brain activity? It has been suggested that one measures the outcome of treatment rather than a specific mechanism [8]. Thus, it might be useful to assess brain activity during multiple points of time during a treatment protocol: e.g. after each session, or even during treatment sessions, or at follow up. For this kind of research especially the EEG as research tool seems to be well suited. In contrast to more invasive and expensive measures such as PET or fMRI, the EEG is a relatively noninvasive, inexpensive and even portable measure of cortical activity [16, 56]. Furthermore, in such a design a combination with psychophysiological research measures might be useful to compare different levels of change (behavioral, physiological, and neural). This kind of research could also be helpful for answering questions regarding optimal length of treatment protocol and session as well as optimal time between sessions.

There are several methodological problems associated with the measurement of neural correlates of psychotherapy of PTSD. Several general methodological confounders of studies of brain function in PTSD may have influenced results and complicate comparisons of different studies. These confounders are method of measurement of brain activity (e.g. fMRI, PET, SPECT, EEG), type of trauma, comorbidity, current medication, dominant symptoms (e.g. intrusions, numbing, avoidance), or dominant emotions elicited by trauma reminders (e.g. anxiety, anger, guilt, shame) [122]. Therefore, comparative research strategies (e.g. the comparison of different imaging techniques in one sample of subjects, or the comparison of different subgroups of PTSD subjects) could help to disentangle the common and distinct effects of these methodological variations. Furthermore, neuroimaging research on PTSD and non-PTSD individuals might assess individuals in the immediate aftermath of a traumatic event, which might avoid some confounds of the assessment of chronic PTSD (e.g. medication history and alcohol abuse). It is also necessary to compare different groups of psychological disorders and PTSD to demonstrate if findings of brain alterations are specific to PTSD. Furthermore, research regarding the conceptualization of PTSD is necessary to increase validity of the PTSD concept [123, 124].

Another important factor for psychotherapy research is that the paradigm used during measurement of brain activity will influence the resulting brain activation patterns. For example, Peres et al. [119] reported a post-therapy increase in left hemisphere function using a script-driven paradigm in which the subject is instructed to read aloud about the traumatic event. It is conceivable that this paradigm involves verbal activity which depends on left hemisphere function. In contrast, the study of Rabe et al. [36] that used standardized pictures with no necessary verbal response found a reduction in right hemisphere activation after CBT. Thus, although the direction of changed asymmetry was similar in both studies, differences in experimental paradigms may have a great impact on the observed results and may explain some inconsistent findings of treatment-related changes in PTSD. Future studies should compare the effects of different paradigms (symptom provocation, baseline) which may increase validity of the results. It should be noted that several other methodological variables are probably influencing results of studies of psychological treatment such as number of sessions, amount of therapists and therapeutic setting (group vs. individual therapy) [8, 125].

References

[1] Davidson RJ. Anterior electrophysiological asymmetries, emotion, and depression: conceptual and methodological conundrums. *Psychophysiology* 1998;35(5):607-14.

[2] Davidson RJ. What does the prefrontal cortex "do" in affect: perspectives on frontal EEG asymmetry research. *Biological Psychology* 2004;67(1-2):219-33.

[3] Heller W, Koven NS, Miller GA. Regional brain activity in anxiety and depression, cognition/emotion interaction, and emotion regulation. In: Hugdahl K, Davidson RJ, editors. *The Asymmetrical Brain.* Cambridge, MA, US: MIT Press; 2003. p 533-64.

[4] Karl A, Schaefer M, Malta LS, Dorfel D, Rohleder N, Werner A. A meta-analysis of structural brain abnormalities in PTSD. *Neuroscience and Biobehavioral Reviews* 2006;30(7):1004-31.

[5] Karl A, Werner A. The use of proton magnetic resonance spectroscopy in PTSD research-Meta-analyses of findings and methodological review. *Neuroscience and Biobehavioral Reviews* 2009.

[6] Liberzon I, Sripada CS. The functional neuroanatomy of PTSD: a critical review. *Progress in Brain Research* 2008;167:151-69.

[7] Francati V, Vermetten E, Bremner JD. Functional neuroimaging studies in posttraumatic stress disorder: review of current methods and findings. *Depression and Anxiety* 2007;24(3):202-18.

[8] Roffman JL, Marci CD, Glick DM, Dougherty DD, Rauch SL. Neuroimaging and the functional neuroanatomy of psychotherapy. *Psychological Medicine* 2005;35(10):1385-98.

[9] Demaree HA, Everhart DE, Youngstrom EA, Harrison DW. Brain lateralization of emotional processing: historical roots and a future incorporating "dominance". *Behavioral and Cognitive Neuroscience Reviews* 2005;4(1):3-20.

[10] Coan JA, Allen JJ. The state and trait nature of frontal EEG asymmetry in emotion. In: Hugdahl K, Davidson RJ, editors. *The Asymmetrical Brain.* Cambridge, MA, U.S.: MIT Press; 2003. p 565-615.

[11] Harmon-Jones E, Allen JJ. Anger and frontal brain activity: EEG asymmetry consistent with approach motivation despite negative affective valence. *Journal of Personality and Social Psychology* 1998;74(5):1310-6.

[12] Sutton SK, Davidson RJ. Prefrontal brain asymmetry: A biological substrate of the behavioral approach and inhibition systems. *Psychological Science* 1997;8:204-10.

[13] Davidson RJ. Cerebral asymmetry, emotion, and affective style. In: Davidson RJ, Hugdahl K, editors. *Brain Asymmetry.* Cambridge: MIT Press; 1995. p 361-87.

[14] Davidson RJ. Well-being and affective style: Neural substrates and biobehavioural correlates. Philosophical Transactions of the Royal Society of London Series B: *Biological Sciences* 2004;359(1449):1395-411.

[15] Cook IA, O'Hara R, Uijtdehaage SH, Mandelkern M, Leuchter AF. Assessing the accuracy of topographic EEG mapping for determining local brain function. *Electroencephalography and Clinical Neurophysiology* 1998;107(6):408-14.

[16] Davidson RJ, Jackson DC, Larson CL. Human electroencephalography. In: Cacioppo JT, Tassinary LG, Berntson GG, editors. *Handbook of Psychophysiology.* New York: Cambridge University Press; 2000. p 27-52.

[17] Hagemann D. Individual differences in anterior EEG asymmetry: methodological problems and solutions. *Biological Psychology* 2004;67(1-2):157-82.

[18] Davidson RJ, Irwin W. The functional neuroanatomy of emotion and affective style. *Trends in Cognitive Science* 1999;3(1):11-21.

[19] Bruder GE. Frontal and Parietotemporal Asymmetries in Depressive Disorders: Behavioral, Electrophysiologic, and Neuroimaging Findings. In: Hugdahl K, Davidson RJ, editors. *The Asymmetrical Brain.* Cambridge, MA, US: MIT Press; 2003. p 719-42.

[20] Wiedemann G, Pauli P, Dengler W, Lutzenberger W, Birbaumer N, Buchkremer G. Frontal brain asymmetry as a biological substrate of emotions in patients with panic disorders. *Archives of General Psychiatry* 1999;56(1):78-84.

[21] Davidson RJ, Marshall JR, Tomarken AJ, Henriques JB. While a phobic waits: regional brain electrical and autonomic activity in social phobics during anticipation of public speaking. *Biological Psychiatry* 2000;47(2):85-95.

[22] Thibodeau R, Jorgensen RS, Kim S. Depression, anxiety, and resting frontal EEG asymmetry: a meta-analytic review. *Journal of Abnormal Psychology* 2006;115(4):715-29.

[23] Bruder GE, Fong R, Tenke CE, Leite P, Towey JP, Stewart JE et al. Regional brain asymmetries in major depression with or without an anxiety disorder: a quantitative electroencephalographic study. *Biological Psychiatry* 1997;41(9):939-48.

[24] Reid SA, Duke LM, Allen JJ. Resting frontal electroencephalographic asymmetry in depression: inconsistencies suggest the need to identify mediating factors. *Psychophysiology* 1998;35(4):389-404.

[25] Heller W, Nitschke JB. The puzzle of regional brain activity in depression and anxiety: The importance of subtypes and comorbidity. *Cognition and Emotion* 1998;12:421-47.

[26] Heller W, Nitschke JB, Etienne MA, Miller GA. Patterns of regional brain activity differentiate types of anxiety. *Journal of Abnormal Psychology* 1997;106(3):376-85.

[27] Bremner JD, Staib LH, Kaloupek D, Southwick SM, Soufer R, Charney DS. Neural correlates of exposure to traumatic pictures and sound in Vietnam combat veterans with and without posttraumatic stress disorder: a positron emission tomography study. *Biological Psychiatry* 1999;45(7):806-16.

[28] Rauch SL, Shin LM. Functional neuroimaging studies in posttraumatic stress disorder. *Annals of the New York Academy of Sciences* 1997;821:83-98.
[29] Rauch SL, van der Kolk BA, Fisler RE, Alpert NM, Orr SP, Savage CR et al. A symptom provocation study of posttraumatic stress disorder using positron emission tomography and script-driven imagery. *Archives of General Psychiatry* 1996;53(5):380-7.
[30] Pagani M, Hogberg G, Salmaso D, Tarnell B, Sanchez-Crespo A, Soares J et al. Regional cerebral blood flow during auditory recall in 47 subjects exposed to assaultive and non-assaultive trauma and developing or not posttraumatic stress disorder. *European Archives of Psychiatry and Clinical Neuroscience* 2005;255(5):359-65.
[31] Herskovits EH, Gerring JP, Davatzikos C, Bryan RN. Is the spatial distribution of brain lesions associated with closed-head injury in children predictive of subsequent development of posttraumatic stress disorder? *Radiology* 2002;224(2):345-51.
[32] Lanius RA, Williamson PC, Densmore M, Boksman K, Neufeld RW, Gati JS et al. The nature of traumatic memories: a 4-T FMRI functional connectivity analysis. *The American Journal of Psychiatry* 2004;161(1):36-44.
[33] McCaffrey RJ, Lorig TS, Pendrey DL, McCutcheon NB, Garrett JC. Odor-induced EEG changes in PTSD Vietnam veterans. *Journal of Traumatic Stress* 1993;6:213-24.
[34] Metzger LJ, Paige SR, Carson MA, Lasko NB, Paulus LA, Pitman RK et al. PTSD arousal and depression symptoms associated with increased right-sided parietal EEG asymmetry. *Journal of Abnormal Psychology* 2004;113(2):324-9.
[35] Shankman SA, Silverstein SM, Williams LM, Hopkinson PJ, Kemp AH, Felmingham KL et al. Resting electroencephalogram asymmetry and posttraumatic stress disorder. *Journal of Traumatic Stress* 2008;21(2):190-8.
[36] Rabe S, Beauducel A, Zollner T, Maercker A, Karl A. Regional brain electrical activity in posttraumatic stress disorder after motor vehicle accident. *Journal of abnormal psychology* 2006;115(4):687-98.
[37] Davidson RJ, Schwartz GE, Pugash E, Bromfield E. Sex differences in patterns of EEG asymmetry. *Biological Psychology* 1976;4(2):119-38.
[38] Miller A, Fox NA, Cohn JF, Forbes EE, Sherrill JT, Kovacs M. Regional patterns of brain activity in adults with a history of childhood-onset depression: gender differences and clinical variability. *American Journal of Psychiatry* 2002;159(6):934-40.
[39] Bremner JD, Mletzko T, Welter S, Quinn S, Williams C, Brummer M. Effects of phenytoin on memory, cognition and brain structure in post-traumatic stress disorder: a pilot study. *Journal of Psychopharmacology* 2005;19(2):159-65.
[40] Vermetten E, Vythilingam M, Southwick SM, Charney DS, Bremner JD. Long-term treatment with paroxetine increases verbal declarative memory and hippocampal volume in posttraumatic stress disorder. *Biological Psychiatry* 2003;54(7):693-702.
[41] Lindauer RJ, Vlieger EJ, Jalink M, Olff M, Carlier IV, Majoie CB et al. Effects of psychotherapy on hippocampal volume in out-patients with post-traumatic stress disorder: a MRI investigation. *Psychological Medicine* 2005;35(10):1421-31.
[42] Farrow TF, Hunter MD, Wilkinson ID, Gouneea C, Fawbert D, Smith R et al. Quantifiable change in functional brain response to empathic and forgivability judgments with resolution of posttraumatic stress disorder. *Psychiatry Research* 2005;140(1):45-53.

[43] Felmingham K, Kemp A, Williams L, Das P, Hughes G, Peduto A et al. Changes in anterior cingulate and amygdala after cognitive behavior therapy of posttraumatic stress disorder. *Psychological Science* 2007;18(2):127-9.

[44] Lansing K, Amen DG, Hanks C, Rudy L. High-resolution brain SPECT imaging and eye movement desensitization and reprocessing in police officers with PTSD. *The Journal of Neuropsychiatry and Clinical Neurosciences* 2005;17(4):526-32.

[45] Levin P, Lazrove S, van der Kolk B. What psychological testing and neuroimaging tell us about the treatment of Posttraumatic Stress Disorder by Eye Movement Desensitization and Reprocessing. *Journal of Anxiety Disorders* 1999;13(1-2):159-72.

[46] Lindauer RJ, Booij J, Habraken JB, van Meijel EP, Uylings HB, Olff M et al. Effects of psychotherapy on regional cerebral blood flow during trauma imagery in patients with post-traumatic stress disorder: a randomized clinical trial. *Psychological Medicine* 2008;38(4):543-54.

[47] Pagani M, Hogberg G, Salmaso D, Nardo D, Sundin O, Jonsson C et al. Effects of EMDR psychotherapy on 99mTc-HMPAO distribution in occupation-related post-traumatic stress disorder. *Nuclear Medicine cCommunications* 2007;28(10):757-65.

[48] Rabe S, Zoellner T, Beauducel A, Maercker A, Karl A. Changes in brain electrical activity after cognitive behavioral therapy for posttraumatic stress disorder in patients injured in motor vehicle accidents. *Psychosomatic Medicine 2008*;70(1):13-9.

[49] Maercker A, Zollner T, Menning H, Rabe S, Karl A. Dresden PTSD treatment study: randomized controlled trial of motor vehicle accident survivors. *BMC Psychiatry* 2006;6:29.

[50] Brewin CR, Dalgleish T, Joseph S. A dual representation theory of posttraumatic stress disorder. *Psychological Review* 1996;103(4):670-86.

[51] Ehlers A, Clark DM. A cognitive model of posttraumatic stress disorder. *Behaviour Research and Therapy 2000*;38(4):319-45.

[52] Linley PA, Joseph S. Positive change following trauma and adversity: a review. *Journal of Traumatic Stress* 2004;17(1):11-21.

[53] Tedeschi RG, Calhoun LG. Posttraumatic Growth: Conceptual Foundations and Empirical Evidence. *Psychological Inquiry* 2004;15(1):1-18.

[54] Urry HL, Nitschke JB, Dolski I, Jackson DC, Dalton KM, Mueller CJ et al. Making a life worth living: neural correlates of well-being. *Psychological Science* 2004;15(6):367-72.

[55] Rabe S, Zollner T, Maercker A, Karl A. Neural correlates of posttraumatic growth after severe motor vehicle accidents. *Journal of Consulting and Clinical Psychology* 2006;74(5):880-6.

[56] Jackson DC, Mueller CJ, Dolski I, Dalton KM, Nitschke JB, Urry HL et al. Now you feel it, now you don't: frontal brain electrical asymmetry and individual differences in emotion regulation. *Psychological Science* 2003;14(6):612-7.

[57] Ochsner KN, Bunge SA, Gross JJ, Gabrieli JD. Rethinking feelings: an FMRI study of the cognitive regulation of emotion. *Journal of Cognitive Neuroscience* 2002;14(8):1215-29.

[58] Davidson RJ. Affective style and affective disorders: Perspectives from affective neuroscience. *Cognition and Emotion* 1998;12:307-30.

[59] Allen JJ, Urry HL, Hitt SK, Coan JA. The stability of resting frontal electroencephalographic asymmetry in depression. *Psychophysiology* 2004;41(2):269-80.

[60] Davidson RJ, Pizzagalli D, Nitschke JB, Putnam K. Depression: perspectives from affective neuroscience. *Annual Review of Psychology* 2002;53:545-74.

[61] Yehuda R. Advances in understanding neuroendocrine alterations in PTSD and their therapeutic implications. *Annals of the New York Academy of Sciences* 2006;1071:137-66.

[62] Blackhart GC, Minnix JA, Kline JP. Can EEG asymmetry patterns predict future development of anxiety and depression? A preliminary study. *Biological Psychology* 2006;72(1):46-50.

[63] Nitschke JB, Heller W, Palmieri PA, Miller GA. Contrasting patterns of brain activity in anxious apprehension and anxious arousal. *Psychophysiology* 1999;36(5):628-37.

[64] Buckley TC, Kaloupek DG. A meta-analytic examination of basal cardiovascular activity in posttraumatic stress disorder. *Psychosomatic Medicine* 2001;63(4):585-94.

[65] Orr SP, Metzger LJ, Pitman RK. Psychophysiology of post-traumatic stress disorder. *The Psychiatric Clinics of North America* 2002;25(2):271-93.

[66] Pole N. The psychophysiology of posttraumatic stress disorder: a meta-analysis. *Psychological Bulletin* 2007;133(5):725-46.

[67] Blanchard EB, Hickling EJ, Buckley TC, Taylor AE, Vollmer A, Loos WR. Psychophysiology of posttraumatic stress disorder related to motor vehicle accidents: replication and extension. *Journal of Consulting and Clinical Psychology* 1996;64(4):742-51.

[68] Veazey CH, Blanchard EB, Hickling EJ, Buckley TC. Physiological responsiveness of motor vehicle accident survivors with chronic posttraumatic stress disorder. *Applied Psychophysiology and Biofeedback* 2004;29(1):51-62.

[69] Davidson RJ. Affective style, psychopathology, and resilience: brain mechanisms and plasticity. *American Psychologist* 2000;55(11):1196-214.

[70] Helgeson VS, Reynolds KA, Tomich PL. A meta-analytic review of benefit finding and growth. *Journal of Consulting and Clinical Psychology* 2006;74(5):797-816.

[71] Liberzon I, Martis B. Neuroimaging studies of emotional responses in PTSD. *Annals of the New York Academy of Sciences* 2006;1071:87-109.

[72] Brewin CR. A cognitive neuroscience account of posttraumatic stress disorder and its treatment. *Behaviour Research and Therapy* 2001;39(4):373-93.

[73] Brewin CR. What is it that a neurobiological model of PTSD must explain? *Progress in Brain Research* 2008;167:217-28.

[74] Coan JA, Allen JJ. Frontal EEG asymmetry as a moderator and mediator of emotion. *Biological Psychology* 2004;67(1-2):7-49.

[75] Brewin CR, Holmes EA. Psychological theories of posttraumatic stress disorder. *Clinical Psychology Review* 2003;23(3):339-76.

[76] Foa EB, Rothbaum B. Treating the trauma of rape. New York: Guilford Press; 1998.

[77] Cabeza R, Nyberg L. Imaging cognition II: An empirical review of 275 PET and fMRI studies. *Journal of Cognitive Neuroscience 2000*;12(1):1-47.

[78] Hugdahl K, Davidson RJ. *The Asymmetrical Brain.* Cambridge, MA, US: MIT Press; 2003.

[79] Maguire EA. Neuroimaging studies of autobiographical event memory. *Philosophical Transactions of the Royal Society of London* 2001;356(1413):1441-51.

[80] Piefke M, Fink GR. Recollections of one's own past: the effects of aging and gender on the neural mechanisms of episodic autobiographical memory. *Anatomy and Embryology* 2005;210(5-6):497-512.

[81] Astur RS, St Germain SA, Tolin D, Ford J, Russell D, Stevens M. Hippocampus function predicts severity of post-traumatic stress disorder. *Cyberpsychology and Behavior* 2006;9(2):234-40.

[82] Bremner JD, Vythilingam M, Vermetten E, Southwick SM, McGlashan T, Staib LH et al. Neural correlates of declarative memory for emotionally valenced words in women with posttraumatic stress disorder related to early childhood sexual abuse. *Biological Psychiatry* 2003;53(10):879-89.

[83] Geuze E, Vermetten E, Ruf M, de Kloet CS, Westenberg HG. Neural correlates of associative learning and memory in veterans with posttraumatic stress disorder. *Journal of Psychiatric Research 2008*;42(8):659-69.

[84] Shin LM, Shin PS, Heckers S, Krangel TS, Macklin ML, Orr SP et al. Hippocampal function in posttraumatic stress disorder. *Hippocampus* 2004;14(3):292-300.

[85] Holmes EA, Brewin CR, Hennessy RG. Trauma films, information processing, and intrusive memory development. *Journal of Experimental Psychology: General* 2004;133(1):3-22.

[86] Joseph S, Linley PA. Positive adjustment to threatening events: An organismic valuing theory of growth through adversity. *Review of General Psychology* 2005;9(3):262-80.

[87] Tomarken AJ, Davidson RJ, Wheeler RE, Doss RC. Individual differences in anterior brain asymmetry and fundamental dimensions of emotion. *Journal of Personality and Social Psychology* 1992;62(4):676-87.

[88] Park CL, Helgeson VS. Introduction to the special section: growth following highly stressful life events--current status and future directions. *Journal of Consulting and Clinical Psychology* 2006;74(5):791-6.

[89] Zoellner T, Maercker A. Posttraumatic growth in clinical psychology - a critical review and introduction of a two component model. *Clinical Psychology Review* 2006;26(5):626-53.

[90] Frazier P, Tennen H, Gavian M, Park C, Tomich P, Tashiro T. Does self-reported posttraumatic growth reflect genuine positive change? *Psychological Science* 2009;20(7):912-9.

[91] Sears SR, Stanton AL, Danoff-Burg S. The yellow brick road and the emerald city: benefit finding, positive reappraisal coping and posttraumatic growth in women with early-stage breast cancer. *Health Psychology* 2003;22(5):487-97.

[92] Siegel K, Schrimshaw EW, Pretter S. Stress-related growth among women living with HIV/AIDS: examination of an explanatory model. *Journal of Behavioral Medicine* 2005;28(5):403-14.

[93] Kline JP, Allen JJ, Schwartz GE. Is left frontal brain activation in defensiveness gender specific? *Journal of Abnormal Psychology* 1998;107(1):149-53.

[94] Tomarken AJ, Davidson RJ. Frontal brain activation in repressors and nonrepressors. *Journal of Abnormal Psychology* 1994;103(2):339-49.

[95] Lane RD, Merikangas KR, Schwartz GE, Huang SS, Prusoff BA. Inverse relationship between defensiveness and lifetime prevalence of psychiatric disorder. *The American Journal of Psychiatry* 1990;147(5):573-8.

[96] Cacioppo JT. Feelings and emotions: roles for electrophysiological markers. *Biological Psychology* 2004;67(1-2):235-43.

[97] Etkin A, Pittenger C, Polan HJ, Kandel ER. Toward a neurobiology of psychotherapy: basic science and clinical applications. *The Journal of Neuropsychiatry and Clinical Neurosciences* 2005;17(2):145-58.

[98] Moras K. The value of neuroscience strategies to accelerate progress in psychological treatment research. *Canadian Journal of Psychiatry* 2006;51(13):810-22.

[99] Peres JF, McFarlane A, Nasello AG, Moores KA. Traumatic memories: bridging the gap between functional neuroimaging and psychotherapy. *The Australian and New Zealand Journal of Psychiatry* 2008;42(6):478-88.

[100] Freud S. Entwurf einer Psychologie [Draft of a psychology]. In: Freud S, editor. *Aus den Anfängen der Psychoanalyse* [From the beginnings of psychoanalysis]. Frankfurt: Fischer; 1950.

[101] Kandel ER. A new intellectual framework for psychiatry. *The American Journal of Psychiatry* 1998;155(4):457-69.

[102] Cappas NM, Andres-Hyman R, Davidson L. What Psychotherapists Can Begin to Learn from Neuroscience: Seven Principles of a Brain-Based Psychotherapy. *Psychotherapy: Theory, Research, Practice, Training* 2005;42(3):374-83.

[103] van der Kolk BA. Clinical implications of neuroscience research in PTSD. *Annals of the New York Academy of Sciences* 2006;1071:277-93.

[104] Foa EB. Psychological processes related to recovery from a trauma and an effective treatment for PTSD. *Annals of the New York Academy of Sciences* 1997;821:410-24.

[105] Goldapple K, Segal Z, Garson C, Lau M, Bieling P, Kennedy S et al. Modulation of cortical-limbic pathways in major depression: treatment-specific effects of cognitive behavior therapy. *Archives of General Psychiatry* 2004;61(1):34-41.

[106] Furmark T, Tillfors M, Marteinsdottir I, Fischer H, Pissiota A, Langstrom B et al. Common changes in cerebral blood flow in patients with social phobia treated with citalopram or cognitive-behavioral therapy. *Archives of General Psychiatry* 2002;59(5):425-33.

[107] Seminowicz DA, Mayberg HS, McIntosh AR, Goldapple K, Kennedy S, Segal Z et al. Limbic-frontal circuitry in major depression: a path modeling metanalysis. *NeuroImage* 2004;22(1):409-18.

[108] Bryant RA, Felmingham K, Whitford TJ, Kemp A, Hughes G, Peduto A et al. Rostral anterior cingulate volume predicts treatment response to cognitive-behavioural therapy for posttraumatic stress disorder. *Journal of Psychiatry and Neuroscience* 2008;33(2):142-6.

[109] Bryant RA, Felmingham K, Kemp A, Das P, Hughes G, Peduto A et al. Amygdala and ventral anterior cingulate activation predicts treatment response to cognitive behaviour therapy for post-traumatic stress disorder. *Psychological Medicine* 2008;38(4):555-61.

[110] Bruder GE, Tenke CE, Stewart JW, McGrath PJ, Quitkin FM. Predictors of therapeutic response to treatments for depression: a review of electrophysiologic and dichotic listening studies. *CNS spectrums* 1999;4(8):30-6.

[111] Bruder GE, Stewart JW, McGrath PJ, Deliyannides D, Quitkin FM. Dichotic listening tests of functional brain asymmetry predict response to fluoxetine in depressed women and men. *Neuropsychopharmacology* 2004;29(9):1752-61.
[112] Bruder GE, Sedoruk JP, Stewart JW, McGrath PJ, Quitkin FM, Tenke CE. Electroencephalographic alpha measures predict therapeutic response to a selective serotonin reuptake inhibitor antidepressant: pre- and post-treatment findings. *Biological Psychiatry* 2008;63(12):1171-7.
[113] Grawe K. Psychologische Therapie [*Psychological Therapy]*. Göttingen: Hogrefe; 2000.
[114] Grawe K. Neuropsychotherapie [*neuropsychotherapy]*. Göttingen: Hogrefe; 2004.
[115] Greenberg LS, Pascual-Leone A. Emotion in psychotherapy: a practice-friendly research review. *Journal of Clinical Psychology* 2006;62(5):611-30.
[116] Wampold B. The Great Psychotherapy Debate. Models, Methods, and Findings. Mahwah, NJ: Lawrence Earlbaum; 2001.
[117] Imel Z, Wampold B. The Importance of Treatment and the Science of Common Factors in Psychotherapy In: Brown SD, Lent RW, editors. *Handbook of counseling psychology*. New York: John Wiley & Sons Inc.; 2008. p 249-62.
[118] Kazdin AE. Understanding how and why psychotherapy leads to change. *Psychother Res* 2008:1-11.
[119] Peres JF, Newberg AB, Mercante JP, Simao M, Albuquerque VE, Peres MJ et al. Cerebral blood flow changes during retrieval of traumatic memories before and after psychotherapy: a SPECT study. *Psychological Medicine* 2007;37(10):1481-91.
[120] Benish SG, Imel ZE, Wampold BE. The relative efficacy of bona fide psychotherapies for treating post-traumatic stress disorder: a meta-analysis of direct comparisons. *Clinical Psychology Review* 2008;28(5):746-58.
[121] Bisson J, Andrew M. Psychological treatment of post-traumatic stress disorder (PTSD). *Cochrane database of systematic reviews (Online)* 2007(3):CD003388.
[122] Lanius RA, Bluhm R, Lanius U, Pain C. A review of neuroimaging studies in PTSD: heterogeneity of response to symptom provocation. *Journal of Psychiatric Research* 2006;40(8):709-29.
[123] McNally RJ. Progress and controversy in the study of posttraumatic stress disorder. *Annual Review of Psychology* 2003;54:229-52.
[124] Rosen GM, Lilienfeld SO. Posttraumatic stress disorder: an empirical evaluation of core assumptions. *Clinical Psychology Review* 2008;28(5):837-68.
[125] Dougherty DD, Rauch SL, Rosenbaum JF. Essentials of neuroimaging for clinical practice. Washington: *American Psychiatric Association;* 2004.

In: Neurobiology of Post-Traumatic Stress Disorder
Editors: L. Sher and A. Vilens, pp. 175-200
ISBN: 978-1-61470-593-2

Chapter XII

Developmental Traumatic Experiences, PTSD and Substance Abuse Vulnerability: The Neurobiological Link

Gilberto Gerra[1], Lorenzo Somaini[2], Amir Zaimovic[3], Maria Lidia Gerra[4], Icro Maremmani[5], Mario Amore[4] and Roberto Ciccocioppo[6]

[1]Health and Human Development Branch, United Nations Office on Drugs and Crime, Wien, Austria
[2]Addiction Treatment Centre, Local Health Unit, Biella, Italy
[3]Addiction Treatment Centre, Local Health Unit, Parma, Italy
[4]University of Parma, Parma, Italy
[5]University of Pisa, Pisa, Italy
[6]University of Camerino, Camerino, Italy

Abstract

The diagnosis of PTSD dates back to the 80s and is characterized by the observation that as a consequence of a series of traumatic events, it is possible to develop a psychiatric syndrome. Recent epidemiological data show a correlation between stress-related disorders, as in the case of PTSD, and substance abuse disorders. A large majority of drug dependent patients have been reported experiencing one or more PTSD criterion of traumatic experiences. For a better understanding of this comorbidity, it is important to consider shared neurobiological risk factors and relevant stressful life events as possible vulnerability conditions to both PTSD and substance use disorders. The main reason given by subjects with PTSD for the abuse of psychotropic drugs, including illicit drugs, prescription drug and alcohol seems to be obtaining temporary relief and improve impulse control. Yet, the onset of drugs use may also anticipate the development of PTSD syndrome, determining an increased individual susceptibility to PTSD. Common neurotransmitter abnormalities i.e. low plasma levels of serotonin, elevated plasma levels of catecholamines, a similar derangement of the HPA axis, a generalized over-activity of the sympathetic nervous system and morphological and functional alterations of the central nervous system have been reviewed in this chapter.

PTSD and Substance Abuse Epidemiology

Epidemiological data show frequent associations between stress-related disorder, such as post-traumatic stress disorder (PTSD), and substance use disorders. We, therefore, find it important to review the possible neurobiological changes underlying this comorbidity.

A large majority of drug-dependent patients have been reported to experience one or more PTSD criteria for traumatic experiences. More than one third were found to meet criteria for current PTSD and half of them for lifetime PTSD. The assessment of trauma and PTSD in substance misusers and, conversely, the assessment of substance abuse in victims of PTSD appear to be essential in clinical practice [1].

On this basis, other studies that have examined the association between traumatic exposure, PTSD and substance use in a large sample of out-of-treatment Afro-American drug users have reported that 40% of the respondents experienced a traumatic event, with almost half of these developing PTSD. In this sample of subjects, early onsets of marijuana and heroin use, alcohol dependence and opiate dependence were each associated with exposure to a traumatic event for males, while early onset alcohol use and alcohol dependence were associated with females' exposure to a traumatic event [2].

Again, very recently, a history of being abused has been reported to increase the likelihood that a woman will abuse alcohol and other drugs [3]. Some specificity with respect to the type of substance abuse has been also documented, as the prevalence of PTSD proves to be higher in drug than in alcohol dependence. Additionally, PTSD seems to be an independent risk factor for an unfavourable outcome of substance use disorders [4]. In this connection, other findings suggest that it is not PTSD per se, but the frequent co-occurrence of PTSD and associated psychopathologies that largely accounts for the greater problem severity of substance use disorders in patients with PTSD [5].

Lastly, post-traumatic stress disorder is frequently associated with other psychiatric diagnoses, of which substance use disorder one of the most prevalent. Comorbid disorders in general, and substance use disorders (SUD) in particular, frequently complicate the course and outcome of PTSD, and *vice versa*. For a better understanding of this comorbidity, so that new treatment can be developed, it is important to explore the relevance of stressful life events as a factor in vulnerability to PTSD as well as to substance use disorders and the neurobiological basis of the association of these two types of disorder [6].

Adverse childhood experiences have been demonstrated to play a crucial negative role in development: in particular, impaired parental bonding seems to be a major risk factor for mental diseases, substance abuse and dependence later in life [7, 8, 9, 10]. Accordingly, all types of childhood maltreatment, including emotional abuse, have been reported as serious risk factors for substance use during adolescence [11] and emotional neglect was found to be associated with a greater severity of substance abuse, particularly among female subjects [12]. The individuals affected by substance use disorders have been previously reported to perceive high maternal and paternal control and low maternal care, a pattern characteristic of an "affectionless control" rearing style [13], that was also reported in the history of pathological gamblers [14], depressed subjects [15], alcoholics [16], young offenders [17] and, in general, psychiatric disorders [18]. In a retrospective evaluation, disturbed parenting was frequently found in the history of narcotic addicts [19], with the perception of parents as cold, indifferent, controlling and intrusive [20]. Psychometric investigation in a large sample of high school students showed an inverse correlation between parental care and aggressiveness/social maladaptation scores, in turn related to experimenting with illicit drugs [21].

Similarly, the lack of adequate parental care was very recently identified as a consistent predictor of adult affective disorders [22] and a prospective investigation showed that child abuse and neglect were associated with an increased risk of major depression [23], again suggesting that disruption in early rearing conditions may significantly contribute to mood problems, negative affects and behavioural under-control, as possible risk factors in substance abuse susceptibility. As previously noted by Roy in cocaine addicts [24], adverse childhood experiences, particularly poor child-parent relationships, appear to negatively influence personality development, possibly contributing to a stable dysfunction of brain monoamines, with an inverse correlation between emotional neglect during infancy and CSF metabolites of serotonin and dopamine in adults. Nevertheless, hypothalamic-pituitary-adrenal axis and autonomic nervous system hyper-reactivity, which were observed in the adults affected by affective and addictive disorders [25, 26, 27, 28, 29], could be a persistent consequence of childhood abuse and neglect; these may contribute to a diathesis for adulthood psychopathological conditions [30, 31].

A variety of clinical studies have evaluated the long-term effects of early developmental stress, such as child abuse and neglect or parental loss. Severe early stress and maltreatment have been found to induce a neurobiological cascade of events that have the potential to alter brain development and to increase the risk of developing depression, symptoms of attention-deficit/hyperactivity, borderline personality disorder, dissociative identity disorder and substance abuse [32]. In particular, compared to levels in subjects who had experienced neither loss nor a separation, parental loss by death was found to be associated with higher cortisol throughout the day, suggesting that affective deprivation induced by bereavement during childhood may have lasting effects on the HPA axis, even in the absence of psychopathology [33]. Accordingly, infants who experienced frequent emotional withdrawal by their mothers (either as a result of maternal depression, or their mother's strategic use of withdrawal as a control tactic) showed high baseline levels of cortisol, with sensitization to later stress, cognitive deficits, and social-emotional problems [34].

Taking into account all this evidence, a possible neurobiological derangement, involving both catecholaminergic (i.e., dopamine, noradrenaline, serotonin) function and the hypothalamus-pituitary-adrenal (HPA) axis function, may be hypothesized to underlie substance dependence and depression vulnerability in humans [35, 36], possibly as a consequence of chronic exposure to stress during childhood. For example, lower basal levels of homovanillic acid (HVA), and higher basal levels of prolactin (PRL), adrenocorticotropic hormone (ACTH) and cortisol with respect to controls has been reported in addicted individuals. In particular, these neurobiological changes, revealing a possible derangement of the dopamine system and stress response axis, have been found to be associated with a childhood history of adverse experiences and a low perception of parental care among cocaine addicts and heroin addicts [37, 38]. Childhood experiences of neglect and poor parent-child attachment may be responsible for a complex neurobiological dysregulation of the HPA axis and of the dopamine systems, which are known to play a crucial role in addictive and affective disorder susceptibility.

Link between PTSD and Substance Abuse: Possible Mechanisms

A complex relationship between PTSD and substance abuse exists and a few hypotheses could be suggested to explain the neurobiological link between these two psychiatric disorders. One hypothesis is that PTSD and SUD may simply share common vulnerability conditions affecting the brain and creating a predisposition to psychiatric and behavioural disorders. In this case, genetic factors and environmental factors, which both influences brain function and affect behavioural and personality traits, could make individuals simultaneously more sensitive to stress or trauma and more susceptible to substance abuse through parallel processes. The hypothesis that exposure to traumatic events *per se* increases the risk for substance use disorders has not been demonstrated. A modestly higher risk for nicotine dependence might be an exception. Post-traumatic stress disorder might be a causal risk factor for nicotine and drug use disorders or, alternatively, the co-occurrence of PTSD and these disorders might be influenced by shared risk factors other than traumatic exposure [39].

A second possibility is that traumatic events and stressful developmental conditions might be able to induce stable neurobiological changes which in their turn could contribute not only to PTSD symptoms or stress-related behavioural disorders, but also to substance abuse vulnerability. Neurotransmitter pathway derangement triggered by stress might also be responsible for trauma-related disorders and simultaneously for a proneness to experiment with illicit drugs and abuse alcohol, and for the continuous use and development of dependence. In this case, traumatic or stressful experiences appear to play a crucial role with respect to pre-existing vulnerability by inducing a risk-prone neurobiological pattern. In this connection, increased exposure to adverse experiences in childhood was found to be related to high-risk behaviours, the diagnosis of a substance use disorder, exposure to trauma in adulthood and a diagnosis of posttraumatic stress disorder [40].

Alternatively, a third hypothesis linking PTSD and SUD is based on the possibility that substances are used or abused in an attempt to self-medicate against the negative symptoms of PTSD. PTSD patients are affected by anxiety, impulsiveness related to increased stress reactivity, reduction of memory performance, high emotionality in response to trauma-related stimuli and over-representation of traumatic memory, particularly with conditioned fear responses or retrieval of traumatic memories [41, 42]. Psychotropic substances, including illicit drugs, prescription drugs and alcohol, could be used by PTSD patients to cope with stress, obtain temporary relief and improve impulse control.

Most of the data published so far support a pathway whereby PTSD precedes substance abuse or dependence. Substances are initially used to modify PTSD symptoms. With the development of dependence, physiological arousal resulting from substance withdrawal may exacerbate PTSD symptoms, so contributing to a relapse into substance use. Preclinical work has led to the proposal that in PTSD, corticotropin-releasing hormone and noradrenergic systems may interact, so that the stress response is progressively raised. Patients may use sedatives, hypnotics, or alcohol in an effort to interrupt this progressive increase [43]. This view has been supported by other authors, who suggest that the avoidance of trauma reminders and associated distress may be achieved by using drugs and alcohol. Alternatively, a substance-abusing lifestyle might predispose such individuals to experience traumatic events [1].

Accordingly, the onset of drug use and/or alcohol abuse may also precede the development of the entire PTSD syndrome in the clinical history, possibly contributing to an increase in individual susceptibility to PTSD. In this case, a history of adverse and stressful experiences could have primarily induced substance abuse, while the neurobiological derangement induced by stress and psychotropic substances would be responsible for the development of PTSD.

Finally, an undefined area of the interactions between stress, traumatic experiences and drug use should be considered in a large proportion of these cases, without any possibility of recognizing a well-defined sequence in the clinical history and related neurobiological changes. In these cases, many patients do not fulfil the entire set of criteria for DSM PTSD, although they show clear evidence of stressful and traumatic implications in substance abuse pathogenesis. The concomitant symptoms of PTSD among drug users significantly influence the clinical picture and the outcome. Young people with current SUD-PTSD have been found to report significantly higher levels of substance use in negative situations, as well as emotion-focused coping, compared to the current SUD-only group. Severe PTSD was a significant predictor of negative situational drug use, and emotion-focused coping was found to mediate this relationship [44].

The Neurobiology of Developmental Traumatic Experiences and Risk in PTSD/ Substance Use Disorders

Hypothalamus-Pituitary-Adrenal Axis

Early childhood traumatic experiences, such as childhood maltreatment, have been found to be associated with an enhanced risk of adolescent and adult alcohol and substance use disorders (defined as DSM-IV alcohol or substance abuse or dependence). Maltreated children and adolescents show a dysregulation of the stress response systems and suffer from altered brain development, which may lead to an enhanced vulnerability to psychopathology, particularly post-traumatic stress disorder (PTSD) and depression. These negative affect disorders may, in turn, increase the risk that an adolescent or a young adult will develop alcohol or substance use disorders [45].

Stress-related hypothalamic-pituitary-adrenal dysregulation seems to be associated with the development of substance use disorders. On one hand, pre-existing hyperactivity of HPA axis and higher cortisol seems to increase the risk of developing substance use disorder and depression-related illness in adolescents [46]. Seemingly, studies on females with MDD frequently report a hyperactivity of the HPA axis in this patient cohort. On the other, evidence from a variety of HPA axis dysregulations in PTSD suggests that gene variants, timing of the stressful experience as well as the type of the trauma may influence the outcome [47]. For example not only hyperactivity of the HPA axis system but also low baseline cortisol levels have been associated with PTSD. It is unclear whether it plays an etiological role or is an epiphenomenon of some other process. It is possible that reduced baseline cortisol reflects the net result of input to the hypothalamus from cortical and sub-cortical regions of the brain linked with increased vigilance, sensitization to trauma because of prior traumatic experiences, or genetic factors [48]. Exposure to a long-lasting/persistent stimulation of the HPA axis could have induced a sensitization of the system devoted to stress response in vulnerable individuals. Studies using low-dose dexamethasone suppression and metapyrone testing, two pharmacological agents that alter the availability of stress hormones by exerting feedback on the HPA axis, revealed that hypocortisolism in PTSD occurs in the context of heightened sensitivity of the HPA axis to negative glucocorticoid feedback [6]. Findings of increased glucocorticoid receptor binding and function support the assumption of increased negative feedback sensitivity of the HPA axis in PTSD [6]. At the level of central nervous system (CNS), marked and sustained increases in CRF concentrations have been measured in the cerebrospinal fluid (CSF) of patients with PTSD [8, 9]. Evidence of blunted ACTH responses to CRF stimulation in PTSD supports the hypothesis that PTSD involves high levels of hypothalamic CRF activity and corresponding down-regulating of pituitary CRF receptors [6]. In addition, reduced volume of the hippocampus, the major brain region inhibiting the HPA axis, is a cardinal feature of PTSD [10]. Taken together, the specific constellation of neuroendocrine findings in PTSD reflects sensitization of the HPA axis to exposure to stressors [6, 42].

HPA function dysregulation could represent the most important neurobiological link between developmental traumatic experiences (or stress-related disorders such as PTSD) and substance use disorders. A significant body of evidence, in fact, underlines the involvement of HPA function in vulnerability to substance abuse. The hypothalamo-pituitary-adrenal axis has been found to be involved in several aspects of cocaine self-administration. Corticosterone seems to be crucial to the acquisition of cocaine use since its self-administration is associated to high levels of hormones that are found in stress. High circulating levels of corticosterone also raise sensitivity to low doses of cocaine, possibly due to a sensitization-associated phenomenon, suggesting that exposure to stress may increase individual vulnerability to cocaine [49]. In addition, HPA function seems to be involved in craving intensity [50, 51] and in stress-induced relapses [52, 53].

HPA axis abnormalities seem to contribute to the development of alcohol use disorders and may simultaneously be a result of alcohol misuse. For example, non-physiological interaction between the HPA mechanisms and serotonergic systems seems to contribute to suicidal behaviour and vulnerability to substance abuse in adolescents [54]. Besides the HPA axis function, the serotonin function, in its turn, may be modulated by stressful stimuli, such as adverse childhood experiences and maternal deprivation [33]. Analogously with PTSD, as shown in studies on the offspring of alcohol- and drug-abusing parents, interactions between pre-existing alterations in frontal-limbic and HPA functions reflect vulnerability to addiction [55]. Associations between altered HPA-axis activity and family history and alcohol dependence have been documented [56].

Some evidence was found for HPA axis hypoactivity at awakening in adolescents with an early onset of cannabis use compared to late onset users, possibly indicating an increased risk that early onset users will seek stimulation to restore arousal levels by using substances [57].

HPA sensitization to stressful factors in the history of these individuals could be responsible for their low cortisol levels, as previously reported in PTSD.

CRF, which is over-secreted in subjects exposed to trauma and adverse childhood experiences, appears to be activated during the stress-induced reinstatement of drug-taking as well as in acute withdrawal from all major drugs of abuse. In addicted individuals the CRF system has been also hypothesized to contribute to allostatic changes leading to increased vulnerability to relapse during prolonged abstinence [58, 59, 60]. Analogously with PTSD, escalation in drug intake, and during withdrawal produces an activation of the brain extrahypothalamic CRF system of the extended amygdala [61].

Genetic variants affecting the affinity of the glucocorticoid receptor (low-affinity) have been demonstrated to modify HPA axis responsiveness at several levels. As a result of these genetic variants, HPA axis reactivity will be changed, exposing the brain to suboptimal cortisol levels during challenges, affecting individual stress responsiveness and coping style [62] and, possibly, becoming involved in susceptibility both to substance use disorders and trauma-related disorders.

Impaired HPA response to stressful stimuli, such as experimentally induced aggressiveness, has been repeatedly reported in abstinent heroin dependent individuals [63], in subjects with 3,4-methylenedioxy-methamphetamine ("Ecstasy") use history [64], in heroin addicts under methadone medication [65] and in adolescents at risk for substance abuse [24] suggesting that reduced ability to cope with stress could be attributable to a pre-morbid psychobiological condition, in association with increased sympathetic arousal.

Exposure to stress and alcohol cues have both been demonstrated to produce a significantly enhanced and persistent craving. Blunted stress-induced cortisol responses were observed in the alcohol-dependent subjects compared to the social drinker group, demonstrating that stress and cue exposure are able to induce a persistent negative emotion-related alcohol craving state in abstinent alcoholics accompanied by dysregulated HPA and physiological arousal responses [66].

The deleterious effects of early life stress, child maltreatment, and accumulated adversity on the brain CRF system and the hypothalamic-pituitary-adrenal axis function are probably involved in the pathogenesis of both SUD and PTSD. On the other hand, adaptation of the noradrenergic, mesolimbic dopamine, glutamate, and gamma-amino-butyric acid (GABA) pathways have also been associated with a rise in the stress-related risks of addiction. Furthermore, the effects of regular and chronic drug use on alterations in these stress and motivational systems may impact stress regulation, impulse control, and the perpetuation of compulsive drug seeking and relapse susceptibility [67].

Nor-epinephrine and Dopamine changes

Autonomic nervous system hyper-reactivity, presumably due to CRF hypersecretion, is a persistent consequence of childhood abuse that may contribute to the diathesis for psychopathological conditions in adults [31]. A sustained hyperactivity of the sympathetic branch of the autonomic nervous system, as shown by heart rate, blood pressure, skin conductance level, and other psycho-physiological parameters has been described in PTSD patients. Accordingly, increased urinary excretion of norepinephrine and epinephrine, and their metabolites, has been documented in combat veterans, abused women, and children with

PTSD. In addition, patients with PTSD exhibit increased heart rate, blood pressure, and NE responses to challenge, such as traumatic reminders. Decreased platelet α2 receptor binding further suggests NE hyperactivity in PTSD [68, 69]. Serial sampling revealed sustained increases in CSF NE concentrations and increased CSF NE responses to psychological stressors in PTSD [70, 71]. Taking a wider view, increased CNS NE (re)activity plausibly contributes to features of PTSD, including hyperarousal, increased startle, and encoded fear memories [69].

Interestingly, prospective studies have shown that increased heart rate and peripheral adrenaline excretion at the time of exposure to trauma predict the subsequent development of PTSD [72, 73], again suggesting that changes in the sympathetic system could favour trauma or chronic stress exposure and contribute to a condition of vulnerability.

The involvement of the noradrenergic afferents of the extended amygdala in anxiety, stress-induced relapse and altered reward processing reflects a common function in the stress modulation of drug-seeking [74], attributing to the sympathetic system, which could have been dysregulated by adverse experiences, a primary role among the biological correlates of addictive behaviour. Persistent neurobiological changes, including CRF and NE hyper-activation in the extended amygdala, form a powerful motivational background for a relapse into drug and alcohol dependence. Due to CRF and NE hyperactivity, the amygdala could well play a key role not only in mediating emotional memories and driving drug-seeking motivation, but also in providing a basis for withdrawal-associated negative reinforcement, so facilitating the progressive involvement in the addiction spiral [75]. In particular, in this context the function of hyper-secretion of NE seems to represent a strong link between PTSD and SUD.

With many similarities respect to PTSD, a condition of hyperarousal and increased responses in cathecolamines to stressful conditions have been found among individuals affected by substance use disorders [64, 65, 76] and healthy adolescents with an aggressive temperament [77]. Accordingly, anxious adolescents showed increased levels of NE in response to experimental psychological stress [78]. A common neurobiological pattern of alterations to sympathetic reactivity in the developmental age group could underlie a sensitivity to trauma and vulnerability to substance abuse.

Starting in the late 1960s and early 1970s, the NE system was identified as a key element in the regulation of both natural and drug-induced rewards. Pre-clinical studies have indicated that NE is critically involved in mediating behavioural sensitization, drug discrimination and reinstatement of the search for drugs [79]. NE signalling from the nucleus of the tractus solitarius to the extended amygdala, along with increased CRF transmission, have been reported to be critical for the expression of acute opiate withdrawal as well as stress-induced relapses into drug-seeking for opiates, cocaine, ethanol, and nicotine. NE and CRF transmission in the extended amygdala are also implicated in the increased anxiety that occurs during prolonged abstinence from chronic opiates, cocaine, ethanol and cannabinoids [74].

Genetically defined levels of catecholamines has been reported to be related with the risk of personality disorders and substance abuse, possibly concurring also to PTSD vulnerability. High levels of catecholamines related to low-activity Mono-Amino-Oxidasis (MAO) gene variant appear to be associated with the most antisocial and severe forms of addiction to drugs [80].

Dopamine

A large body of evidence indicates that DA plays a critical role in reward processing, but DA has also been implicated in fear conditioning. Moreover, evidence in humans indicates that exposure to stressors increases mesolimbic DA release, which, in its turn, may have an impact on HPA axis responses. Dysregulation of the HPA axis and and alteration in mesocorticolimbic dopamine neurotransmission have also been described in children exposed to adverse experiences or stressful life events or those reconding with particularly neglected parental care and attention [81], This makes it predictable that dysregulation of these systems plays a role both in addictive/affective disorders and in PTSD. In line with this view an increase in urinary excretion of DA and DA metabolites in patients with PTSD has been reported. Moreover, genetic variations in the DA system have been implicated in moderating risk for PTSD [42].

Continuous dopamine system over stimulation due to chronic stress could significantly affect the mesolimbic pathway, inducing an exhausted/impaired response to natural rewards and a sensitization to stressful experiences. Recent studies have shown low levels of homovanillic acid (HVA), the dopamine metabolite, in addicted individuals and demonstrated an inverse relationship between HVA levels and markers of adverse childhood experiences such as neglect [37, 82].

Although drugs of abuse possess various different neuropharmacological profiles, activation of the mesocorticolimbic system, particularly the ventral tegmental area, nucleus accumbens, amygdala and prefrontal cortex dopaminergic pathways, constitutes a shared mechanism by which various drugs of abuse mediate their acute reinforcing effects [83, 84]. An impaired dopaminergic system against a background of developmental traumatic experiences would constitute a substantive risk condition for susceptibility to substance use disorders.

Some time ago, the propensity of individual rats to develop drug intake was reviewed in relation to the dopaminergic hyperactivity that characterizes drug-prone animals [85]; stress-induced increases in dopamine concentrations in the nucleus accumbens were found to be associated with amphetamine self-administration [86]. Accordingly, in human studies, adolescents have been reported to be more sensitive to nicotine reward, because of their overdeveloped excitatory glutamatergic system that facilitates dopamine, and their underdeveloped GABAergic systems that inhibit dopamine [87].

The plasticity of the dopamine system in responding to stressful and adverse experiences has been demonstrated by studies on concentrations of dopamine D2 receptors. Experimental animals individually housed and raised in isolation showed falls in concentrations of D2 receptors, which only recovered in individuals with a dominant social status [88]. Submissive animals that remained with a low concentration of D2 receptors showed a significantly higher proneness to cocaine self-administration. In this case too, a history of adverse stressful events and problematic social relationships appears to affect the dopamine system and influence mood and behavioural reactions.

Serotonin

Indirect evidence suggests a role for 5-HT in the pathophysiology of PTSD, including symptoms of impulsiveness, hostility, aggression, depression, and suicidal tendencies. An important link between 5-HT neurotransmission and PTSD is provided by data showing the therapeutic efficacy of the selective serotonin reuptake inhibitors (SSRI). In addition, biochemical data demonstrate that PTSD patients have altered 5-HT serum levels, a low density of platelet 5-HT uptake sites and an altered responsiveness to CNS serotoninergic challenge [68, 89]. Taken together, these data indicate that altered 5-HT transmission may contribute to symptoms of PTSD such as hyper vigilance, a tendency to be easily startled, impulsiveness, and intrusive memories. A significant body of evidence associates low serotoninergic function with impulsiveness, aggression (including extreme violence), and depression [90, 91], indicating that the impairment of the serotonin system could be involved in a variety of psychiatric disorders.

Impulse control inabilities, abnormal learning processing and impaired coping with stress seem to be related to a serotonin deficit in addicted individuals too. For example, Impaired responses to d-fenfluramine challenges, indicative of a low level of reactivity of the serotonin system, have previously been reported in heroin addicts [92], particularly those characterized by a novelty-seeking temperament [93] and mood disorders [94].

Accordingly, diminished central serotonergic responsiveness to serotoninergic agonists, has been described in adults with an impulsive personality or alcohol use disorders. As a result, dysregulation of the central serotonergic function seems to contribute not only to impulsive-aggressive behaviour, but also to alcohol use disorders [95].

Recently, gene variants affecting the serotonin transporter and modulating serotonin system function have been reassociated with drug abuse vulnerability, novelty-seeking, aggressiveness and poor school achievement. The short-short (SS) genotype frequency (serotonin transporter promoter polymorphism: 5-HTTLPR) was significantly higher among users compared with abstinent subjects. When evaluated over the entire sample, mean total scores for aggressiveness, novelty-seeking scores and school failure frequency were significantly higher in SS than in LL subjects [96]. Accordingly, low level of expression of the gene encoding the 5-HTT transporter, due to "S" promoter polymorphism, has been found to be associated with smoking behaviour among adolescents, together with an increased risk of developing nicotine dependence [97], with heroin dependence, particularly in the subgroup of antisocial heroin addicts [98], and with alcoholism [99].

In addition to the risk conditions induced by genetic variants - stress in general, and adverse early life experiences in particular - have been considered to have an impact on the development of the serotonergic system, possibly inducing a dysregulation of this inhibitory mechanism in the brain [100]. Monkeys that suffered from maternal deprivation in childhood have been reported to be fearful, more aggressive, less exploratory, and subject to binge drinking when exposed to alcohol, with a significant reduction of the serotonin transporter concentrations that are related to stressful early experiences [101].

Taken together, these findings suggest that, with reference to serotonin dysfunctions too, a possible vulnerability pattern may be shared by SUD and PTSD, related to individual psychobiological traits and personality characteristics already evident at a developmental age.

GABA/Benzodiazepine Receptor System

Alterations in the gamma-aminobutyric acid (GABA) receptor complex and GABA neurotransmission have been reported to influence the reinforcing effects of alcohol, drugs of abuse and benzodiazepines. Chronic modulation of the GABA(A)-benzodiazepine receptor complex seems to play a major role in central nervous system dysregulation during alcohol abstinence. In addition, the GABA(B) receptor agonist baclofen has shown promise in relapse prevention in a number of substance dependence disorders [102].

Accordingly, patients with PTSD exhibit low numbers of platelet BZ binding sites [30]. Single photon emission computed tomography and PET imaging studies revealed decreased BZ receptor binding in the cortex, hippocampus and thalamus of patients with PTSD. A low capability of the inhibitory GABAergic system could be the basis for an increased sensitivity to trauma and stressful events, adrenergic hyper-arousal and vulnerability in SUD/PTSD subjects.

Glutamate/NMDA Receptor System

The glutamate/NMDA receptor system has been implicated in synaptic plasticity, learning, and memory, including the well-studied phenomenon of long-term potentiation, the extended excitation of neural circuits, leading to long-lasting enhancement in the communication between neurons. These processes appear to be essential both in the consolidation of trauma memories in PTSD [103] and in the conditioning mechanism in addictive behaviour [67, 104]. In addition to its role in learning and memory, overexposure to glutamate is associated with excitotoxicity, and, plausibly, could contribute to a loss of neurons in the hippocampus and pre-frontal cortex in PTSD.

In experimental animals both prenatal and postnatal environmental factors have been demonstrated to affect the concentration of glutamate. Pups exposed to gestational stress and control pups raised by a foster mother, whether gestationally stressed or not, had higher levels of dopamine and NMDA receptors in cortical areas. These findings suggest that the normal expression of DA and glutamate receptors is influenced by in utero experiences and by lactation. Particularly for cortical D [2] receptors and NMDA receptors in the cerebral cortex and the nucleus accumbens, early stressful conditions such as adoption could be crucial in influencing brain functions [105].

Repeated maternal deprivation has been found associated with alterations in the NMDA channel activity in the caudate-putamen of adult rats, and may be responsible for addictive behaviour vulnerability [106]. In this case too, the NMDA system appears to be modulated by early adverse experiences and social deprivation, confirming that brain pathway plasticity may underlie trauma-related disorders and SUD.

The glutamatergic system plays a crucial role in alcohol addiction, by offering targets for the action of ethanol via its antagonism to the N-methyl-D-aspartate (NMDA) subtype of the glutamate receptor. The modulation of the glutamatergic function by ethanol contributes both to the euphoric and the dysphoric consequences of ethanol intoxication. Adaptations within glutamatergic systems appear to contribute to ethanol tolerance and dependence and to both acute and protracted ethanol withdrawal. Alterations in the NMDA receptor function have been reported to contribute to subjective responses to ethanol and, therefore, to the risk of developing alcoholism, too [107]; they have been hypothesized in subjects with a family history of alcohol dependence and comorbid depression [108]. In this connection, a shared condition of NMDA vulnerability that predisposes to conditioned emotional reactions and sensitization to stress, as well as glutamate system changes induced by adverse childhood experiences, could help to explain the high rate of alcohol abuse comorbidity in PTSD.

Endogenous Opioids

The available data suggest that a dysfunction of the opioid reward system is significantly linked to opiate abuse vulnerability and that heroin use alters mesolimbic dopamine neurotransmission tone in the mesolimbic proenkephalin function [109].

Alterations in endogenous opioids have been postulated to be involved in symptoms of numbing, stress-induced analgesia, and dissociation in PTSD. Endogenous opioids exert inhibitory influences on the HPA axis, whereas naloxone, a mu opiate receptor antagonist, increases HPA axis activity by blocking an inhibitory opioidergic influence on hypothalamic CRF secretion. Interestingly, patients with PTSD have been reported to exhibit an exaggerated HPA axis response to naloxone. Naloxone has also been shown to reverse the analgesia of PTSD patients after exposure to traumatic reminders. Lastly, PTSD patients exhibit increased CSF β-endorphin levels, suggesting an increased activation of the endogenous opioid system. Interestingly, the opiate receptor antagonist, naltrexone, has been reported to be effective in treating symptoms of dissociation and flashbacks in traumatized patients.[42, 69, 110].

The reward pathways for different addictive drugs converge to form a common pathway in which beta-endorphin is a modulating element. Beyond the role of the endogenous opioid system in the regulation of stress responses, beta-endorphins have been shown to play a major role in regulating drug abuse reward, too. This may occur via interactions with the mesolimbic dopaminergic system and also by effects that have an impact on learning and memory processing [111, 112].

Considering the contribution of the endogenous opioid systems to the regulation of these functions, one could speculate that disinhibition, alterations in their receptor activities (i.e., μ- or δ-opioid systems) might contribute to impulse control disorders [113] and could heighten the derangement in neuromodulation that underlies susceptibility both to substance abuse and to PTSD.

Structural and Functional Neuroanatomy

Strong similarities appear to characterize SUD and PTSD in terms of brain morphology and functional neuroanatomy. Changes in brain structure and function have been identified in patients with PTSD using brain imaging methods [114, 115, 116]. Brain regions that are altered in patients with PTSD include the hippocampus, amygdala, and prefrontal cortical regions, including the anterior cingulate and orbitofrontal cortex [114]. Accordingly, these areas have been found to be involved in addictive disorders, with stable changes in reward perception, decision-making processes, motivational patterns and abnormal associative learning [61, 117]. As reported for neuroendocrine changes, it is difficult to establish if continuous stressful stimuli or exposure to drugs are responsible for these changes or, alternatively, vulnerability to trauma and psychotropic agents is due to pre-existing structural and functional changes in the brain. Recent evidence demonstrated exposure-related structural abnormalities in the hippocampus and amygdala in long-term heavy cannabis users [118], suggesting that drug use could have contributed to the vulnerability of PTSD neurobiological correlates.

Hippocampal volume reduction in PTSD may reflect the toxic effects over time of repeatedly increased glucocorticoid exposure or heightened glucocorticoid sensitivity, though recent evidence also suggests that a small hippocampus might represent a pre-existing vulnerability factor for developing PTSD [119]. Accordingly, exposure to early abuse and early stress has is both cases been associated with the emergence of epileptiform electroencephalogram (EEG) abnormalities, alterations in the corpus callosum area, and reduced volume or synaptic density of the hippocampus [120]. The development of reward-based learning and decision-making, and of the neural circuitry underlying these processes, appears to be negatively influenced by adverse child-rearing environments characterized by abuse and other forms of maltreatment. Adolescents whose mothers displayed more punishing responses to their positive affective behaviour during experimental tasks showed significant changes in the left dorsal anterior cingulate cortex, bilateral orbito-frontal cortex and amygdala volumes [121].

Alternatively, previous dysfunctions in these brain areas could be attributable to individual vulnerability, as a pre-existing condition with respect to both trauma-stress and drug exposure. Emerging data on the neural mechanisms of impulse control highlight brain regions involved in emotion and decision-making in young subjects, including the ventromedial prefrontal cortex, anterior cingulate cortex and amygdala [122]. Attention-related dysfunctions in the ventrolateral prefrontal cortex were seen in ADHD patients, and reward-related dysfunctions in the orbitofrontal cortex were seen in conduct disorder patients. Together with the pattern of paralimbic dysfunctions demonstrated in children with conduct disorder during sustained attention, these findings support theories of abnormalities in orbitofrontal-paralimbic motivation networks in individuals with conduct disorder and ventrolateral fronto-cerebellar attention network dysfunction in individuals with ADHD [123].

The functional role of the amygdala in mediating both stress responses and emotional learning suggests that changes in this region and its connected circuitry may be implicated in the pathophysiology of PTSD. Although there is no clear evidence for structural alterations of the amygdala in PTSD, functional imaging studies have revealed hyper-responsiveness of the amygdala in PTSD during the presentation of traumatic scripts, cues and other reminders [124, 116]. PTSD patients also show increased amygdala responses to general emotional stimuli that are not associated with the trauma, such as emotional faces [123]. It is striking that, the amygdala too seems to be sensitized to subliminally presented threatening cues in PTSD [42, 103, 125, 126].

Similarly, the brain reward system involved in the development of addiction includes key elements of the extended amygdala, with increased reactivity to drug-related stimuli [104], and also to subliminally presented cues [42, 103, 125, 126]. To increase the body of evidence pertinent to the link with PTSD, addicted patients have been reported to over-react not only to drug-related stimuli, but also to emotional and stressful images [127, 128]; these data suggest that drugs and traumatic experiences probably share the same brain pathways in inducing conditioned behaviours. Considering these findings, PTSD may be maintained by second-order conditioning where trauma-relevant cues come to serve as unconditioned stimuli, thus generalizing enhanced emotional responses to many previously neutral cues and impeding extinction [129], in full analogy with the associative learning without extinction established by drugs [130]. Extinction is a form of inhibitory learning that suppresses a previously conditioned response. Both fear and drug-seeking are conditioned responses that can lead to maladaptive behaviour when expressed inappropriately, manifesting as anxiety disorders and addiction, respectively. Recent evidence indicates that the medial prefrontal cortex (mPFC) is critical for the extinction of both fear and drug-seeking behaviours [131].

Still on this topic, patients with PTSD exhibit a low-volume volumes frontal cortex [132], including a low volume for the anterior cingulate cortex (ACC) [133, 134]. Functional imaging studies have found low levels of activation of the medial prefrontal cortex (mPFC) in PTSD patients responding to stimuli such as traumatic scripts [135, 136], combat pictures and sounds [137], trauma-unrelated negative narratives [138], fearful faces [139], emotional stroop [140], and others. Considering frontal cortex neuro-anatomy in this relationship to substance dependence, reduced medial OFC grey matter might reflect long-term adaptations within the reward-learning circuit that underlies pathological decision-making [141].

In the case of the frontal cortex too, early life stress has been demonstrated to affect the adult anterior cingulated cortex, leading to a significant reduction in its volume [142], and possibly playing a role in susceptibility to SUD and PTSD.

Genetic Factors

Gene variants have been demonstrated to significantly influence vulnerability to PTSD, besides contributing to the risk of experimenting with psychotropic substances and developing drug dependence [143].

Evidence from family and twin studies has long suggested an inheritable contribution in the development of PTSD. In addition, there is evidence for inheritable contributions to some of the neurobiological endophenotypes of PTSD, such as low hippocampal volume [144] or unusually high amygdala reactivity [145]. One study has linked a polymorphism in the DA transporter gene to PTSD risk. An excess of the SLC6A39 repeat allele was present in those with PTSD. This finding suggests that genetically determined changes in DA reactivity may contribute to PTSD among trauma survivors [146]. Several studies have investigated polymorphisms in the D2 receptor in relation to PTSD risks, though results have been inconsistent [147]. Lastly, there is considerable evidence linking a low expression variant of the serotonin transporter, the so-called "s" allele of the serotonin-transporter-linked polymorphic region (5-HTTLPR), to stress responsiveness and to the risk of developing depression as a result of life stress [42].

Similarly, the DA transporter gene (9 repeat variant) has been found to be involved in externalizing behaviour in children [148] at risk of SUD and addiction to drugs with antisocial comorbidity [149]; the D2 receptor gene (allele A1) has been widely studied in addiction to drugs and alcohol [150]; 5HT transporter polymorphism S (short) has been associated with addictive behaviour [98, 99] possibly affecting stress responses and anxiety/dysphoria, which are factors relevant to initial vulnerability, the process of addiction, and relapses [151].

Nevertheless, the hypothesis that adverse experiences in childhood, not only individual gene variants, may contribute to central monoamine dysfunctions in addicted individuals was previously put forward by Roy [24], who reported an inverse correlation between emotional neglect during infancy and CSF metabolites of serotonin and dopamine in the adult. In this connections, recent studies have focused on the possible gene-environment (or gene-stress/trauma) interactions in investigating the effects of stressful developmental experiences in individuals carrying specific gene variants.

A specific gene x environment interaction has been reported in borderline personality disorder patients, so suggesting a synergetic effect linking serious life events and the serotonin transporter gene (5-HTTLPR S/L) polymorphism in the development of impulsiveness [152], which could be a crucial element in both SUD and PTSD. Accordingly, bulimic 5HTTLPR S-allele carriers who were previously abused display elevations on dissocial behaviour levels, again corroborating previous observations on the phenomenological correlates of traumatic stress in 5HTTLPR S allele carriers [153].

Considering the pathogenesis of addictive disorders, adverse childhood experiences and genotype appear to combine in inducing vulnerability conditions and psychobiological characteristics. The association between 5-HT transporter polymorphism and psycho stimulant use has been found to be mediated by the mother-child relationship and parental attachment perception; both of these are active as intervene as environmental and genetic factors involved in proneness to substance use disorders, particularly in aggressive-antisocial individuals [82]. Similarly, as another example of gene-environment interaction in susceptibility to substance abuse, adolescents with the S allele of the 5-HTT gene and with family relationships assessed as "neutral" or "bad" displayed as risk of high alcohol intoxication frequency of as much as 12 to 14 times the normal value; this pattern was not attributed to genetic and environmental factors alone [154]. Besides this, in individuals carrying a dopamine receptor (D2) gene variant (allele A1), only patients in the subgroup with stressful life events in their clinical history were found to be at risk of alcoholism [155].

Reports speak of genotypes as raising or as lowering the risks induced by traumatic experiences. The high activity gene variant coding for MAO enzyme, which is able to induce low levels of norepinephrine, has been found to protect children from the behavioural and psychiatric consequences of adverse childhood experiences and maltreatment, including antisocial personality disorder and substance abuse [156].

More recently, reduced maternal care perception identified as a key intermediate factor in the association between serotonin transporter Short-Short (SS) polymorphism and drug use, suggesting that genetic factors and problematic parental behavior may co-determinate drug use susceptibility. During checks on “neglect” levels as assessed by the Child Experience of Care and Abuse Questionnaire (CECA-Q scores) the relative risk associated with the SS genotype shows sharply lower levels, suggesting a crucial role for traumatic childhood experiences in the proneness to experimenting with drug use [157].

Considering the link between SUD and PTSD neurobiology previously described, this type of gene-stress interaction, already observed in studies concerning addictive behaviour and related vulnerability, could be crucial for PTSD, too.

Thus, studies in developmental traumatology may prove to be critical in the effort to attempt to link the neurobiology of maltreatment-related PTSD with the neurobiology of alcohol and substance use disorders and in developing early strategies for the prevention of adolescent and adult alcohol and substance use disorders.

Conclusion

PTSD is frequently associated with other psychiatric disorders, in particular with Substance Use Disorder. Furthermore, SUD can complicate the clinical course of PTSD, which in turn influences the clinical course of SUD. These two diseases have deep and important interrelationships and few hypotheses can explain the neurobiological link that binds these psychiatric disorders. Of course, genetic and environmental factors may cause individuals to be more predisposed to stress and trauma more vulnerable to both the development of PTSD and SUD with a parallel process. In this sense also alterations both in specific neurotransmitters such as dopamine, serotonin, CRF and endogenous opioids, and in the autonomic nervous system, with high levels of catecholamines on the one hand, and deep derangement HPA axis and the morphological and structural changes over some areas of the central nervous system which can alter the functions of executive control and decision making on the other hand, seem to represent the common neurobiological substrate of the two diseases. Another evidence of this neurobiological link is also the observation of a higher incidence of developing SUD in those who suffer from PTSD, who often resort to the use of substances in order to obtain temporary relief and impulse control, and the presence of more or less overt forms of PTSD in those suffering from SUD. For a better understanding of this comorbidity and for the development of new treatments, it is important to explore the relevance of stressful life events as a vulnerability factor for both PTSD and SUD.

References

[1] Reynolds M, Mezey G, Chapman M, Wheeler M, Drummond C, Baldacchino A: Co-morbid post-traumatic stress disorder in a substance misusing clinical population. *Drug Alcohol Depend.* 2005; 77(3):251-8.

[2] Johnson SD, Striley C, Cottler LB: The association of substance use disorders with trauma exposure and PTSD among African American drug users. *Addict Behav.* 2006; 31(11):2063-73.

[3] Covington SS: Women and addiction: a trauma-informed approach. *J Psychoactive Drugs.* 2008; Suppl 5:377-85.

[4] Driessen M, Schulte S, Luedecke C, Schaefer I, Sutmann F, Ohlmeier M, et al.,. : TRAUMAB-Study Group. Trauma and PTSD in patients with alcohol, drug, or dual dependence: a multi-center study. *Alcohol Clin Exp Res.* 2008; 32(3):481-8.

[5] Cacciola JS, Koppenhaver JM, Alterman AI, McKay JR: Posttraumatic stress disorder and other psychopathology in substance abusing patients. *Drug Alcohol Depend.* 2009; 101(1-2):27-33.

[6] Martínez-Raga J, Didia-Attas J, Ruiz M, Knecht CC, Cepeda S, San Molina L, et al.,: Post-traumatic stress disorder and substance use disorder: epidemiology, nature and neurobiology. *Vertex.* 2005; 16(63):325-31.

[7] Canetti, L., Bachar, E., Galili-Weisstub, E., De-Nour, A.K., Shalev, A.Y: Parental bonding and mental health in adolescence. *Adolescence.* 1997: 32(126):381-394.

[8] Newcomb MD, Felix-Ortiz M: Multiple protective and risk factors for drug use and abuse: cross-sectional and prospective findings. *J Pers Soc Psychol.* 1992; 63(2):280-96.

[9] Petraitis J, Flay BR, Miller TQ: Reviewing theories of adolescent substance use: organizing pieces in the puzzle. *Psychol Bull.* 1995; 117(1):67-86.

[10] Brook JS, Nomura C, Cohen P: Prenatal, perinatal, and early childhood risk factors and drug involvement in adolescence. *Genet Soc Gen Psychol Monogr.* 1989; 115(2):221-41.

[11] Moran PB, Vuchinich S, Hall NK: Associations between types of maltreatment and substance use during adolescence. *Child Abuse Negl.* 2004; 28(5):565-74.

[12] Hyman SM, Garcia M, Sinha R: Gender specific associations between types of childhood maltreatment and the onset, escalation and severity of substance use in cocaine dependent adults. *Am J Drug Alcohol Abuse.* 2006; 32(4):655-64.

[13] Torresani S, Favaretto E, Zimmermann C: Parental representations in drug-dependent patients and their parents. Compr Psychiatry. 2000; 41(2):123-9.

[14] Grant, J.E., Kim, S.W: Parental bonding in pathological gambling disorder. *Psychiatr Q.* 2002: 73(3): 239-247.

[15] Narita T, Sato T, Hirano S., Gota, M, Sakado K, Uehara T: Parental child-rearing behavior as measured by the Parental Bonding Instrument in a Japanese population: factor structure and relationship to a lifetime history of depression. *J Affect Disord.* 2000: 57(1-3): 229-234

[16] Joyce PR, Sellman D, Wells E, Frampton CM, Bushnell JA, Oakley-Browne M, et al.: Parental bonding in men with alcohol disorders: a relationship with conduct disorder. *Aust N Z J Psychiatry.* 1994: 28(3): 405-411.

[17] Chambers JA, Power KG, Loucks N, Swanson V: Psychometric properties of the Parental Bonding Instrument and its association with psychological distress in a group of incarcerated young offenders in Scotland. *Soc Psychiatry Psychiatr Epidemiol.* 2000: 35(7): 318-25.

[18] Scinto A, Marinangeli MG, Kalyvoka A, Daneluzzo E, Rossi A: The use of the Italian version of the Parental Bonding Instrument (PBI) in a clinical sample and in a student group: an exploratory and confirmatory factor analysis study. *Epidemiol Psichiatr Soc.* 1999: 8(4): 276-283.

[19] Bernardi E, Jones M, Tennant C : Quality of parenting in alcoholics and narcotic addicts. *Br J Psychiatry.* 1989 ; 154: 677-682.

[20] Schweitzer RD, Lawton PA: Drug abusers perceptions of their parents. *Br J Addict.* 1989; 84(3):309-14.

[21] Gerra G, Angioni L, Zaimovic A, Moi G, Bussandri M, Bertacca S, et al.: Substance use among high-school students: relationships with temperament, personality traits, and parental care perception. *Subst Use Misuse.* 2004; 39(2):345-67.

[22] Lancaster G, Rollinson L, Hill J: The measurement of a major childhood risk for depression: comparison of the Parental Bonding Instrument (PBI) 'Parental Care' and the Childhood Experience of Care and Abuse (CECA) 'Parental Neglect'. *J Affect Disord.* 2007; 101(1-3):263-7.

[23] Widom CS, DuMont K, Czaja SJ: A prospective investigation of major depressive disorder and comorbidity in abused and neglected children grown up. *Arch Gen Psychiatry*. 2007; 64(1):49-56.
[24] Roy A: Self-rated childhood emotional neglect and CSF monoamine indices in abstinent cocaine-abusing adults: possible implications for suicidal behavior. *Psychiatry Res*. 2002; 112(1):69-75
[25] Gold PW, Loriaux DL, Roy A, Kling MA, Calabrese JR, Kellner CH: Responses to corticotropin-releasing hormone in the hypercortisolism of depression and Cushing's disease: pathophysiologic and diagnostic implications. *New England Journal of Medicine*. 1986; 314: 1329–1335.
[26] Holsboer F, Gerken A, von Bardeleben U, Grimm W, Beyer H, Muller OA et al.: Human corticotropin-releasing hormone in depression. *Biological Psychiatry* 1986; 21: 601–611.
[27] Holsboer F, Lauer CJ, Schreiber W, Krieg J-C: Altered hypothalamic–pituitary–adrenocortical regulation in healthy subjects at high familial risk for affective disorders. *Neuroendocrinology* 1995; 62 340–347.
[28] Krishnan KRR, Rayasam K, Reed DR, Smith M, Chappell P, Saunders WB: The corticotropin-releasing factor stimulation test in patients with major depression: relationship to dexamethasone suppression test results. *Depression* 1993; 1: 133–136.
[29] Sher L: The role of the hypothalamic-pituitary-adrenal axis dysfunction in the pathophysiology of alcohol misuse and suicidal behavior in adolescents. *Int J Adolesc Med Health*. 2007; 19(1):3-9.
[30] Arborelius L, Owens MJ, Plotsky PM, Nemeroff CB: The role of corticotropin-releasing factor in depression and anxiety disorders. *J Endocrinol*. 1999; 160(1):1-12
[31] Heim C, Newport DJ, Heit S, Graham YP, Wilcox M, Bonsall R: Pituitary-adrenal and autonomic responses to stress in women after sexual and physical abuse in childhood. *JAMA*. 2000; 284(5):592-7.
[32] Teicher MH, Andersen SL, Polcari A, Anderson CM, Navalta CP; Developmental neurobiology of childhood stress and trauma. *Psychiatr Clin North Am*. 2002; 25(2):397-426,
[33] Nicolson NA: Childhood parental loss and cortisol levels in adult men. *Psychoneuroendocrinology*. 2004; 29(8):1012-8.
[34] Bugental DB, Martorell GA, Barraza V: The hormonal costs of subtle forms of infant maltreatment. *Horm Behav*. 2003; 43(1):237-44.
[35] Oswald LM, Wong DF, McCaul M, Zhou Y, Kuwabara H, Choi L, et al.: Relationships among ventral striatal dopamine release, cortisol secretion, and subjective responses to amphetamine. *Neuropsychopharmacology*. 2005; 30(4):821-32
[36] Duval F, Mokrani MC, Monreal-Ortiz JA, Fattah S, Champeval C, Schulz P, et al.: Cortisol hypersecretion in unipolar major depression with melancholic and psychotic features: dopaminergic, noradrenergic and thyroid correlates. *Psychoneuroendocrinology* 2006; 31 (7): 876–888.
[37] Gerra G, Leonardi C, Cortese E, Zaimovic A, Dell'agnello G, Manfredini M, et al.: Childhood neglect and parental care perception in cocaine addicts: relation with psychiatric symptoms and biological correlates. *Neurosci Biobehav Rev*. 2009; 33(4):601-10.

[38] Gerra G, Leonardi C, Cortese E, Zaimovic A, Dell'Agnello G, Manfredini M, et al.: Adrenocorticotropic hormone and cortisol plasma levels directly correlate with childhood neglect and depression measures in addicted patients. *Addict Biol.* 2008; 13(1):95-104.
[39] Breslau N, Davis GC, Schultz LR: Posttraumatic stress disorder and the incidence of nicotine, alcohol, and other drug disorders in persons who have experienced trauma. *Arch Gen Psychiatry.* 2003; 60(3):289-94.
[40] Lu W, Mueser KT, Rosenberg SD, Jankowski MK: Correlates of adverse childhood experiences among adults with severe mood disorders. *Psychiatr Serv.* 2008; 59(9):1018-26.
[41] Vermetten E: Stress, trauma, and post-traumatic stress disorder.*Tijdschr Psychiatr.* 2009; 51(8):595-602.
[42] Heim C, Nemeroff CB: Neurobiology of posttraumatic stress disorder. *CNS Spectr.* 2009; 14(1 Suppl 1):13-24.
[43] Jacobsen LK, Southwick SM, Kosten TR: Substance use disorders in patients with posttraumatic stress disorder: a review of the literature. *Am J Psychiatry.* 2001; 158(8):1184-90.
[44] Staiger PK, Melville F, Hides L, Kambouropoulos N, Lubman DI: Can emotion-focused coping help explain the link between posttraumatic stress disorder severity and triggers for substance use in young adults? *J Subst Abuse Treat.* 2009; 36(2):220-6.
[45] De Bellis MD: Developmental traumatology: a contributory mechanism for alcohol and substance use disorders. *Psychoneuroendocrinology.* 2002; 27(1-2):155-70.
[46] Rao U, Hammen CL, Poland RE: Mechanisms underlying the comorbidity between depressive and addictive disorders in adolescents: interactions between stress and HPA activity. *Am J Psychiatry.* 2009; 166(3):361-9.
[47] Shea A, Walsh C, Macmillan H, Steiner M: Child maltreatment and HPA axis dysregulation: relationship to major depressive disorder and post traumatic stress disorder in females. *Psychoneuroendocrinology.* 2005; 30(2):162-78.
[48] Marshall RD, Garakani A: Psychobiology of the acute stress response and its relationship to the psychobiology of post-traumatic stress disorder. *Psychiatr Clin North Am. 2002*; 25(2):385-95.
[49] Goeders NE: Stress and cocaine addiction. J Pharmacol Exp Ther. 2002; 301(3):785-9.
[50] Sinha R, Catapano D, O'Malley S: Stress-induced craving and stress response in cocaine dependent individuals. *Psychopharmacology (Berl).* 1999; 142(4):343-51.
[51] Elman I, Lukas SE, Karlsgodt KH, Gasic GP, Breiter HC: Acute cortisol administration triggers craving in individuals with cocaine dependence. *Psychopharmacol Bull.* 2003; 37(3):84-9.
[52] Sinha R: Modeling stress and drug craving in the laboratory: implications for addiction treatment development. *Addict Biol.* 2009; 14(1):84-98
[53] Sinha R, Garcia M, Paliwal P, Kreek MJ, Rounsaville BJ: Stress-induced cocaine craving and hypothalamic-pituitary-adrenal responses are predictive of cocaine relapse outcomes. *Arch Gen Psychiatry.* 2006; 63(3):324-31.
[54] Sher L, Oquendo MA, Grunebaum MF, Burke AK, Huang YY, Mann JJ: CSF monoamine metabolites and lethality of suicide attempts in depressed patients with alcohol dependence.*Eur Neuropsychopharmacol.* 2007; 17(1):12-5.

[55] Lovallo WR: Cortisol secretion patterns in addiction and addiction risk. *Int J Psychophysiol.* 2006; 59(3):195-202
[56] Dai X, Thavundayil J, Santella S, Gianoulakis C: Response of the HPA-axis to alcohol and stress as a function of alcohol dependence and family history of alcoholism. *Psychoneuroendocrinology.* 2007; 32(3):293-305.
[57] Huizink AC, Ferdinand RF, Ormel J, Verhulst FC: Hypothalamic-pituitary-adrenal axis activity and early onset of cannabis use. *Addiction.* 2006; 101(11):1538-9.
[58] Koob GF: Stress, corticotropin-releasing factor, and drug addiction. *Ann N Y Acad Sci.* 1999; 897:27-45.
[59] Bruijnzeel AW, Gold MS: The role of corticotropin-releasing factor-like peptides in cannabis, nicotine, and alcohol dependence. *Brain Res Brain Res Rev.* 2005; 49(3):505-28.
[60] Valdez GR, Koob GF: Allostasis and dysregulation of corticotropin-releasing factor and neuropeptide Y systems: implications for the development of alcoholism. *Pharmacol Biochem Behav.* 2004; 79(4):671-89.
[61] Koob G, Kreek MJ: Stress, dysregulation of drug reward pathways, and the transition to drug dependence. *Am J Psychiatry.* 2007; 164(8):1149-59.
[62] Derijk RH: Single nucleotide polymorphisms related to HPA axis reactivity. *Neuroimmunomodulation.* 2009; 16(5):340-52.
[63] Gerra G, Zaimovic A, Moi G, Bussandri M, Bubici C, Mossini M, et al.: Aggressive responding in abstinent heroin addicts: neuroendocrine and personality correlates. *Prog Neuropsychopharmacol Biol Psychiatry.* 2004; 28(1):129-39.
[64] Gerra G, Zaimovic A, Ampollini R, Giusti F, Delsignore R, Raggi MA, et al.: Experimentally induced aggressive behavior in subjects with 3,4-methylenedioxy-methamphetamine ("Ecstasy") use history: psychobiological correlates. *J Subst Abuse.* 2001; 13(4):471-91.
[65] Gerra G, Zaimovic A, Raggi MA, Giusti F, Delsignore R, Bertacca S, et al.: Aggressive responding of male heroin addicts under methadone treatment: psychometric and neuroendocrine correlates. *Drug Alcohol Depend.* 2001; 65(1):85-95.
[66] Sinha R, Fox HC, Hong KA, Bergquist K, Bhagwagar Z, Siedlarz KM: Enhanced negative emotion and alcohol craving, and altered physiological responses following stress and cue exposure in alcohol dependent individuals. *Neuropsychopharmacology.* 2009; 34(5):1198-208.
[67] Sinha R: Chronic stress, drug use, and vulnerability to addiction. *Ann N Y Acad Sci.* 2008; 1141:105-30.
[68] Vermetten E, Bremner JD: Circuits and systems in stress. II. Applications to neurobiology and treatment in posttraumatic stress disorder. *Depress Anxiety;* 2002:16:14-38.
[69] Strawn JR, Geracioti TD: Noradrenergic dysfunction and the psychopharmacology of posttraumatic stress disorder. *Depress Anxiety.* 2008; 25:260-271.
[70] Geracioti TD Jr, Baker DG, Ekhator NN, West SA, Hill KK, Bruce AB, et al.: CSF norepinephrine concentrations in posttraumatic stress disorder. *Am J Psychiatry.* 2001; 158:1227-1230.
[71] Geracioti TD Jr, Baker DG, Kasckow JW, Strawn JR, Jeffrey Mulchahey J, Dashevsky BA, et al: Effects of trauma-related audiovisual stimulation on cerebrospinal fluid

norepinephrine and corticotropin-releasing hormone concentrations in post-traumatic stress disorder. *Psychoneuroendocrinology*. 2008; 33:416-424.

[72] Yehuda R, McFarlane AC, Shalev AY: Predicting the development of posttraumatic stress disorder from the acute response to a traumatic event. *Biol Psychiatry*. 1998; 44:1305-1313.

[73] Delahanty DL, Nugent NR: Predicting PTSD prospectively based on prior trauma history and immediate biological responses. *Ann N Y Acad Sci*. 2006;1071:27-40.

[74] Smith RJ, Aston-Jones G: Noradrenergic transmission in the extended amygdala: role in increased drug-seeking and relapse during protracted drug abstinence. B*rain Struct Funct*. 2008; 213(1-2):43-61.

[75] Koob GF: Brain stress systems in the amygdala and addiction. *Brain Res*. 2009.

[76] Gerra G, Zaimovic A, Giusti F, Baroni MC, Delsignore R, Raggi MA et al.: Pivagabine effects on neuroendocrine responses to experimentally-induced psychological stress in humans. *Behav Brain Res*. 2001; 122(1):93-101.

[77] Gerra G, Zaimovic A, Giucastro G, Folli F, Maestri D, Tessoni A, et al.: Neurotransmitter-hormonal responses to psychological stress in peripubertal subjects: relationship to aggressive behavior. *Life Sci*. 1998; 62(7):617-25.

[78] Gerra G, Zaimovic A, Zambelli U, Timpano M, Reali N, Bernasconi S, et al.: Neuroendocrine responses to psychological stress in adolescents with anxiety disorder. *Neuropsychobiology*. 2000; 42(2):82-92.

[79] Sofuoglu M, Sewell RA: Norepinephrine and stimulant addiction. *Addict Biol*. 2009; 14(2):119-29.

[80] Gabel S, Stadler J, Bjorn J, Shindledecker R, Bowden CL: Monoamine oxidase and homovanillic acid in boys with predispositions to substance abuse. *Alcohol Clin Exp Res*. 1994; 18(5):1137-42.

[81] Uhart M, Wand GS: Stress, alcohol and drug interaction: an update of human research. *Addict Biol*. 2009; 14(1):43-64.

[82] Gerra G, Leonardi C, Cortese E, Zaimovic A, Dell'Agnello G, Manfredini M, et al.: Homovanillic acid (HVA) plasma levels inversely correlate with attention deficit-hyperactivity and childhood neglect measures in addicted patients. *J Neural Transm*. 2007; 114(12):1637-47.

[83] Feltenstein MW, See RE: The neurocircuitry of addiction: an overview. *Br J Pharmacol*. 2008; 154(2):261-74.

[84] Dayan P: Dopamine, reinforcement learning, and addiction. *Pharmacopsychiatry*. 2009; 42 Suppl 1:S56-65.

[85] Lucas LR, Angulo JA, Le Moal M, McEwen BS, Piazza PV: Neurochemical characterization of individual vulnerability to addictive drugs in rats. *Eur J Neurosci*. 1998; 10(10):3153-63.

[86] Rougé-Pont F, Piazza PV, Kharouby M, Le Moal M, Simon H: Higher and longer stress-induced increase in dopamine concentrations in the nucleus accumbens of animals predisposed to amphetamine self-administration. A microdialysis study. *Brain Res*. 1993; 602(1):169-74.

[87] O'Dell LE: A psychobiological framework of the substrates that mediate nicotine use during adolescence. *Neuropharmacology*. 2009; 56 Suppl 1:263-78.

[88] Morgan D, Grant KA, Gage HD, Mach RH, Kaplan JR, Prioleau O, et al.: Social dominance in monkeys: dopamine D2 receptors and cocaine self-administration. *Nat Neurosci.* 2002; 5(2):169-74.

[89] Ressler KJ, Nemeroff CB: Role of serotonergic and noradrenergic systems in the pathophysiology of depression and anxiety disorders. *Depr Anxiety.* 2000; 12:2-19.

[90] Carver CS, Johnson SL, Joormann J: Serotonergic function, two-mode models of self-regulation, and vulnerability to depression: what depression has in common with impulsive aggression. *Psychol Bull.* 2008; 134(6):912-43.

[91] Krakowski M: Violence and serotonin: influence of impulse control, affect regulation, and social functioning. *J Neuropsychiatry Clin Neurosci.* 2003; 15(3):294-305.

[92] Gerra G, Zaimovic A, Moi G, Bussandri M, Delsignore R, Caccavari R, et al.: Neuroendocrine correlates of antisocial personality disorder in abstinent heroin-dependent subjects. *Addict Biol.* 2003; 8(1):23-32.

[93] Gerra G, Zaimovic A, Timpano M, Zambelli U, Begarani M, Marzocchi GF, et al.: Neuroendocrine correlates of temperament traits in abstinent opiate addicts. *J Subst Abuse.* 2000; 11(4):337-54.

[94] Gerra G, Zaimovic A, Zambelli U, Delsignore R, Baroni MC, Laviola G, et al.: Neuroendocrine correlates of depression in abstinent heroin-dependent subjects. *Psychiatry Res.* 2000; 96(3):221-34.

[95] Soloff PH, Lynch KG, Moss HB: Serotonin, impulsivity, and alcohol use disorders in the older adolescent: a psychobiological study. *Alcohol Clin Exp Res.* 2000; 24(11):1609-19.

[96] Gerra G, Garofano L, Castaldini L, Rovetto F, Zaimovic A, Moi G, et al.: Serotonin transporter promoter polymorphism genotype is associated with temperament, personality traits and illegal drugs use among adolescents. *J Neural Transm.* 2005; 112(10):1397-410.

[97] Gerra G, Garofano L, Zaimovic A, Moi G, Branchi B, Bussandri M, et al.: Association of the serotonin transporter promoter polymorphism with smoking behavior among adolescents. *Am J Med Genet B Neuropsychiatr Genet.* 2005; 135B(1):73-8.

[98] Gerra G, Garofano L, Santoro G, Bosari S, Pellegrini C, Zaimovic A, et al.: Association between low-activity serotonin transporter genotype and heroin dependence: behavioral and personality correlates. *Am J Med Genet B Neuropsychiatr Genet.* 2004; 126B(1):37-42.

[99] Hallikainen T, Saito T, Lachman HM, Volavka J, Pohjalainen T, Ryynanen OP, et al.: Association between low activity serotonin transporter promoter genotype and early onset alcoholism with habitual impulsive violent behavior. *Mol Psychiatry.* 1999; 4(4):385-8.

[100] Van Goozen SH, Fairchild G: Neuroendocrine and neurotransmitter correlates in children with antisocial behavior. *Horm Behav.* 2006; 50(4):647-54.

[101] Ichise M, Vines DC, Gura T, Anderson GM, Suomi SJ, Higley JD, et al.: Effects of early life stress on [11C]DASB positron emission tomography imaging of serotonin transporters in adolescent peer- and mother-reared rhesus monkeys. *J Neurosci.* 2006; 26(17):4638-43.

[102] Malcolm RJ: GABA systems, benzodiazepines, and substance dependence. *J Clin Psychiatry.* 2003; 64 Suppl 3:36-40.

[103] Davis M, Ressler K, Rothbaum BO, Richardson R: Effects of D-cycloserine on extinction: translation from preclinical to clinical work. *Biol Psychiatry.* 2006; 60:369-375.

[104] Koob GF: Neuroadaptive mechanisms of addiction: studies on the extended amygdala. *Eur Neuropsychopharmacol.* 2003; 13(6):442-52.

[105] Barros VG, Berger MA, Martijena ID, Sarchi MI, Pérez AA, Molina VA, et al.: Early adoption modifies the effects of prenatal stress on dopamine and glutamate receptors in adult rat brain. *J Neurosci Res.* 2004; 76(4):488-96.

[106] Sircar R, Mallinson K, Goldbloom LM, Kehoe P: Postnatal stress selectively upregulates striatal N-methyl-D-aspartate receptors in male rats. *Brain Res.* 2001; 904(1):145-8.

[107] Petrakis IL, Limoncelli D, Gueorguieva R, Jatlow P, Boutros NN, Trevisan L, et al.: Altered NMDA glutamate receptor antagonist response in individuals with a family vulnerability to alcoholism. *Am J Psychiatry.* 2004; 161(10):1741-2.

[108] Phelps LE, Brutsche N, Moral JR, Luckenbaugh DA, Manji HK, Zarate CA Jr: Family history of alcohol dependence and initial antidepressant response to an N-methyl-D-aspartate antagonist. *Biol Psychiatry.* 2009; 65(2):181-4.

[109] Nikoshkov A, Drakenberg K, Wang X, Horvath MC, Keller E, Hurd YL: Opioid neuropeptide genotypes in relation to heroin abuse: dopamine tone contributes to reversed mesolimbic proenkephalin expression. *Proc Natl Acad Sci U S A.* 2008; 105(2):786-91.

[110] Newport DJ, Nemeroff CB: Neurobiology of posttraumatic stress disorder. *Curr Opin Neurobiology.* 2000; 10:211-218.

[111] Roth-Deri I, Green-Sadan T, Yadid G: Beta-endorphin and drug-induced reward and reinforcement. *Prog Neurobiol.* 2008; 86(1):1-21.

[112] Racz I, Schürmann B, Karpushova A, Reuter M, Cichon S, Montag C, et al.: The opioid peptides enkephalin and beta-endorphin in alcohol dependence. *Biol Psychiatry.* 2008; 64(11):989-97.

[113] Olmstead MC, Ouagazzal AM, Kieffer BL: Mu and delta opioid receptors oppositely regulate motor impulsivity in the signaled nose poke task. *PLoS One.* 2009; 4(2):e4410.

[114] Bremner JD, Elzinga B, Schmahl C, Vermetten E: Structural and functional plasticity of the human brain in posttraumatic stress disorder. *Prog Brain Res.* 2008; 167:171-186

[115] Rauch SL, Shin LM, Phelps EA: Neurocircuitry models of posttraumatic stress disorder and extinction: human neuroimaging research--past, present, and future. *Biol Psychiatry.* 2006; 60:376-382

[116] Liberzon I, Sripada CS: The functional neuroanatomy of PTSD: a critical review. *Prog Brain Res.* 2008; 167:151-169.

[117] Makris N, Oscar-Berman M, Jaffin SK, Hodge SM, Kennedy DN, Caviness VS, et al.: Decreased volume of the brain reward system in alcoholism. *Biol Psychiatry.* 2008; 64(3):192-202.

[118] Yücel M, Solowij N, Respondek C, Whittle S, Fornito A, Pantelis C, et al.: Regional brain abnormalities associated with long-term heavy cannabis use. *Arch Gen Psychiatry.* 2008; 65(6):694-701.

[119] Pitman RK, Gilbertson MW, Gurvits TV, et al. Clarifying the origin of biological abnormalities in PTSD through the study of identical twins discordant for combat exposure. *Ann N Y Acad Sci.* 2006; 1071:242-254.

[120] Teicher MH, Tomoda A, Andersen SL: Neurobiological consequences of early stress and childhood maltreatment: are results from human and animal studies comparable? *Ann N Y Acad Sci.* 2006; 1071:313-23.

[121] Whittle S, Yap MB, Yücel M, Sheeber L, Simmons JG, Pantelis C, et al.: Maternal responses to adolescent positive affect are associated with adolescents' reward neuroanatomy. *Soc Cogn Affect Neurosci.* 2009.

[122] Boes AD, Bechara A, Tranel D, Anderson SW, Richman L, Nopoulos P: Right ventromedial prefrontal cortex: a neuroanatomical correlate of impulse control in boys. *Soc Cogn Affect Neurosci.* 2009; 4(1):1-9.

[123] Rubia K, Smith AB, Halari R, Matsukura F, Mohammad M, Taylor E, et al.: Disorder-specific dissociation of orbitofrontal dysfunction in boys with pure conduct disorder during reward and ventrolateral prefrontal dysfunction in boys with pure ADHD during sustained attention. *Am J Psychiatry.* 2009; 166(1):83-94.

[124] Shin LM, Rauch SL, Pitman RK: Amygdala, medial prefrontal cortex, and hippocampal function in PTSD. *Ann N Y Acad Sci.* 2006; 1071:67-79.

[125] Gelpin E, Bonne O, Peri T, Brandes D, Shalev AY: Treatment of recent trauma survivors with benzodiazepines: a prospective study. *J Clin Psychiatry.* 1996; 57:390-394.

[126] Rasmusson AM, Hauger RL, Morgan CA, Bremner JD, Charney DS, Southwick SM: Low baseline and yohimbine-stimulated plasma neuropeptide Y (NPY) levels in combat-related PTSD. *Biol Psychiatry.* 2000; 47:526-539.

[127] Moeller SJ, Maloney T, Parvaz MA, Dunning JP, Alia-Klein N, Woicik PA, et al.: Enhanced choice for viewing cocaine pictures in cocaine addiction. *Biol Psychiatry.* 2009; 66(2):169-76.

[128] Zijlstra F, Veltman DJ, Booij J, van den Brink W, Franken IH: Neurobiological substrates of cue-elicited craving and anhedonia in recently abstinent opioid-dependent males. *Drug Alcohol Depend.* 2009; 99(1-3):183-92.

[129] Wessa M, Flor H: Failure of extinction of fear responses in posttraumatic stress disorder: evidence from second-order conditioning. *Am J Psychiatry.* 2007; 164(11):1684-92.

[130] Di Chiara G: Drug addiction as dopamine-dependent associative learning disorder. *Eur J Pharmacol.* 1999; 375(1-3):13-30.

[131] Peters J, Kalivas PW, Quirk GJ: Extinction circuits for fear and addiction overlap in prefrontal cortex. *Am J Med Genet B Neuropsychiatr Genet.* 2008; 147B(1):128-30.

[132] Rauch SL, Shin LM, Segal E, Pitman RK, Carson MA, McMullin K, et al.: Selectively reduced regional cortical volumes in post-traumatic stress disorder. *Neuroreport.* 2003; 14:913-916.

[133] Yamasue H, Kasai K, Iwanami A, Ohtani T, Yamada H, Abe O, et al.: Voxel-based analysis of MRI reveals anterior cingulate gray-matter volume reduction in posttraumatic stress disorder due to terrorism. *Proc Natl Acad Sci U S A.* 2003; 100:9039-9043.

[134] Woodward SH, Kaloupek DG, Streeter CC, Martinez C, Schaer M, Eliez S: Decreased anterior cingulate volume in combat-related PTSD. *Biol Psychiatry.* 2006; 59:582-587.

[135] Shin LM, Orr SP, Carson MA, Rauch SL, Macklin ML, Lasko NB, et al.: Regional cerebral blood flow in the amygdala and medial prefrontal cortex during traumatic imagery in male and female Vietnam veterans with PTSD. *Arch Gen Psychiatry.* 2004; 61:168-176.

[136] Britton JC, Phan KL, Taylor SF, Fig LM, Liberzon I: Corticolimbic blood flow in posttraumatic stress disorder during script-driven imagery. *Biol Psychiatry.* 2005; 57:832-840.

[137] Bremner JD, Narayan M, Staib LH, Southwick SM, McGlashan T, Charney DS: Neural correlates of memories of childhood sexual abuse in women with and without posttraumatic stress disorder. *Am J Psychiatry.* 1999; 156:1787-1795.

[138] Lanius RA, Williamson PC, Hopper J, Densmore M, Boksman K, Gupta MA, et al.: Recall of emotional states in posttraumatic stress disorder: an fMRI investigation. *Biol Psychiatry.* 2003; 53:204-210.

[139] Shin LM, Wright CI, Cannistraro PA, Wedig MM, McMullin K, Martis B, et al.: A functional magnetic resonance imaging study of amygdala and medial prefrontal cortex responses to overtly presented fearful faces in posttraumatic stress disorder. *Arch Gen Psychiatry.* 2005; 62:273-281.

[140] Bremner JD, Vermetten E, Vythilingam M, Afzal N, Schmahl C, Elzinga B, et al.: Neural correlates of the classic color and emotional stroop in women with abuse-related posttraumatic stress disorder. *Biol Psychiatry.* 2004; 55:612-620.

[141] Tanabe J, Tregellas JR, Dalwani M, Thompson L, Owens E, Crowley T, et al.: Medial orbitofrontal cortex gray matter is reduced in abstinent substance-dependent individuals. *Biol Psychiatry.* 2009; 65(2):160-4.

[142] Cohen RA, Grieve S, Hoth KF, Paul RH, Sweet L, Tate D, et al.: Early life stress and morphometry of the adult anterior cingulate cortex and caudate nuclei. *Biol Psychiatry.* 2006; 59(10):975-82.

[143] Kreek MJ, Nielsen DA, LaForge KS: Genes associated with addiction: alcoholism, opiate, and cocaine addiction. *Neuromolecular Med.* 2004; 5(1):85-108.

[144] Gilbertson MW, Shenton ME, Ciszewski A, Kasai K, Lasko NB, Orr SP et al.: Smaller hippocampal volume predicts pathologic vulnerability to psychological trauma. *Nat Neurosci.* 2002; 5:1242-1247.

[145] Hariri AR, Mattay VS, Tessitore A, Kolachana B, Fera F, Goldman D, et al.: Serotonin transporter genetic variation and the response of the human amygdala. *Science.* 2002; 297:400-403.

[146] Segman RH, Cooper-Kazaz R, Macciardi F, Goltser T, Halfon Y, Dobroborski T, et al.: Association between the dopamine transporter gene and posttraumatic stress disorder. *Mol Psychiatry.* 2002; 7:903-907.

[147] Broekman BF, Olff M, Boer F. The genetic background to PTSD. *Neurosci Biobehav Rev.* 2007;31:348-62.

[148] Young SE, Smolen A, Corley RP, Krauter KS, DeFries JC, Crowley TJ, et al.: Dopamine transporter polymorphism associated with externalizing behavior problems in children. *Am J Med Genet.* 2002; 114(2):144-9.

[149] Gerra G, Garofano L, Pellegrini C, Bosari S, Zaimovic A, Moi G, et al.: Allelic association of a dopamine transporter gene polymorphism with antisocial behaviour in heroin-dependent patients. *Addict Biol.* 2005; 10(3):275-81.

[150] Noble EP: The DRD2 gene in psychiatric and neurological disorders and its phenotypes. *Pharmacogenomics.* 2000; 1(3):309-33.
[151] Oroszi G, Goldman D: Alcoholism: genes and mechanisms. *Pharmacogenomics.* 2004; 5(8):1037-48.
[152] Wagner S, Baskaya O, Lieb K, Dahmen N, Tadić A: The 5-HTTLPR Polymorphism modulates the association of serious life events (SLE) and impulsivity in patients with Borderline Personality Disorder. *J Psychiatr Res.* 2009.
[153] Steiger H, Richardson J, Joober R, Israel M, Bruce KR, Ng Ying Kin NM, et al.: Dissocial behavior, the 5HTTLPR polymorphism, and maltreatment in women with bulimic syndromes. *Am J Med Genet B Neuropsychiatr Genet.* 2008; 147B(1):128-30.
[154] Nilsson KW, Sjöberg RL, Damberg M, Alm PO, Ohrvik J, Leppert J, et al.: Role of the serotonin transporter gene and family function in adolescent alcohol consumption. *Alcohol Clin Exp Res.* 2005; 29(4):564-70.
[155] Conner BT, Noble EP, Berman SM, Ozkaragoz T, Ritchie T, Antolin T, et al.: DRD2 genotypes and substance use in adolescent children of alcoholics. *Drug Alcohol Depend.* 2005; 79(3):379-87.
[156] Caspi A, McClay J, Moffitt TE, Mill J, Martin J, Craig IW, et al.: Role of genotype in the cycle of violence in maltreated children. *Science.* 2002; 297(5582): 851-854.
[157] Gerra G., Zaimovic A., Castaldini L., Garofano L., Manfredini M., Somaini L., et al.: Relevance of perceived childhood neglect, 5-HTT gene variants and hypothalamus-pituitary-adrenal axis dysregulation to substance abuse susceptibility. *Am J Med Genet B Neuropsychiatr Genet.* 2010; 153B(3): 715-22

In: Neurobiology of Post-Traumatic Stress Disorder
Editors: L. Sher and A. Vilens, pp. 201-216
ISBN: 978-1-61470-593-2

Chapter XIII

Posttraumatic Stress Disorder (PTSD) and Alzheimer's Disease (AD)

Magda Tsolaki
Aristotle University of Thessaloniki, Thessaloniki, Greece

Abstract

Posttraumatic Stress Disorder and Alzheimer's Disease have many similarities and only some differences. The main similarities are that both disorders have as risk factor a severe traumatic event (100% in PTSD and 80% in AD), the prevalence F/M ratio of PTSD is the same as in AD, (2:1), stress hormones and other neurotransmitters are involved in both diseases, the areas of brain which are involved in both disorders are about the same (hippocampus, amygdale, prefrontal cortex, anterior cingulated gyrus) and finally the areas that are activated in fMRI are about the same in both conditions. Perhaps there are some differences in genes, but this area needs more research and the age of onset is the point which is different in vast majority of patients with AD, because the mean age in AD patients is 75 years old. This is the reason which leads to different therapeutic strategies.

Introduction

The concept of trauma currently occupies a central position in interdisciplinary dialogue. Using the concept of psychological trauma as a bridge, many authors attempt an interdisciplinary dialogue with neurology, psychiatry, biology and all neurosciences in general.

Research on stress is multifaceted and undertaken at many levels of analysis. Discipline differences in definition and measurement have contributed to divergent findings.

The actions of stress on cognition involve multiple mechanisms and different time courses. Chronic stress is associated with hippocampal damage and impaired memory in animals [1] and humans [2]. In addition, stress has been shown to decrease and survival proliferation of new cells [3].

There is also a theory that there are vulnerability and resilience genes for brain diseases.

It is well known that young or adult patients with PTSD suffer from memory problems and old persons may present severe cognitive disorders after severe psychological stress, which can lead to dementia.

Memory problems are the main symptoms- the same brain target in both diseases, -hippocampus is the first target- , increased levels of GCs were noticed also in both PTSD and AD and prevalence F/M ratio of PTSD is the same as in AD, 2:1. The possible only difference is the age of onset which possibly means different plasticity of the aging brain and therefore different response. During the following discussion more similarities will be shown and discussed.

1. Definitions

Posttraumatic Stress Disorder (PTSD) is an anxiety disorder resulting from experiencing or witnessing an extreme traumatic stressor that involved potential loss of life or serious injury to the self or others. PTSD is described in DSM-IV as a disorder associated with exposure to a traumatic event outside the range of usual human experience, e.g., rape, combat, or natural disaster and consists of a predictable constellation of distressing behavioural symptoms and physiological features. These symptoms are characterized by persistent re-experiencing of the traumatic event, avoidance of stimuli associated with the traumatic event, amnesia and increased arousal. Furthermore, sleep disturbances, memory problems, inappropriate irritability, difficulty in concentration, and exaggerated startle responses are often observed in PTSD patients [4]. This disorder is usually related with young or adult patients.

Alzheimer's disease (AD) is a progressive neurodegenerative disorder first affecting memory and attention functions and then gradually affecting all cognitive functions with behavioural impairment and eventually causing death [5]. It is today the most common cause of dementia, and accounts for 60–70% of all cases. The mean age of onset is 75 years old [6, 7]. The strongest evidence for amyloid being a cause –and not just a consequence- of the disease comes from genetics. However, the heterogeneity of the etiologic factors makes it difficult to define the clinically most important factor determining the onset and progression of the disease specially in Late Onset AD which accounts for 95-98% of all cases. US National Institute of Aging makes a special effort to pick up grants that refute the amyloid hypothesis.

2. Epidemiology

PTSD

The estimated lifetime prevalence of PTSD among adult Americans is 7,8% with women twice as likely as men to have PTSD [8]. However, prevalence estimates of PTSD have varied widely in the literature. Recent study which examined data from 24 studies found that the proportion of PTSD cases with delayed PTSD was 24.8% (95% CI = 22.6% to 27.2%) after adjusting for differences in study methodology, demographic features, and event-related characteristics. Military combat exposure, Western cultural background, and lower cumulative PTSD incidence were associated with delayed PTSD. Participants with initial subthreshold PTSD were at increased risk of developing delayed PTSD. Delayed PTSD was found among about a quarter of PTSD cases and represents exacerbations of prior symptoms [9].

AD

The prevalence of dementia worldwide is approximated to about 24 million people with women twice as likely as men to have AD and if the current trend with 4.6 million new cases of dementia each year continuous, the population of patients with dementia will nearly double every 20 years until 2040 [10]. Dementia is a psychosocial tragedy to the growing elderly population and a financial burden for the society [11]. As AD is the most frequent type of dementia the social and financial burden is mainly because of it.

3. Neurochemistry

Several studies suggest that the duration of exposure to corticosteroids or stress, the dose of corticosteroids, but also the vulnerability of the hippocampus, amygdala and prefrontal cortex, areas targeted by elevated stress hormones, mediate the damaging effects of elevated corticosteroids. According to one theory, there is a correlation of increased *corticosteroids (*GCs) and the ageing hippocampus. A second theory suggests that the damaging effects of stress are mediated by the stress-induced increases in *glutamate and calcium. Insulin deficiency* as well as oxidative stress, disturbed protein metabolism and failure of hippocampal neurogenesis in the adult brain have been also proposed as mediators of the stress-associated damage and consequent cognitive decline. Appreciation of *norepinephrine*'s role in modulating the neural circuitry mediating cognition and affect should help elucidate the pathophysiology of a variety of neuropsychiatric disorders and the development of novel treatments [12].

PTSD

Decreased baseline *cortisol* levels in patients with PTSD many years after the original trauma do not mean the possibility of higher levels of cortisol release at the time of the original trauma. Also we do not know what happens to cortisol during the stress of traumatic reminders in patients with PTSD, since all studies were conducted in a baseline unperturbed state. One consistent observation seems to be that the return of corticosterone to baseline after a stress is delayed in aging.

The PTSD patients showed in a recent study widespread impairment of GABA(A)-ergic function, which was reversed toward facilitation in both hemispheres in one-half of the patients, marked increase of *glutamatergic function* in the right hemisphere, and right-sided impairment of short-latency afferent inhibition. Illness duration and avoidance symptoms but not anxiety correlated with right-lateralized dysfunctions of cortical excitability. Although the neurobiological complexity of each transcranial magnetic stimulation variable makes these results theoretical, the pattern of cortical excitability accompanying PTSD symptoms suggests a bilateral decrease of the GABA(A)-ergic function. This prevails in the right hemisphere, in association with a relative prevalence of the glutamatergic tone, a new finding that current neuroimaging investigations cannot provide due to the lack of reliable glutamate tracers. Results might help to disclose new pathophysiological aspects of PTSD symptoms, providing a rationale for future neuromodulatory strategies of treatment [13].

The body's principal adaptive responses to stress stimuli are mediated by an intricate stress system, which includes the hypothalamic-pituitary-adrenocortical axis and the sympathoadrenal system. Dysregulation of the system, caused by the cumulative burden of repetitive or chronic environmental stress challenges (allostatic load) contributes to the development of a variety of illnesses including hypertension, atherosclerosis, and the *insulin-resistance*-dyslipidemia syndrome, as well as certain disorders of immune function [14].

AD

There are data, obtained in the aged human population which showed that long-term exposure to high endogenous levels of *GCs* is associated with both memory impairment and a 14% smaller volume of the hippocampus. It has been suggested that chronic elevations of GC hormones result in morphological changes in hippocampal neurons. Eighty-eight effect sizes and 1642 participants from 28 studies were meta-analyzed for the effects of stress on memory performance and GC activation. Analyses showed that stress was associated with *GCs* activation and declarative memory decline. It was concluded that, although the relationship between stress, glucocorticoids, and memory loss was empirically supported, there were other factors, such as stress condition and gender, as well as individual differences within groups, that influenced the association between these variables, and warrant further examination [15]. GC treatment also induced accumulation of transiently expressed EGFP-tau in the neuronal perikarya. Together with previous evidence showing that Abeta can activate cyclin-dependent kinase 5 and GSK3, these results uncover a potential mechanism through which GC may contribute to AD neuropathology [16]. Having all this information in mind we measured serum cortisol levels in 53 patients with AD, 32 patients with Mild Cognitive Impairment and 22 controls, age, gender and education-matched. About half of the patients had a Severe Psychological Stress (SPS) just before the onset of cognitive symptoms. We found that there was no significant difference between persons with and without SPS [17]. Our results are one to two years after the onset of dementia and perhaps the levels of cortisol were decreased. It could be extremely harmful if cortisol levels remained high for long duration.

Glutamate is a neurotransmitter, which participates in normal function but also contributes to hippocampal pathology, such as neuronal loss produced by ischemia, seizures and prolonged psychosocial stress. Experimental manipulations or natural processes (aging) can cause glutamatergic dysfunctions modelling corresponding states in AD [18]. Under or over stimulation may result in impaired development of glutamate-mediated transmission [19]. Episodes of severe stress life may cause cognitive problems, possibly through reduction of glutamatergic systems in old age. Enhanced glutamate responses after stress have been reported to increase the vulnerability of the aging brain to neuronal damage [20].

Insulin plays a role in the structural responses of the brain to stressors. Lack of insulin decreases neuron number in dentate gyrus and leads to remodeling of dendrites of CA3 neurons that is further accelerated by repeated restraint stress. Stress and hyperglycemia both increase oxidative stress in the brain and this is likely to contribute, over time, to impaired neural function in chronic stress and diabetes. Acute cortisol rises have been reported to suppress hippocampal glucose uptake. Thus the acute memory impairing effects of GCs might be related to their ability to impair glucose uptake, as it is known that glucose supplementation improves memory, particularly in the elderly [21].

4. Anatomy

PTSD

Studies in animals and humans support the hypothesis that stress can lead to damage to the *Hippocampus* (HC). Repeated stress causes atrophy of dendrites in the CA3 region and both acute and chronic stress suppress neurogenesis of dentate gyrus granule neurons. A structure closely linked in the initial processing of a fearful stimulus, prior to lateral amygdalar activation, is the *hippocampus*, which is involved in memory and can therefore provide contextual information about a stimulus. Understanding the conditioned fear response in patients with PTSD may provide a tool for uncovering the neurobiological processes that underlie PTSD.

The *amygdala* plays a key role in consolidating the emotional significance of events. Thus, this region has played a crucial role in our understanding of conditioned fear processing, a process that is important in the pathophysiology of PTSD. The activation of neuromodulatory systems affecting the basolateral complex of the *amygdala* (BLA) and its projections to other brain regions involved in processing different kinds of information plays a key role in enabling emotionally significant experiences as in PTSD to be well remembered [22]. In fact, laboratory animal studies have clearly demonstrated that the central nucleus of the amygdala is one of the critical neuroanatomical loci in the development of the fear-potentiated startle (one of the hyperarousal symptoms of PTSD).

Of the cerebrocortical regions, the *prefrontal cortex* appears to play a seminal role in the working memory component of explicit memory and may play a counter-regulatory role in the stress response through inhibitory effects upon the amygdala. Alterations in prefrontal cortical activity may explain the deficits in explicit memory function often seen in PTSD. The *prefrontal cortex* (PFC) is a higher brain region that regulates thought, behaviour, and emotion using representational knowledge, operations often referred to as working memory. Cognitive and neurophysiological studies have shown that HC-PFC processing is impaired, and amygdaloid processing is enhanced, by stress and in anxiety and mood disorders, including Major Depressive Disorder, Post Traumatic Stress Disorder and AD.

The *anterior cingulate gyrus* is responsible for the maintenance of social mores, fear-related behaviour, and selective attentional processing (as evidenced by the Stroop task). Anterior cingulate dysfunction may therefore play a role in the re-experiencing phenomena reported in

PTSD. A recent study showed enhanced *anterior cingulated* responses in PTSD contrast with reports of reduced activity for threat stimuli, suggesting that the latter may be specific to processing of threat-related content. Activation of rostral and dorsal anterior cingulate, left amygdala and posterior parietal networks in response to salient nontreatening stimuli may reflect generalized hypervigilance [23].

Also involved in conditioned fear processing and its extinction is the *orbitofrontal cortex.* A conditioned fear response occurs when a conditioned stimulus (e.g. a light) is repeatedly paired with an unconditioned stimulus (e.g. a shock) such that the fear response occurs to the conditioned stimulus alone. Extinction (i.e. loss of the fear response to the conditioned stimulus) eventually occurs when the conditioned stimulus is repeatedly shown without the accompanying unconditioned stimulus. Because the orbitofrontal cortex is crucial to the process of extinction, orbitofrontal dysfunction probably contributes to the tenacity of the hyperarousal symptoms such as the exaggerated startle response in patients with PTSD. Finally, explicit memory buffers in the various sensory association cortices and the motor cortex appear to be activated when PTSD patients are exposed to traumatic cues, and thus are likely to play a role in the pathophysiology of re-experiencing.

AD

Atrophy of human *HC* is noticed in Cushing's Syndrome, Recurrent Depressive Illness, Post Traumatic Stress Disorder, Schizophrenia, Aging, Prodromal AD and AD [24].

A recent study revealed that surface inward-deformation in MCI and AD is most prominent in the anterior hippocampal segment and the basolateral complex of *the amygdala* [25].

In a recent study significantly higher activation (intensity and number of voxels) was observed in *dorsolateral-prefrontal-cortex (DLPFC)*, of MCI subjects an prodromal stage of Alzheimer's Disease. The overactivity seen in DLPFC of MCI may represent a compensatory mechanism that enables them to perform normally [26].

Apathy which is the most common psychiatric disorder in AD was associated with Gray Matter density loss in the *anterior cingulate* and frontal cortex bilaterally, the head of the left caudate nucleus and in bilateral putamen [27].

5. Risk Factors

PTSD

The only environmental risk factor is extreme traumatic stressor. But the question is why not all people who are exposed to similar traumatic stressors do not develop PTSD? The answer to this critical question may be our genes [28].

AD

Many approaches-theories as amyloid cascade, oxidative stress, inflammation, specific protein mutations, long term response to injury or infection and defects in normal brain maintenance –such as the clearance of defective proteins- have been undertaken to understand risk factors of dementia and AD. Many approaches-theories as amyloid cascade, oxidative stress, inflammation, specific protein mutations, long term response to injury or infection and defects in normal brain maintenance –such as the clearance of defective proteins- have been undertaken to understand risk factors of dementia and Alzheimer's Disease (AD). There are many risk factors for AD: Age, mutations, a relative with Down syndrome, vascular risk factors-hypertension, diabetes mellitus, hyperlipidemia, smoking, obesity, hyperomocystinaimia-, depression, helicobacter of pylori etc [29, 30, 31].

We conducted a study where we describe that the vast majority of patients with dementia report a very stressful event, just before the onset of symptoms of dementia [32].

6. Genetics

PTSD

Several keys to future genetic studies are highlighted, thorough analyses of the gene by environment (GxE) interactions that govern one's vulnerability to anxiety disorder(s), the effectiveness of individual treatment strategies, and the severity of symptoms may lead to more effective prophylactic (e.g. social support) and treatment strategies [33].

AD

Last twenty years many mutations in APP gene, PS1 and PS2 genes in chromosomes 21, 14 and 1 respectively are recognized in familial cases of AD [34, 35]. The only high risk gene accepted in AD is allele ε4 of Apolipoprotein E (heterozygotes and homozygotes [36, 37].

7. Clinical and Laboratory Examinations

PTSD

Many of the hallmark symptoms of PTSD (e.g. nightmares, flashbacks, amnesia for the traumatic event, dissociative episodes, exaggerated startle response) represent, at least in part, disturbances in neurocognitive processing. In particular, sensory input and memory processing appear to be awry in PTSD. Not only are pharmacological probes being used to investigate psychiatric illness at the cellular level, but new tools are also becoming available to investigate brain function in psychiatric disorders. Some of the methods used to investigate neurocognitive processing in patients with PTSD include neuropsychological testing, sensory evoked potentials, electroencephalography, polysomnography, and various modalities for functional brain imaging, including single photon emission computed tomography (SPECT), positron emission tomography (PET), and functional magnetic resonance imaging (fMRI) [38].

AD

Neuropsychological testing, evoked potentials [39] and structure and functional brain Imaging are the main clinical and laboratory exams for AD and others dementias also. In 2004 PIB amyloid imaging was published for the first time.

8. Neuroimaging

The knowledge about the development and maintenance of the PTSD has increase significantly in the past ten years with important insights coming from imaging studies. Through these insights PTSD has changed from “traumatic neurosis” to a biologically based psychological disorder. Neuroimaging findings suggest that four brain areas are of special interest: the hyperactive amygdale, the HC with volume reduction, as well as the cingulated gyrus and orbitofrontal cortical regions, which may not be able to inhibit the hyperactive amygdala to trauma related stimuli [40].

AD

Event Related Potentials

The disorganization of the AERP waveform in MCI patients –prodromal AD- is a potential basis upon which a neurophysiologic methodology for identifying and "staging" MCI can be sought. It was also found that delayed N200 latency not only identifies memory changes better than the MMSE, but also may be a potential predictor of the MCI patients who convert to AD.

MRI

It seems to be a difference of onset regions between early and late onset of AD where elderly AD patients tend to present with involvement of the medial temporal lobes (MTL) associated with a marked memory loss whereas early onset patients predominantly have the decrease of rCBF in posterior cortical association regions [41]. This difference has also been shown in voxel based morphometric studies with MRI [42].

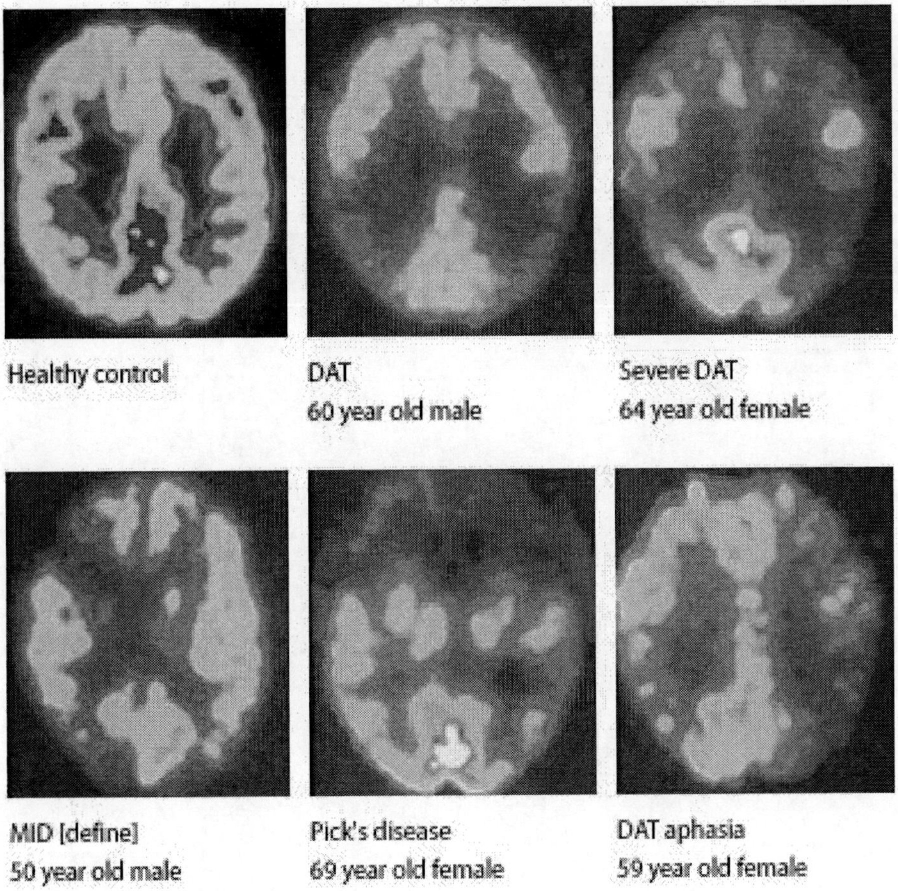

Figure 1. **PET images of brains showing dementia.** DAT = dementia of the Alzheimer type. MID = multi-infarct dementia.

PET and SPECT

Functional imaging with PET or SPECT has a high sensitivity and can detect subtle pathophysiologic changes in the brain before structural changes are present [43]. Although severe degeneration in the posterior cingulate cortex was demonstrated in autopsy brains from end-stage Alzheimer's disease patients [44], non-invasive and repeatable imaging techniques permit investigations of the earliest functional changes in living human subjects that can be followed longitudinally. SPECT has been established as a valuable clinical examination for the diagnosis of dementia disorders. The methods are used both for verification of dementia as well as for discrimination of different types of dementia. The diagnostic accuracy of PET and SPECT to distinguish patients with AD from healthy elderly has in an evidence-based review, been found to be comparable to the accuracy of a histopathologically confirmed diagnosis. Concerning the discrimination between different types of dementia i.e. AD, Lewy body dementia and frontotemporal dementia, both PET and SPECT have a higher sensitivity than clinical evaluation (figure 1) [45].

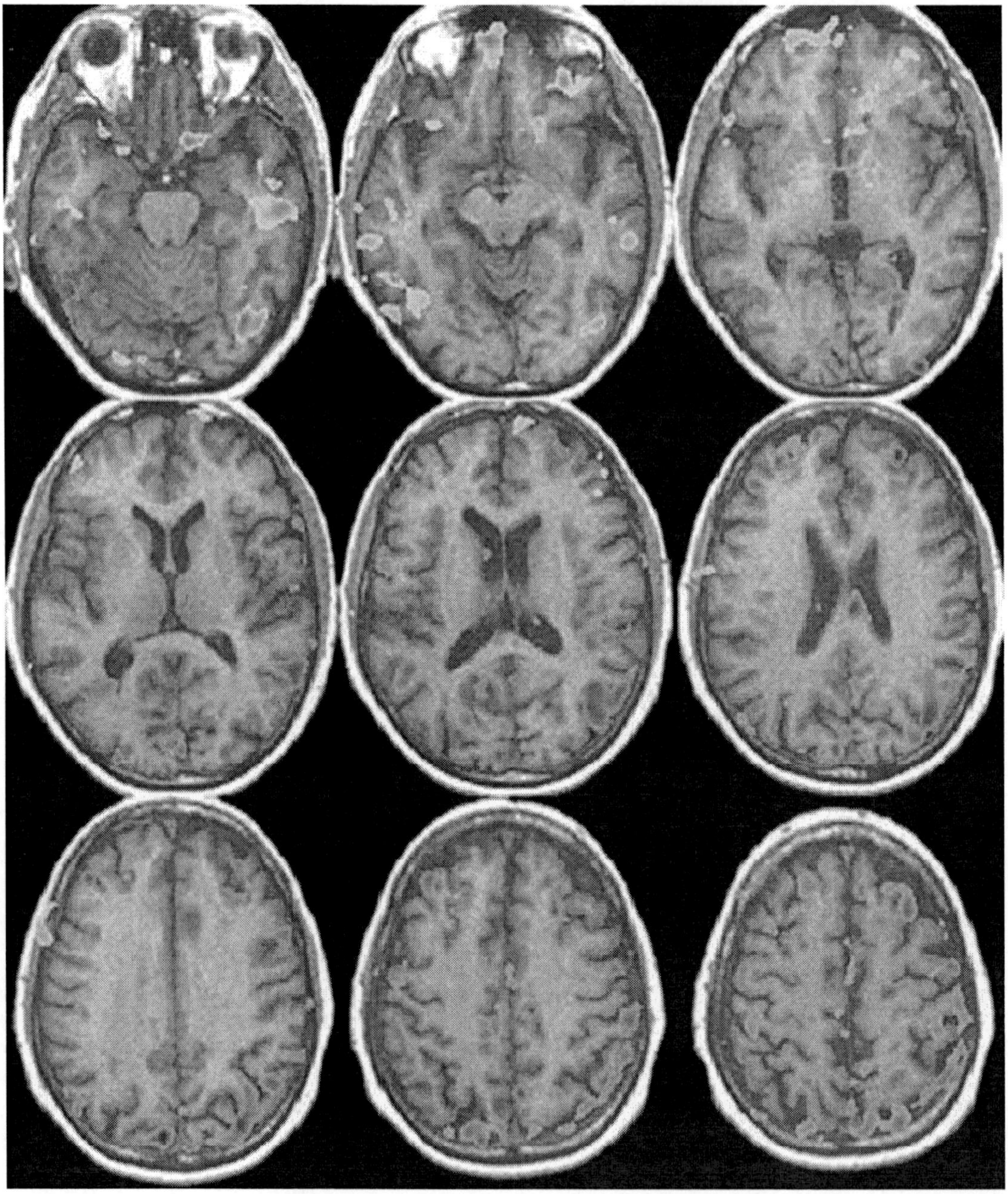

Figure 2. Statistical parametric mapping of activation differences observed between Alzheimer's disease (AD) and control groups is displayed on highresolution T1-weighted (3DFSPGR) axial images. Greater activation seen in the AD group compared with the control group is displayed in red to yellow scale, as seen in bilateral frontal and temporal regions. Greater activation in the predominantly temporal frontal, fusiform gyrus, precuneus, precentral gyrus, and postcentral control group compared with the AD group is displayed in blue to green scale, as seen in predominantly temporal frontal, fusiform gyrus, precuneus, precentral gyrus, and postcentral gyrus.

fMRI

fMRI of working memory (WM) in patients with Mild Cognitive Impairment and probable.

Alzheimer's disease showed more activation than the controls in the right superior frontal gyrus, bilateral middle temporal, middle frontal, anterior cingulate, and fusiform gyri. Activation in the right parahippocampal gyrus, left inferior frontal gyrus, bilateral cingulate and lingual gyri, right lentiform nucleus, right fusiform gyrus, and left supramarginal gyrus in the AD group was less than in the MCI group. The WM task evoked activation in widely distributed regions, consistent with previous fMRI studies. AD and MCI patients showed an increased extent of activation and recruitment of additional areas [46] (Figure 2).

PTSD

Event Related Potentials

There are studies using event-related potentials (ERPs) in post-traumatic stress disorder (PTSD) which demonstrated reduced P3 amplitude during target detection and working memory (WM) processes [47].

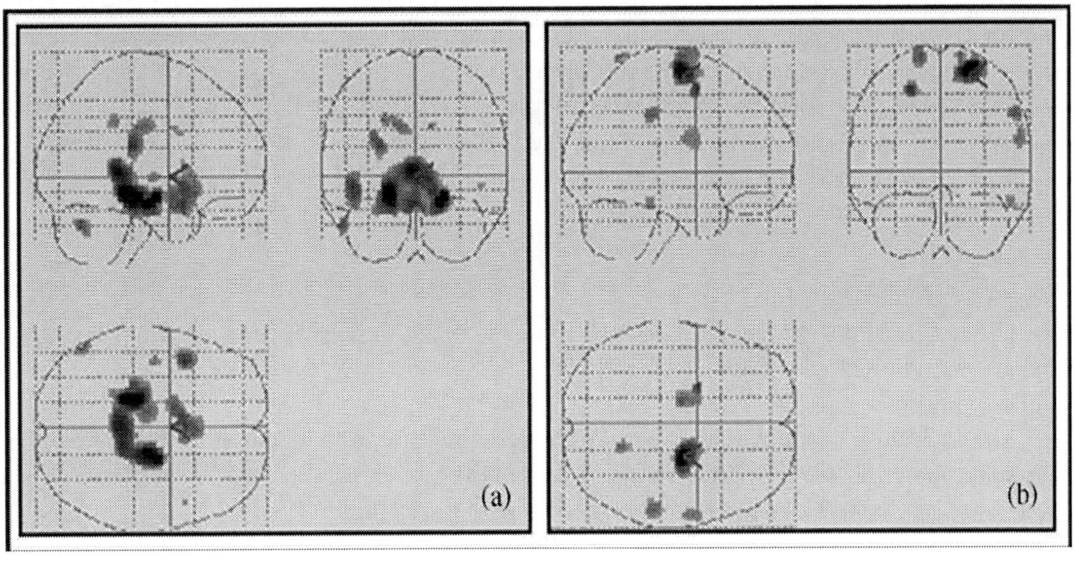

Figure 3. Maximum intensity projection images of (a) increased rCBF levels and (b) decreased rCBF levels in PTSD patients (nZ23) compared with normal controls (nZ64) [48].

MRI

Stress-related hippocampal atrophy appears to be a consequence of increased exposure to excitatory amino acids and GCs.

PET

An early report noted increased right parahippocampal activation in a PET study of Vietnam veterans with PTSD when presented with traumatic stimuli. One 1999 PET study reported decreased right hippocampal activation during script-driven guided mental imagery of traumatic experiences in women with childhood abuse-related PTSD when compared to sexually abused women without PTSD [49].

SPECT

In a SPECT study comparing Vietnam combat veterans with PTSD to combat veterans without PTSD and to noncombatant controls, only the PTSD group exhibited left amygdala activation in response to exposure to combat sounds [50].

Therefore, PTSD patients exhibit alterations in cerebral perfusion compared with age-matched normal subjects without re-exposure to accident-related stimuli. Yong An Chung et al found that PTSD patients showed alterations in cerebral blood perfusion under resting conditions compared with normal controls over a number of brain regions, in particular increased rCBF levels in limbic regions and decreased rCBF levels in the superior frontal gyrus and parietal and temporal regions. This result indicates that PTSD patients have abnormal cerebral perfusion levels in limbic regions and the frontal and temporal cortex under resting conditions (as displayed in Figure 3).

This finding supports the hypothesis of the involvement of limbic regions, which are thought to regulate emotion and memory, in the pathophysiology of PTSD.

Conclusion

The results of the present review concerning the epidemiology, neurochemistry, anatomy, risk factors, clinical and laboratory findings in PTSD and AD revealed that there are some similar points.

Although there is still little or no information about the relationship between PTSD and AD and none at all regarding the mode of presentation, course, treatment, complications, or patterns of service use in either disease when there is comorbidity.

Many caregivers in both clinical studies French [51] and Greek [32] but also in other studies have reported that the onset of symptoms of dementia occurred after a severe psychological stress, like a loss of a loved person, retirement, loss of their property etc. These observations suggest a possible relationship between stress factors, and cognitive decline. For the establishment of the possible role of posttraumatic stress in the occurrence of dementia more extensive and thorough studies should be performed. More concern of family and doctors during the period following the stressful event, in order to provide emotional support to the elderly, might be a field of intervention targeting the prevention of dementia. The conclusion of this review and the benefit is that severe psychological stress in the elderly is also another risk factor for dementia as a cause of PTSD. Therefore, the education of all people young and elderly in order to face severe psychological stress should include directions for the appropriate treatment of these conditions.

References

[1] Coburn-Litvak PS, Pothakos K, Tata DA, McCloskey DP, Anderson BJ. Chronic administration of corticosterone impairs spatial reference memory before spatial working memory in rats. *Neurobiol Learn Mem.* 2003;80:11-23.

[2] Starkman MN, Giordani B, Berent S, Schork MA, Schteingart DE. Elevated cortisol levels in Cushing's disease are associated with cognitive decrements. *Psychosom Med.* 2001;63:985-93

[3] Huang GJ, Herbert J. Serotonin modulates the suppressive effects of corticosterone on proliferating progenitor cells in the dentate gyrus of the hippocampus in the adult rat. *Neuropsychopharmacology.* 2005;30:231-41

[4] Yehuda R.. Post-traumatic stress disorder. *N Engl J Med* 2002*;346*: 108–14.

[5] [1] McKhann G, Drachman D, Folstein M, Katzman R, Price D, Stadlan EM.. Clinical diagnosis of Alzheimer's disease: report of the NINCDSADRDA Work Group under the auspices of Department of Health and Human Services Task Force on Alzheimer's Disease. *Neurology* 1984*;34*:939-944.

[6] Brookmeyer R, Gray S, Kawas C., Projections of Alzheimer's disease in the United States and the public health impact of delaying disease onset. *Am J Public Health* 1998*;88*:1337-1342.

[7] Fratiglioni L, Launer LJ, Andersen K, Breteler MM, Copeland JR, Dartigues JF, et al. Incidence of dementia and major subtypes in Europe: A collaborative study of population-based cohorts. Neurologic Diseases in the Elderly Research Group. *Neurology* 2000.*;54*:S10-5.

[8] Olszewski TM, Varrasse JF. The neurobiology of PTSD: Implications for nurses. *J Psychosoc Nurs Ment Health Serv.* 2005;43:40-47.

[9] Smid GE, Mooren TT, van der Mast RC, Gersons BP, Kleber RJ. Delayed posttraumatic stress disorder: systematic review, meta-analysis, and meta-regression analysis of prospective studies. *J Clin Psychiatry.* 2009 Jul 14.

[10] Ferri CP, Prince M, Brayne C, Brodaty H, Fratiglioni L, Ganguli M, et al. Global prevalence of dementia: a Delphi consensus study. *Lancet.* 2005*;366:*2112-2117.

[11] Wimo A, Winblad B, Engedal K, Soininen H, Verhey F, Waldemar G, et al.. An economic evaluation of donepezil in mild to moderate Alzheimer's disease: results of a 1-year, double-blind, randomized trial. *Dement Geriatr Cogn Disord* 2003*;15*:44-54.

[12] Miller G. Forgetting and Remembering. Learning to forget. *Science*, 2004; 304: 5667: 34-36.

[13] Rossi S, De Capua A, Tavanti M, Calossi S, Polizzotto NR, Mantovani A, Falzarano V, Bossini L, Passero S, Bartalini S, Ulivelli M. Dysfunctions of cortical excitability in drug-naïve posttraumatic stress disorder patients. *Biol Psychiatry.* 2009 1;66(1):54-61.

[14] Vanitallie TB. Stress: a risk factor for serious illness. *Metabolism.* 2002 ;51(6 Suppl 1):40-5.

[15] Sauro MD, Jorgensen RS, Pedlow CT. Stress, glucocorticoids, and memory: a meta-analytic review. *Stress.* 2003; 6(4):235-45.

[16] Sotiropoulos I, Catania C, Riedemann T, Fry JP, Breen KC, Michaelidis TM, Almeida OF. Glucocorticoids trigger Alzheimer disease-like pathobiochemistry in rat neuronal cells expressing human tau. *J Neurochem.* 2008;107(2):385-97

[17] Tsolaki M., Vasiliades G., Patousi A., Touliou A., Kazis A. Cortisol serum levels in Post Traumatic Stress Dementia *2004;19th Medical Conference of North Greece,* May.

[18] Bonhaus DW, Perry WB, McNamara JO. Decreased density, but not number, of N-methyl-D-aspartate, glycine and phencyclidine binding sites in hippocampus of senescent rats. *Brain Res* 1990;532:82-6.

[19] Seligman ME, Maier SF. Failure to escape traumatic shock. *J Exp Psychol* 1967;74:1-9.

[20] Stein-Behrens BA, Elliott EM, Miller CA, Schilling JW, Newcombe R, Sapolsky RM. Glucocorticoids exacerbate kainic acid-induced extracellular accumulation of excitatory amino acids in the rat hippocampus. *J Neurochem* 1992;58:1730-5.

[21] Gold PE. Sweet memories. *Am. Sci.* 1987;75:151-155.

[22] McGaugh JL. The amygdala modulates the consolidation of memories of emotionally arousing experiences. *Annu Rev Neurosci.* 2004; 27: 1-28.

[23] Bryant RA, Felmingham KL, Kemp AH, Barton M., Peduto AS, REnnie C., Gordon E., Williams LM. Neural networks of information processing in posttraumatic stress disorder: a functional magnetic resonance imaging study. *Biol Psychiatry* 2005;58:111-118.

[24] Tsolaki M, Kounti F, Karamavrou S. Severe Psychological Stress in Elderly Individuals: A Proposed Model of Neurodegeneration and Its Implications. *Am J Alzheimers Dis Other Demen.* 2009;24(2):85-94.

[25] [1] Qiu A, Fennema-Notestine C, Dale AM, Miller MI; Alzheimer's Disease Neuroimaging Initiative. Regional shape abnormalities in Mild Cognitive Impairment and Alzheimer's disease. *Neuroimage.* 2009 Jan 22.

[26] Gigi A, Babai R, Penker A, Hendler T, Korczyn AD. Prefrontal Compensatory Mechanism May Enable Normal Semantic Memory Performance in Mild Cognitive Impairment (MCI). *J Neuroimaging.* 2009 May 20.

[27] Bruen PD, McGeown WJ, Shanks MFVenneri A. Neuroanatomical correlates of neuropsychiatric symptoms in Alzheimer's disease. *Brain.* 2008;131(Pt 9):2455-63.

[28] de Quervain DJ, Kolassa IT, Ertl V, Onyut PL, Neuner F, Elbert T, Papassotiropoulos A. A deletion variant of the alpha2b-adrenoceptor is related to emotional memory in Europeans and Africans. *Nat Neurosci.* 2007;10(9):1137-9.

[29] Work Group on Alzheimer's Disease and other Dementias. Practice guideline for the Treatment of Patients With Alzheimer's Disease and Other Dementias. Available online at :http ://www.psychiatryonline.com/pracGuide/loadGuidelinePdf.aspx?file= AlzPG101007).

[30] Waldemar G, Dubois B, Emre M, Georges J, McKeith IG, Rossor M, et al. Recommendations for the diagnosis and management of Alzheimer's disease and other disorders associated with dementia: EFNS guideline. *Eur J Neurol* 2007;14(1): e1-26.

[31] Patterson C, Feightner J, Garcia A, Hsiung R, MacKnight C, Sadovnick D. Diagnosis and treatment of dementia: 1. Risk assessment and primary prevention of Alzheimer disease. *Can Med Assoc J* 2008;178:548-556.

[32] Tsolaki M., Papaliagkas V., Kounti F., Messini C., Boziki M., Anogianakis G., Vlaikidis N. Severe stress events and dementia: A study of an elderly Greek demented population. *Psychiatry Research* 2009.

[33] Norrholm SD, Ressler KJ. Genetics of anxiety and trauma-related disorders. *Neuroscience.* 2009 Jun 18.

[34] Mullan M, Crawford F, Axelman K, Houlden H, Lilius L, Winblad B, et al. A pathogenic mutation for probable Alzheimer's disease in the APP gene at the n-terminus of beta-amyloid. *Nat Genet* 1992;1:345-347.

[35] Sherrington R, Rogaev EI, Liang Y, Rogaeva EA, Levesque G, Ikeda M, et al. Cloning of a gene bearing missense mutations in early-onset familial Alzheimer's disease. *Nature* 1995;375:754-760.

[36] Corder EH, Saunders AM, Strittmatter WJ, et al. Gene dose of apolipoprotein E type 4 allele and the risk of Alzheimer's disease in late onset families. *Science* 1993; 261: 921–23.

[37] Tsolaki M., Boziki M., Molyva D., Pitsalidis C., Papassotiropoulos A., Kazis A. APOE alleles in Greek normal elderly and patients with Alzheimer's Disease.Nevrologia *(Greek Journal)* 2006; 15(2): 96-108.

[38] Newport, D. J., Nemeroff, C. B. Neurobiology of posttraumatic stress disorder. *Current Opinion in Neurobiology,* 2000;10:211–218.

[39] Papaliagkas V, Kimiskidis V, Tsolaki M, Anogianakis G..Usefulness of event-related potentials in the assessment of mild cognitive impairment.. *BMC Neurosci.* 2008 Nov 5;9:107.

[40] Jatzko A, Schmitt A., Kordon A., Braus DF. Neuroimaging findings in posttraumatic stress disorder: review of the literature. *Fortschr Neurol Psychiatr.* 2005; 73:377-91.

[41] Kemp PM, Holmes C, Hoffmann SM, Bolt L, Holmes R, Rowden J, et al. Alzheimer's disease: differences in technetium-99m HMPAO SPECT scan findings between early onset and late onset dementia. *J Neurol Neurosurg Psychiatry* 2003*;74*:715-719.

[42] Ishii K, Kawachi T, Sasaki H, Kono AK, Fukuda T, Kojima Y, et al.. Voxel-based morphometric comparison between early- and late-onset mild Alzheimer's disease and assessment of diagnostic performance of z score images. *American journal of neuroradiology* 2005*;26*:333-340.

[43] Mosconi L, De Santi S, Li Y, Li J, Zhan J, Tsui WH, et al. Visual rating of medial temporal lobe metabolism in mild cognitive impairment and Alzheimer's disease using FDG-PET. *European journal of nuclear medicine and molecular imaging.* 2006;33:210-221.

[44] Brun, A., Gustafson, L.,. Distribution of cerebral degeneration in Alzheimer's disease. A clinico-pathological study*, Arch. Psychiatr. Nervenkrankh.* 1976;223, 15– 33.

[45] Kantarci K, Jack CR Jr. Neuroimaging in Alzheimer disease: an evidence-based review. *Neuroimaging clinics of North America* 2003*;13*:197-209.

[46] Yetkin, F. Z., Rosenberg, R N., Weiner, M F., Purdy, P D., Cullum C. M. FMRI of working memory in patients with mild cognitive impairment and probable Alzheimer's disease. *Eur Radiol* 2006; *16*: 193–206.

[47] Veltmeyer MD, Clark CR, McFarlane AC, Moores KA, Bryant RA, Gordon E. Working memory function in post-traumatic stress disorder: an event-related potential study. *Clin Neurophysiol.* 2009;120(6):1096-106.

[48] Chung YA, Kim SH, Chung SK, Chae JH, Yang DW, Sohn HS, Jeong J.. Alterations in cerebral perfusion in posttraumatic stress disorder patients without re-exposure to accident-related stimuli. *Clinical Neurophysiology* 2006;*117*:637–642.

[49] Bremner J, Narayan M, Staib L, Southwick S, McGlashan T, Charney D., Neural correlates of memories of childhood sexual abuse in women with and without posttraumatic stress disorder. *Am J Psychiatry*, 1999; *156*:1787-1795.

[50] Liberzon I, Taylor SF, Amdur R, Jung TD, Chamberlain KR, Minoshima S, Koeppe RA, Fig LM. Brain activation in PTSD in response to trauma related stimuli. *Biol Psychiatry* 1999;*45*:817–26.

[51] Charles, E., Bouby-Serieys, V., Thomas, P., Clement, J.P., Links between life events, traumatism and dementia; an open study including 565 patients with dementia. *Encephale.* 2006;32,746–752.

In: Neurobiology of Post-Traumatic Stress Disorder
Editors: L. Sher and A. Vilens, pp. 217-227
ISBN: 978-1-61470-593-2

Chapter XIV

Traumatic Stress, PTSD and Complex Partial Seizure-like Symptoms

Petr Bob
Charles University, Prague, Czech Republic

Abstract

Recent reported findings indicate that stress experiences are related to psychological and neurobiological processes that may have lasting consequences and significantly influence brain functions. Stress also causes atypical memory consolidation leading to traumatic memory that in patients with PTSD has certain specific psychological and neurophysiological characteristics. Together these findings suggest that specific memory impairment and intrusive autobiographical memories of stressful experiences, and increased sensitivity to stress that occur in PTSD are related to dissociation. Cognitive and emotional dysregulation related to traumatic stress and dissociation is also likely linked to deficits in inhibitory functions and increased limbic excitability that may lead to temporo-limbic seizure-like activity. These findings strongly suggest that stress-activated limbic kindling may be involved in the pathogenesis of PTSD and may explain efficacy of antiepileptic drugs in the treatment of PTSD. Future research needs to determine whether patients with PTSD who have heightened level of psychosensory seizure-like symptoms are more likely to benefit from treatment interventions that use antiepileptic drugs.

Introduction

Recent research findings in clinical neurosciences provide evidence that stress experiences and especially traumatic stress are related to psychological and neurobiological processes that may have lasting consequences and significantly influence brain functions and morphology [1-5]. These maladaptive responses frequently exert negative influences on brain structures and systems involved in stress response and lead to an increased probability of relapses, recurrences, residual symptomatology, symptoms of traumatic stress, PTSD and dissociative symptoms [6-11].

A basic feature of this stress activation presents corticotropin-releasing hormone (CRH) that is closely related to locus ceruleusnorepinephrine autonomic neurons of the hypothalamus and brainstem, and regulates the peripheral activities of the hypothalamic-pituitary-adrenal (HPA) axis, the sympathetic nervous activity and also immune responses [12-14]. The CRH also directly stimulates IL-6, IL-1, IL-12 and substance P production and also upregulates TNFalpha [14-16]. On the other hand the stress activation of the immune system predominantly produces various cytokines such as IL-6, IL-1 and TNF-alpha that coordinate and integrate the brain-endocrine-immune response, and stimulate the hypothalamus, amygdala, and pituitary, which in turn stimulate CRH secretion and activate both the HPA axis and the sympathetic nervous system [14,17,18].

Other basic features of the stress activation are influences of proinflammatory cytokines, such as IL-6, IL-1 and TNF-alpha on deficits in neurogenesis and BDNF expression that may explain cognitive and memory deficits related to traumatic stress and dissociation [19-21]. The relationship between BDNF and cognition probably is mediated by influence of stress, and especially chronic stress that increases release of glucocorticoids from the adrenal gland, cause cell death or atrophy of vulnerable neurons through an inhibitory influence on BDNF synthesis in the hippocampus and the cortisol action [22]. Together recent findings indicate that BDNF plays an important role in stress response, modification of synaptic plasticity, transmission and memory formation especially in the hippocampus and neocortex, that have specific consequences in development of PTSD, depression, schizophrenia, epilepsy, neurodegenerative disorders and pain sensitization [23,24]. Additionaly, there is evidence that decreased BDNF expression may significantly influence reparative processes and neurogenesis of hippocampal neurons with resulting hippocampal atrophy that frequently occurs in patients with PTSD [2,25].

Traumatic Stress and Memory Consolidation

According to recent evidence, the hippocampus plays particularly important role in memory consolidation [26,27]. Memory consolidation presents fixation of new information in long-term memory by the process of long-term potentiation which can be significantly influenced by stress experiences that may affect memory processing [26,28]. Historical and recent findings indicate that repeated stress and especially traumatic stress experiences may also disturb mental integrity and lead to dissociation of memory and mental experience [29-31]. This clinical experience of psychological fragmentation of memory contents is in agreement with experimental evidence of the neural dissociability of the memory processes [32-36]. The evidence supports the view that memory systems concerned with encoding emotion and context are dissociable at psychological, physiological, and anatomical levels [35,37]. This process leads to memories that are stored without a contextual or spatiotemporal frame. Neurophysiologic processes that lead to consolidation of stressful emotional experiences therefore produce memories, which are often fragmentary, temporally, and spatially disorganized, mainly because they originate from entirely unrelated events [35,38].

Disturbed temporal memory related to stress conditions is evident from studies focused on episodic and autobiographical memories. For example, results in a group of 87 children aged 7–15 years, who were exposed to a traumatic event requiring hospitalization, indicate that specifically children, who showed temporal disorganization, but not absence of emotion or dissociative amnesia, in narrative themes were more likely to report concurrent subsyndromal PTSD symptoms at 4–7 weeks post-trauma [39].

Similarly other results provide evidence demonstrating that exposure to a significant psychological stressor preserves or even enhances memory for emotional aspects of an event, and simultaneously disrupts memory for non-emotional aspects of the same event [38]. These results are consistent with other findings of differential effects of stress on brain systems responsible for encoding and retrieving emotional memories in the amygdala and non-emotional memories in the hippocampal formation, and consistent with the view that memories formed under high levels of stress are not qualitatively the same as those formed under ordinary emotional circumstances but display typical forms of spatiotemporal disorganization, fragmentation, and incompleteness [35,38,40].

Traumatic Stress and Dissociation

Recent reported findings suggest that consolidation of traumatic memory in patients with PTSD has certain specific psychological and neurophysiological characteristics [8]. In subjective experience traumatic memory is not acceptable for conscious awareness because of coupled strong negative emotions [6]. Extremely negative emotion experienced during the traumatic event may lead to atypical memory consolidation characterized by consolidation process predominantly on an implicit (subliminal) level that produces a dissociated state of memory that does not fit into a conscious cognitive scheme [35,40,41]. This consolidation on an implicit level likely may be caused by inescapable stress, which may block the induction of long-term potentiation in the medial prefrontal cortex (PFC) and hippocampus. This blocking of higher order behavior mediated by the hippocampus and PFC allows more automatic responses dependent on subcortical structures, mainly the amygdala [26,41-43]. These findings are also in accordance with neuroimaging data which suggest that characteristic changes in the perfusion of limbic brain structures, such as the amygdala and the hippocampus, coincide with high arousal and/or anxiety during traumatic recall [43]. High anxiety and arousal are thought to intensely focus the attention, and this attentional shift may produce fragmented memories, personality fragmentation [41,43], psychological automatisms, and lowering of mental functioning as described by Janet [6,31].

These data also implicate important consequences for psychotherapy, and especially for therapy of PTSD that must enable memory reconsolidation in safe and nondangerous conditions that can lead to neurobiological reprocessing of memory traces. This reconditioning and reconsolidation is therefore possible only by reexperiencing the traumatic memory in a new and safe situation during a psychotherapeutic process that enables integration of the dissociated mental state related to previous traumatic experience [41]. During this process, implicitly consolidated traumatic memory in subcortical structures, mainly in the amygdala, is probably transformed from automatic into a higher level of conscious experience by long-term potentiation in the higher level structures of the central nervous system, such as the medial prefrontal cortex and hippocampus.Without reconsolidation, traumatic recall in patients with PTSD cannot be processed in an integrated mode of consciousness, and released traumatic memory remains dissociated as a discontinuous experience with amnesic gaps [7,41,44,45].

Together these findings suggest that specific memory impairment and intrusive autobiographical memories of stressful experiences, and increased sensitivity to stress that occur in PTSD are related to dissociation [7,31,44,45]. The memories reliving a dissociated traumatic event are often accompanied to increased sensitivity, aversive feelings and also to physical sensations representing somatoform components of dissociation. These symptoms of somatoform dissociation are frequently related to defensive reactions leading to analgesia and anesthesia, painful symptoms or passive coping strategies as a response to inescapable and threatening stress [6,7,31,44,45].

Stress, Sensitization and Epileptic like Brain Activity

Recent evidence indicates that various environmental factors such as perinatal damage, hypoglycemia, childhood stressful experiences, and other influences interacting with genetic basis may significantly affect persisting sensitivity to stress and other stimuli in later times [5,9,22,46,47]. This state of increased sensitivity and consequent alterations in cognitive biases that present important and critical conditions in pathogenesis of dissociative states and PTSD [5,9,11,46-48]. According to recent findings this increased sensitivity and imbalance of neural systems related to stress and dissociation could be explained by sensitization that leads to increased response to certain stimuli, which at the beginning of this process were subthreshold and caused no or low responses [49].

Recent findings suggest that repeated stressful events leading to an increase in responsiveness to stress stimuli and increased vulnerability to stressors may determine sensitization process that have lasting consequences with kindling-like progression [9,11]. According to several reported findings this specific form of sensitization with progressively increasing response of groups of neurons due to repetitive subthreshold stimulation may later lead to epileptic or epileptiform activity [9]. Process of kindling was firstly reported by Sevillano in 1961 and later elaborated by Goddard who comprehensively described this process and using stereotactically implanted electrodes into the amygdala and other brain regions found that repetitive subthreshold stimuli elicit only little changes in EEG activity or behavior whether later applied stimuli in a state of increased sensitivity result in local after-discharges, increased synchronization or seizures [50,51]. Recent findings suggest that the kindling related focal after-discharges within the amygdala and other brain structures may cause local changes in synchronization and seizure-like activity which could be important in pathogenesis of PTSD [9,11]. These findings potentially may explain treatment resistance to usual psychotropic medication in several PTSD patients and clinical importance of appropriate anticonvulsant medication even in patients who do not display seizures or epileptiform changes on scalp EEG [52,53].

The kindling-model of stress-related sensitization [9,11,46,47] seems to be in agreement with suggestive evidence that stress may influence significantly increased occurrence of EEG abnormalities that have been reported in significantly traumatized patients mainly in the frontotemporal region, which consisted of spikes, sharp waves, or paroxysmal slowing, predominantly in the left hemisphere [4,5,29,48,54].

Stress-related sensitization has been proposed to cause changes in GABA postsynaptic receptors that may lead to overstimulation of neurons mainly in the limbic system, resulting in limbic system irritability manifesting as markedly increased prevalence of seizure-like symptoms suggestive of temporal lobe epilepsy [4,5,48]. Recent data strongly suggest that traumatic stress may determine limbic irritability and temporal-limbic seizure-like activity [4,5,29,48,55] and close link between limbic irritability measured by questionnaire LSCL-33 (Limbic System Checklist) and several pathophysiological changes in cerebellar vermis has been reported [4,5,48,56]. These findings suggest that cognitive and emotional dysregulation related to traumatic stress and atypical memory consolidation leading to dissociation likely could be closely linked to deficits in inhibitory functions and increased limbic excitability (or irritability) that may also lead to temporo-limbic seizure-like activity.

Conclusion

Together recent clinical findings suggest that epileptiform activity in temporolimbic structures may cause psychiatric symptoms that may occur independently on conditions of common neurological diagnostic criteria for epilepsy. This evidence indicates that this epileptic-like process may emerge in the form of symptoms similar to several symptoms of temporal lobe epilepsy (the so-called complex partial seizure-like symptoms) such as somatic, sensory, behavioral and memory symptoms (See Table 1) that may occur also in nonepileptic conditions [4,5,9,29,48,57-59]. Clinical importance of these symptoms in PTSD patients suggests study by Roca and Freeman [60] who reported relevance of psychosensory complex partial seizure-like symptoms for the study of chronic PTSD and found that their presence is associated with significantly more severe PTSD symptoms, dissociative symptoms, aggression, and overall psychopathology.

At this point it is needed to emphasize a clear distinction between epileptic and epileptiform discharges and relationship between ictal and interictal symptomatology. While epilepsy is defined as a chronic condition characterized by spontaneous, recurrent seizures and seizure is defined as a clinical event associated with a transient, hypersynchronous neuronal discharge (i.e. word epileptic is a descriptive term used to denote the presence of epilepsy), epileptiform discharges is an interpretative term used in electroencephalography that applies to distinctive waves or complexes distinguishable from the background activity, which resemble the waveforms recorded in a proportion of human subjects suffering from an epileptic disorder [61]. Epileptiform patterns include spike and sharp waves, alone or accompanied by slow waves, occurring singly or in bursts lasting at most a few seconds. The term epileptiform typically refers to interictal paroxysmal activity and not to the EEG activity seen during an actual seizure, which is called an electrographic seizure [61]. In principle this could implicate that epileptiform activity may occur in two significant conditions. The first presents interictal epileptiform activity in epileptic patients whereas the second condition presents cases in which epileptiform activity may occur independently of any seizure in patient's history (e.g. in patients with psychiatric disorders).

Table 1. Items of Structured clinical interview for complex partial seizure-like symptoms (also called Iowa Interview for Partial Seizure-like Symptoms) by Roberts et al. (1992) [57]

1. Do you sometimes smell things which other people can't smelf, such as feces, urine, rot, body odor, or smoke? Be sure in responding to this that the smells you report have no apparent cause (e.g., smelling kitty litter when you don't own a cat).
2. Do you sometimes have a bad taste in your mouth, such as a metallic taste, which comes and goes for no reason?
3. Do you sometimes sense movement in peripheral vision, but when you turn to look you can not see anything?
4. Do you sometimes see things in your peripheral vision, such as stars, bugs, snakes, worms or threads?
5. Do you sometimes see mice or cockroaches run across the floor, but when you turn to look, you do not see them?
6. Do you sometimes feel as though bugs are crawling on you or that something is brushing up against your skin, such as a cobweb?
7. Do you sometimes go numb in a part of your body for no apparent reason?
8. Do you sometimes get a ringing, buzzing, rushing or tapping noise in your ears which comes and goes for no reason?
9. Do you sometimes answer the telephone only to find that it had not actually been ringing?
10. Do you sometimes get severe headaches that are so bad you become nauseated or want to throw up?
11. Do you sometimes get a pain in your head which you would not classify as 'headache'?
12. Do you sometimes have marked urinary urgency, but fail to produce any urine when going to the bathroom?
13. Do you sometimes have trouble with the pronunciation of words with the effect that you appear a bit intoxicated even though you are not?
14. Is it a common problem of yours that you will suddenly have trouble thinking of words you should know and were able to say moments before?
15. Do you sometimes find that you have uttered a sentence which doesn't make any sense and involves words other than those you wished to say?
16. Do you sometimes become quite suddenly and intensely confused and perplexed and then have the feeling pass in a few minutes?
17. Do you sometimes have an overwhelming feeling that things are weird, strange, or wrong, sort of like entering the twilight zone?
18. Do you sometimes feel that familiar places or persons are somehow not familiar or the way they should be?
19. Do you sometimes get the feeling that you have experienced something or been someplace before even though you know you have not?
20. Do you have clear cut gaps in your memory during which you cannot remember anything that happened over a period of 5 minutes or more?
21. Do you sometimes find that you have missed major sections of TV shows you have been watching, like someone has spliced a section out of a movie?
22. Have you ever found yourself driving without remembering how you got there or where you are going?
23. Do people often tell you about things you have done or said for which you have no memory at all?

24. Do you have staring spells where you become sort of hypnotized by a bright or shiny object?
25. Do people often tell you that there are times when you are staring and have a blank look on your face?
26. Do you feel that your memory or concentration is getting substantially worse every year? (no=0; yes=5)
27. Do you sometimes lose consciousness or just black out?
28. Are you regularly so depressed that you think seriously about suicide? (no=0; yes=5)
29. Do you sometimes become abruptly more depressed than you were a few minutes or seconds earlier with no apparent cause?
30. Are you often inclined to panic or become very anxious for no reason?
31. Do you sometimes become extremely and intensely angry for no reason?
32. Do people tell you that you have become very angry and you do not remember?
33. Do people tell you that you sometimes have an intensely angry expression on your face while asleep?
34. Do you sometimes feel an irresistible urge to sleep during the day and then sleep so soundly that no one can arouse you?
35. Do you sometimes wake up to realize that you have been sweating so much that the bed sheets are soaked?

Note: The items are rated by the patient for frequency of occurrence on the Likert scale (0, 1, 2, 3, 4, 5): 0 Never or not in the past year; 1 Two a three times in the past year; 2 At least once a month; 3 At least once a week; 4 Several times a week; 5 At least once a day. Score is a sum of all items (70 or higher is a criterion for epilepsy spectrum disorder- Roberts et al., 1992).

These reported findings are in agreement with clinical evidence that many patients with PTSD are unresponsive or display only moderate responses with frequent side effects to first-line serotonin reuptake inhibitor treatment and several studies suggest that antiepileptic drugs may be an effective treatment alternative or efficient adjunctive treatment for PTSD [52,53]. Recent clinical evidence from case reports, double-blind controlled studies and placebo-controlled trials on the efficacy and tolerability of antiepileptic drugs in PTSD have shown that lamotrigine, topiramate, and tiagabine seem to be effective for PTSD treatment [52,53]. Other antiepileptic drugs that seem to be promising in open-label trials in PTSD patients include carbamazepine, valproate, gabapentin, vigabatrin, phenytoin, and levetiracetam [52]. These findings also suggest that stress-activated limbic kindling may be involved in the pathogenesis of PTSD and may explain efficacy of antiepileptic drugs in the treatment of PTSD due to their antikindling effect. Future research needs to determine whether patients with PTSD who have heightened level of complex partial seizure-like symptoms or dissociative symptoms are more likely to benefit from treatment interventions that use antiepileptic drugs. On the other side various psychotherapeutic approaches to PTSD that in principle seem to be based on various techniques that enable reconsolidation of traumatic memories also present necessary approach to psychological and neurobiological understanding of PTSD and its therapy.

References

[1] Bremner JD: Neuroimaging in posttraumatic stress disorder and other stress-related disorders. *Neuroimaging Clin N Am* 2007;17:523-38.

[2] Bremner JD, Elzinga B, Schmahl C, Vermetten E: Structural and functional plasticity of the human brain in posttraumatic stress disorder. *Prog Brain Res* 2008;167:171-186.

[3] Bremner JD: Does stress damage the brain? *Biol Psychiatry* 1999;45:797-805.

[4] Teicher MH, Tomoda A, Andersen SL: Neurobiological consequences of early stress and childhood maltreatment: are results from human and animal studies comparable? *Ann N Y Acad Sci* 2006;1071:313-323.

[5] Teicher MH, Andersen SL, Polcari A, Anderson CM, Navalta CP, Kim DM: The neurobiological consequences of early stress and childhood maltreatment. *Neurosci Biobehav Rev* 2003;27:33-44.

[6] Bob P: Pain, dissociation and subliminal self-representations. *Conscious Cogn* 2008;17:355-369.

[7] Bryant RA: Does dissociation further our understanding of PTSD? *J Anxiety Disord* 2007;21:183-191.

[8] Heim C, Nemeroff CB: Neurobiology of posttraumatic stress disorder. *CNS Spectr* 2009;14:13-24.

[9] Kraus JE: Sensitization phenomena in psychiatric illness: lessons from the kindling model. *J Neuropsychiatry Clin Neurosci* 2000;12:328-343.

[10] Ozer EJ, Best SR, Lipsey TL, Weiss DS: Predictors of posttraumatic stress disorder and symptoms in adults: a meta-analysis. *Psychol Bull* 2003;129:52-73.

[11] Post RM, Weiss SR, Smith M, Li H, McCann U: Kindling versus quenching. Implications for the evolution and treatment of posttraumatic stress disorder. *Ann N Y Acad Sci* 1997;821:285-295.

[12] Chrousos GP: The hypothalamic-pituitary-adrenal axis and immune-mediated inflammation. *N Engl J Med* 1995;332:1351-1362.

[13] Elenkov IJ, Wilder RL, Chrousos GP, Vizi ES: The sympathetic nerve- an integrative interface between two supersystems: the brain and the immune system. *Pharmacol Rev* 2000;52:595-638.

[14] Elenkov IJ, Chrousos GP: Stress hormones, proinflammatory and antiinflammatory cytokines, and autoimmunity. *Ann N Y Acad Sci* 2002;966:290-303.

[15] Leu SJ, Singh VK: Stimulation of interleukin-6 production by corticotropin-releasing factor. *Cell Immunol* 1992;143:220-227.

[16] Lotz M, Vaughan JH, Carson DA: Effect of neuropeptides on production of inflammatory cytokines by human monocytes. *Science* 1988;241:1218-1221.

[17] Besedovsky H, del RA, Sorkin E, Dinarello CA: Immunoregulatory feedback between interleukin-1 and glucocorticoid hormones. *Science* 1986;233:652-654.

[18] Kovacs KJ, Elenkov IJ: Differential dependence of ACTH secretion induced by various cytokines on the integrity of the paraventricular nucleus. *J Neuroendocrinol* 1995;7:15-23.

[19] Bob P, Raboch J, Maes M, Susta M, Pavlat J, Jasova D et al.: Depression, traumatic stress and interleukin-6. *J Affect Disord* 2009 doi:10.1016/j.jad.2009.03.017 (in press).

[20] Maes M, Yirmyia R, Noraberg J, Brene S, Hibbeln J, Perini G et al.: The inflammatory & neurodegenerative (I&ND) hypothesis of depression: leads for future research and new drug developments in depression. *Metab Brain Dis* 2009;24:27-53.

[21] Schulte-Herbruggen O, Nassenstein C, Lommatzsch M, Quarcoo D, Renz H, Braun A: Tumor necrosis factor-alpha and interleukin-6 regulate secretion of brain-derived neurotrophic factor in human monocytes. *J Neuroimmunol* 2005;160:204-209.

[22] Savitz J, Solms M, Ramesar R: The molecular genetics of cognition: dopamine, COMT and BDNF. *Genes Brain Behav* 2006;5:311-328.

[23] Binder DK, Scharfman HE: Brain-derived neurotrophic factor. *Growth Factors* 2004; 22:123-131.

[24] Thomas K, Davies A: Neurotrophins: a ticket to ride for BDNF. *Curr Biol* 2005;15: R262-R264.

[25] Bremner JD: Stress and brain atrophy. *CNS Neurol Disord Drug Targets* 2006;5:503-512.

[26] Lynch MA: Long-term potentiation and memory. *Physiol Rev* 2004;84:87-136.

[27] Morris RG, Moser EI, Riedel G, Martin SJ, Sandin J, Day M et al.: Elements of a neurobiological theory of the hippocampus: the role of activity-dependent synaptic plasticity in memory. *Philos Trans R Soc Lond B Biol Sci* 2003;358:773-786.

[28] McGaugh JL: Memory- a century of consolidation. *Science* 2000;287:248-251.

[29] Bob P: Dissociation and neuroscience: history and new perspectives. *Int J Neurosci* 2003;113:903-914.

[30] Kihlstrom JF: Dissociative disorders. *Annu Rev Clin Psychol* 2005;1:227-253.

[31] Spiegel D: Trauma, dissociation, and memory. *Ann N Y Acad Sci* 1997;821:225-237.

[32] LeDoux JE: Brain mechanisms of emotion and emotional learning. *Curr Opin Neurobiol* 1992;2:191-197.

[33] LeDoux JE: Emotional memory systems in the brain. *Behav Brain Res* 1993;58:69-79.

[34] LeDoux JE: Emotion, memory and the brain. *Sci Am* 1994;270:50-57.

[35] Nadel L, Jacobs WJ: Traumatic memory is special. *Curr Dir Psychol Sci* 1998;7:154-157.

[36] Phillips RG, LeDoux JE: Differential contribution of amygdala and hippocampus to cued and contextual fear conditioning. *Behav Neurosci* 1992;106:274-285.

[37] Bechara A, Tranel D, Damasio H, Adolphs R, Rockland C, Damasio AR: Double dissociation of conditioning and declarative knowledge relative to the amygdala and hippocampus in humans. *Science* 1995;269:1115-1118.

[38] Payne JD, Jackson ED, Ryan L, Hoscheidt S, Jacobs JW, Nadel L: The impact of stress on neutral and emotional aspects of episodic memory. *Memory* 2006;14:1-16.

[39] Kenardy J, Smith A, Spence SH, Lilley PR, Newcombe P, Dob R et al.: Dissociation in children's trauma narratives: an exploratory investigation. *J Anxiety Disord* 2007;21:456-466.

[40] Brewin CR: Autobiographical memory for trauma: update on four controversies. *Memory* 2007,15:227-248.

[41] Bob P: Hypnotic abreaction releases chaotic patterns of electrodermal activity during dissociation. *Int J Clin Exp Hypn* 2007;55;435-456.

[42] Maroun M, Richter-Levin G: Exposure to acute stress blocks the induction of long-term potentiation of the amygdala-prefrontal cortex pathway in vivo. *J Neurosci* 2003;23:4406-4409.

[43] Vermetten E, Douglas BJ: Functional brain imaging and the induction of traumatic recall: a cross-correlational review between neuroimaging and hypnosis. *Int J Clin Exp Hypn* 2004;52:280-312.

[44] Nijenhuis ER, van der HO, Kruger K, Steele K: Somatoform dissociation, reported abuse and animal defence-like reactions. *Aust N Z J Psychiatry* 2004;38:678-686.

[45] van der Hart O, Nijenhuis ER, Steele K: Dissociation: An insufficiently recognized major feature of complex posttraumatic stress disorder. *J Trauma Stress* 2005;18:413-423.

[46] Post RM, Weiss SR, Pert A: Implications of behavioral sensitization and kindling for stress-induced behavioral change. *Adv Exp Med Biol* 1988;245:441-463.

[47] Post RM, Weiss SR, Li H, Leverich GS, Pert A: Sensitization components of post-traumatic stress disorder: implications for therapeutics. *Semin Clin Neuropsychiatry* 1999;4:282-294.

[48] Teicher MH, Glod CA, Surrey J, Swett C, Jr.: Early childhood abuse and limbic system ratings in adult psychiatric outpatients. *J Neuropsychiatry Clin Neurosci* 1993;5:301-306.

[49] Post RM, Weiss SR: A speculative model of affective illness cyclicity based on patterns of drug tolerance observed in amygdala-kindled seizures. *Mol Neurobiol* 1996, 13: 33-60.

[50] Goddard GV: Development of epileptic seizures through brain stimulation at low intensity. *Nature* 1967;214:1020-1021.

[51] McNamara JO, Morrisett R, Nadler JV: Recent advances in understanding mechanisms of the kindling model. *Adv Neurol* 1992;57:555-560.

[52] Berlin HA: Antiepileptic Drugs for the Treatment of Post-traumatic Stress Disorder. *Curr Psychiatry Rep* 2007;14:185-189.

[53] Johannessen LC: Antiepileptic drugs in non-epilepsy disorders: relations between mechanisms of action and clinical efficacy. CNS Drugs 2008;22:27-47.

[54] Spiegel D: Neurophysiological correlates of hypnosis and dissociation. *J Neuropsychiatry Clin Neurosci* 1991;3:440-445.

[55] Bob P, Susta M, Prochazkova-Vecerova A, Kukleta M, Pavlat J, Jagla F et al.: Limbic irritability and chaotic neural response during conflicting stroop task in the patients with unipolar depression. *Physiol Res* 2006;55 Suppl 1:S107-S112.

[56] Anderson CM, Teicher MH, Polcari A, Renshaw PF: Abnormal T2 relaxation time in the cerebellar vermis of adults sexually abused in childhood: potential role of the vermis in stress-enhanced risk for drug abuse. *Psychoneuroendocrinology* 2002;27: 231-244.

[57] Roberts RJ, Gorman LL, Lee GP, Hines ME, Richardson ED, Riggle TA et al.: The phenomenology of multiple partial seizure-like symptoms without stereotyped spells: an epilepsy spectrum disorder? *Epilepsy Res* 1992;13:167-177.

[58] Silberman EK, Post RM, Nurnberger J, Theodore W, Boulenger JP: Transient sensory, cognitive and affective phenomena in affective illness. A comparison with complex partial epilepsy. *Br J Psychiatry* 1985;146:81-89.

[59] Varney NR, Garvey MJ, Cook BL, Campbell DA, Roberts RJ: Identification of treatment-resistant depressives who respond favorably to carbamazepine. *Ann Clin Psychiatry* 1993;5:117-122.

[60] Roca V, Freeman TV: Psychosensory symptoms in combat veterans with posttraumatic stress disorder. *J Neuropsychiatry Clin Neurosci* 2002;14:185-189.

[61] Chatrian GE, Bergamini L, Dondey M, et al.: A glossary of terms most commonly used by clinical electroencephalographers. *Electroencephalogr Clin Neurophysiol* 1974;37:538-548.

In: Neurobiology of Post-Traumatic Stress Disorder
Editors: L. Sher and A. Vilens, pp. 229-257
ISBN: 978-1-61470-593-2

Chapter XV

Brain Adenosine Signaling in Psychological Trauma and Comorbid Depression

Thomas R. Minor, Traci N. Plumb, Christopher J. Schell and Anh K. Pham
University of California, Los Angeles, California, USA

Abstract

This chapter reviews human and animal literature on the psychological and biological consequences of maintaining and intense, prolonged fear response. This type of psychological trauma alters neural regulatory mechanisms in the CNS that renders fear- and anxiety-related circuitry hyper-responsive such that individual overreact to relatively mild stressors. The metabolic challenge associated with maintaining a state of hyperactivity engages a compensatory signaling mechanism involving adenosine. We review evidence from the learned helplessness rat model of PTSD and comorbid depression that enhanced brain adenosine signaling mediates the transition from a state of anxiety and agitation to a depression-like state termed conservation-withdrawal. We also present new evidence that the conservation-withdrawal aspects of helplessness are mediated in the striatum where adenosine activates specific A_{2A} receptors to uncouple dopamine's motivational signal from ongoing behavior.

I. Introduction

Dire war conditions in Iraq underscore the need for basic and applied research directed at decreasing the prevalence of post-traumatic stress disorder (PTSD) among troops returning from duty and developing effective treatments for symptoms of the disorder. The prevalence of PTSD in active duty soldiers is estimated at 16.6% from a screen of over 2800 veterans returning from Iraq [1, 2]. This alarming prevalence rate has not been seen since the Vietnam War, where 14.7% of soldiers met criteria for lifetime PTSD [3]. The two wars are remarkably similar in that they represent *wars without fronts* where allies and enemies are difficult to discern [3]. Current conditions also are exacerbated by a lack of adequate training for the law-enforcement role played by soldiers in the Iraqi theater, and the use of insidious IED munitions (improvised explosive devises), which add dimensions of terror and unpredictability to everyday life. The uncertainty associated with these conditions often exaggerates emotional output under conditions of threat and can seriously compromise an individual's perception of control in a combat situation [4, 5]. The loss of perceived control and prediction further compromises the ability to regulate negative emotional output. This combination of conditions provides "the prefect storm" for an unusually high incidence of PTSD among American troops in Iraq.

This chapter reviews research from our laboratory on the psychological and neurobiological processes associated with the development of PTSD-like symptoms and comorbid depression in a rat model of psychological trauma. This research emphasizes the brain metabolic consequences of an emotional reaction of extreme fear or terror. PTSD was referred to as combat fatigue in wars past. Our work suggests in a very real sense that PTSD symptoms arise from fatigue associated with excessive emotional output. The critical metabolic challenge and fatigue occur in the brain following unusually high and prolonged neural activation. Metabolic challenge and neural fatigue have the paradoxically effect of making fear-processing centers in the brain more reactive [6]. In turn, overactive of these systems engages compensatory mechanisms that decrease activity. This chapter focuses on one of these compensatory signaling mechanisms involving the purine nucleoside adenosine and the consequences of activation of this system for emotional health.

II. Modeling Psychological Trauma

According to the Diagnostic and Statistical Manual for Mental Disorders, 4th edition (DSM-IV), post-traumatic stress disorder is an Axis I anxiety disorder where the individual has direct personal experience of a traumatic event or series of events that are deemed life threatening and have the possibility to inflict serious, mortal harm [7, 8, 9, 10]. The threat of injury or death then elicits a strong sense of fear, horror, and helplessness perceived by the individual as inescapable. Extreme psychological trauma further overwhelms the person's biological and psychological defenses against stress, disrupting the individual's locus of control and predictability within the situation [11].

Three different symptoms are evident in the aftermath of the trauma: The individual frequently re-experiences the trauma, avoids reminders of the event, and suffers symptoms of hyperarousal (12, 7, 8, 9, 10]. The first symptom is often unprovoked and unwanted, as the individual re-experiences the event through nightmares, memories, and flashbacks [13, 8, 12, 10]. Active avoidance of reminders of the event includes people who were present during the experience, the location of the trauma, and even event-related thoughts [14, 8, 12, 10]. Hyperarousal is primarily somatic/ physical remnant of trauma in which the person suffers from impaired concentration, mild to severe insomnia, irritability, and an exaggerated startle response [15, 8, 12, 10]. Among the three core symptoms, an individual also may experience diminished interest in daily activities, dysthymia, and restricted affect and emotion [16]. These symptoms are evident for at least 1 month post-trauma, although they often times persists for several years [9, 12, 10, 17]. Substance abuse and generalized anxiety disorder are also common long-term problems and can complicate treatment and recovery [16].

Major depression is the most common comorbid disorder with PTSD [16, 12, 9]. Data from the National Comorbidity Survey (NCS) show that 49% of women and 48% of men with PTSD also had lifetime major depressive disorder [18, 16]. Furthermore, the onset of the two disorders can co-exist and as a result make the traumas more difficult to distinguish [16]. This is remarkable yet informative when considering the symptoms of major depression, symptoms which are both distinct and related to those in PTSD [7, 19, 20]. An individual with depression usually experiences at least five of these symptoms within a two-week period: (a) depressive mood and/or feelings of sadness and emptiness; (b) diminished interest in many activities; (c) weight loss or gain with erratic changes in appetite; (d) fatigue and loss of energy almost every day; (e) increased agitation; (f) insomnia; (g) trouble concentrating; (h) and often times, increased ideation and thoughts of suicide [7, 19, 20]. The two categories of (1) fatigue and (2) sadness have come to be those of most interest in our laboratory. However, our focus has primarily been geared toward the aspects of fatigue rather than the emotional state of the animal, as the physiological response of fatigue is a measurable component of helplessness; measuring sadness in rodents is not an option.

Something is always lost when animals are used to model a human psychiatric disorder. The complexity of human cognition and emotions are never well represented. Nonetheless, the obvious advantages of being able to rapidly produce the disorder, as well as the ability to analyze the psychological and neurobiological components of the disorder make the endeavor worthwhile. We have use the learned helplessness procedure to this end, both because of its face and construct validity in the conditions producing psychological trauma and a reasonable overlap in PTSD-like symptomology.

A. Learned Helplessness Procedure

This procedure is a traditional method for analyzing the effects of acute, traumatic stress and modeling related symptoms of post-traumatic stress disorder and comorbid major depression in rats [5, 21, 22, 23, 24]. The experiment consists of two phases in which rats are exposed to a series of escapable electric shocks, yoked inescapable shocks, or simple apparatus restraint in the pretreatment phase [25]. The restrained group provides a baseline from which any effect of stressor controllability can be assessed during later testing. Although the nature of the test varies with the interests of the experimenter, the traditional measure of helplessness has been performance in a shuttle-escape task conducted 24 hr after stress pretreatment. Our shuttle-escape task consist of five trials during which a rat must cross from one side of a two-compartment shuttlebox to the other in order to terminate electric footshock (FR-1 trials), followed by 25 trials during which a rat must run from one side of the shuttlebox to the other, then return to terminate shock (FR-2 trials). Both trial types are presented with an average intertrial interval of 60 seconds; however, 3 minutes intervene between trial types, which maximize escape deficits in inescapably shocked rats without adversely affecting control subjects [26]. Trials terminate automatically if the appropriate response requirement is not met within 40 sec. of shock onset.

We also have measured behavioral freezing during the FR-1 trials of the shuttle-escape test in recent experiments. Freezing is a species-specific defense response of the rat [27] and is an excellent measure of fearfulness during testing. This response is greatly exaggerated in rats pre-exposed to inescapable shock [28, 29] and is ameliorated by manipulations that improve test escape performance and reduce other symptoms of PTSD. Some of these data will be presented toward the end of the chapter.

Escape latencies typically do not differ among groups during FR-1 trials. However, large performance differences occur when the response requirement is made more difficult (FR-2 trials). Rats pre-exposed to escapable shock in the pretreatment phase perform as efficiently as restrained controls during escape testing 24 hours later. By contrast, rats pre-exposed to yoked inescapable shock show severe impairment, with near-maximum escape latencies during the FR-2 trials.

This general pattern among groups holds for a wide variety of behavioral and biological stress indexes. Moreover, because escapably and inescapably shocked rats receive the same pattern, intensity, and durations of shock during pretreatment, the differential performance of these two groups in the test phase provides unequivocal evidence that some psychological variable related to behavioral control, or lack thereof, modulates the impact of the shock stressor [30].

A simpler version of this procedure often is used when testing the efficacy of experimental compounds at alleviating stress-induced pathology. Because rats exposed to escapable shock do not show behavioral or physiological impairment, this condition adds an unnecessary expense to basic pharmaceutical research. Thus, much of the research discussed here compares rats exposed to variable-duration inescapable shocks during the pretreatment phase with the restrained control.

B. Intense, Chronic Fear and Helplessness

A considerable effort has been made over the years at both the empirical and theoretical level to understand the psychology and neurobiology of helplessness. Early explanations [see reviews: 5, 6] of the phenomenon emphasized the deleterious impact of uncontrollable stressors on psychological well-being, physiological function, and disease processes. Extensive analysis of the psychological dimension of stressor controllability eventually indicated that the lack of control over a stressor is neither necessary nor sufficient for a helplessness effect [31]. Moreover, the availability of an escape or coping response turned out to be only one of several conditions that mitigated the impact of a shock stressor [32, 31, 33, 34]. More recent analyses clearly implicate the prolonged, intense fear that is elicited during pretreatment with inescapable shock as the culprit. As such, the learned helplessness procedure has morphed from a model of major depression to a model of PTSD. Whether test behavior is more anxiety-like or more depression-like depends on when and how the animal is tested. Both emotional states are evident at different times during the traditional shuttle-escape test. Whereas inescapably shocked rats begin testing in a highly anxious/agitated state, they rapidly transition into a depression-like state following exposure to relatively mild footshock stress. Thus we have used this paradigm to model PTSD and comorbid depression.

The conditions that modulate fear during a shock stressor in the helplessness paradigm are very similar to those that regulate emotional outflow in human in the face of a severe threat [4]. Exposure to unsignaled, inescapable shock is a low-information condition in which periods of danger and safety are difficult to discriminate [35, 6, 5, 36]. In the absence of clear safety information, the maximum level of fear supported by the shock intensity remains relatively constant throughout the entire pretreatment. This session is long (about 2 hours) and shocks used to induce helplessness are intense. Thus the rat remains in an intense, chronic fear state for an extended period of time. The animal simply is not designed to remain in this highly catabolic state for such a prolonged period. Fear conditioning is maximized [35, 26, 29], fearfulness is sensitized [28, 6] and transient, nonassociative changes in brain and endocrine function are provoked that render the rat highly vulnerable to subsequent stresses for an extended period of time [5, 37].

Escapable shock is far less debilitating because it is less fear provoking. Stimuli generated in the act of escaping acquire fear-inhibitory (safety) properties due to their consistent association with shock termination and a shock-free period. These safety cues dramatically reduce the total amount of time spent in fear during the pretreatment session [32, 35, 31]. Less fear is conditioned to the pretreatment context and the nonassociative consequences of excessive fear are prevented as a result.

Inescapably shocked rats enter the test phase of the experiment in an anxious, agitated state relative to escapably shocked or restrained controls [29]. Re-exposure to mild footshock on the initial escape trials provokes intense behavioral activation, rapid and exaggerated fear acquisition [28] and excessive and unregulated neural activation [38, 39, 40, 41, 42, 43, 44].

We have argued that the most substantial impact of this initial activation response is on brain metabolic processes [36, 6]. The challenge to energetic homeostasis recruits regulatory process that attempt to restore homeostasis, apparently at the expense of ongoing behavioral processes. At this point, inescapably shocked rats show a dramatic shift from a state of active coping to a depression-like state of dissociation and behavior quiescence.

As discussed in detail below, we have linked this transition in behavioral and emotional state to an increase in brain adenosine signaling. Adenosine is liberated in extracellular space during excessive neural activation or other compromises to energy homeostasis [46, 47, 48, 49, 50, 51].

III. Adenosine Neuromodulation

A. Pharmacology and Function

Figure 1 shows a generic synaptic cleft in the CNS as a means of summarizing main features of adenosine neuromodulation. Adenosine participates in a feedback signaling pathway that regulates excitable tissue with respect to available energy [52, 53, 54]. Adenine nucleotides (ATP: adenosine triphosphate) and nucleosides (adenosine) serve as endpoints in intracellular energy metabolism, with expendable energy represented by the number of high-energy phosphate bonds. Research over the past decade has shown that purines also are liberated into extracellular space to convey information about intracellular energy state. External binding sites for adenosine (P_1 receptors) and ATP (P_2 receptors) are widely distributed in the CNS and at neuroeffector junctions in the periphery [55, 53]. Although there are a number of complexities to this form of homeostatic regulation, extracellular ATP usually signals a state of high or adequate energy and enhances membrane excitability. Conversely, adenosine is liberated during energy failure and inhibits membrane excitability (among other effects).

Extracellular ATP can exert a paracrine modulatory function by binding to stereospecific set of P2 receptors [56]. Activation of these receptors alters vascular tone and has a general excitatory effect on neural transmission. This type of signaling has been reviewed in detail elsewhere [57, 58, 59, 60, 61, 62, 63] and will not be considered here.

There are two main sources of extracellular adenosine. First, ATP utilization increases within a cell as workload increases. Dephosphorylation of ATP and other nucleotides is a source of intracellular adenosine in the nanomolar range. The increase intracellular adenosine translated into a rise in extracellular levels as most cells are equipped with bidirectional transporters [64] which extrude the nucleoside into extracellular space. Second, adenosine is produced by the extracellular degradation of ATP by several enzyme families of ectonucleotidases [58, 65]. This process begins with sequestration of ATP into synaptic vesicles. ATP is actively transported into transmitter-containing vesicles [66, 67] where it is co-released during synaptic activation. The nucleotide is known to be released as a cotransmitter in this manner with norepinephrine, serotonin, dopamine, glutamate, and γ-aminobutyric acid (GABA) in the CNS [68] and probably is sequestered with all transmitters, albeit at various concentrations.

The overall production of adenosine resulting from the synaptic release of ATP is a function of the level of neural activation, and therefore the concentration of ATP released as a cotransmitter, as well as the level of enzymatic activity in the ectonucleotidase cascade. These enzymes respond to changes in pH as pCO_2 concentrations vary with cellular respiration. Under normal circumstances, low pCO_2 and a relatively alkaline pH associated with neural quiescence inhibit ectonucleotidases. ATP concentrations increase in the synaptic cleft with little adenosine production. As the level of neural activation increases, pH becomes more acidic with an increase in pCO_2. The inhibition of ectonucleotidases is removed and extracellular adenosine concentrations increase [69, 70, 71, 72, 65].

Adenosine exerts its homeostatic and regulatory actions by interacting with four G-protein coupled stereospecific P1 receptors: A_1, A_{2A}, A_{2B}, & A_3 (see reviews: 73, 74). A_1 receptors are widely distributed in the brain and mediate adenosine's inhibitory actions by coupling with a G_i protein that inhibits adenylyl cyclase [75]. A_2 receptors mediate adenosine's excitatory actions by coupling a G_s protein that excites adenylyl cyclase [76, 77]. The A_{2B} subtype is a low-affinity receptor that is widely distributed in most brain regions. The high-affinity A_{2A} subtype has a much more limited distribution, being localized primarily on enkephaline-containing GABAergic neurons in the striopallidal tract of the striatum [78, 79]. Limited concentrations of A_{2A} receptors also are found in thalamus [80, 81, 82], nucleus tractus solitarius [83, 84], and cholinergic neurons of the pontine reticular formation [85, 86]. A_3 receptors (not shown in Figure 1) are found primarily in the periphery, with high concentrations in testes and mast cells, and are not heavily expressed in brain. These receptors play an important role in regulating inflammatory reactions [87, 88]. The receptor-mediated actions of adenosine are mimicked by a number of synthetic adenosine analogs that vary in their affinity for subtypes of the adenosine receptor. Moreover, methylxanthine stimulants (caffeine; theophylline) are nonselective high-affinity antagonists of the adenosine receptor, and derive their immediate stimulant properties by disinhibiting neurons from adenosine regulation during states of fatigue or depressed mood [89, 5, 53, 24].

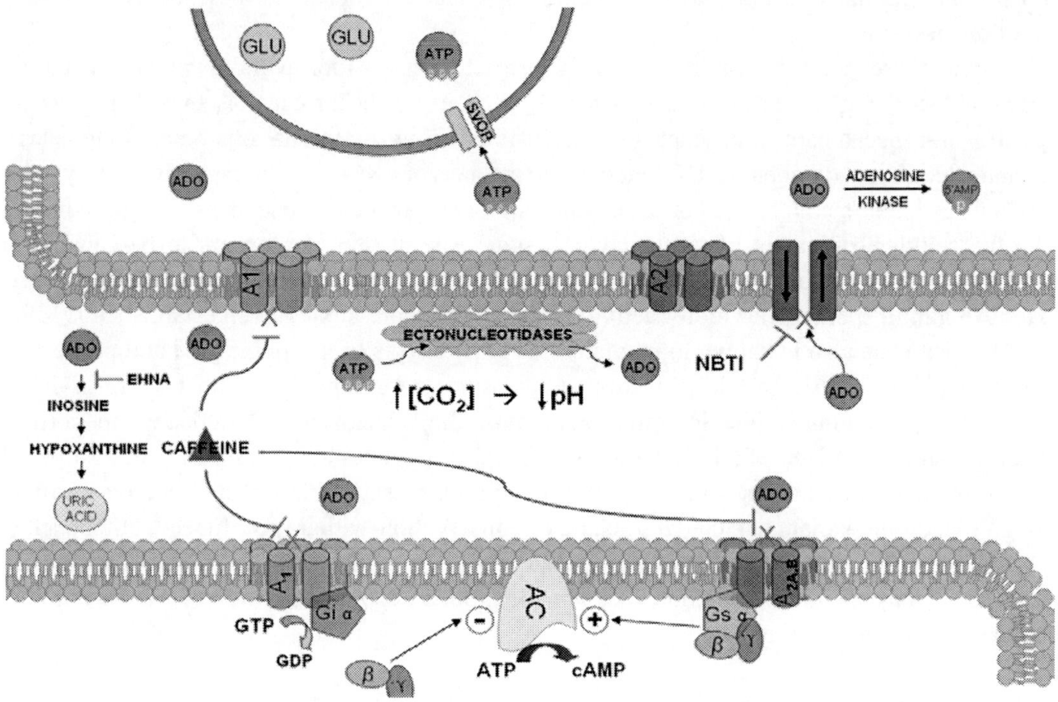

Figure 1. Regulation of adenosine in the synaptic cleft. Build up of adenosine in the synaptic cleft occurs by 1) conversion of ATP to ADO by ectonucleotidases when triggered by increased levels of CO2 or 2) through a bidirectional transporter after release from synaptic vesicles and dephosphorylation from ATP. Synaptic levels of ADO can be decreased by 1) Conversion of ADO into Inosine, hypoxanthine, and Uric Acid, which can be blocked by Inosine Deaminase. 2) ADO binding to Adenosine receptors A1, A_{2A}, A_{2B}. Caffeine acts as ADO receptor antagonist to increase ADO levels. 3) The bidirectional transporter also decreases ADO levels; this pathway can be blocked by EHNA. ADO receptor binding results in release of G protein ϒ and β units; this regulates levels of cAMP in postsynaptic space through binding to Adenylyl Cyclase.

B. Adenosine Signaling and Conservation-Withdrawal

Engle and Schmale [90] argued that periods of intense activation, stress or anxiety are automatically and unconditionally followed by a transition to a state termed *conservation-withdrawal*. The sensory unresponsiveness, behavioral depression, and cognitive dullness that characterize the state were considered to be adaptive mechanisms for husbanding limited resources and facilitating the eventual recovery of energy homeostasis.

This proposition was presented, in part, as an explanation for the learned helplessness effect. Sensory unresponsiveness, cognitive dullness, and behavioral depression are reasonably good descriptors of the depression-like component of helplessness. Moreover, the symptoms of conservation-withdrawal are a good description of the fatigue component of major depression [36]. Finally, the idea that conservation-withdrawal follows a period of intense catabolic output is consistent with the shift in behavior and emotional state that occurs in inescapably shock rats at the time of testing.

Despite these positive features of the construct, conservation-withdrawal has had little impact of theorizing in the helplessness paradigm. Part of the problem is that there was no known neural mechanism to produce such an instinctive, unconditional outcome. Nonetheless, the characterization of adenosine's function in the nervous system corresponds closely with Engel and Schmale's concept of conservation-withdrawal; indeed, adenosine's function appears to be the cellular equivalent of the behavioral state. The nucleoside appears to be capable of producing the global characteristics of conservation-withdrawal. The ability of adenosine and its analogs to inhibit spontaneous motor activity (behavioral depression) is well established [91, 92, 93, 94]. Adenosine also is known to mediate the spinal effects of morphine [95] and participate in antinociception [96], certainly a form of sensory unresponsiveness that is relevant to helplessness [97]. Finally, the inhibitory action of the nucleoside is implicated in cortical spreading depression [98, 99, 100, 101] and tonically regulates hippocampus [102, 103]. An increase in adenosine's receptor-mediated action could be responsible for the impaired learning or cognitive dullness that characterizes a state of conservation-withdrawal. Indeed, Hitchcock et al. [104] recently reported that deficits in spatial learning produced by the administration of an adenosine agonist or the cholinergic receptor antagonist scopolamine are reversed by highly selective A_1 receptor antagonists.

IV. Adenosine Signaling in Conservation-Withdrawal and Helplessness

A. Effect of Methylxanthine Stimulants

Caffeine and theophylline decrease sensory reaction times, increase attention and memory retrieval, and promote behavioral activation, particularly under conditions of fatigue or depressed mood [26, 106]. These stimulant actions are derived from antagonizing adenosine's inhibitory action in the CNS: Caffeine and theophylline bind readily to the major subtypes of brain adenosine receptors, influence cellular effector systems in a manner opposite to that of adenosine, and enhance most forms of excitatory transmission [107, 52, 26, 95].

This drug profile suggested that methylxanthine stimulants should be highly effective in reversing the sensory unresponsiveness, cognitive dullness, and behavioral depression associated with conservation-withdrawal. If a rise in adenosine inhibition in the CNS is directly responsible for these aspects of helplessness, then xanthine stimulants should be uniquely beneficial to rats preexposed to inescapable shock. Conversely, enhanced noradrenergic [108, 109] and cholinergic [108] transmission have been implicated in behavioral depression following inescapable shock. If these neural changes are directly responsible for behavioral impairment, then the ability of the methylxanthines to enhance both forms of transmission should exacerbate escape deficits in inescapably shocked rats.

We injected rats with caffeine (10 mg/kg) or vehicle either before exposure to inescapable shock or simple restraint, before shuttle-escape testing, or both in a factorial design as a preliminary test of this hypothesis [110]. Caffeine had no beneficial effect when injected prior to pretreatment in the inescapable stressor; indeed, it produced slightly greater impairment of later shuttle-escape performance. By contrast, caffeine treatment prior to shuttle-escape testing (or prior to both pretreatment and testing) completely eliminated the deficits in escape performance typically observed following pretreatment with inescapable shock. Caffeine treatment had no effect on performance by restrained controls. This pattern of results indicates that the ability of xanthine stimulants to reverse the helplessness effect is not an artifact of drug state dependency, or an untoward effect of the drugs on escape latencies.

A comparison of pretest treatment with caffeine and theophylline yielded classic dose-response curves for the ability of these adenosine receptor antagonists to reverse the helplessness effect. These stimulants have very similar affinities for central adenosine receptors and were equally effective in eliminating FR-2 escape deficits. Some improvement in test performance occurred for all methylxanthine doses, with asymptotic improvement occurring at a dose of 7 mg/kg. Amphetamine, another psychomotor stimulant had no beneficial effects on escape performance.

B. Effect of Adenosine Analogs

The possibility that adenosine mediates the conservation-withdrawal aspects of helplessness is supported by a striking degree of overlap among the effects of inescapable shock and those of adenosine and its analogs. Both inescapable shock [108, 111, 112] and adenosine [113, 91] are well known for their ability to depress motor activity. Suggestive parallels also are found in the similar effects of these events on vegetative functions such as feeding and weight loss [114], hypothermia [115, 116], and bitter-taste finickiness [39, 107]. The reinforcing efficacy of electrical brain stimulation also is similarly depressed by an adenosine analog [117] and inescapable shock [108], suggesting that adenosine might be involved in the anhedonia produced by the traumatic stressor [6]. Another well-documented effect of inescapable shock is the production of opiate antinociception [97]. This stress-induced analgesia is reversed by naloxone, is cross tolerant with morphine, and has both spinal and supraspinal components. Adenosine also produces potent antinociception with spinal and supraspinal components. Several findings suggest that endogenous adenosine is released as a transmitter via opiate receptor activation to directly mediate the effects of morphine in the spinal dorsal horn [118]. Although the contribution of adenosine to stress analgesia has not been examined, the similarities in the effects of stress and adenosine on nociception are highly suggestive of a critical role. Finally, adenosine analogs and inescapable shock have very similar effects on hippocampal electrophysiology [119, 102, 109, 103] which may contribute to their detrimental effect on performance in spatial learning tasks [105, 109].

If a stress-induced rise in endogenous adenosine is the proximate cause of impaired test performance in inescapably shocked rats, then it should be possible to mimic the effects of inescapable preshocks by administering an exogenous adenosine agonist to non-stressed rats just before testing. Even though the neural conditions that antedate adenosine receptor activation in drug-treated and preshocked rats probably differ substantially, similar impairment should be observed if adenosine is the last link in the causal chain that produces escape deficits.

We tested this prediction by exposing one group of rats to inescapable tailshocks and 10 other groups to simple restraint in tubes 24 hr before standard shuttle-escape testing [120].

Twenty minutes prior to the test the shocked group and one restrained group received an i.p. injection of vehicle and served to define the standard boundaries of the helplessness effect during escape testing. The remaining restrained groups received an i.p. injection of a dose of one of three synthetic adenosine analogs: a) N-ethyl-carboximidoadenosine -- *NECA* (0.01-1.0 μM/kg), an uptake resistant analog with very high affinity for both A_1 and A_2 receptors; b) R(-) N^6-(2-Phenylisopropyl)adenosine -- *R-PIA* (0.01-1.0 μM/kg), a selective A_1 agonist; and c) S(+) N^6-(2-Phenyl-isopropyl)adenosine -- *S-PIA* (0.1-10.0 μM/kg), the less active enantiomer of R-PIA.

Adenosine analogs had little effect on FR-1 escape trials. Performance during the more difficult FR-2 trials varied as a function of drug dose and the specific binding characteristic of the adenosine analog. All NECA doses significantly retarded FR-2 escape. R-PIA was 10 to 12 times less effective than NECA at mimicking the effects of inescapable shock; no deficit occurred at the lowest dose. S-PIA was 12 to 15 times less effective than R-PIA. The relative efficacy of these drugs at mimicking the effects of earlier inescapable shock suggests that activation of A_2 (or a combination of A_1 & A_2) receptors produce symptoms of helplessness or conservation-withdrawal.

We obtained preliminary evidence for a central role of adenosine by comparing the efficacy of theophylline and 8-sulphophenel-theophylline (8-SPT) at reversing shock-induced and NECA-induced escape deficits. Whereas theophylline readily binds to adenosine receptors in both the CNS and periphery, 8-SPT is a polarized molecule that does not cross the blood-brain barrier. Thus, if adenosine and its analogs act centrally to impair escape performance, then only theophylline should be effective in reversing shock-induced and NECA-induced escape deficits.

As predicted, pretest treatment with 7 mg/kg of theophylline completely reverses deficits in escape performance in rat exposed to inescapable shock 24 hours earlier. Treatment with 7 mg/kg of 8-SPT has no such beneficial effect. Similarly, restrained rats show deficits in escape performance when treated with 0.1 μM/kg of NECA shortly before testing. The NECA-induced deficit is reversed by pretest treatment with theophylline, but not 8-SPT.

C. Manipulating Endogenous Adenosine

Extracellular adenosine concentrations – and access to extracellular adenosine receptors – are regulated by two mechanisms (see Figure 1). Adenosine is converted to inactive inosine and eventually to uric acid via a degradation pathway involving adenosine deaminase [121, 122, 123]. The nucleoside also is removed from the synaptic cleft via equilibrase nucleoside transporters (ENT1 & ENT2), as well as an active transport system involving Na+ exchange [124]. Adenosine is rapidly converted to 5'AMP via adenosine kinase once inside the cell, which stops active gradient transport of the nucleoside.

Disabling either of these regulatory mechanisms functionally increases synaptic adenosine concentrations and its action at extracellular receptors. Thus, if brain adenosine signaling mediates the escape deficits produced by initial exposure to inescapable shock, then enhancing the small and otherwise nondebilitating endogenous concentrations produced by prior restraint stress by either blocking degradation or uptake transport should disrupt test performance.

Woodson et al. [125] demonstrated that inhibiting adenosine deaminase with erythro-9-(2-hydroxy-3-nonyl)adenine (EHNA) mimics the effects of inescapable shock on shuttle-escape performance in rats. Microinfusion of EHNA into the brain ventricles of previously restrained rats impaired test escape performance in a dose-dependent manner. This effect of EHNA, as well as the effect of earlier exposure to inescapable shock, was reversed by pretest treatment with adenosine receptor antagonist caffeine. A subthreshold dose of EHNA also interacted in synergy with pre-exposure to an ineffective number of inescapable shocks to maximally impair shuttle-escape performance at the time of testing.

Similarly, increasing endogenous adenosine concentrations by blocking the reuptake transporter yields a similar pattern of results [126]. Microinfusion of the nucleoside transport blocker S-(4-nitrobenzyl)-6-theoinosine) (NBTI) into the right lateral ventricle of rats undergoing restraint stress 24 hours earlier produces a large deficit in shuttle-escape performance, which is comparable to that produced by prior exposure to inescapable shock. NBTI impairs escape performance in a dose-dependent manner. Moreover, a subthreshold dose of the adenosine uptake inhibitor acts synergistically with an ineffective number of inescapable preshocks maximize deficits in test escape performance, suggesting that NBTI and inescapable shock are acting on the same neural mechanism.

Analysis of the receptor mediating shock-induced and NBTI-induced escape deficits using highly selective receptor antagonists supported earlier evidence for an A_2 receptor. Both types of deficits were reversed by the nonselective adenosine receptor antagonist caffeine, and the highly selective A_{2A} receptor antagonist CSC (8-(3-chloro-styrl)caffeine) in a dose-dependent manner. The highly selective A_1 (DPCPX: 8-Cyclopentyl-1,3-Dipropylxanthine) and A_{2B} (CGS 21680) receptor antagonists failed to improve performance in rats pre-exposed to inescapable shock or receiving i.c.v. infusion of NBTI shortly before escape testing. These data strongly suggest that activation of A_{2A} receptors mediate deficits in escape performance, regardless of whether those deficits occur because of prior traumatic stress or by enhancing endogenous adenosine signaling through pharmacological means. As discussed in detail later, A_{2A} receptors limited and unique distribution in the CNS. Thus, these data have important implications for the anatomical locus and mechanism by which adenosine signaling mediates behavioral depression.

D. Excessive Neural Excitation, Metabolic Challenge, and Adenosine Signaling

Petty and Sherman [41] reported several years ago that small pools of GABA are depleted in frontal cortex in rats pre-exposed to inescapable shock. They speculated that loss of this form of neural inhibition might lead to over-activation of glutamate neurons in frontal cortex, leading to at least some of the behavioral impairment falling under the rubric of learned helplessness. Petty, McChesney & Kramer [127] infused glutamate directly into the frontal cortex in an attempt to mimic the effect of neuronal over-activation. Deficits in escape performance occurred 1 and 72 hours after injection. Moreover, injection of gamma-aminobutyric acid (GABA) into the frontal cortex prevented escape impairments in the helplessness procedure—a finding that suggests glutamate involvement given the opposing roles of GABA and glutamate in the central nervous system. Other studies have shown that chronic antidepressant treatment alters radioligand binding to the NMDA glutamate receptor subtype in mice and that these changes are restricted to the cerebral cortex [128, 129].

More recently, Amy Hunter from our laboratory argued that excessive activity in frontal cortex glutamate neurons is more likely to be associated with the PTSD-like symptoms shown by inescapably shocked rats than those related to conservation-withdrawal or behavior depression [39]. The enhanced neophobia [29], enhanced context fear conditioning [28], and exaggerated initial reaction to the footshock stressor during the first few FR-1 escape trials is more PTSD-like than depression-like. This enhanced behavioral reaction to test stimuli was occasioned by an exaggerated neural response as well. Excessive activation involving the biogenic amines, and in particular norepinephrine, was identified as a critical factor early on [130, 131, 132, 40].

Acute stress is associated with increased glutamate efflux in frontal cortical regions. Behavioral stressors including restraint, forced swimming, and tail pinch increase extracellular levels of endogenous glutamate in the frontal cortex [133, 134]. Pre-exposure to stress increases glutamate uptake and release from frontal cortical synaptosomes *in vitro* [135]. Exposure to a pharmacological stressor has been shown to increase glutamate efflux in a fashion similar to other stressors. The 'inverse' benzodiazepine agonist, FG7142, has been shown to be anxiogenic in both humans and laboratory animals [136, 137, 138]. Importantly, systemic administration of FG7142 produces increased levels of glutamate in the prefrontal cortex that are blocked by pretreatment with the anxiolytic diazepam [139].

Glutamate also is specifically associated with enhanced sensory responding such as is observed during the beginning of the test phase in the learned helplessness paradigm. In other paradigms glutamate mediates related behaviors such as fear-potentiated startle [140, 141], increases in sensory reactivity to noise, and facilitated conditioning after exposure to stress [43]. These findings are consistent with the argument that increased glutamate may underlie the heightened behavioral responding that immediately precedes behavioral depression in the helplessness procedure.

Frontal cortex is implicated as a critical brain structure in both PTSD and major depression. Frontal cortex shows a dramatic increase activity as measured by single photon emission tomography in combat veterans listing to sounds of combat positron, but not in age-matched controls. By contrast, this area shows substantially diminished neuronal activity as measured by positron-emission or single-photon emission tomography in depressed individuals [142, 143]. Glutamate processes are among those that are altered in the frontal cortex in depression. For instance, depressed patients show significant reductions in the glutamate pool as measured by ^{11}C-glucose positron emission tomography [144]. Moreover, depressed suicide victims versus sudden death controls show decreased glutamate binding in the frontal cortex [145].

Hunter et al. [39] speculated on the basis of these data that the generation of escape deficits in the helplessness paradigm is a multi-step process. Excessive activation in frontal cortex glutamate neurons (and probably other brain regions and transmitter systems as well) is associated with the PTSD-like symptoms of exaggerated reactivity and fearfulness shown by inescapably shocked rats. The unregulated nature of this activation eventually leads to metabolic fatigue in frontal cortex or its projection regions. Adenosine is produced as a compensatory homeostatic mechanism. The nucleosides inhibitory action in the nervous system mediates the transition to a state of conservation-withdrawal.

As a test of this hypothesis, rats received bilateral infusion of glutamate into frontal cortex. Escape deficits occurred 1 and 72 hours later as in the Petty et al. experiments. Treatment with caffeine shortly before escape testing eliminated the escape deficits 1 hour post injection, but not 72 hours later, indicating that a non-purine mechanism is responsible for the long-term deficit. Moreover, NMDA lesions of the frontal context prior to exposure to inescapable shock completely eliminated evidence of impaired escape performance 24 hours later. This later result was taken as evidence that adenosine signaling in a projection region, rather than frontal cortex *per se*, is responsible for escape deficits.

We also have provided evidence that metabolic fatigue is a critical component of neural over-activation and compensatory adenosine signaling in the helpless paradigm. We produced metabolic inhibition in these experiment by treating previously restrained rats with 2-deoxyglucose (2-DG). This compound readily crosses the blood-brain barrier and neural membranes to interfere with normal respiration by competitively inhibiting glucose transport and disrupting cytosolic glycolysis [146]. Because membrane transport occurs in direct proportion to metabolic activity, 2-DG induces marked glucoprivation in the most active brain cells.

Administering 2-DG in this manner impaired subsequent escape performance in a dose-dependent manner. However, the performance deficit in this case was not a direct consequence of metabolic inhibition. Instead, the challenge to metabolic homeostasis produced by 2-DG resulted in compensatory adenosine signaling, as the escape deficits was completely reversed by caffeine treatment. Moreover, preexposure to an ineffective number of inescapable shock acted synergistically with pretest treatment of a subthreshold dose of 2-DG to maximize the performance deficit during escape testing. This latter result suggests that inescapable shock and 2-DG affect the same mechanism.

IV. Role of Striatum in Behavioral Depression

A systematic analysis of the adenosine receptor subtype mediating escape deficits in the helplessness paradigm clearly implicated the A_{2A} receptor. This receptor subtype has a very limited distribution in brain, being expressed primarily on the dendritic spines of enkephalin-containing GABAergic neurons in the ventral striatum. Adenosine signaling in this region is linked to dopamine signaling in a manner that appears to functionally regulate the integration of motivation with ongoing behavior. A brief overview of striatum function and the specifics of adenosine signaling are given below.

A. The Psychobiology of the Ventral Striatum

The basal ganglia (striatum, globus pallidus, substantia nigra, & subthalamic nucleus) are a cluster of nuclei in the brain whose function has been associated mainly with motor control. If damage is sustained to the basal ganglia or its pathways, the cortex cannot properly direct motor control and motor abilities will be impaired, resulting in common motor disorders including Parkinson's disease and Huntington's disease [147]. Recent evidence suggests, however, that the nuclei of the basal ganglia may not only control motor function but also play a role in the integration of emotion and motivation with ongoing behavior.

The striatum, consisting of the caudate nucleus, putamen, nucleus accumbens, and olfactory tubercle, is the most prominent and well known component of the basal ganglia. The striatum is divided into dorsal and ventral sections, each with its own function. Heimer and Wilson [148] were the first to differentiate the dorsal striatum and its sensory-motor connections from the ventral striatum, with pathways to allocortical regions, such as the hippocampus and amygdala, and neocortical regions such as the prefrontal cortex.

The dorsal striatum (caudate-putamen complex) performs the traditional sensory-motor functions that are typically associated with the basal ganglia [148]. The caudate-putamen complex receives the majority of its input from the cerebral cortex. This information is then sent as inhibitory output by GABAergic neurons through the substantia nigra and global pallidus to the motor nuclei of the thalamus. The thalamus then relays the information to the motor cortex where motor movement is initiated [149]. The ventral striatum (nucleus accumbens) is the site of integration of dopaminergic inputs from the mesencephalon and excitatory glutamatergic inputs from the limbic cortices, amygdala and hippocampus [36]. Medium-sized spiny GABAergic neurons make up more than 90% of the striatal neuronal population, where they project either to the globus pallidus via the indirect pathway or to the substantia nigra via the direct pathway [150]. Figure 2 outlines the circuit structure of the indirect and direct pathways between the cerebral cortex and the basal ganglia.

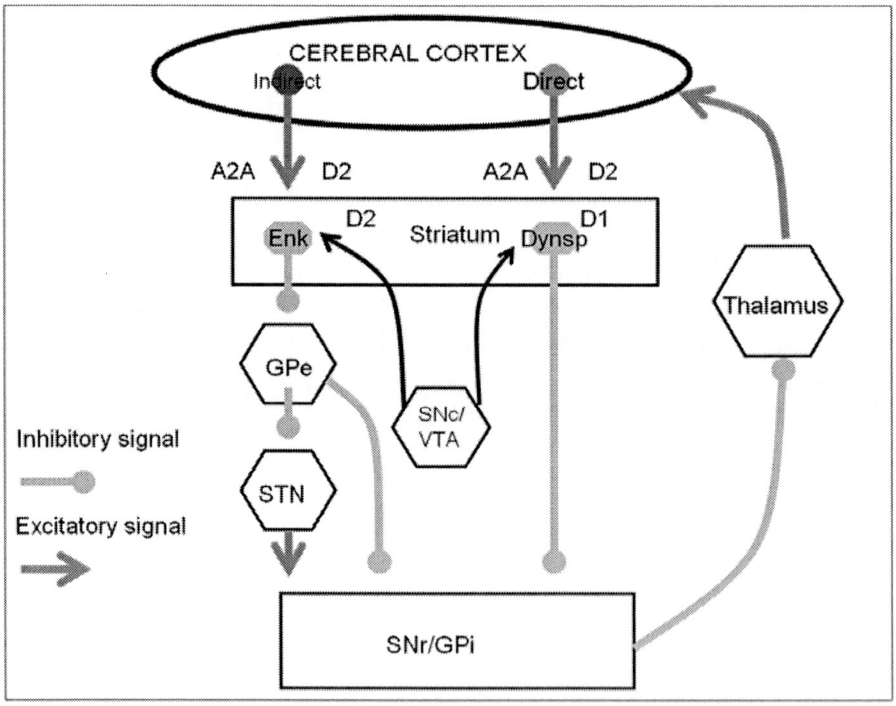

Figure 2. Hypomotoric state. Basal ganglia circuitry acting through closely tied adenosine and dopamine receptors in the striatum to regulate GABAergic striatal pallidal neurons. Enkaphalinergic A_{2A} receptors interacts closely with D_2 receptors to regulate the indirect pathway, connecting through the external segment of globus pallidus , subthalamic nucleus, and internal segment of globus pallidus to result in overall inhibition of activity, while the direct pathway sends an overall disinhibitory signal to the thalamus.

The ventral striatum consists mainly of the nucleus accumbens which is broken down into two regions, the shell and core [151, 152], each with efferent and afferent pathways connecting to various structures of the limbic system and prefrontal cortex. The ventral hippocampus and amygdala project mainly to the shell, but innervate only small portions of the core [153, 154]. The core subregion receives inputs primarily from the dorsal medial prefrontal cortex, parahippocampal cortex, and the anterior basolateral amygdala nucleus [155] and parallels the dorsal striatum in its outputs. Both subregions play significant but distinct roles in instrumental learning and Pavlovian conditioning. Stimulus-response learning appears to be controlled mainly by the core, whereas the integration of basic biological drives and the limbic system originates in the shell [156, 157]. Perhaps the most important distinction between the shell and core of the nucleus accumbens is the shell's dopaminergic pathway to and from the ventral tegmental area (VTA). This pathway is of particular importance to the aspect of motivation and reward-related behaviors [152].

The ventral striatum, unlike the dorsal striatum, receives inputs from limbic structures as well as inputs from certain regions of the prefrontal cortex related to cognitive function [158]. The limbic system and cerebral cortex have been termed the "emotive brain" and "cognitive brain," respectively [159]. Both the "emotive brain" and "cognitive brain" provide information to the basal ganglia, more specifically the ventral striatum, which then translates into behavioral motor responses [160].

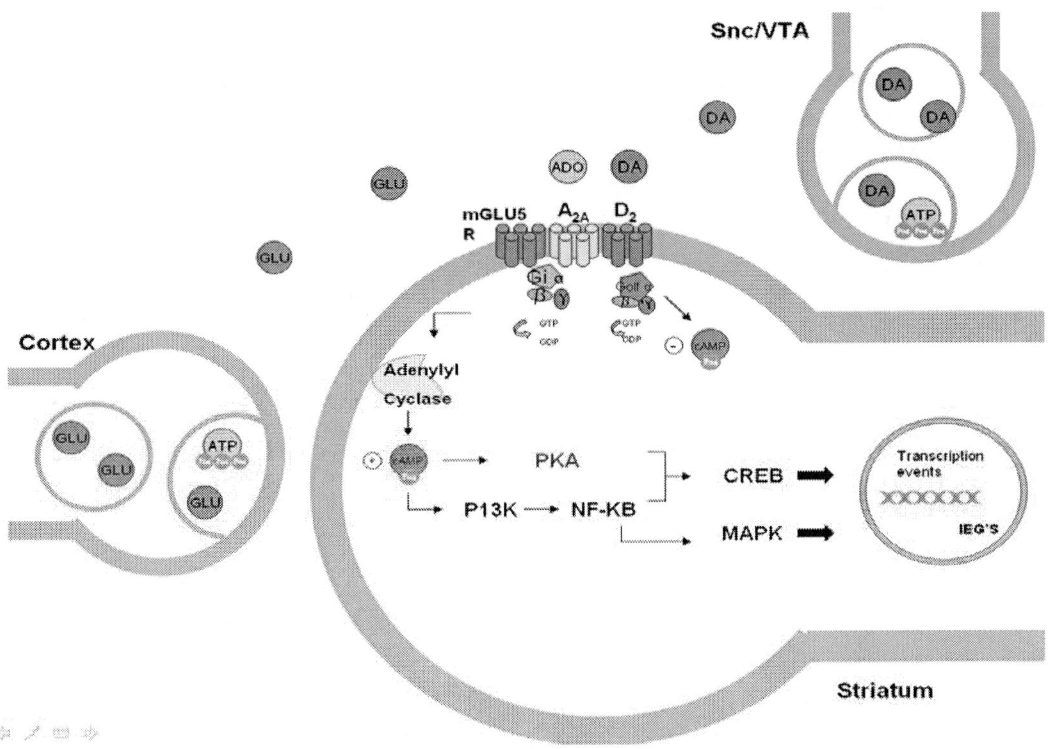

Figure 3. The striatum receives glutamatergic inputs from the cortex and dopaminergic inputs from the ventral tegmental area (VTA). Glutamate ($mGLU_5$) and dopamine (D_2) receptors bond with adenosine receptors (A_{2A}) in the ventral striatum to form a metabotropic heteromer. Activation of the A_{2A} receptor decreases the affinity for bonding in the D_2 receptor.

This coupling of cognitive and emotive systems provides the ventral striatum with functionally strong associations to the motivational aspects of behavior. While the dorsal striatum predominately engages in sensorimotor functions, the ventral striatum appears to act as the functional interface between the limbic system and motor system [160, 161]. Ferré et al. [162] described the ventral striatum as the site of goal-directed behavior and the conversion of motivation and action. As such, the striatum is a plausible center for common behavioral disorders such as obsessive-compulsive disorder and depression as these are most likely caused by a dysfunction of this conversion.

B. Adenosine, Dopamine, and Metabotropic Heteromers

The current theory of goal-directed behavior describes a complex system of glutamatergic and dopaminergic inputs to the ventral striatum [162]. Glutamatergic input carries information about motivationally relevant stimuli to the nucleus accumbens resulting in behavioral responses. Dopaminergic input then provides a means to decide the most appropriate course of action depending on the predictive value of the reward. This excitatory input modulates changes in the neuronal pathways in the nucleus accumbens so that the motivational value of the stimuli can be remembered for future use [163, 164]. Without this goal-directed behavior, motivation will cease and motor retardation will ensue, resulting in symptoms homologous to behavioral depression. This dissociation of motivation and action is dependent on the relationship between dopamineric input from the VTA and adenosine signaling in the ventral striatum [36].

Figure 3 reviews the relationship between adenosine and dopamine in the striatum. Adenosine A_{2A} receptors are located mainly in the ventral striatum on dendritic spines of GABAergic enkephalinergic neurons. These spines contain heteromers, or receptor complexes, consisting of receptors for adenosine (A_{2A}), dopamine (D_2), and glutamate ($mGLU_5$) [162]. The postsynaptic changes at the glutamatergic synapses within the striatum are modulated by these metabotropic heteromers. A_{2A} and D_2 are G protein coupled, and these complexes engage in antagonistic and synergistic cross-talk [162, 165]. A_{2A} receptors modulate the binding characteristics of D_2 receptors, where activation of adenosine A_{2A} receptors decreases the affinity for D_2 agonist bonding. The synergistic interaction of A_{2A} and $mGLU_5$ allows A_{2A} receptors to override D_2 receptor antagonism during co-activation of $mGLU_5$ [166, 167, 162]. This antagonistic interaction of A_{2A} and D_2 receptors is likely to be responsible for the uncoupling of motivation from ongoing behavior often seen in behavioral depression [36].

V. Adenosine A_{2A} Signaling in the Ventral Medial Striatum

Stephanie Furst and Traci Plumb currently are assessing the role of adenosine A_{2A} signaling in the striatum in the helplessness effect. In a preliminary study they surgically implanted rats with bilateral guide cannulas in the ventral medial striatum. Rats were subsequently exposed to inescapable shock or restraint. Rats from each of these groups received microinfusion of the highly selective A_{2A} receptor antagonist CSC or vehicle into the ventral medial striatum. Standard shuttle-escape testing occurred 10 minutes later.

Figure 4 shows mean percent freezing in each group during the initial FR-1 trials. Again, freezing is a species-typical defensive response of a rat and is an excellent measure of fear. Groups did not differ in FR-1 escape latencies during these trials; thus, any difference in freezing among groups is not an artifact of differential shock exposure.

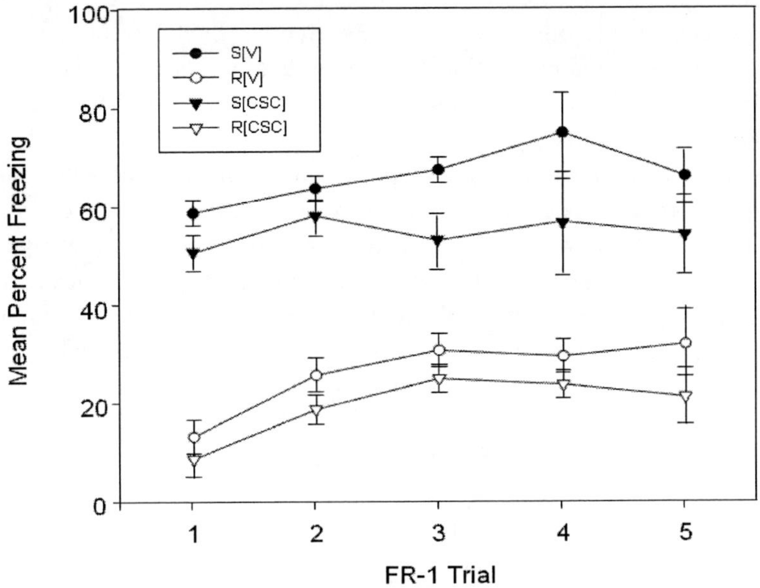

Figure 4. Percent freezing in all groups during initial FR-1 trials. Rats were exposed to inescapable shock (s) or restraint (r) as a measure of traumatic stress and mild stress respectively. They were then injected with CSC, a highly selective adenosine A_{2A} antagonist, or a vehicle (v). Rats exposed to traumatic stress exhibited freezing at a higher rate during the initial trials than those exposed to mild stress. CSC had little effect on time spent freezing.

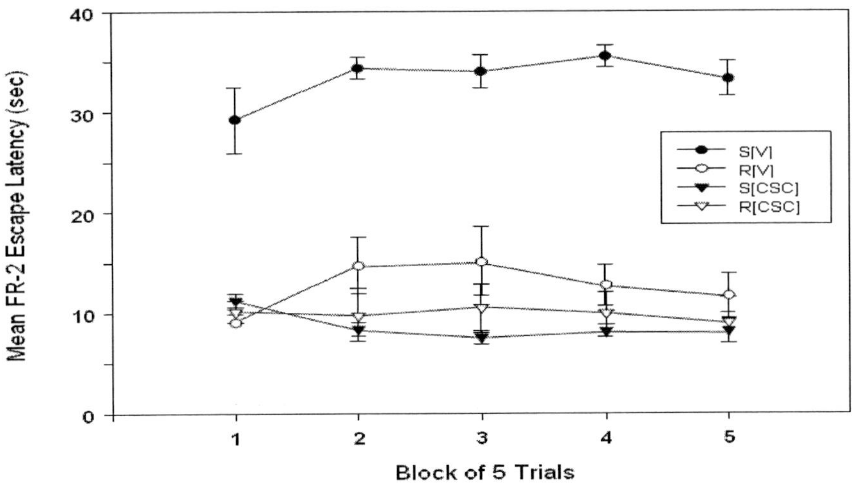

Figure 5. Mean escape latencies in all groups during FR-2 trials. The standard learned helplessness effect is observed between Groups S[V] and R[V] where performance was drastically impaired in those

exposed to inescapable shock. Injection of CSC in the ventral striatum eliminated the learned helplessness effect.

A comparison of shocked (S[V]) and restrained (R[V]) vehicle controls shows the exaggerated fearfulness that is typically observed following traumatic stress. Inescapably shocked rats spend dramatically more time freezing from the first shock trial on, and reach a high asymptote, than do restrained controls. Blocking adenosine A_{2A} receptors in the ventral striatum had no appreciable effect on the outcome. Even though ventral medial striatum is associated with emotional and motivational components of ongoing behavior, this region is not specifically associated with fear or freezing. Thus, the absence of an effect of CSC on this index of traumatic stress is not surprising.

Figure 5 shows the mean FR-2 escape latencies in each of the four groups. The differences in performance between Groups S[V] and R[V] represent the standard learned helplessness effect. The shift from easy (FR-1) to difficult (FR-2) escape contingencies dramatically impaired performance in vehicle-treated inescapably shocked rats relative to restrained controls. These rats not only showed near maximum escape latencies on across training, but many rats also failed to escape on a majority of trials. Importantly, the devastating effect of prior traumatic stress on escape performance was completely reversed by micro infusion of the adenosine A_{2A} receptor antagonist CSC. The antagonist had a slight, but not statistically significant, beneficial effect on the performance of restrained controls.

Although control injection sites still need to be examined in this project, the available data strongly suggest that enhanced adenosine signaling and the activation of A_{2A} receptors in the ventral medial striatum mediate deficits in escape performance in the helplessness procedure. Enhanced adenosine signaling may be a consequence of overactivation of descending glutamatergic projections from frontal cortex. Simultaneous activation of A_{2A} and $mGLU_5$ receptors dramatically reduces the affinity of D2 receptors in the heteromeric complex to dopamine. Loss of dopaminergic input from the VTA essentially uncouples the motivational signal from ongoing behavior. The symptoms of helplessness, behavioral depression or conservation-withdrawal ensue.

VI. Conclusion

PTSD is an anxiety disorder with its origin in psychological trauma. The traumatic event elicits a strong emotion reaction of fear or terror. The perceived absence of control or prediction in the situation exacerbates negative emotional outflow and increases the likelihood that PTSD will become a life-long problem. The disorder is occasioned by a number of comorbid problems, the most prevalent and serious of which is major depression.

We model these aspects of the disorder by exposing rats to a series of uncontrollable and unpredictable shocks in the learned helplessness paradigm. This manipulation engenders an intense and prolonged fear reaction [35, 5, 31]. The emotional reaction compromises a number of regulatory mechanisms that renders brain fear-processing pathways hyper-responsive [168, 130, 41, 40, 132]. The end consequence is that inescapably shock rats enter the test phase of the experiment in a highly fearful, agitated state [29, 28]. The metabolic demands of the underlying neural activation associated with attempting to cope with a threatening situation and maintaining this highly catabolic emotion state rapidly compromises energy homeostasis. At this point, the animal transitions from a state of anxiety and agitation to one of conservation-withdrawal.

A now sizable literature is reviewed in the foregoing pages indicating that this transition in emotional state in mediated by an increase in brain adenosine signaling. This signaling mechanism links neural excitability to the available energy supply [122, 49, 50]. Neural over-activation and other forms of metabolic challenge are met with a very potent compensatory inhibition. The substrates of active coping are compromised as a result.

A systematic analysis of the adenosine receptor subtype mediating conservation-withdrawal and helplessness [126] implicated the A_{2A} receptor. This subtype has a very limited distribution in the nervous system and is confined primarily to the dendritic spines of enkephaline-containing GABAergic neuron in the indirect pathway of the striatum. As such, these receptors are strategically located to produce component of a conservation-withdrawal reaction related to behavioral depression.

A_{2A} receptors form trimeric heteromers with dopamine D_2 and $mGLU_5$ receptors. Simultaneous activation of the adenosine and metabotropic glutamate components of the receptor complex functionally uncouples the dopamine signal from VTA from ongoing behavior. The loss of this motivational signal produces a hypomotoric state. The animal has difficulty initiating and maintaining active movement [38]. Escape deficits ensue.

References

[1] Hoge CW, Terhakopian A, Castro CA, Messer SC, Engel CC: Association of posttraumatic stress disorder with somatic symptoms, health care visits, and absenteeism among Iraq War Veterans. *Am J Psychiatry* 2007; 164: 150-153.

[2] Marx BP: Posttraumatic stress disorder and Operations Enduring Freedom and Iraqi Freedom: Progress in a time of controversy. *Clin Psychol Rev* 2009; 1-3.

[3] Dohrenwend BP, Turner B J, Turse NA, Adams BG, Koenen KC, Marshall R: *The Psychological Risks of Vietnam for U.S. Veterans: A Revisit with New Data and Methods.* Science 2006; 313: 979-982.

[4] Başoğlu M, Mineka S, Paker M, Aker T, Livanon M, Gök S: Psychological preparedness for trauma as a protective factor in survivors of torture. *Psychol Med,* 1997; 27(6): 1421-1433.

[5] Minor TR, Dess, N.K., Overmier, JB: Inverting the traditional view of "learned helplessness", in *Fear, avoidance, and phobias: A fundamental analysis.* Edited by Denny MR, Hillsdale, N.J, Lawrence-Erlbaum Associates 1991, p. 87-133.

[6] Minor TR, Hunter AM: Stressor controllability and learned helplessness research in the United States: sensitization and fatigue processes. *Integr Physiol Behav Sci* 2002; 37(1): p. 44-58.

[7] American Psychological Association: *Diagnostic and Statistical Manual of Mental Disorders (DSM-IV).* 4th ed. Washington, DC: American Psychiatric Association, 1994.

[8] Yehuda R: Post-Traumatic Stress Disorder. *N Engl J Med* 2002; *346*(2): 108-114.

[9] Ozer EJ, Best SR, Lipsey TL, Weiss DS: Predictors of posttraumatic stress disorder and symptoms in adults: A meta-analysis. *Psychol Bull* 2003; *129*(1): 52-73.

[10] Bisson JI: Post-traumatic stress disorder. *BMJ* (Clinical research ed. 2007; 334(7597): 789-93.

[11] Hood SK, Carter MM. A: Preliminary Examination of Trauma History, Locus of Control, and PTSD Symptom Severity in African American Women. *J Black Psychology* 2008; *34*(2): 179-191.

[12] McNally RJ Progress and controversy in the study of posttraumatic stress disorder. *Ann Review of Psychol* 2003; *54:* 229-52.

[13] American Psychological Association: D*iagnostic and Statistical Manual of Mental Disorders (DSM-IV),* Criteria B. 4th ed. Washington, DC: American Psychiatric Association, 1994.

[14] American Psychological Association: *Diagnostic and Statistical Manual of Mental Disorders (DSM-IV),* Criteria C. 4th ed. Washington, DC: American Psychiatric Association, 1994.

[15] American Psychological Association: *Diagnostic and Statistical Manual of Mental Disorders (DSM-IV),* Criteria D 4th ed. Washington, DC: American Psychiatric Association, 1994.

[16] Brady K T, Killeen T K, Brewerton T, Lucerini S: Comorbidity of Psychiatric Disorders and Posttraumatic Stress Disorder. *J Clin Psychiatry* 2000; 61: 22-32.

[17] Pole N, Neylan TC, Otte C, Metzler TJ, Best SR, Henn-Haase C: Associations Between Childhood Trauma and Emotion-Modulated Psychophysiological Responses to Startling Sounds: A Study of Police Cadets. *J Abnorm Psychol* 2007; 116(2): 352-361.

[18] Kessler RC, DuPont RL, Berglund P, Wittchen HU: Impairment in pure and comorbid generalized anxiety disorder and Major depression at 12 months in two national surveys. *Am J Psychiatry* 1999; 156(2): 1915-23.

[19] Kessler RC, Berglund P, Demler O, Jin R, Koretz D, Merikangas KR: The epidemiology of major depressive disorder: results from the National Comorbidity Survey Replication (NCS-R). *J American Medical Association 2003*; *289*(23): 3095-105.

[20] Hasin DS, Goodwin RD, Stinson FS, Grant B: Epidemiology of major depressive disorder: results from the National Epidemiologic Survey on Alcoholism and Related Conditions 2005; *Archives of general psychiatry* 2005; *62*(10): 1097-106.

[21] Maier SF, Watkins L: Cytokines for psychologists: Implications of bidirectional immune-to-brain communication for understanding behavior, mood, and cognition. *Psychological Review* 1998; 105: 83-107.

[22] Anisman H, Merali Z: Anhedonic and anxiogenic effects of cytokine exposure. Adv Exp Med Biol 1999; 461: 199-233.

[23] Overmier JB, Hellhammer DH: The learned helplessness model of human depression, in An inquiry into schizophrenia and depression: *Animal models of psychiatric disorders.* Edited by Simon P, Soubrie P, Widlocher D. Basel, Switzerland, AG Karger, 1988.

[24] Seligman ME: *Learned Helplessness: On depression development and death.* New York: W H Freeman, 1975.

[25] Maier SF, Albin RW, Testa TJ: Failure to learn to escape in rats previously exposed to inescapable shock depends on the nature of the escape response. *J Comp Physio Psych* 1973; 85: 581-592.

[26] Minor TR, LoLordo VM: Escape deficits following inescapable shock: the role of contextual odor. J Exp Psychol: *Anim Beh Process* 1984; 10:168-81.

[27] Bolles RC, Fanselow MS: A perceptual-defensive-recuperative model of fear and pain. *Behav Brain Sci* 1980; 3: 291-323.

[28] Maier SF: Role of fear in mediating shuttle escape learning deficit produced by inescapable shock. *J Exp Psychol Anim Behav Process* 1990; 16(2): 137-49.

[29] Minor TR: Conditioned fear and neophobia following inescapable shock. Anim Learn Beh 1990; 18: 212-226.

[30] Overmier JB, Seligman ME: Effects of inescapable shock upon subsequent escape and avoidance responding. *J Comp Physiol Psychol* 1967; 63(1): 28-33.

[31] Minor TR, Trauner MA, Lee CY, Dess NK: Modeling signal features of escape response: effects of cessation conditioning in "learned helplessness" paradigm. *J Exp Psychology Anim Behav Process* 1990; 16: 123-136.

[32] Mineka S, Cook M, Miller S: Fear conditioned with escapable and inescapable shock: effects of a feedback stimulus. *J Exp Psych: Anim Behav Process* 1984; 10: 307-324.

[33] Rosellini RA, DeCola JP, Warren DA: The effects of feedback on contextual fear depends upon the minimum intertrial interval. *Learning and Motivation* 1986; 17: 229-242.

[34] Rosellini RA, DeCola JP, Warren DA: Predictability and controllability: Differential effects upon contextual fear. *Learning and Motivation* 1987; 18: 392-420.

[35] Jackson RL, Minor TR: Effects of signaling inescapable shock on subsequent escape learning: implications for theories of coping and "learned helplessness". *J Exp Psychol Anim Behav Process* 1988; 14(4): 390-400.

[36] Minor TR, Huang Q, Witt AE: Cytokine-purine interactions in traumatic stress, behavioral depression, and sickness. *CSN & Neurological Disorders-Drug Targets* 2006; 5(5): 547-560.

[37] Simson PE, Weiss JM: Altered activity of the locus coeruleus in an animal model of depression. *Neuropsychopharmacol* 1988; 1: 287-295.

[38] Anisman H, Remington G, Sklar LS: Effect of inescapable shock on subsequent escape performance: Catecholaminergic and cholinergic mediation of response initiation and maintenance. *Psychopharmacol* 1979; 61: 107-124.

[39] Hunter AM, Balleine BW, Minor TR: Helplessness and escape performance: glutamate-adenosine interactions in the frontal cortex. *Behav Neurosci* 2003; 17(1): 123-135.

[40] Minor TR, Pelleymounter MA, Maier SF: Uncontrollable shock, forebrain norepinephrine, and stimulus selection during choice-escape leaning. *Psychobiology* 1988; 16: 135-145.

[41] Petty F, Sherman AD: Regional aspects of the prevention of learned helplessness by desiprimine. *Life Sci* 1980; 26: 1447-1452.

[42] Anisman H, Zacharko RM: Behavioral and neurochemical consequences associated with stressors, in *Stress-induced Analgesia*. Edited by Kelly DD, Annals of the New York Academy of Sciences 1986; 467: 205-225.

[43] Shors TJ, Servatius RJ: Stress-induced sensitization and facilitated learning require NMDA receptor activation. Neuroreport 1995; 6: 677-680.

[44] Bodnar RJ, Kelly DD, Glusman M: 2-deoxy-D-glucose analgesia: influence of opiate and nonopiate factors. *Pharmacol Biochem Behav*1979; 21: 749-754.

[45] Servatius RJ, Beck KD: Facilitated acquisition of the classically conditioned eyeblink response in male rats after systemic IL-1beta. *Integr Physiol Behav Sci* 2003; 38(3): 169-178.

[46] Headrick JP, Bendall MR, Faden AI, Vink R: Dissociation of adenosine levels from bioenergetic state in experimental brain trauma: Potential role in secondary injury. *J Cereb Blood Flow and Metab* 1994; 14: 853-861.

[47] Hoehn K, White TD: Role of excitatory amino acid receptors in K+ and glutamate-evoked release of endogenous adenosine from rat cortical slices. *J Neurochem* 1990; 54: 256-265.

[48] Lloyd HG, Lindstrom K, Fredholm BB: Intracellular formation and release of adenosine from rat hippocampal slices evoked by electrical stimulation or energy depletion. *Neurochem Int* 1993; 23: 173-185.

[49] Newby AC: Adenosine and the concept of "retaliatory metabolites". *Trends Biochem Sci* 1984; 9: 42-44.

[50] Stone TW: Physiological roles of adenosine and adenosine 5'triphosphate in the nervous system. *Neurosci* 1981; 6: 523-555.

[51] Santos MS, Moreno AJ, Carvalho AP: Relationships between ATP depletion, membrane potential, and the release of neurotransmitters in rat nerve terminals. An in vitro study under conditions that mimic anoxia, hypoglycemia, and ischemia. Stroke 1996; 27(5): 941-950.

[52] Weiner MF, Lovitt R: Conservation-withdrawal versus depression. *Gen Hosp Psychiatry* 1979; 1(4): 347-349.

[53] Svenningsson P, Lindskog M, Rognoni F: Activation of adenosine A2A and dopamine D1 receptors stimulates cyclic AMP-dependent phosphorylation of DARPP-32 in distinct populations of striatal projection neurons. *Neuroscience* 1998;. 84(1): 223-228.

[54] Bshesh K, Zhao B, Spight D: The A2A receptor mediates an endogenous regulatory pathway of cytokine expression in THP-1 cells. *J Leukoc Biol* 2002; 72(5): 1027-1036.

[55] Rosin DL, Hettinger BD, Lee A, Linden J: Anatomy of adenosine A2A receptors in brain: morphological substrates for integration of striatal function. *Neurology* 2003; 61(11): 10-11.

[56] Burnstock G: A basis for distinguishing two types of purinergic receptors. *Cell Membrane Receptors for Drugs and Hormones* 1978; 107-118.

[57] Burnstock G: Pathophysiology and therapeutic potential of purinergic signaling. *Pharmacol Rev* 2006; 58(1): 58-86.

[58] Burnstock G: Historical review: ATP as a neurotransmitter. *Trends in Pharmocl Sci* 2006; 27(3): 166-176.

[59] Burnstock G: Purinergic signalling and disorders of the central nervous system. *Nat Rev Drug Discov* 2008; 7: 575-590.

[60] Burnstock G, Verkhratsky A: Evolutionary origins of the purinergic signaling system. *Acta Physiol* (osf) 2009; 195(4): 415-417.

[61] Gourine AV, Wood JD, Burnstock G: Purinergic signaling in autonomic control. *Trends Neurosci* 2009; 32(4): 241-248.

[62] Gulbransen BD, Sharkey KA: Purinergic neuron-to-glia signaling in the enteric nervous system. *Gastroenterolgy* 2009; 136 94): 1349-1358.

[63] Zimmermann J: Correspondence (letter to the editor): Brachytherapy is an Established Therapeutic Option. *Dtsch Arztebl Int* 2008; 105(50): 882.

[64] Geiger JD, Fyda DM: Adenosine transport in nervous tissues, in *Adenosine in the Central Nervous System*. Edited by Stone TW, London, Academic Press 1991: 1–23.

[65] Vorhoff t, Zimmermann H, Pelletier J, Sevigny J, Braun N: Cloning and characterization of the ecto-nucleotidases NTPDase3 from rat brain: Predicted secondary structure and relation to the other members of the E-NTPDase family and actin. *Purinergic Signal* 2005; 1(3): 259-270.

[66] Sawada K, Echigo N, Juge N, Miyahi T, Otsuka M, Omote H, Yamamoto A, Moriyama Y. *Proc Ntal Acad Sci USA* 2008; 105(15): 5683-5686.

[67] Yao J, Bajjalieh SM: SVOP is a nucleotide binding protein. *PLoS One* 2009; 4(4): 5315.

[68] Burnstock G: Cotransmission. *Curr Opin Pharmacol* 2004; 4(1): 47-52.

[69] Cunha RA, Sebastioa AM, Ribeiro JA: Inhibition of ATP of hippocampal synaptic transmission requires localized extracelllar catabolism by ecto-nucleotidases into adenosine and channeling to adenosine A1 receptors. *J Neurosci* 1998; 18(6): 1987-1985.

[70] Dulla CG, Dobelis P, Pearson T, Frenguelli BG, Staley KJ, Masino SA: Adenosine and ATP link Pco_2 to cortical excitability via pH. *Neuron* 2005; 48: 1011-1023.

[71] Dunwiddie TV, Diao L, Proctor WR: Adenine nucleotides undergo rapid, quantitative conversion to adenosine in the extracellular space in rat hippocampus. J Neurosci 1997; 17(20): 7673-7682.

[72] Langer d, Hammer K, Kosalka P, Schrader J, Robson S, Zimmermann H: Distribution of ectonucleotidases in the rodent brain revisited. *Cell Tissue Res* 2008; 334: 199-217.

[73] Haas HL, Selbach O: Functions of neuronal adenosine receptors. Naunyn Schmiedebergs Arch Pharmacol 2000; 362(4-5): 375-381.

[74] Phillis JW: Adenosine and adenine nucleotides as regulators of cerebral blood flow: Roles of acidosis, cell swelling, and KATP channels. Crit Rev Neurobiol 2004; 16(4): 237-270.

[75] Linden J: Structure and function of A1 adenosine receptors. *Faseb J* 1991; 5(12): p. 2668-2676.

[76] Bruns RF, Lu GH, Pugsley TA: Characterization of the A2 adenosine receptor labeled by [3H]NECA in rat striatal membranes. *Mol Pharmacol* 1986; 29(4): 331-46.

[77] Sebastião AM, Ribeiro JA: Adenosine A2 receptor-mediated excitatory actions on the nervous system. Progress in Neurobiology 1996; 48: 167–189.

[78] Calon F, Dridi M, Hornykiewicz O, Bédard PJ, Rajput AH, Di Paolo T: Increased adenosine A2A receptors in the brain of Parkinson's disease patients with dyskinesias. *Brain* 2004: 127(Pt 5): 1075-1084.

[79] Svenningsson P, Le Moine C, Fisone G, Fredholm BB: Distribution, biochemistry and function of striatal adenosine A2A receptors. *Prog Neurobiol* 1999; 59(4): 355-396.

[80] Ishiwata K, Mishina M, Kimura Y, Oda K, Sasaki T, Ishii K: First visualization of adenosine A(2A) receptors in the human brain by positron emission tomography with [11C]TMSX. *Synapse* 2005; 55(2): 133-136.

[81] Mishina M, Ishiwata K, Kimura Y, Naganawa M, Oda K, Kobayashi S, Katayama Y, Ishii K: Evaluation of distribution of adenosine A2A receptors in normal human brain measured with [11C]TMSX PET. *Synapse* 2007; 61(9): 778-784.

[82] Weaver DR: A2a adenosine receptor gene expression in developing rat brain. Brain Res Mol Brain Res 1993; 20(4): 313-327.

[83] Castillo-Meléndez M, Krstew E, Lawrence AJ, Jarrott B: Presynaptic adenosine A2a receptors on soma and central terminals of rat vagal afferent neurons. *Brain Res* 1994; 652(1): 137-144.

[84] Scislo TJ, O'Leary DS: Vasopressin V1 receptors contribute to hemodynamic and sympathoinhibitory responses evoked by stimulation of adenosine A2a receptors in NTS. *Am J Physiol Heart Circ Physiol* 2006; 290(5): 1889-1898.

[85] Coleman CG, Baghdoyan HA, Lydic R: Dialysis delivery of an adenosine A2A agonist into the pontine reticular formation of C57BL/6J mouse increases pontine acetylcholine release and sleep. *J Neurochem* 2006; 96(6): 1750-1759.

[86] Ferre S, Fredholm BB, Morelli M: Adenosine-dopamine receptor-receptor interactions as an integrative mechanism in the basal ganglia. *Trends Neurosci* 1997; 20(10): 482-487.

[87] Gessi S, Merighi S, Varani K, Leung E, Mac Lennan S, Borea PA: The A3 adenosine receptor: an enigmatic player in cell biology. *Pharmacol Ther* 2008; 117(1): 123-140.

[88] Linden J, Taylor HE, Robeva AS, Tucker AL, Stehle JH, Rivkees SA, Fink JS, Reppert SM: Molecular cloning and functional expression of a sheep A3 adenosine receptor with widespread tissue distribution. *Mol Pharmacol* 1993; 44(3): 524-532.

[89] Dantzer R: Cytokine-induced sickness behavior: where do we stand? *Brain Behav Immun* 2001; 15(1): 7-24.

[90] Engel G L, Schmale AH: Conservation-withdrawal: A primary regulatory process for organismic homeostasis. Ciba Foundation Symposium 1972; 8: 57–75.

[91] Mohta M, Sethi AK, Tyagi A: Psychological care in trauma patients. *Injury* 2003; 34(1): 17-25.

[92] Weiner MF: Conservation-withdrawal and mental retardation in medical and surgical patients. *Psychosomatics* 1983; 24(1): 41-43.

[93] Huang QJ, Jiang H, Hao XL, Minor TR: Brain IL-1 beta was involved in reserpine-induced behavioral depression in rats. *Acta Pharmacol Sin* 2004: 25(3): 293-296.

[94] El Yacoubi M, Costentin J, Vaugeois JM: Adenosine A2A receptors and depression. *Neurology* 2003; 61(11 Suppl 6): 282-287.

[95] Huang Q, Minor TR: Brain IL-1beta mediates reserpine-induced behavioral depression in rats. *Chinese Journal of Behavioral and Medical Science* 2003; 12(5): 491.

[96] Kent S, Rodriguez F, Kelly KW, Dantzer R: Anorexia induced by microinjection of interleukin-1β in the ventromedial hypothalamus of the rat. *Physiol Beh* 1994; 56: 1031-1036.

[97] Milusheva E, Sperlagh, B, Kiss B, Szporny L, Pasztor E, Papasova M, Vizi ES: Inhibitory effect of hypoxic condition on acetylcholine release is partly due to the effect of adenosine release from tissue. Brain Research Bulletin 1990; 24: 369-373.

[98] Burns RF: Role of adenosine in supply/demand balance. *Nucleos Nucleotides* 1991; 10: 931-933.

[99] Ikemoto K: [Human striatal D-neurons and their significance]. *Nihon Shinkei Seishin Yakurigaku Zasshi* 2002; 22(4): 131-135.

[100] Fuxe K, Agnati LF, Jacobsen K: Receptor heteromerization in adenosine A2A receptor signaling: relevance for striatal function and Parkinson's disease. *Neurology* 2003; 61(11 Suppl 6): 19-23.

[101] Overmier JB, Hellhammer DH: The learned helplessness model of human depression, in An inquiry into schezophrenia and depression: Animal models of psychiatric

disorders. Edited by Simon P, Soubrie P, Widlocher D. Basel, Switzerland, AG Karger, 1988.

[102] Selye H: The general adaptation syndrome and the disease of adaptation. *J Clin Endocr* 1946; 6: 117.

[103] Kern JA, Lamb RJ, Reed JC: Dexamethasone inhibition of interleukin 1 beta production by human monocytes. Posttranscriptional mechanisms. *J Clin Invest* 1988; 81(1): 237-244.

[104] Hitchcock JM, Chaney SF, Zwolshen JM, Ketteler HJ, Peet NP, Soreson SM: Potential cognition-enhancing activity of the adenosine A1 receptor antagonist MDL 102,234 in spatial learning and long-term potentiation models. Society for Neuroscience Abstracts 1992; 18: 867.

[105] White TD, Hoehn, K: Release of adenosine and ATP from nervous tissue, in *Adenosine in the Nervous System*. Edited by Stone TW. San Diego, CA, Academic Press, 1991, 173-196.

[106] Maier SF, Watkins LR: Intracerebroventricular interleukin-1 receptor antagonist blocks the enhancement of fear conditioning and interference with escape produced by inescapable shock. Brain Research 1995: 695: 279-286.

[107] Minor TR, Huang Q, Foley EA: Cytokine-purine interactions in behavioral depression in rats. *Integr Physiol Behav Sci* 2003; 38(3): 189-202.

[108] Field TM: Coping with separation stress in infants and young children. In *Stress and Coping*. Edited by Fields TM, McCabe PM, Schneiderman N. Hillsdale, NJ, Lawrence Erlbaum Associates, 1985.

[109] Avitsur R, Pollak Y, Yirmiya R: Administration of interleukin-1β into the hypothalamic nucleus induces febrile and behavioral effects. *Neuroimmunomodulation* 1997; 4: 258-265.

[110] Minor TR, Chang, WC, Winslow, JL: Stress and adenosine I: Effects of methylxanthine and amphetamine stimulants on learned helplessness in rats, *Behav Neurosci 1994a*; 108(2): 254-264.

[111] Burnstock G: A basis for distinguishing two types of purinergic receptors. *Cell Membrane Receptors for Drugs and Hormones* 1978; 107-118.

[112] Maier SF, Nguyen KT, Deak T: Stress, learned helplessness, and brain interleukin-1 beta. *Adv Exp Med Biol* 1999; 461: 235-249.

[113] Field T, Reite M: Children's responses to separation from mother during the birth of another child. *Child Dev,* 1984; 55(4): 1308-1316.

[114] Dragunow M: Purinergic mechanisms in epilepsy. *Progress in Neurobiology* 1988; 31: 154-158.

[115] Daly JW: Adenosine agonists and antagonists. In *Purines in cellular signaling : Targets for more drugs.* Edited by Jacobs KA, Daly JW Manganiello V. New York, NY, Springer-Verlag, 1990.

[116] Slisli Y, de Beaurepaire R: Interleukin-1beta and calcitonin, but not corticotropin-releasing factor, alter sleep cycles when injected into the rat hypothalamic lateral paraventricular area. *Neurosci Lett* 1999; 265(1): 29-32.

[117] Hestad KA, Tonseth S, Stoen CD Ueland T, Aukrust P: Raised plasma levels of tumor necrosis factor alpha in patients with depression: normalization during electroconvulsive therapy. *J ECT* 2004; 19(4): 183-188.

[118] Haung Q, Jiang H, Minor TR: Reserpine-induced depression in rats: Variations in brain intereukin-1β define early versus late behavioral deficits. *Biological Psychiatry*, submitted.
[119] Minor TR, Winslow JL, Chang WC: Stress and adenosine: II. Adenosine analogs mimic the effect of inescapable shock on shuttle-escape performance in rats. *Behav Neurosci* 1994b; 108(2): 265-276.
[120] Minor TR, Dess NK, Ben-David E, Chang WC: Individual differences in vulnerability to inescapable shock in rats. *J Experimental Psychol Anim Behav Process* 1994b; 20: 402-412.
[121] Arch RS, Newsholme EA: Activities and some properties of 5'-nucleotidase, adenosine kinase and adenosine deaminase in tissues from vertebrates and invertebrates in relation to the control of the concentration and the physiological role of adenosine. *J Biochem* 1978: 174: 965-977.
[122] Geiger JD, Padua RA, Nagy JI: Adenosine deaminase regulation of purine actions. In J. W. Phillis (Ed.), *Adenosine and adenine nucleotides as regulators of cellular function.* Boca Raton, FL 1991; CRC Press: 71-84.
[123] Newby AC: The interaction of inhibitors with adenosine metabolizing enzymes in intact isolated cells. *Biochem Pharmacol* 1981; 30: 2611-2615.
[124] Noji T, Karasawa, A., Ksuka H: Adenosine uptake inhibitors. *Eur J Pharmacol* 2004; 495: 1-16.
[125] Woodson JC, Minor TR, Job RF: Inhibition of adenosine deaminase by erythro-9-(2-hydroxy-3-nonyl)adenine (EHNA) mimics the effect of inescapable shock on escape learning in rats. *Behav Neurosci* 1998; 112(2): 399-409.
[126] Minor TR, Rowe M, Cullen PK, Furst S: Enhancing brain adenosine signaling with the nucleoside transport blocker NBTI (S-(4-nitrobenzyl)-6-theoinosine) mimics the effects of inescapable shock on later shuttle-escape performance in rats. Behav Neurosci 2008; 122(6): 1236-1247.
[127] Petty F, McChesney C, Kramer G: Intracortical glutamate injection produces helpless-like behavior in the rat. *Pharmacol Biochem Behav* 1985; 22(4): 531-533.
[128] Nowak G, Trullas R, Layer RT, Skolnick P, Paul IA: Adaptive changes in the N-methyl-D-aspartate receptor complex after chronic treatment with Imipramine and 1-aminocyclopropanecarboxylic acid. *J Pharmacol Exp ther* 1993; 265(3): 1380-1386.
[129] Paul IA, Layer RT, Skolnick P, Nowak G: Adaptation of the NMDA receptor in rat cortex following chronic electroconvulsive shock or Imipramine. *Eur J Pharmacol* 1993; 147(3): 305-311.
[130] Cassens G, Roffman M, Kuruc A: Alterations in brain norepinephrine metabolism induced by environmental stimuli previously paired with inescapable shock. *Science* 1980; 209(4461): 1138-1140.
[131] Weiss JM, Goodman PA, Losito BG, Corrigan S, Charry JM, Bailey WH: Behavioral depression produced by an uncontrollable stressor: Relationship to norepinephrine, dopamine, and serotonin levels in various regions of rat brain. *Brain Res Rev* 1981; 3: 167-205.
[132] Weiss JM, Simson PE: Mechanisms underlying stress-induced behavioral depression. In Stress and Coping. Edited by Field T, McCabe P, Schneiderman N. Hillsdale, *Lawrence-Erlbaum Associates,* 1985; 99-116.

[133] Moghaddam B: Stress preferentially increases extraneuronal levels of excitatory amino acids in the prefrontal cortex: comparison to hippocampus and basal ganglia. *J Neurochem* 1993; 60(5): 1650-1657.

[134] Bagley J, Moghaddam B: Temporla dynamics of glutamate efflux in the prefrontal cortex and in the hippocampus following repeated stress: effects of pretreatment with saline or diazepam. *Neuroscience* 1997; 77(1): 65-73.

[135] Gilad GM, Gilad VH, Wyatt RJ, Tizabi Y: Region-selective stress-induced increase of glutamate uptake and release in rat forebrain. *Brain Res* 1990; 5252: 335-338.

[136] Corda MG , Blaker WD, Mendelson WB, Guidotti A, Costa E: Beta-Carbolines enhance shock-induced suppression of drinking in rats. *Proc Natl Acad Sci USA* 1983; 80(7): 2072-2076.

[137] Dorow R, Horowski R, Paschelke G, Amin M: Severe anxiety induced by FG 7142, a beta-carboline ligand for benzodiazepine receptors. *Lancet* 1983, 2(8341): 98-99.

[138] Ongini e, Barzaghi C, Marzanatti M: Intrinsic and antagonistic effects of beta-carboline FG 7142 on behavioral and EEG actions of benzodiazepines and pentobarbital in cats. *Eur J Phamarcol* 1983, 95(1-2): 125-129.

[139] Karreman M, Moghaddam B: Effect of pharmacological stress on glutamate efflux in the prefrontal cortex. *Brain Res* 1996; 716(1-2): 180-182.

[140] Davis M, Walker DL, Lee Y: Neurophysiology and neuropharamocology of startle and its affective modulation, in Startle Modification. Edited by Dawson ME, Schell AM, Böhmelt, AH, United Kingdom, Cambridge University Press 1999: 95-113.

[141] Davis TN: The centrosome on centre stage. *Trends Cell Biol* 1997.

[142] Baxter, LR: Pet studies of cerebral function in major depression and obsessive-compulsive disorder: The emerging prefrontal cortex consensus. Annals of Clinical Psychiatry: The Official Journal of the American Academy of Clinical Psychiatrists 1991; 3(2): 103-109.

[143] Soares JC, Mann: The functional neuroanatomy of mood disorders. *J Psychiatr Res* 1997; 31(4): 393-342.

[144] Kishimoto H, Yamada K, Iseki E, Kosaka K & Okoshi T: Brain imagine of affective disorders and schizophrenia. *Psychiatry Clin Neurosci* 1998; 53 suppl: 212-214.

[145] Nowak G, Ordway GA, Paul IA: Alterations in the N-methyl-D-aspartate (NMDA) receptor complex in the frontal cortex of suicide victims. Brain Res 1995; 675(1-2): 157-164.

[146] Huang Q, Haoxingling, Minor TR: Reserpine increases brain IL-1beta expression in rat brain. *Chinese Journal of Immunology* 2003; 19(11): 765-767.

[147] Albin RL, Young AB, Penney JB: The functional anatomy of basal ganglia disorders. *Trends Neurosci* 1989; 12: 366–375.

[148] Heimer L, Wilson RD: The subcortical projections of the allocortex: similarities in the neural associations of the hippocampus, the piriform cortex, and the neocortex. In *Golgi centennial symposium.* Edited by Santinni M. New York, NY, Raven press, 1975, 177–193.

[149] Kemp JM, and Powell TPS: The connexions of the striatum and globus pallidus: Synthesis and speculation. *Phil Trans R Soc Lond* 1971; 262(B): 441-457.

[150] Herrero MT, Barcia C, Navarro JM: Functional anatomy of thalamus and basal ganglia. *Childs Nerv Syst* 2002; 18(8): 386-404.

[151] Zaborszky L, Alheid GF, Beinfeld MC, Eiden LE, Heimer L, Palkovits M: Cholecystokinin innervation of the ventral striatum: a morphological and radioimmunological study. *Neurosci* 985; 14: 427–453.

[152] Meredith GE, Baldo BA, Andrezjewski, ME, Kelley AE: The structural basis for mapping behavior onto the ventral striatum and its subdivisions. *Brain Struct and Funct* 2008; 213(1-2); 17-27.

[153] Groenewegen HJ, Berendse HW, Meredith GE, Haber SN, Voorn P, Wolters JG, Lohman AHM: Functional anatomy of the ventral, limbic system-innervated striatum. In *The mesolimbic dopamine system: from motivation to action.* Edited by Willner P, Scheel-Krüger J. Chichester, Wiley, 1991, 19–59.

[154] Ragsdale CW Jr, Graybiel AM: Fibers from the basolateral nucleus of the amygdale selectively innervate striosomes in the caudate nucleus of the cat. *J Comp Neurol* 1988; 269: 506–522.

[155] Groenewegen HJ: The ventral striatum as an interface between the limbic and motor systems. *CNS Spect* 2007; 12(12): 887-892.

[156] Corbit LH, Muir JL, Balleine BW: The role of the nucleus accumbens in instrumental conditioning: evidence of a functional dissociation between accumbens core and shell. *J Neurosci* 2001; 21: 3251-3260.

[157] Parkinson JA, Olmstead MC, Burns LH, Robbins TW, Everitt BJ: Dissociation in effects of lesions of the nucleus accumbens core and shell on appetitive pavlovian approach behavior and the potentiation of conditioned reinforcement and locomotor activity by D-amphetamine. J eurosci 1999; 19: 2401-2411.

[158] Alexander GE, Crutcher MD, DeLong MR: Basal ganglia-thalamocortical circuits: parallel substrates for motor, oculomotor, "prefrontal" and "limbic" functions. *Prog Brain Res* 1990; 85: 19-146.

[159] Konorski J: *Integrative activity of the brain: An interdisciplinary approach.* University of Chicago Press, Chicago 1967.

[160] Mogenson GJ, Jones DL, Yim CY: From motivation to action: Functional interface between the limbic system and the motor system. *Prog Neurobiol* 1980; 14: 69–97.

[161] Graybiel AM: Input-output anatomy of the basal ganglia. In Symposium Lecture, Proc Soc Neurosci, Toronto, Canada 1976.

[162] Ferré S, Diamond I, Goldberg SR, Yao L, Hourani SMO, Huang ZL, Urade Y, Kitchen I: Adenosine A_{2A} receptors in ventral striatum, hypothalamus and nociceptive circuitry: Implications for drug addiction, sleep and pain. *Prog Neurobiol* 2007; 8: 332-347.

[163] Schultz W: Getting formal with dopamine and reward. *Neuron* 2002; 36: 241–263.

[164] Horvitz JC: Dopamine gating of glutamatergic sensorimotor and incentive motivational nput signals to the striatum. *Behav Brain Res* 2002; 137: 65–74.

[165] Agnati LF, Ferre´ S, Lluis C, Franco R, Fuxe K: Molecular mechanisms and therapeutical implications of intramembrane receptor/receptor interactions among heptahelical receptors with examples from the striatopallidal GABA neurons. *Pharmacol Rev* 2003; 55: 509–550.

[166] Ferre, S, Karcz-Kubicha M, Hope BT, Popoli P, Burgueno J, Gutierrez MA, Casado V, Fuxe K, Goldberg SR, Lluis C, Franco R, Ciruela F: Synergistic interaction between adenosine A2A and glutamate mGlu5 receptors: implications for striatal neuronal function. *Proc Natl Acad Sci USA* 2002; 99: 11940–11945.

[167] Ferre´,S, Borycz J, Goldberg SR, Hope BT, Morales M, Lluis C, Franco R, Ciruela F, Cunha R: Role of adenosine in the control of homosynaptic plasticity in striatal excitatory synapses. *J Integr Neurosci* 2005; 4: 445–464.
[168] Anisman H, Sklar LS: Catecholamine depletion in mice upon reexposure to stress: mediation of the escape deficits produced by inescapable shock. *J Comp Physiol Psychol* 1979; 93: 610-625.

In: Neurobiology of Post-Traumatic Stress Disorder
Editors: L. Sher and A. Vilens, pp. 259-269
ISBN: 978-1-61470-593-2

Chapter XVI

Traumatic Stress, Depression and Suicide

Leo Sher
Columbia University and New York State Psychiatric Institute,
New York, New York, USA

Abstract

The psychological consequences of trauma exposure are well recognized throughout history. The study of post-traumatic stress disorder (PTSD) dates back more than 100 years. Before 1980, post-traumatic syndromes were recognized by different names, including railway spine, shell shock, traumatic (war) neurosis, concentration-camp syndrome, and rape-trauma syndrome. The symptoms described in these syndromes overlap considerably with what we now recognize as PTSD. Individuals exposed to severe traumatic stress often develop comorbid PTSD and major depressive disorder (MDD). I developed a new concept suggesting that that some or all individuals diagnosed with comorbid PTSD and MDD have a separate psychobiological condition that can be termed "post-traumatic mood disorder" (PTMD). This concept is supported by a significant number of clinical, translational and other studies suggesting that patients suffering from comorbid PTSD and MDD are different clinically and biologically from individuals with PTSD alone or MDD alone. Patients with comorbid PTSD and MDD are characterized by greater impairment compared to individuals with PTSD alone or MDD alone. Neurobiological evidence supporting the concept of PTMD includes the findings related to dopaminergic, serotonergic, and hypothalamic-pituitary-adrenal axis function and other observations. It is crucial to develop interventions to prevent and to treat PTMD and measures to prevent and to treat suicidal ideation and behavior in individuals with PTMD.

Traumatic Stress

The psychological consequences of trauma exposure are well recognized throughout history. The study of post-traumatic stress disorder (PTSD) dates back more than 100 years. Before 1980, post-traumatic syndromes were recognized by different names, including railway spine, shell shock, traumatic (war) neurosis, concentration-camp syndrome, and rape-trauma syndrome [1]. The symptoms described in these syndromes overlap considerably with what we now recognize as PTSD. Nearly 100 years ago, the French psychiatrist J. Rogues de Fursak suggested that clinicians should distinguish: a. traumatic psychoses proper, resulting either from a localized cerebral lesion or from a general concussion of the brain; *b.* emotional psychoses, due to emotional shock with or without a traumatism which is more or less slight and not necessarily involving the brain; to which should be added *c.* the cases of war psychoses or so-called shell shock, which, as we shall see, are nothing but emotional psychoses [2]. He suggested that bad heredity, an emotionally unstable constitution, an accident producing an insignificant injury and an intense emotion; transient state of confusion with dream manifestations systematized upon the accident of which the patient has preserved an exact recollection lead to the development of "Shell Shock."

Despite multiple observations over many decades, observations, the scientific study of traumatic responses is a recent enterprise. After World Wars I and II, studies of psychological distress among combatants and concentration camp survivors set the stage for the recognition of the long-term negative effects of trauma exposure.

Since its official recognition as a psychiatric diagnosis nearly 30 years ago, there has been remarkable progress in understanding PTSD and developing and testing evidence-based treatments. PTSD is a common and severe psychiatric disorder precipitated by exposure to a psychologically distressing event. PTSD first appeared in the Diagnostic and Statistical Manual of Psychiatric Disorders (DSM-III) in 1980 [1], arising from studies of the Vietnam war, and of civilian victims of natural and man-made disasters [3,4]. According to the most recent edition of the Diagnostic and Statistical Manual of Psychiatric Disorders (DSM-IV-TR) [5], the essential feature of PTSD is the development of characteristic symptoms following exposure to an extreme traumatic stressor characterized by: direct personal experience of an event that involves actual or threatened death or serious injury, or other threat to one's physical integrity; or witnessing an event that involves death, injury, or a threat to the physical integrity of another person; or learning about unexpected or violent death, serious harm, or threat of death or injury experienced by a family member or other close associate. The person reacts to this event with fear and helplessness, and tries to avoid being reminded of it.

PTSD holds an important place in psychiatric thinking for several reasons. First it is one of the few disorders (besides disorders induced by substances or general medical conditions) in which an etiology is specified, despite the effort of DSM-IV to be atheoretical. As a consequence it has become a paradigm for understanding the role of the environment in psychopathology. Throughout the history of the concept of PTSD there has been a tension between an emphasis on factors relating to the host and factors relating to the nature of the trauma in the understanding of the etiology of this disorder.

PTSD is characterized by the presence of three distinct, but co-occurring, symptom clusters [1,5]. *Reexperiencing symptoms* describe spontaneous, often insuppressible intrusions of the traumatic memory in the form of images or nightmares that are accompanied by intense physiological distress. *Avoidance symptoms* involve restricting thoughts and distancing oneself from reminders of the event, as well as more generalized emotional and social withdrawal. *Hyperarousal symptoms* reflect more overt physiological manifestations, such as insomnia, irritability, impaired concentration, hypervigilance, and increased startle responses. These symptoms must be severe enough to impair social, occupational, or interpersonal function and co-occur for at least 1 month. The impairment from PTSD is amplified by poor coping strategies, substance abuse, co-occurring mood and anxiety disorders, lack of social support, and the accelerated development of stress-related medical conditions. A large body of research indicates that there is a correlation between PTSD and suicide [1,5]. There is evidence that traumatic events such as sexual abuse, combat trauma, rape, and domestic violence generally increase a person's suicide risk.

PTSD is a disorder of forgetting perhaps even more than of remembering [6,7]. It is the inability to forget the trauma that leads to the pathology and suffering in PTSD. Forgetting is a critical component of recovery. Of course, if we could not forget, our brains would rapidly be cluttered with information and observations and perhaps more limited in cognitive control functions for other activities.

Much of the early research on PTSD was based on the assumption that PTSD is a natural consequence of trauma exposure [2-6]. However, a growing body of research suggests that many persons exposed to traumatic events do not develop PTSD, or they recover quickly from stress symptomatology experienced in the immediate aftermath of trauma exposure. Although most people (50-90%) encounter trauma over a lifetime, only about 8% develop full PTSD. Vulnerability to PTSD presumably stems from an interaction of biological diathesis, early childhood developmental experiences, and trauma severity. Predictor models have consistently found that childhood trauma, chronic adversity, and familial stressors increase risk for PTSD as well as risk for biological markers of risk for PTSD after a traumatic event in adulthood. This effect of childhood trauma, which is not well understood, may be a marker for both traumatic experiences and attachment problems.

War, terrorism, and natural disasters create large populations in distress [6]. Not all distress amounts to mental disorder, but fear, worry, insomnia, and changes in health-risk behaviors all contribute to the health burden of mass violence and are targets for early public health intervention. Substantial evidence supports essential principles of immediate and midterm mass trauma interventions that are now described as psychological first aid [8].

About 30 armed conflicts are occurring now around the globe involving more than 25 countries [9,10]. For those in the United States and the United Kingdom, Iraq and Afghanistan are the present teachers of lessons long known and too often forgotten. Epidemiological surveys conducted during the current conflicts in Iraq and Afghanistan suggest that as many as 13% to 17% of service members screen positive for PTSD [11].

It is important to note that one of the most serious consequences of PTSD is its association with the heavy use of alcohol and/or other drugs [2-6]. Substance abuse has immediate consequences in the form of increased accidents, injuries, and long-term effects in terms of occupational and familial instability and early mortality. Furthermore, substance abuse, in itself, is often a risk factor for additional traumatic exposures either through accidents or interpersonal violence. The use of alcohol and/or other drugs is associated with suicidal and homicidal behavior. Impulsivity in combination with a mood disorder or PTSD increases the risk of suicidal behavior. Individuals exposed to severe traumatic stress often develop comorbid PTSD and major depressive disorder (MDD).

Major Depressive Disorder

MDD is a significant public health problem [3,5]. MDD is relatively common and its recurrent nature profoundly disrupts patients' lives. MDD afflicts one in eight Americans during their lifetimes and costs the U.S. economy more than $80 billion annually in medical treatment and lost productivity [12]. Many people suffer from dysthymia which is a milder form of depression than MDD. About 10-15% of patients with MDD die from suicide [3]. Overall, patients with MDD die at a younger age, especially from cardiovascular disease, than the general population. They show substantial impairment on numerous measures of social functioning, including measures of marital and occupational functioning. MDD is associated with high rates of alcohol and drug abuse, which may be a form of self-medication. About 25% of patients with MDD abuse alcohol or illegal drugs, percentages that are far higher than the general population rate of 8% [3,5]. The rate of occupational, social, and physical disabilities of patients with mood disorders is equal to or greater than that for patients with other major chronic illnesses, including hypertension, diabetes, coronary artery disease, lung disease, gastrointestinal disease, and arthritis. Patients with MDD become medically ill more often and show greater physical disability and sometimes higher death rates when ill [3]. The DSM-IV criteria for major depression stipulate that a patient has a persistent, distinct feeling of depression and/or loss of pleasure or interest in usual activities that lasts for at least two weeks [5]. In order to complete the diagnosis, at least four associated symptoms must also be present (decrease or increase in appetite, insomnia or hypersomnia, loss of energy, psychomotor agitation or retardation, feelings of worthlessnessin or appropriate guilt, diminished ability to concentrate or indecisiveness, and recurrent thoughts of death or suicidal behavior). It is worth noting that patients with MDD and/or PTSD often feel sad because they experience a personal sense of guilt or inadequacy. Cognitive dysfunction is another way in which patients are affected by PTSD and/or MDD [2-6].

The Concept of Post-traumatic Mood Disorder

PTSD and major depression are frequently comorbid [13-15]. Brown and colleagues [14] assessed lifetime anxiety and mood disorders comorbidity in a community sample of outpatients and found that PTSD was the anxiety disorder most likely to be associated with MDD, with 69% of individuals with PTSD also meeting criteria for MDD. Findings from the National Comorbidity Survey indicated that 78% of individuals meeting criteria for both disorders reported that their PTSD preceded the MDD [15].

Many research groups reported that individuals with comorbid PTSD and MDD were characterized by greater severity of symptoms and the higher level of impairment in social and occupational functioning compared to individuals with PTSD alone [13,16-23]. One study found that comorbidity of PTSD and MDD was associated with more severe symptoms as well as higher levels of disability on all indices (global dysfunction, distress, social impairment, and occupational disability) [23]. Another research group observed that a severity of overall symptoms was three- to fivefold greater in subjects with comorbid PTSD and MDD compared to those with PTSD alone [18]. A study of military veterans found that adjusting for sociodemographic differences and physical illness comorbidity, patients with comorbid PTSD and MDD reported more severe depression, lower social support, more frequent outpatient health care visits, and were more likely to report suicidal ideation than MDD patients [13]. A large-scale epidemiological survey found that the comorbid group was five times more likely to manifest functional impairment compared to those diagnosed with PTSD alone [16]. A recent study reported that patients with comorbid PTSD and MDD had higher objective MDD, impulsivity, and hostility scores and were significantly more likely to have made a suicide attempt compared to subjects with MDD alone [24]. We found that depressed subjects with comorbid PTSD tended towards earlier age of first hospitalization and a higher number of hospitalizations compared to depressed individuals without comorbid PTSD [25].

Maes et al. [26] reported that there is an association between PTSD with concurrent major MDD and lower affinity of alpha-2 adrenoreceptors, as well as higher plasma tyrosine availability to the brain, not found in patients suffering from PTSD alone. Woodward et al. [27] found that patients with comorbid PTSD and MDD exhibited less slow wave sleep and less facial (mentalis) electromyographic activity, compared with PTSD patients without comorbid MDD. Patients with comorbid PTSD and MDD did not exhibit the classic rapid eye movement sleep architectural modifications associated with unipolar depression, despite the fact that several other psychophysiological indices of dysphoria were detectable in their sleep.

Cortisol response to placebo or fenfluramine was examined in depressed patients with or without comorbid PTSD and in a control group of healthy volunteers [28]. Depressed patients with comorbid PTSD had the lowest plasma cortisol; depressed patients without comorbid PTSD had the highest plasma cortisol; and healthy volunteers had intermediate levels. This suggests differences in hypothalamic-pituitary-adrenal axis function among these three groups. We compared the effect of age on cortisol levels in depressed patients with or without comorbid PTSD and in healthy volunteers [29,30]. We found that cortisol levels increased with age in depressed patients with PTSD alone; they did not increase in depressed patients with comorbid PTSD or in healthy volunteers. We also observed that the number of previous major depressive episodes was a predictor of the cortisol response to fenfluramine administration in depressed patients without PTSD, but not in depressed patients with comorbid PTSD. We also found that depressed subjects with comorbid PTSD had higher cerebrospinal fluid (CSF) homovanillic acid (HVA) levels compared with depressed subjects without comorbid PTSD or healthy volunteers [25]. The dexamethasone suppression-corticotrophin releasing hormone (DEX-CRH) test has shown to be a more sensitive test to assess HPA-axis dysregulation in MDD and therefore may provide a useful test tool to probe HPA-axis regulation in MDD and PTSD [35,36]. To evaluate the effect of PTSD on HPA-axis regulation, the response to a DEX-CRH test between male veterans with PTSD and male veterans, who had been exposed to similar traumatic events during their deployment, without PTSD was compared [37]. The DEX-CRH test did not reveal HPA-axis abnormalities in PTSD patients as compared to trauma controls. PTSD patients with a co-morbid MDD showed an attenuated ACTH response compared to PTSD patients without co-morbid MDD, suggesting the presence of subgroups with different HPA-axis regulation within the PTSD group. Altered sensitivity of the CRH receptors at the pituitary or differences in AVP secretion might explain these differences in response.

Davidson et al. [31] observed genetic shared liability for PTSD and MDD, with familial loading for MDD predicting chronic PTSD in trauma survivors was found. Koenen et al. [32] analyzed data from about 7,000 members of the Vietnam Era Twin Registry and found substantial genetic overlap between PTSD and MDD and suggested that genes implicated in the etiology of MDD are strong candidates for PTSD and vice versa.

Subjects with PTSD without comorbid MDD and patients with PTSD with comorbid MDD were examined using the script-driven symptom-provocation paradigm adapted to functional magnetic resonance imaging [33]. This study found different patterns of brain activation in subjects with PTSD without comorbid MDD compared to patients with PTSD with comorbid MDD. PTSD patients with comorbid MDD relative to both healthy controls and trauma-exposed control subject showed reduced norepinephrine transporter expression in the left thalamus [34]. PTSD patients without comorbid MDD did not show differences in norepinephrine transporter expression relative to healthy controls and trauma-exposed control subject, i.e., PTSD patients showed reduced left thalamic norepinephrine transporter expression only when having comorbid MDD.

In summary, clinical and neurobiological evidence strongly suggests that patients suffering from comorbid PTSD and MDD differ clinically and biologically from individuals with PTSD alone or MDD alone. I have previously suggested that some or all individuals diagnosed with comorbid PTSD and MDD have a separate psychobiological condition that can be termed "post-traumatic mood disorder" (PTMD) [38,39].

Suicidal Behavior in PTMD

Suicidal act refers to the occurrence of suicide attempts, which can be defined as self-directed injurious acts with at least some intent to end one's own life [3,40]. Suicidal behavior ranges from fatal acts (completed suicide), to highly lethal and failed suicide attempts (where high intention and planning are evident, and survival is fortuitous), and to low-lethality attempts (usually impulsive attempts that are triggered by a social crisis). Multiple risk factors are associated with suicidal behavior, which has been conceptualized as the outcome of the interaction between an individual's diathesis for suicidal acts and triggers for suicidal behavior. The diathesis refers to the propensity for manifesting suicidal behavior and may be considered trait-related and independent of psychiatric diagnosis. In contrast, triggers are precipitants or stressors that determine the timing and probability of suicidal acts. Thus, triggers may be considered state-related. In this regard, risk factors for suicidal behavior may be categorized according to whether they affect the diathesis or the triggers. Both PTSD and MDD are associated with suicidal behavior and there are observations suggesting that individuals with comorbid PTSD and MDD (i.e., PTMD) have higher suicidality compared to persons with MDD only [3,38-40].

Multiple neurobiological factors including serotonin, glutamate, gamma-aminobutyric acid, norepinephrine, neuropeptide Y, corticotrophin-releasing hormone, dopamine, opioids, and thyroid hormones are involved in the pathophysiology of PTSD which is enormously complex [2-4,6,38,39]. Some neurobiological alterations observed in PTSD may be related to suicidal behavior. Serotonergic and hypothalamic-pituitary adrenal (HPA) axis abnormalities are associated with both PTSD and suicidal behavior and may underlie both conditions.

The Concept of PTMD and Suicidal Behavior

The experience of a potentially traumatic event is a correlate of suicidal behavior and MDD [40-43]. PTMD is associated with suicidal behavior. Considerable evidence suggests that child abuse increases the risk for PTSD, MDD, and suicidal behavior in adolescents and adults [44-48]. About one-third of those who have been abused or neglected during childhood develop PTSD, and a history of childhood abuse more than doubles the odds of developing major depressive episode [45-46]. Many victims of childhood abuse develop comorbid PTSD and MDD, i.e., they develop PTMD. This suggests that an association of PTMD with suicidal behavior may frequently be related to childhood abuse.

Risk factors which may contribute to both PTMD and suicidal behavior in adolescents include: social and educational disadvantage; childhood and family adversity; psychopathology; individual and personal vulnerabilities; exposure to stressful life events and circumstances; and social, cultural and contextual factors [49-51]. Frequently, suicidal behaviors in adolescents appear to be a consequence of adverse life sequences in which multiple risk factors from these domains combine to increase risk of suicidal behavior.

I have recently proposed a model of suicidal behavior in war veterans with PTMD [52]. The model consists of the following components: 1) genetic factors; 2) prenatal development; 3) biological and psychosocial influences from birth to mobilization/deployment; 4) mobilization/pre-deployment stress; 5) combat stress, traumatic brain injury, and physical injury; 6) post-deployment stress; 7) biological and psychosocial influences after the deployment; 8) trigger (precipitant) of a suicidal act; and 9) suicidal act. The first four components determine vulnerability to combat stress. The first seven components determine predisposition to suicidal behavior, a key element that differentiates PTMD patients who are at high risk from those at lower risk. Suicidal behavior in PTMD can be attributed to the coincidence of a trigger with a predisposition for suicidal behavior.

Conclusion

PTSD, MDD, and PTMD are neither normal variations in mood nor appropriate reactions to severe stress. These disorders are distinguished from normal reactions by the duration and intensity of the patient's suffering and the degree of his or her functional impairment. It is also crucial for physicians, and the general public, to realize that these disorders do not represent a failure of "will power" or some other form of moral weakness.

Trauma-related mental health pathology is a critical medical and social problem. It is crucial to develop interventions to prevent and to treat PTMD and measures to prevent and to treat suicidal ideation and behavior in persons with PTMD. Priorities for intervening to reduce suicidal behavior in traumatized individuals lie with interventions focused upon the improved recognition, treatment and management of individuals with PTSD, MDD or PTMD.

References

[1] Friedman MJ, Keane TM, Resick PA. (eds). *Handbook of PTSD: Science and Practice. New* York: The Guilford Press, 2007.
[2] De Fursac JR, transl. by Rosanoff AJ. Traumatic and emotional psychoses. So called Shell Shock. *American Journal of Insanity* 1918; 75: 19-51.
[3] Cutler JL, Marcus ER (editors). *Psychiatry.* Philadelphia, WB Saunders, 1999.
[4] Hageman I, Andersen HS, Jorgensen MB. Post-traumatic stress disorder: a review of psychobiology and pharmacotherapy. *Acta Psychiatr Scand* 2001; 104: 411–422.
[5] American Psychiatric Association. Diagnostic and Statistical Manual of Mental Disorders, Fourth Edition, Text Revision. Washington DC, *American Psychiatric Association, 2000*: 463–468.
[6] Ursano RJ, Benedek DM, Fullerton CS. Posttraumatic Stress Disorder: Neurobiology, *Psychology, and Public Health. Psychiatric Times,* 2008; 25(3). http://www.consultantlive.com/pathologic-processes/article/10168/1147506?pageNumber=2 Accessed August 7, 2008
[7] Kuhl BA, Dudukovic NM, Kahn I, Wagner AD. Decreased demands on cognitive control reveal the neural processing benefits of forgetting. *Nat Neurosci* 2007;10:908-914.
[8] Hobfoll SE, Watson P, Bell CC, et al. Five essential elements of immediate and mid-term mass trauma intervention: empirical evidence. *Psychiatry* 2007;70: 283-315.
[9] *Project Plowshares Armed Conflicts Report 2007 Summary.* Available at: http://www.ploughshares.ca/ libraries/ACRText/Summary2006.pdf. Accessed August 5, 2008.
[10] http://www.globalsecurity.org/military/world/war/index.html Accessed August 8, 2008
[11] Hoge CW, Castro CA, Messer SC, et al. Combat duty in Iraq and Afghanistan, mental health problems and barriers to care. *N Engl J Med* 2004;351:13-22.
[12] Greenberg PE, Kessler RC, Birnbaum HG, Leong SA, Lowe SW, Berglund PA, Corey-Lisle PK. The economic burden of depression in the United States: how did it change between 1990 and 2000? *J Clin Psychiatry* 2003;64(12):1465-75.
[13] Campbell DG, Felker BL, Chuan-Fen L et al. Prevalence of PTSD-Depression comorbidity: Implications for clinical practice guidelines and primary care interventions. *J Gen Intern Med* 2007;22:711-8.
[14] Brown TA, Campbell LA, Lehman CL, Grisham JR, Mancill RB. Current and lifetime comorbidity of the DSM-IV anxiety and mood disorders in a large clinical sample. *J Abnorm Psychol 2001*; 110: 585-599.
[15] Kessler RC, Sonnega A, Bromet E, Hughes M, Nelson CB. Posttraumatic stress disorder in the National Comorbidity Survey. *Arch Gen Psychiatry* 1995; 52: 1048-1060.
[16] Mollica RF, McInnes K, Sarajlic N, Lavelle J, Sarajlic I, Massagli MP. Disability associated with psychiatric comorbidity and health status in Bosnian refugees living in Croatia. *J Am Med Assoc* 1999; 282: 433-439.
[17] Shalev AY, Freedman S, Peri T, Brandes D, Sahar T, Orr SP, Pitman RK. Prospective study of posttraumatic stress disorder and depression following trauma. *Am J Psychiatry.* 1998; 155: 630-637.

[18] Karam EG. Comorbidity of posttraumatic stress disorder and depression. In: Fullerton C, Ursano RJ (editors). *Posttraumatic Stress Disorder: Acute and Long-Term Responses to Traumas and Disaster. Progress in Psychiatry Series,* No. 51. American Psychiatric Press, Washington, DC, 1997; pp. 77-90.

[19] Moore LJ, Boehnlein JK. Posttraumatic stress disorder, depression, and somatic symptoms in U.S. Mien patients. *J Nerv Ment Dis* 1991; 179: 728-733.

[20] Skodol AE, Schwartz S, Dohrenwend BP, Levav I, Shrout PE, Reiff M. PTSD symptoms and comorbid mental disorders in Israeli war veterans. *Br J Psychiatry.* 1996; 169: 717-725.

[21] Koren D, Arnon I, Klein E. Acute stress response and posttraumatic stress disorder in traffic accident victims: a one-year prospective, follow-up study. *Am J Psychiatry.* 1999; 156: 367-373.

[22] Mintz J, Mintz LI, Arruda MJ, Hwang SS. Treatments of depression and the functional capacity to work. *Arch Gen Psychiatry* 1992; 49:761-768.

[23] Momartin S, Silove D, Manicavasagar V, Steel Z. Comorbidity of PTSD and depression: association with trauma exposure, symptom severity and functional impairment in Bosnian refugees resettled in Australia. *J Affect Disord* 2004; 80: 231–248.

[24] Oquendo M, Brent DA, Birmaher B, Greenhill L, Kolko D, Stanley B, Zelazny J, Burke AK, Firinciogullari S, Ellis SP, Mann JJ. Posttraumatic stress disorder comorbid with major depression: factors mediating the association with suicidal behavior. *Am J Psychiatry* 2005; 62: 560-566.

[25] Sher L, Oquendo MA, Li S, Burke AK, Grunebaum MF, Zalsman G, Huang YY, Mann JJ. Higher cerebrospinal fluid homovanillic acid levels in depressed patients with comorbid post-traumatic stress disorder. *Eur Neuropsychopharmacol* 2005; 15: 203-209.

[26] Maes M, Lin AH, Verkerk R, Delmeire L, Van Gastel A, Van der Planken M, Scharpe S. Serotonergic and noradrenergic markers of post-traumatic stress disorder with and without major depression. *Neuropsychopharmacology* 1999; 20: 188–197.

[27] Woodward SH, Friedman MJ, Bliwise DL. Sleep and depression in combat-related PTSD inpatients. *Biol Psychiatry* 1996; 39: 182–192.

[28] Oquendo MA, Echavarria G, Galfalvy HC, Grunebaum MF, Burke A, Barrera A, and others. Lower cortisol levels in depressed patients with comorbid post-traumatic stress disorder. *Neuropsychopharmacology* 2003; 28: 591– 598.

[29] Sher L, Oquendo MA, Galfalvy HC, Cooper TB, Mann JJ. Age effects on cortisol levels in depressed patients with and without a history of post-traumatic stress disorder, and healthy volunteers. *J Affect Disord* 2004; 82: 53-59.

[30] Sher L, Oquendo MA, Galfalvy HC, Cooper TB, Mann JJ. The number of previous depressive episodes is positively associated with cortisol response to fenfluramine administration. In: Yehuda R, McEwen B (editors). Biobehavioral Stress Response: Protective and Damaging Effects. *Ann N Y Acad Sci* 2004; 1032: 283-286.

[31] Davidson JR, Tupler LA, Wilson WH, Connor KM. A family study of chronic post-traumatic stress disorder following rape trauma. J Psychiatr Res 1998; 32: 301–309.

[32] Koenen KC, Fu QJ, Ertel K, Lyons MJ, Eisen SA, True WR, Goldberg J, Tsuang MT. Common genetic liability to major depression and posttraumatic stress disorder in men. *J Affect Disord* 2008;105(1-3):109-15.

[33] Lanius RA, Frewen PA, Girotti M, Neufeld RW, Stevens TK, Densmore M Neural correlates of trauma script-imagery in posttraumatic stress disorder with and without comorbid major depression: a functional MRI investigation. *Psychiatry Res.* 2007;155(1):45-56.

[34] Czermak C, Ding Y, Henry S, B.Planeta-Wilson B, S.Kasserman S, J.Frost J, W.Williams W, J.Krystal J, R.Carson R, A.Neumeister A. Norepinephrine transporter imaging in posttraumatic stress disorder. In: *46th Annual Meeting of the American College of Neuropsychopharmacology,* Boca Raton, Florida, December 9-13, 2007. Abstracts. Program No. 133.

[35] Sher L, Cooper TB, Mann JJ, Oquendo MA. Modified dexamethasone suppression – corticotropin-releasing hormone stimulation test: a pilot study of young healthy volunteers and implications for alcoholism research in adolescents and young adults. *Int J Adolesc Med Health* 2006;18(1):133-137.

[36] Ising M, Kunzel HE, Binder EB, Nickel T, Modell S, Holsboer F. The combined dexamethasone/CRH test as a potential surrogate marker in depression. *Prog Neuropsychopharmacol Biol Psychiatry.* 2005;29(6):1085-93.

[37] de Kloet C, Vermetten E, Lentjes E, Geuze E, van Pelt J, Manuel R, Heijnen C, Westenberg H. Differences in the response to the combined DEX-CRH test between PTSD patients with and without co-morbid depressive disorder. *Psychoneuroendocrinology* 2008;33(3):313-20.

[38] Sher L. The concept of post-traumatic mood disorder. *Med Hypotheses* 2005; 65(2):205-10.

[39] Sher L. Posttraumatic mood disorder: a new concept. *Can J Psychiatry* 2005;50(2):127-8.

[40] Kilpatrick DG, Saunders BE, Veronen LJ, Best CL, Von JM. Criminal victimization: Lifetime prevalence, reporting to police, and psychological impact. *Crime & Delinquency* 1987; 33: 479–89.

[41] Waldrop AE, Hanson RF, Resnick HS, Kilpatrick DG, Naugle AE, Saunders BE. Risk factors for suicidal behavior among a national sample of adolescents: Implications for prevention. *J Traum Stress* 2007; 20: 869-879.

[42] Belik S-L, Cox BJ, Stein MB, Asmundson GJG, Sareen J. Traumatic events and suicidal behavior: Results from a national mental health survey. *J Nerv Ment Dis* 2007; 195(4): 342-349.

[43] Joiner TEJr, Sachs-Ericsson NJ, Wingate LR, Brown JS, Anestis MD, Selby EA. Childhood physical and sexual abuse and lifetime number of suicide attempts: A persistent and theoretically important relationship. *Behav Res Ther* 2007; 45: 539–547.

[44] Oquendo M, Brent DA, Birmaher B, Greenhill L, Kolko D, Stanley B, Zelazny J, Burke AK, Firinciogullari S, Ellis SP, Mann JJ. Posttraumatic stress disorder comorbid with major depression: factors mediating the association with suicidal behavior. *Am J Psychiatry.* 2005;162(3):560-566.

[45] Widom CS: Posttraumatic stress disorder in abused and neglected children grown up. *Am J Psychiatry* 1999; 156:1223–1229.

[46] MacMillan HL, Fleming JE, Streiner DL, Lin E, Boyle MH, Jamieson E, Duku EK, Walsh CA, Wong MY-Y, Beardslee WR: Childhood abuse and lifetime psychopathology in a community sample. *Am J Psychiatry* 2001; 158:1878–1883.

[47] Makhija N, Sher L. Childhood abuse, adult alcohol use disorders and suicidal behavior. *QJM* 2007; 100(5): 305-309.
[48] Arnow BA. Relationships between childhood maltreatment, adult health and psychiatric outcomes, and medical utilization. *J Clin Psychiatry*. 2004;65 Suppl 12:10-5.
[49] Beautrais AL. Risk factors for suicide and attempted suicide among young people. *Aust N Z J Psychiatry*. 2000;34(3):420-36.
[50] Sher L, Zalsman G. Alcohol and adolescent suicide. *Int J Adol Med Health* 2005; 17(3):197-203.
[51] Beautrais AL. Suicide and serious suicide attempts in youth: a multiple-group comparison study. *Am J Psychiatry* 2003;160(6):1093-9.
[52] Sher L. A concept of suicidal behavior in war veterans with posttraumatic mood disorder. *Med Hypotheses* 2009;73(2):215-219.

In: Neurobiology of Post-Traumatic Stress Disorder
Editors: L. Sher and A. Vilens, pp. 271-289
ISBN: 978-1-61470-593-2

Chapter XVII

Pharmacotherapy of PTSD: An Evidence-based Review

Lakshmi N. Ravindran[1], Murray B. Stein[2] and Arun V. Ravindran[1]
[1]University of Toronto, Toronto, Ontario, Canada
[2]University of California San Diego, La Jolla, California, USA

Abstract

Post-traumatic stress disorder (PTSD) is a common psychiatric illness with significant morbidity and impact on function. Several classes of pharmacological agents, including antidepressants, anticonvulsants, atypical antipsychotics, and several novel agents, have been investigated for the treatment of PTSD. This chapter reviews the published literature with an emphasis on randomized controlled trials. The evidence for benefit as monotherapy appears to be strongest for antidepressants, specifically SSRIs and SNRIs, with some benefit from trycyclics and MAOIs. Promising preliminary results have been reported for mirtazapine, prazosin and some anticonvulsant such as lamotrigine. Atypical antipsychotics seem to be most useful as adjunctive agents. Limited efficacy data exists on the efficacy of the combination of medication and psychotherapy but this strategy would likely benefit refractory illness. The general dearth of large, well-designed RCTs on treatments for PTSD must be noted.

Introduction

PTSD, which occurs as a consequence of physical and psychological trauma, is characterized by intrusive re-experiencing of the event, avoidance behaviour, and symptoms of hyperarousal. It became recognized as an official psychiatric condition in 1980, with its inclusion in the DSM-III [1]. The diagnosis and treatment of PTSD is often complicated because of the complex presentation and frequent co-morbidity with other psychiatric conditions, including depressive and other anxiety disorders and substance use disorders.

While both psychological and pharmacological strategies have been investigated in PTSD, this chapter will focus on the latter form of treatment.

Pharmacological agents that have demonstrated efficacy in treating patients with PTSD include several classes of antidepressants, anticonvulsants, atypical antipsychotics, and benzodiazepines, with several novel agents in development. This review focuses mainly on randomized controlled trials (RCTs) of the above agents, which are mostly short-term. The few reports on augmentation and combination strategies (for treatment resistance), and prophylaxis and maintenance, are outlined with some proposed predictors of treatment response. Finally, recommendations for lines of treatment, derived using standard methodology for evaluating levels of evidence [2], are provided (see Table 1), together with summary tables of published RCTs and meta-analyses (see Tables 2-5).

Pharmacotherapy

I. Antidepressants

A. Selective Serotonin Reuptake Inhibitors (SSRIs)

SSRIs are currently the most investigated agents in the pharmacological treatment of PTSD. In part, the use of SSRIs has been supported by neurobiological findings, in particular from neuroimaging and pharmacogenetic studies, which suggest a role for serotonergic dysfunction in the pathophysiology of PTSD [3,4].

Table 1. Summary of recommendations

Line of Treatment	*Agent*
First-line	SSRIs (most evidence for fluoxetine, sertraline, paroxetine; likely all agents effective) SNRIs (venlafaxine)
Second-line (as augmentation)	Anti-Adrenegic Agent (specifically prazosin) Atypical antipsychotics (most evidence for olanzapine, risperidone, and quetiapine)
Third-line (based on limited evidence and/or side effect profile)	TCAs (imipramine) MAOIs (phenelzine) Mirtazapine Valproate

Table 2. Meta-analyses of RCTS in PTSD – active agents superior to placebo

Meta-Analysis	Comparison	Number of Studies Included	Duration (weeks)
Ipser et al., 2006	SSRIs vs. placebo	17 studies on SSRIs 2 studies on TCAs	5 to 28 4 to 8

Gao et al., 2006	Augmentation with atypicals or placebo	8 studies	5 to16
Pae et al., 2008	Augmentation with atypicals or placebo	7 studies	5 to 16
Adamou et al., 2007	Valproate alone or vs. placebo	8 studies	4 to 8, with one 52 week outlier

Table 3. Active agents versus placebo in PTSD – postive studies

Study	**Comparison**	**N**	**Duration (weeks)**
SSRIs and SNRIs			
Marshall et al., 2007	Paroxetine vs. placebo	52	22
Davidson, Rothbaum et al., 2006	Venlafaxine vs. sertraline and placebo	538	12
Davidson, Baldwin et al., 2006	Venlafaxine vs. placebo	329	24
TCAs, MAOIs and RIMAs			
Frank et al., 1988	Imipramine vs. phenelzine and placebo	34	8
Kosten et al., 1991	Imipramine vs. phenelzine and placebo	60	8
Katz et al., 1994	Brofaromine vs. placebo	45	14
Novel Antidepressants			
Davidson et al., 2003	Mirtazapine vs. placebo	29	8
Davis et al., 2004	Nefazodone vs. placebo	41	12
Anticonvulsants			
Hertzberg et al., 1999	Lamotrigine vs. placebo	15	12
Atypical or Novel Antipsychotics			
Padala et al., 2006	Risperidone vs. placebo	20	12
Anti-Adrenergic Agents			
Raskind et al., 2003	Prazosin vs. placebo	10	20
Taylor et al., 2008	Prazosin vs. placebo	13	7
Raskind et al., 2007	Prazosin vs. placebo	40	8
Pharmacological Agents in Development			
Aerni et al., 2004	Hydrocortisone vs. placebo	3	12

Table 4. Active agents versus placebo in PTSD – negative studies

Study	**Comparison**	**N**	**Duration (weeks)**
SSRIs and SNRIs			
Martenyi et al., 2007	Fluoxetine 20 mg or 40 mg vs. placebo	411	12
Friedman et al., 2007	Sertraline vs. placebo	169	12

Table 4. (continued)

TCAs, MAOIs and RIMAs			
Shestazky et al., 1988	Phenelzine vs. placebo	13	9
Baker et al., 1995	Brofaromine vs. placebo	114	12
Novel Antidepressants			
Hertzberg et al., 2001	Bupropion vs. placebo	15	12
Becker et al., 2007	Bupropion vs. Placebo	30	8
Anticonvulsants			
Lindley et al., 2007	Augmentation with topiramate or placebo	40	7
Tucker et al., 2007	Topiramate vs. placebo	38	12
Davidson et al., 2007	Tiagabine vs. placebo	232	12
Connor et al., 2006	Tiagabine vs. placebo	18	24
Davis et al., 2008	Divalproex vs. placebo	85	8
Benzodiazepines, Other Hypnotics and Anxiolytics			
Braun et al., 1990	Alprazolam vs. placebo	10	12
Cates et al., 2004	Clonazepam vs. placebo	6	5
Mellman et al., 2002	Temazepam vs. placebo	22	1
Atypical or Novel Antipsychotics			
Butterfield et al., 2001	Olanzapine vs. placebo	15	10
Rothbaum et al., 2008	Augmentation with risperidone or placebo	25	16
Anti-Adrenergic Agents			
Neylan et al., 2006	Guanfacine vs. placebo	63	8
Pharmacological Agents in Development			
Heresco-Levy et al., 2002	D-cycloserine vs. placebo	11	12
Heresco-Levy et al., 2009	D-serine vs. placebo	22	6

Table 5. Comparator only (no Placebo Arm) RCTS in PTSD

Study	Comparison	N	Duration (weeks)	Result
TCAs, MAOIs and RIMAs				
Onder et al., 2006	Moclobemide vs. fluoxetine and tianeptine	103	12	Comparable efficacy
Novel Antidepressants				
Chung et al., 2004	Mirtazapine vs. sertraline	113	6	Mirtazapine superior

McRae et al., 2004	Nefazodone vs. sertraline	37	12	Comparable efficacy
Saygin et al., 2002	Nefazodone vs. sertraline	60	24	Comparable efficacy

A recent meta-analysis of pharmacotherapy for PTSD by Ipser et al. [5] reported that SSRIs show the greatest efficacy in the treatment of PTSD for both short- and long-term treatment; the analysis included 17 studies of SSRIs, with fluoxetine, sertraline and paroxetine demonstrating the greatest evidence of superiority over placebo. However, results from subsequent RCTs have been more equivocal. A recent large RCT of civilians with PTSD found no difference between fluoxetine and placebo, which the investigators suggested may have been influenced by the predominantly female population or the use of lower than usual doses of fluoxetine [20 mg or 40 mg/day) employed in this study [6]. This is similar to the finding by Friedman et al. [7] that sertraline was no different from placebo nor were there predictors of response [7] in a sample of combat veterans with PTSD. In contrast to these findings, an RCT using an urban, primarily minority sample found paroxetine significantly more effective than placebo in improving PTSD symptoms, during both acute and maintenance treatment [8]. The meta-analysis referred to above also noted that the only controlled trial of citolapram found it comparable to both sertraline and placebo [5]. However, open trials have shown benefits with citalopram, in adults, children and adolescents [9-12], and have also reported moderate efficacy for escitalopram [13] and fluvoxamine [14-17].

Despite the positive findings from several RCTs, the U.S. Institute of Medicine, noting the small effect sizes reported, found the overall evidence for SSRIs in PTSD to be inconclusive [18]. However, though the degree of benefit by SSRIs may vary, the efficacy demonstrated in these studies, particularly over placebo, cannot be dismissed, and is likely applicable to all SSRIs. Further, since many patients with PTSD are also diagnosed with depression and other anxiety disorders, SSRIs have the added advantage of alleviating both conditions [19]. There are also reports that SSRIs may help to improve quality of life in PTSD patients, with the majority of improvement often seen during acute treatment, which is another important clinical consideration [20].

The evidence that extended treatment with SSRIs is beneficial in maximizing symptom improvement, recovery and relapse prevention, is also worth highlighting. Of particular note is the report by Londborg et al. that continuation treatment elicited response in over 50% of acute phase non-responders, which may be a useful consideration in refractory cases when there is at least some response to SSRIs [5]. Guidelines currently suggest that maintenance treatment with SSRIs should be continued for 6-12 months in acute PTSD and 12-24 months in chronic PTSD, while even longer treatment may be useful in cases where residual symptoms or poor psychosocial functioning persist [21,22].

B. Serotonin Norepinephrine Reuptake Inhibitors (SNRIs)

The only published trials of SNRIs in PTSD are with venlafaxine. An open trial compared venlafaxine, sertraline and paroxetine in refugees with PTSD and found all three agents effective in reducing PTSD symptoms [23]. Similarly, an RCT by Davidson et al. [24] found venlafaxine and sertraline superior to placebo in reducing PTSD symptoms, particularly with respect to the hyperarousal and avoidance-numbing clusters of symptoms.

Of note is the finding that improvement in PTSD symptoms was found to be independent of the reduction in depressive symptoms. A second RCT with a duration of six months confirmed the superiority of venlafaxine over placebo in improving both the symptoms of PTSD and quality of life [25]. Re-experiencing and avoidance-numbing symptoms were particularly responsive to the medication.

Though based only on these three trials, venlafaxine is currently considered a suitable first- or second-line alternative to SSRIs. Another possible SNRI option is duloxetine, for which there is no published evidence as yet, but which may be beneficial based on its proposed mechanism of action and preliminary results from trials in depression and other anxiety disorders [26-29]. Further investigations with both these agents are warranted.

C. Tricyclic Antidepressants (TCAs)

One of the earliest classes of antidepressants developed, TCAs are still commonly used in clinical practice. However, they have become mainly second- or third-line choices for the treatment of depression due to their side effects and high toxicity in overdose.

Research into TCAs for the treatment of PTSD is limited with mostly negative results. The meta-analysis by Ipser et al. [5] also noted the lack of significant benefit with amitriptyline or desipramine over placebo in combat veterans with PTSD, though some depressive and anxiety symptoms improved [5]. Two early RCTs also used combat veteran populations with PTSD to compare imipramine, phenelzine (a monoamine oxidase inhibitor) and placebo and found both imipramine and phenelzine were significantly more effective though depressive symptoms remained unchanged [30,31]. Of note, baseline depressive symptoms in these trials were in the mild to moderate range so that significant changes in these symptoms might be harder to demonstrate with the limited sample sizes.

D. Monoamine Oxidase Inhibitors (MAOIs) and Reversible Inhibitors of Monoamine Oxidase A (RIMAs)

MAOIs are another older class of antidepressants. They are thought to affect depression by increasing monoamine availability in the synaptic cleft. Though effective agents, their side effect profile, which includes dietary restrictions and risk of hypertensive crisis, has limited their use to second- or third-line agents.

Trials of MAOIs in PTSD are sparse. Two positive RCTs with phenelzine have been described in the previous section [30,31]. Of note, intrusive thoughts or re-experiencing symptoms were particularly reduced with phenelzine in both studies, and the percentage of responders was higher with phenelzine than with imipramine in the second study [31]. However, a double-blind cross-over RCT found no benefit to phenelzine over placebo, though the small sample may have been insufficient to detect a benefit (i.e., Type II error) [32].

RIMAs are thought to be safer to use than MAOIs due to the selectivity of their action and their reversibility. Among RIMAs, brofaromine and moclobemide have been evaluated in PTSD and results have been mixed. One RCT found brofaromine significantly superior to placebo [33], but a larger RCT found both treatments comparably and significantly effective [34]. Another RCT compared moclobemide to fluoxetine and tianeptine, a serotonin reuptake accelerator, and found similarly significant improvement in PTSD symptoms across groups [35]. However, the study lacked placebo control.

Brofaromine and moclobemide are not commercially available in the U.S., though moclobemide is available in Canada, Europe and other parts of the world. Given the very limited data for moclobemide, its role in the treatment of PTSD is unclear.

E. Novel Antidepressants

i. Bupropion

Bupropion is thought to inhibit the reuptake of norepinephrine and dopamine, but to have little effect on serotonin activity [36]. Besides being used to treat depression, it is often used to aid in smoking cessation. The only published placebo-controlled RCT with bupropion sustained release (SR) in PTSD found no group differences [37]. An RCT that compared bupropion SR to placebo for smoking cessation in chronic PTSD patients found that bupropion significantly reduced smoking behaviour but neither treatment affected PTSD symptoms [38].

ii. Mirtazapine

It has been proposed that mirtazapine blocks the presynaptic α_2-adrenergic receptors and thus enhances the release of both serotonin and norepinephrine, while also blocking the serotonergic 5-HT_2 and 5-HT_3 receptors and the H_1 histaminic receptor [39]. There are two reports of RCTs with mirtazapine in PTSD. In a placebo-controlled trial, Davidson et al. [40] found that mirtazapine significantly improved PTSD symptoms. A randomized open-label study in Korean veterans with PTSD found similar positive effects, with mirtazapine significantly superior to sertraline in reducing PTSD symptoms, though effects on depressive symptoms were comparable [41]. Further investigations of this agent are needed to confirm these initial results.

iii. Nefazodone

Nefazodone is an antidepressant that works via antagonist activity at the post-synaptic 5-HT_{2A} receptor and inhibition of pre-synaptic serotonin and norepinephrine reuptake, as well as displaying relatively potent antagonism of the α_1-adrenergic receptor [42]. Reports of open trials with nefazodone suggest benefits in PTSD. In the only placebo-controlled RCT, nefazodone was significantly superior to placebo in reducing PTSD symptoms, but the sample was small [43]. Two other RCTs compared nefazodone to sertraline and found both agents comparably and significantly effective [44,45].

Nefazodone was recently withdrawn from the markets in Canada, Australia and New Zealand following numerous reports of hepatotoxicity, which may explain the dearth of RCTs. This concern may make consideration of nefazodone in the treatment of PTSD more an academic question than a practical one.

II. Anticonvulsants

Anticonvulsants, with their effect on neuronal kindling, are often used to treat bipolar disorders, but have also shown some efficacy in other depressive and anxiety disorders [46]. Since kindling and sensitization mechanisms have also been proposed in the pathophysiology of PTSD, it has been suggested that anticonvulsants may have a therapeutic role for this disorder [47]. There are several reports from open trials which suggest the benefit of carbamazepine, valproate, lamotrigine, gabapentin, topiramate, tiagabine, as well as levetiracetam in PTSD. However, evidence from RCTs is limited and has reported mixed results.

In an early pilot RCT, Hertzberg et al. found lamotrigine superior to placebo in specifically alleviating the re-experiencing and avoidance-numbing clusters of symptoms, but the small sample size limits the generalizability of the results [48]. The literature reports an open label study that found topiramate significantly effective as monotherapy or augmentation (the majority of the civilian patients received the latter) [49]. However, two short-term RCTs with topiramate, one as monotherapy and the other as augmentation, found no significant difference in global improvement over placebo [50,51]. Davidson et al. followed up on a small RCT of tiagabine [53] with a comparison of efficacy and tolerability of tiagabine versus placebo in a large RCT and found no significant differences [52]. Studies of valproate are few and consist mostly of uncontrolled trials, but a meta-analytic review has suggested that valproate may be useful in treating both PTSD and depressive symptoms [54]. However, in a more recent RCT, no significant differences were noted between divalproex and placebo, which were attributed to the short duration of treatment and chronicity of illness [55].

Due to the limited number of studies and the inconclusive results, the current data do not support the use of anticonvulsant monotherapy as first-line treatment in PTSD. Though more research is needed, lamotrigine seems to be the most promising of the available anticonvulsants, but this may simply be because large definitive RCTs have not yet been conducted. Similarly, few controlled studies exist investigating these agents as adjunctive medications in PTSD.

III. Benzodiazepines, other Hypnotics and Anxiolytics

Benzodiazepine and other sedative-hypnotic agents are frequently used to augment antidepressants in the treatment of PTSD to address sleep difficulties and hyperarousal [56]. However, there are few controlled trials in the literature to support this widespread clinical practice. Available reports have evaluated clonazepam, alprazolam and temazepam.

Braun et al., in an early study with a cross-over design, found no difference between alprazolam and placebo [57]. Similarly results were found in a placebo-controlled trial of clonazepam [58]. A small case series in early PTSD found one week of temazepam treatment effective in reducing symptom severity and preventing development of full-blown PTSD [59]. However, these results were not replicated in a subsequent one-week RCT of temazepam in early PTSD by the same investigators [60]. It must be considered that larger samples and longer duration of treatment might have produced more definitive results. In addition, it is possible that benzodiazepines may be more useful in reducing symptom severity than in symptom prevention.

There are two open label reports of the use of buspirone, an azapirone with anxiolytic properties [61,62]. Although modest benefits were noted for both PTSD and depressive symptoms, the samples were very small and there was a wide range for onset of improvement in the second study (5-29 days). Thus, these trials are of insufficient quality and size on which to base decisions about the usefulness of buspirone in PTSD.

Although these classes of agents are frequently used in clinical practice to treat symptoms of PTSD, the evidence for their benefit is limited. Given the risk of dependence with benzodiazepines in particular and the common co-morbidity of PTSD with substance abuse disorders [63], ample caution should be taken when using them in this population.

IV. Atypical or Novel Antipsychotics

Psychotic symptoms, particularly hallucinations and delusions, are often associated with PTSD diagnoses, but are not always better accounted for by co-morbid psychotic disorders or medical conditions [64,65]. Dopamine dysfunction has been implicated in psychotic disorders, and is also thought to play a role in the hyperarousal symptoms seen in PTSD [66,67]. Thus, it is perhaps not surprising that antipsychotics are increasingly used in clinical practice to treat PTSD. In addition to D_2 receptor blockade (with some off label individual variations), these agents also block the α_1- and α_2-adrenergic, 5-HT_{2A} and H_1 receptors, and are partial agonists of the 5-HT_{1A} receptor [68]. These varied modes of action may contribute to the anxiolytic and sedating effects of these agents, thus making them particularly suited to relieving the hyperarousal and sleep difficulties associated with PTSD, as well as treating any psychotic symptoms that may be present. The range of atypical antipsychotics in use include aripiprazole, clozapine, olanzapine, quetiapine, risperidone and ziprasidone.

Evidence for the use of atypical antipsychotics as monotherapy in PTSD is sparse, with only two RCTs available in the literature. In a small trial, Butterfield et al. found no difference between olanzapine and placebo [69]. The small female sample, high presence of co-morbidities, and shorter course of treatment than is usual (10 weeks), were suggested as possible contributors to the non-significant results. Another RCT found risperidone monotherapy significantly superior to placebo [70]. However, the small sample size was once again a study limitation.

Augmentation studies with atypical antipsychotics are more commonly reported. Two meta-analyses of these agents (8 studies and 7 studies respectively) indicate that such augmentation, specifically with olanzapine and risperidone, is significantly more effective than placebo augmentation in refractory PTSD [71,72]. However, in a recent trial, those who did not reach remission with open label sertraline were randomized to augmentation with risperidone or placebo, and both groups improved similarly [73]. Only one meta-analysis assessed quetiapine augmentation in PTSD, and it only included one trial. But given the effect of quetiapine on multiple neurotransmitters and its proven benefits in other anxiety disorders [71], it may eventually (i.e., given further research) be shown to be of benefit in PTSD, as well.

Studies of aripiprazole were not included in the above meta-analyses but two reports, a retrospective case review and an open-label trial, both reported benefits with aripiprazole augmentation [74,75]. Like another atypical agent, ziprasidone, aripiprazole is thought to have fewer metabolic effects than others in its class, but tolerability and resulting drop-out were significant issues with aripiprazole in the open label study [75].

Given the positive results from several augmentation studies, the not uncommon presence of psychotic symptoms and the frequent treatment refractoriness to antidepressants, atypical antipsychotics likely have a significant role in the pharmacotherapy of PTSD. Monotherapy trials with larger samples and of adequate duration to capture late responders would be useful in delineating the full utility of these agents in this disorder and may also provide objective validation for their in clinical practice to treat PTSD. That being said, the side-effects of atypical antipsychotics, particularly significant metabolic effects, must also be considered and careful monitoring instituted even when doses are low.

V. Anti-Adrenergic Agents

Among anti-adrenergic agents, guanfacine and clonidine (α_2-adrenergic agonists), as well as prazosin (an α_1-adrenergic antagonist) have been most explored. Open label reports suggest these agents may be useful as monotherapy or augmenting agents in treating PTSD symptoms, particularly hyperarousal and nightmares [76-80]. However, controlled trials to replicate these findings are far fewer in number. A single RCT of guanfacine found no superiority over placebo in veterans with chronic PTSD [81]. Results with prazosin are more positive and more substantial. A cross-over RCT in chronic PTSD found prazosin significantly more effective than placebo in reducing nightmares and overall symptom severity [82]. Similar results were found in another cross-over RCT with civilian patients treated for seven weeks, and in a parallel-group RCT with 40 [83,84]. However, sample sizes were limited in all three studies.

Given the promising results with prazosin, larger trials of longer duration would be most useful in providing further guidelines for its use in PTSD. RCTs with other anti-adrenergic agents would also be helpful in clarifying whether agents other than prazosin also have a role in PTSD treatment.

VI. Pharmacological Agents in Development

Promising results have been found with d-cycloserine and hydrocortisone in the treatment of PTSD. D-cycloserine is a partial agonist of the n-methyl-d-aspartate (NMDA) receptor. Animal studies suggest that it may aid in adaptive learning and stress coping, hence research interest in the utility of this medication in anxiety disorders. In human trials, it has been used successfully to augment psychotherapy in social anxiety disorder [85,86] and obsessive-compulsive disorder [87]. A small pilot cross-over RCT found that both d-cycloserine and placebo improved numbing/avoidance symptoms in PTSD, though the active drug also improved some neurocognitive measures more than placebo [88]. More recently, the same group of investigators conducted a similarly-designed study using D-serine, a naturally occurring substance that is a full NMDA receptor agonist [89]. The final analysis did not reveal significant differences between groups with respect to reduction in total Clinician-Administered PTSD Scale (CAPS) scores, although post-hoc analysis suggested an advantage for those patients receiving active drug during the second treatment period. These results with d-cycloserine and D-serine must be considered quite preliminary, and use of these agents in the treatment of PTSD must await the completion of further, more definitive research.

Hydrocortisone is a natural metabolic form of cortisol, a glucocorticoid that is considered a stress indicator. Hypothalamic-pituitary axis dysfunction, with accompanying altered neuroendocrine responses have been hypothesized as being a factor in PTSD [90], with resultant interest in the use of glucocorticoids as a possible treatment option. A pilot cross-over study with hydrocortisone (10 mg daily for one month) found that it improved some PTSD symptoms, particularly re-experiencing and avoidance, more than placebo but the extremely small sample and use of concurrent sleep medications and psychotherapy were limitations that confound interpretation of that trial [91].

As seen above, the results from these preliminary RCTs are far from conclusive, but they do support further investigations.

Pharmacotherapy and Psychotherapy Combinations

Among psychological treatments for PTSD, the greatest efficacy has been reported for trauma-focused cognitive-behavioural therapy (TFCBT), stress management and group cognitive-behavioural therapy [92]. There is also some support for eye movement desensitisation and reprocessing (EMDR), while other therapies, such as hypnotherapy, supportive therapy and psychodynamic therapy, have had inconclusive results. Given the range of both pharmacological and psychological therapies that have been investigated in PTSD, and the generally mixed results with medications (except for SSRIs and SNRIs), it is intuitive to think that combination treatment might produce greater benefits that either treatment alone, especially in refractory case. However, research in this area is limited and diffuse, with few therapies or medications studied consistently. (Research with d-cycloserine to enhance learning with CBT is a special use of pharmacotherapy and psychotherapy combinations, and is discussed above.)

In combination studies with proven psychotherapies, child and adolescent patients were randomized to receive TFCBT combined with sertraline or placebo but both groups improved comparably, indicating little benefit to sertraline [93]. However, Otto et al. [94], in their study of refugees with PTSD resistant to pharmacotherapy, found the combination of CBT and sertraline superior to sertraline alone. Some investigators have explored the effect of treatment order on response. In an RCT reported by Simon et al. [95], partial responders to 8 sessions of exposure therapy were randomized to receive paroxetine or placebo augmentation over 5 additional exposure sessions but little added improvement was noted in either group. Another RCT by the same research group administered 10 weeks of open-label sertraline and then all patients, regardless of response, were randomized to another 5 weeks of sertraline alone or sertraline combined with exposure therapy [96]. Additional benefits were seen in the exposure therapy group but only in partial responders to sertaline monotherapy. In both studies, the initial monotherapy was significantly effective.

In the only combination study with hypnotherapy, patients on SSRIs and supportive therapy received adjunctive hypnotherapy or zolpidem, a hypnotic drug used to treat sleep difficulties, for two weeks [97]. Significant global improvement was noted in the hypnotherapy group for both PTSD symptoms and sleep.

As seen above, the data from combination studies is limited and have reported somewhat conflicting results. It is also of note that in most investigations sample sizes were relatively small and combination treatment was brief (less than six weeks). Since combination treatment is usually a strategy employed in refractory cases, a longer duration of treatment might have produced more robust results, suggesting larger parallel-group studies of longer duration are needed. Overall, the current consensus appears to be that partial responders to either form of monotherapy are likely to benefit from the combination.

There are limited published data on possible predictors of treatment response to pharmacotherapy as only a few pharmacotherapy studies in PTSD have established strong *a priori* hypotheses with respect to possible predictors of response in individuals with PTSD. An early pooled analysis of a sample composed mainly of males with chronic, combat-related PTSD treated with the MAOI brofaromine, found that responders were more likely to have had lower baseline severity of re-experiencing and numbing-avoidance symptoms. Interestingly, a history of sexual trauma was found to be more predictive of placebo response [98]. In a more recent pooled analysis of individuals treated with venlafaxine XR, gender was not found to be a predictor of response to this agent; however, trauma type was found to have a possible affect on certain domains of outcome (resilience and disability) although was not predictive of an effect on specific PTSD symptom clusters [99]. Although there has been some interest in using neuroimaging techniques to assess possible predictors of treatment response with psychotherapy in PTSD, only limited studies have assessed the same with response to pharmacotherapy and results are far from conclusive [100,101]. However, the use of neuroimaging and genetic studies may be a useful strategy in future to assess this issue and better direct treatment. Future RCTs of pharmacological agents would benefit from a more systematic exploration of predictors of response.

Conclusion

PTSD can be a debilitating illness with significant impact on different areas of functioning. A well, given the strong evidence of efficacy for certain forms of psychotherapy in PTSD, these evidence-based therapies should be prioritized. Pharmacotherapy can also play an important role in the treatment of this disorder, and it is salutary that research into effective medications continues to expand and grow. Based on the current evidence, SSRIs and SNRIs, specifically fluoxetine, sertraline, paroxetine and venlafaxine, have the greatest evidence as first-line monotherapy in the treatment of PTSD. There is similar evidence for the use of atypical antipsychotics (namely, olanzapine, risperidone and quetiapine) as second-line augmentation in treatment-refractory PTSD. Trycyclics and MAOIs may have some benefit but their use is limited by adverse affects. Mirazapine, anitconvulsants (e.g., valporate, lamotrigine) and the antiadrenergic agent prazosin have promising preliminary results but due to limited evidence can only be recommended as third-line agents.

As this review found, well-powered, definitive RCTs of both short- and long-term duration are largely lacking and are needed to substantiate current findings and strengthen their generalizability. While many neurobiological findings in PTSD have significant overlap to those reported in depression and other anxiety disorders, some are very specific for PTSD [102]. Thus it is possible that pharmacological agents with specific benefit to PTSD and better tolerability profiles could be developed. Another area of research that warrants more exploration is predictors of response, especially in regard to which medications or psychotherapies are likely to be most beneficial for a particular patient, and in which situations combination therapy might be the optimal choice. Preliminary data suggests that the type of trauma, i.e. combat, violence, abuse, etc., might predict treatment response course and outcome. This needs further systematic investigation.

In clinical practice, the choice of pharmacotherapy and/or psychotherapy to treat PTSD may be influenced by many factors, including response to previous treatments, chronicity and severity of illness, presence of co-morbidities, and patient and clinician preference. The goal of treatment must always be remission of all symptoms, and a return to full functioning. The guidance provided by reliable, well-designed clinical research is an invaluable aid in the achievement of this goal.

References

[1] American Psychiatric Association: *Diagnostic and statistical manual of mental disorders (DSM-III),* 3rd ed. Washington, DC, American Psychiatric Association, 2008.

[2] Yatham LN, Kennedy SH, O'Donovan C, Parikh S, MacQueen G, McIntyre R, Sharma V, Silverstone P, Alda M, Baruch P, Beaulieu S, Daigneault A, Milev R, Young T, Ravindran A, Schaffer A, Connolly M, Gorman CP: Canadian Network for Mood and Anxiety Treatments (CANMAT) guidelines for the management of patients with bipolar disorder: Consensus and controversies. *Bipolar Disord* 2005; 7 (suppl 3): 5-69.

[3] Vermetten E, Vythilingam M, Southwick SM, Charney DS, Bremner JD: Long-term treatment with paroxetine increases verbal declarative memory and hippocampal volume in posttraumatic stress disorder. *Biol Psychiatry* 2003; 54: 693-702.

[4] Szabo ST, de Montigny C, Blier P: Modulation of noradrenergic neuronal firing by selective serotonin reuptake blockers. *Br J Pharmacol* 1999; 126: 568-571.

[5] Ipser J, Seedat S, Stein DJ: Pharmacotherapy for post-traumatic stress disorder – a systematic review and meta-analysis. *SAMJ* 2006: 96: 1088-1096.

[6] Martenyi F, Brown EB, Caldwell CD: Failed efficacy of fluoxetine in the treatment of posttraumatic stress disorder: results of a fixed-dose, placebo-controlled study. *J Clin Psychopharmacol* 2007; 27: 166-170.

[7] Friedman MJ, Marmar CR, Baker DG, Sikes CR, Farfel GM: Randomized, double-blind comparison of sertraline and placebo for posttraumatic stress disorder in a Department of Veterans Affairs setting. *J Clin Psychiatry* 2007; 68: 711-720.

[8] Marshall RD, Lewis-Fernandez R, Blanco C, Simpson HB, Lin SH, Vermes D, Garcia w, Schneier F, Neria Y, Sanchez-Lacay A, Liebowitz MR: A controlled trial of paroxetine for chronic PTSD, dissociation, and interpersonal problems in mostly minority adults. *Depress Anxiety* 2007; 24: 77-84.

[9] Seedat S, Stein DJ, Emsley RA: Open trial of citalopram in adults with post-traumatic stress disorder. *Int J Neuropsychopharmacol* 2000; 3: 135-140.

[10] Seedat S, Lockhat R, Kaminer D, Zungu-Dirwayi N, Stein DJ: An open trial of citalopram in adolescents with post-traumatic stress disorder. *Int Clin Psychopharmacol* 2001; 16: 21-25.

[11] Seedat S, Stein DJ, Ziervogel C, Middleton T, Kaminer D, Emsley RA, Rossouw W: Comparison of response to a selective serotonin reuptake inhibitor in children,

adolescents, and adults with posttraumatic stress disorder. *J Child Adolesc Psychopharmacol* 2002; 12: 37-46.

[12] English BA, Jewell M, Jewell G, Ambrose S, Davis LL: Treatment of chronic posttraumatic stress disorder in combat veterans with citalopram: an open trial. *J Clin Psychopharmacol* 2006; 26: 84-88.

[13] Robert S, Hamner MB, Ulmer HG, Lorberbaum JP, Durkalski VL: Open-label trial of escitalopram in the treatment of posttraumatic stress disorder. *J Clin Psychiatry* 2006; 67: 1522-1526.

[14] De Boer M, Op d, V, Falger PJ, Hovens JE, De Groen JH, Van Duijn H: Fluvoxamine treatment for chronic PTSD: a pilot study. *Psychother Psychosom* 1992; 57: 158-163.

[15] Marmar CR, Schoenfeld F, Weiss DS, Metzler T, Zatzick D, Wu R, Smiga S, Tecott L, Neylan TL: Open trial of fluvoxamine treatment for combat-related posttraumatic stress disorder. *J Clin Psychiatry* 1996; 57 (suppl 8): 66-70.

[16] Davidson JR, Weisler RH, Malik M, Tupler LA: Fluvoxamine in civilians with posttraumatic stress disorder. *J Clin Psychopharmacol* 1998; 18:93-95.

[17] Escalona R, Canive JM, Calais LA, Davidson JR: Fluvoxamine treatment in veterans with combat-related post-traumatic stress disorder. *Depress Anxiety* 2002; 15: 29-33.

[18] Committee on Treatment of Posttraumatic Stress Disorder (Institute of Medicine). Treatment of Posttraumatic Stress Disorder: *An Assessment of the Evidence.* Washington, DC, The National Academies Press, 2007.

[19] Shalev AY, Freedman S, Peri T, Brandes D, Sahar T, Orr SP, Pitman RK: Prospective study of posttraumatic stress disorder and depression following trauma. *Am J Psychiatry* 1998; 155: 630-637.

[20] Rapaport MH, Endicott J, Clary CM: Posttraumatic stress disorder and quality of life: results across 64 weeks of sertraline treatment. *J Clin Psychiatry* 2002; 63: 59-65.

[21] Foa E.B., Davidson J.R., Frances A; The Expert Consensus Guidelines Series: Treatment of Posttraumatic Stress Disorder. J *Clin Psychiatry* 1999; 60 (suppl 16).

[22] Stein DJ, Bandelow B, Hollander E, Nutt DJ, Okasha A, Pollack MH, Swinson RP, Zohar J; World Council on Anxiety: WCA Recommendations for the long-term treatment of posttraumatic stress disorder. *CNS Spectr* 2003; 8 (8 suppl 1): 31-39.

[23] Smajkic A, Weine S, Djuric-Bijedic Z, Boskailo E, Lewis J, Pavkovic I: Sertraline, paroxetine and venlafaxine in refugee posttraumatic stress disorder with depression symptoms. *J Trauma Stress* 2001; 14: 445-452.

[24] Davidson J, Rothbaum BO, Tucker P, Asnis G, Benattia I, Musgnung JJ: Venlafaxine extended release in posttraumatic stress disorder: a sertraline- and placebo-controlled study. *J Clin Psychopharmacol* 2006; 26: 259-267.

[25] Davidson J, Baldwin D, Stein DJ, Kuper E, Benattia I, Ahmed S, Musgnung J, Bennatia I, Rothbaum BO: Treatment of posttraumatic stress disorder with venlafaxine extended release: a 6-month randomized controlled trial. *Arch Gen Psychiatry* 2006; 63: 1158-1165.

[26] Goldstein DJ, Mallinckrodt C, Lu Y, Demitrack MA: Duloxetine in the treatment of major depressive disorder: a double-blind clinical trial. *J Clin Psychiatry* 2002; 63: 225-231.

[27] Raskin J, Goldstein DJ, Mallinckrodt CH, Ferguson MB; Duloxetine in the long-term treatment of major depressive disorder. *J Clin Psychiatry* 2003; 64: 1237-1244.

[28] Hartford J, Kornstein S, Liebowitz M, Pigott T, Russell J, Detke M, Walker D, Ball S, Dunayevich D, Dinkel J, Erickson J: Duloxetine as an SNRI treatment for generalized anxiety disorder: results from a placebo and active-controlled trial. *Int Clin Psychopharmacol* 2007; 22: 167-174.
[29] Endicott J, Russell JM, Raskin J, Detke MJ, Erickson J, Ball SG, Marciniak M, Swindle RW: Duloxetine treatment for role functioning improvement in generalized anxiety disorder: three independent studies. *J Clin Psychiatry* 2007; 68: 518-524.
[30] Frank JB, Kosten TR, Giller EL Jr, Dan E: A randomized clinical trial of phenelzine and imipramine for posttraumatic stress disorder. *Am J Psychiatry* 1988; 145: 1289-1291.
[31] Kosten TR, Frank JB, Dan E, McDougle CJ, Giller EL Jr: Pharmacotherapy for posttraumatic stress disorder using phenelzine or imipramine. *J Nerv Ment Dis* 1991; 179: 366-370.
[32] Shestatzky M, Greenberg D, Lerer B: A controlled trial of phenelzine in posttraumatic stress disorder. *Psychiatry Res* 1988; 24: 149-155.
[33] Katz RJ, Lott MH, Arbus P, Crocq L, Herlobsen P, Lingjaerde O, Lopez G, Loughrey GC, MacFarlane DJ, McIvor R, et al: Pharmacotherapy of post-traumatic stress disorder with a novel psychotropic. *Anxiety* 1994; 1: 169-174.
[34] Baker DG, Diamond BI, Gillette G, Hamner M, Katzelnick D, Keller T, Mellman TA, Pontius E, Rosenthal M, Tucker P, et al: A double-blind, randomized, placebo-controlled, multi-center study of brofaromine in the treatment of post-traumatic stress disorder. *Psychopharmacology (Berl)* 1995; 122: 386-389.
[35] Onder E, Tural U, Aker T: A comparative study of fluoxetine, moclobemide, and tianeptine in the treatment of posttraumatic stress disorder following an earthquake. *Eur Psychiatry* 2006; 21: 174-179.
[36] Davidson JR, Connor KM: Bupropion sustained release: a therapeutic overview. *J Clin Psychiatry 1998*; 59 (suppl 4): 25-31.
[37] Becker ME, Hertzberg MA, Moore SD, Dennis MF, Bukenya DS, Beckham JC: A placebo-controlled trial of bupropion SR in the treatment of chronic posttraumatic stress disorder. *J Clin Psychopharmacol* 2007; 27: 193-197.
[38] Hertzberg MA, Moore SD, Feldman ME, Beckham JC: A preliminary study of bupropion sustained-release for smoking cessation in patients with chronic posttraumatic stress disorder. *J Clin Psychopharmacol* 2001; 21: 94-98.
[39] Stimmel GL, Dopheide JA, Stahl SM: Mirtazapine: an antidepressant with noradrenergic and specific serotonergic effects. *Pharmacotherapy* 1997; 17: 10-21.
[40] Davidson JR, Weisler RH, Butterfield MI, Casat CD, Connor KM, Barnett S, van Meter S: Mirtazapine vs. placebo in posttraumatic stress disorder: a pilot trial. *Biol Psychiatry* 2003; 53: 188-191.
[41] Chung MY, Min KH, Jun YJ, Kim SS, Kim WC, Jun EM: Efficacy and tolerability of mirtazapine and sertraline in Korean veterans with posttraumatic stress disorder: A randomized open label trial. *Hum Psychopharmacol* 2004; 19: 489-494.
[42] Davis R, Whittington R, Bryson HM: Nefazodone. A review of its pharmacology and clinical efficacy in the management of major depression. *Drugs* 1997; 53: 608-636.
[43] Davis LL, Jewell ME, Ambrose S, Farley J, English B, Bartolucci A, Petty F: A placebo-controlled study of nefazodone for the treatment of chronic posttraumatic stress disorder: a preliminary study. *J Clin Psychopharmacol* 2004; 24: 291-297.

[44] McRae AL, Brady KT, Mellman TA, Sonne SC, Killeen TK, Timmerman MA, Bayles-Dazet W: Comparison of nefazodone and sertraline for the treatment of posttraumatic stress disorder. *Depress Anxiety* 2004; 19: 190-196.

[45] Saygin MZ, Sungur MZ, Sabol EU, Cetinkaya P. Nefazodone versus sertraline in treatment of posttraumatic stress disorder. *Bull Clin Psychopharmacol* 2002; 12: 1-5.

[46] Mula M, Pini S, Cassano GB: The role of anticonvulsant drugs in anxiety disorders: a critical review of the evidence. *J Clin Psychopharmacol* 2007; 27: 263-272.

[47] Post RM, Weiss SR, Smith M, Li H, McCann U: Kindling versus quenching. Implications for the evolution and treatment of posttraumatic stress disorder. *Ann N Y Acad Sci* 1997; 821: 285-295.

[48] Hertzberg MA, Butterfield MI, Feldman ME, Beckham JC, Sutherland SM, Connor KM, Davidson JR: A preliminary study of lamotrigine for the treatment of posttraumatic stress disorder. *Biol Psychiatry* 1999; 45: 1226-1229.

[49] Berlant JL. Prospective open-label study of add-on and monotherapy topiramate in civilians with chronic nonhallucinatory posttraumatic stress disorder. *BMC Psychiatry* 2004; 4: 24.

[50] Tucker P, Trautman RP, Wyatt DB, Thompson J, Wu SC, Capece JA, Rosenthal NR: Efficacy and safety of topiramate monotherapy in civilian posttraumatic stress disorder: a randomized, double-blind, placebo-controlled study. *J Clin Psychiatry* 2007; 68: 201-206.

[51] Lindley SE, Carlson EB, Hill K: A randomized, double-blind, placebo-controlled trial of augmentation topiramate for chronic combat-related posttraumatic stress disorder. *J Clin Psychopharmacol* 2007; 27: 677-681.

[52] Davidson JR, Brady K, Mellman TA, Stein MB, Pollack MH: The efficacy and tolerability of tiagabine in adult patients with post-traumatic stress disorder. *J Clin Psychopharmacol* 2007; 27: 85-88.

[53] Connor KM, Davidson JR, Weisler RH, Zhang W, Abraham K: Tiagabine for posttraumatic stress disorder: effects of open-label and double-blind discontinuation treatment. *Psychopharmacology (Berl)* 2006; 184: 21-25.

[54] Adamou M, Puchalska S, Plummer W, Hale AS: Valproate in the treatment of PTSD: systematic review and meta analysis. *Curr Med Res Opin* 2007; 26: 1285-1291.

[55] Davis LL, Davidson JR, Ward LC, Bartolucci A, Bowden CL, Petty F: Divalproex in the treatment of posttraumatic stress disorder: a randomized, double-blind, placebo-controlled trial in a veteran population. *J Clin Psychopharmacol* 2008; 28: 84-88.

[56] Mellman TA, Clark RE, Peacock WJ: Prescribing patterns for patients with posttraumatic stress disorder. *Psychiatr Serv* 2003; 54: 1618-1621.

[57] Braun P, Greenberg D, Dasberg H, Lerer B: Core symptoms of posttraumatic stress disorder unimproved by alprazolam treatment. *J Clin Psychiatry* 1990; 51: 236-238.

[58] Cates ME, Bishop MH, Davis LL, Lowe JS, Woolley TW: Clonazepam for treatment of sleep disturbances associated with combat-related posttraumatic stress disorder. *Ann Pharmacother 2004*; 38: 1395-1399.

[59] Mellman TA, Byers PM, Augenstein JS: Pilot evaluation of hypnotic medication during acute traumatic stress response. *J Trauma Stress* 1998; 11: 563-569.

[60] Mellman TA, Bustamante V, David D, Fins AI: Hypnotic medication in the aftermath of trauma. *J Clin Psychiatry* 2002; 63: 1183-1184.

[61] Duffy J, Malloy P. Efficacy of buspirone in the treatment of posttraumatic stress disorder: an open trial. *Ann Clin Psychiatry* 1994; 6: 33–7.

[62] Wells B, Chu C, Johnson R, and others. Buspirone in the treatment of posttraumatic stress disorder. *Pharmacotherapy* 1991; 11: 340–3.

[63] Kessler RC, Sonnega A, Bromet E, et al. Posttraumatic stress disorder in the National Comorbidity Survey. *Arch Gen Pschiatr* 1995; 52: 1048-1060.

[64] Hamner MB, Frueh BC, Ulmer HG, Arana GW: Psychotic features and illness severity in combat veterans with chronic posttraumatic stress disorder. *Biol Psychiatry* 1999; 45: 846-852.

[65] Sareen J, Cox BJ, Goodwin RD, Asmundson JG: Co-occurrence of posttraumatic stress disorder with positive psychotic symptoms in a nationally representative sample. *J Trauma Stress* 2005; 18: 313-322.

[66] Charney DS, Deutch AY, Krystal JH, Southwick SM, Davis M: Psychobiologic mechanisms of posttraumatic stress disorder. *Arch Gen Psychiatry* 1993; 50: 295-305.

[67] Seedat S, Stein MB, Oosthuizen PP, Emsley RA, Stein DJ: Linking posttraumatic stress disorder and psychosis: a look at epidemiology, phenomenology, and treatment. *J Nerv Ment Dis* 2003; 191: 675-681.

[68] Philip NS, Carpenter LL, Tyrka AR, Price LH: Augmentation of antidepressants with atypical antipsychotics: a review of the current literature. *J Psychiatr Pract* 2008; 14: 34-44.

[69] Butterfield MI, Becker ME, Connor KM, Sutherland S, Churchill LE, Davidson JR: Olanzapine in the treatment of post-traumatic stress disorder: a pilot study. *Int Clin Psychopharmacol* 2001; 16: 197-203.

[70] Padala PR, Madison J, Monnahan M, Marcil W, Price P, Ramaswamy S, Din AU, Wilson DR, Petty F: Risperidone monotherapy for post-traumatic stress disorder related to sexual assault and domestic abuse in women. *Int Clin Psychopharmacol* 2006; 21: 275-280.

[71] Gao K, Muzina D, Gajwani P, Calabrese JR: Efficacy of typical and atypical antipsychotics for primary and comorbid anxiety symptoms or disorders: a review. *J Clin Psychiatry* 2006; 67: 1327-1340.

[72] Pae CU, Lim HK, Peindl K, Ajwani N, Serretti A, Patkar AA, Lee C: The atypical antipsychotics olanzapine and risperidone in the treatment of posttraumatic stress disorder: a meta-analysis of randomized, double-blind, placebo-controlled clinical trials. *Int Clin Psychopharmacol* 2008; 23: 1-8.

[73] Rothbaum BO, Killeen TK, Davidson JR, Brady KT, Connor KM, Heekin MH: Placebo-controlled trial of risperidone augmentation for selective serotonin reuptaker inhibitor-resistant civilian posttraumatic stress disorder. *J Clin Psychiatry* 2008; 69: 520-525.

[74] Worthington JJ 3rd, Kinrys G, Wygant LE, Pollack MH: Aripiprazole as an augmentor of selective serotonin reuptake inhibitors in depression and anxiety disorder patients. *Int Clin Psychopharmacol* 2005; 20: 9-11.

[75] Robert S, Hamner MB, Durkalski VL, Brown MW, Ulmer HG: An open-label assessment of aripiprazole in the treatment of PTSD. *Psychopharmacol Bull* 2009; 42: 69-80.

[76] Horrigan JP: Guanfacine for PTSD nightmares. *J Am Acad Child Adolesc Psychiatry* 1996; 35: 975-976.

[77] Horrigan JP, Barnhill LJ: The suppression of nightmares with guanfacine. *J Clin Psychiatry* 1996; 57: 371.

[78] Porter DM, Bell CC: The use of clonidine in post-traumatic stress disorder. *J Natl Med Assoc* 1999; 91: 475-477.

[79] Raskind MA, Dobie DJ, Kanter ED, Petrie EC, Thompson CE, Peskind ER: The alpha1-adrenergic antagonist prazosin ameliorates combat trauma nightmares in veterans with posttraumatic stress disorder: a report of 4 cases. *J Clin Psychiatry* 2000; 61: 129-133.

[80] Kinzie JD, Leung P: Clonidine in Cambodian patients with posttraumatic stress disorder. *J Nerv Ment Dis* 1989; 177: 546-550.

[81] Neylan TC, Lenoci M, Samuelson KW, Metzler TJ, Henn-Haase C, Hierholzer RW, Lindley SE, Otte C, Schoenfeld FB, Yesavage JA, Marmar CR: No improvement of posttraumatic stress disorder symptoms with guanfacine treatment. *Am J Psychiatry* 2006; 163: 2186-2188.

[82] Raskind MA, Peskind ER, Kanter ED, Petrie EC, Radant A, Thompson CE, Dobie DJ, Hoff D, Rein RJ, Straits-Troster K, Thomas RG, McFall MM: Reduction of nightmares and other PTSD symptoms in combat veterans by prazosin: a placebo-controlled study. *Am J Psychiatry* 2003; 160: 371-373.

[83] Taylor FB, Martin P, Thompson C, Williams J, Mellman TA, Gross C, Peskind ER, Raskind MA: Prazosin effects on objective sleep measures and clinical symptoms in civilian trauma posttraumatic stress disorder: a placebo-controlled study. *Biol Psychiatry* 2008; 63: 629-632.

[84] Raskind MA, Peskind ER, Hoff DJ, Hart KL, Holmes HA, Warren D, Shofer J, O'Connell J, Taylor F, Gross C, Rohde K, McFall ME: A parallel group placebo controlled study of prazosin for trauma nightmares and sleep disturbance in combat veterans with post-traumatic stress disorder. *Biol Psychiatry* 2007; 61: 928-934.

[85] Hofmann SG, Meuret AE, Smits JA, Simon NM, Pollack MH, Eisenmenger K, Shiekh M, Otto MW: Augmentation of exposure therapy with D-cycloserine for social anxiety disorder. *Arch Gen Psychiatry* 2006; 63: 298-304.

[86] Guastella AJ, Richardson R, Lovibond PF, Rapee RM, Gaston JE, Mitchell P, Dadds MR: A randomized controlled trial of D-cycloserine enhancement of exposure therapy for social anxiety disorder. *Biol Psychiatry* 2008; 63: 544-549.

[87] Kushner MG, Kim SW, Donahue C, Thuras P, Adson D, Kotlyar M, McCabe J, Peterson J, Foa EB: D-cycloserine augmented exposure therapy for obsessive-compulsive disorder. *Biol Psychiatry* 2007; 62: 835-838.

[88] Heresco-Levy U, Kremer I, Javitt DC, Goichman R, Reshef A, Blanaru M, Cohen T: Pilot-controlled trial of D-cycloserine for the treatment of post-traumatic stress disorder. *Int J Neuropsychopharmacol* 2002; 5: 301-307.

[89] Heresco-Levy U, Vass A, Bloch B, Wolosker H, Dumin E, Balan L, Deutsch L, Kremer I: Pilot controlled trial of d-serine for the treatment of post-traumatic stress disorder. *Int J Neuropsychopharmacol* 2009; 15: 1-8.

[90] de Kloet CS, Vermetten E, Geuze E, Kavelaars A, Heijnen CJ, Westenberg HGM: Assessment of HPA-axis function in posttraumatic stress disorder: Pharmacological and non-pharmacological challenge tests, a review. *J Psychiatr Res* 2006; 40: 550-67.

[91] Aerni A, Traber R, Hock C, Roozendaal B, Schelling G, Papassotiropoulos A, Nitsch RM, Schnyder U, de Quervain DJ : Low-dose cortisol for symptoms of posttraumatic stress disorder. *Am J Psychiatry* 2004; 161: 1488-1490.

[92] Bisson JI, Ehlers A, Matthews R, Pilling S, Richards D, Turner S. Psychological treatments for chronic post-traumatic stress disorder. Systematic review and meta-analysis. *Br J Psychiatry.* 2007 Feb;190:97-104.

[93] Cohen JA, Mannarino AP, Perel JM, Staron V: A pilot randomized controlled trial of combined trauma-focused CBT and sertraline for childhood PTSD symptoms. *J Am Acad Child Adolesc Psychiatry* 2007; 46: 811-819.

[94] Otto MW, Hinton D, Korbly NB, Chea A, Ba P, Gershuny BS, Pollack MH: Treatment of pharmacotherapy-refractory posttraumatic stress disorder among Cambodian refugees: a pilot study of combination treatment with cognitive-behaviour therapy vs sertraline alone. *Behav Res Ther* 2003; 41: 1271-1276.

[95] Simon NM, Connor KM, Lang AJ, Rauch S, Krulewicz S, LeBeau RT, Davidson JR, Stein MB, Otto MW, Foa EB, Pollack MH: Paroxetine CR augmentation for posttraumatic stress disorder refractory to prolonged exposure therapy. *J Clin Psychiat* 2008; 69: 400-405.

[96] Rothbaum BO, Cahill SP, Foa EB, Davidson JR, Compton J, Connon KM, Astin MC, Hahn CG: Augmentation of sertraline with prolonged exposure in the treatment of posttraumatic stress disorder. *J Trauma Stress* 2006; 19: 625-638.

[97] Abramowitz EG, Barak Y, Ben-Avi I, Knobler HY: Hypnotherapy in the treatment of chronic combat-related PTSD patients suffering from insomnia. *Int J Clin Exp Hypn* 2008; 56: 270-280.

[98] Connor KM, Hidalgo RB, Crockett B, Malik M, Katz RJ, Davidson JR: Predictors of treatment response in patients with posttraumatic stress disorder. *Prog Neuropsychopharmacol Biol Psychiatry* 2001; 25: 337-45.

[99] Rothbaum BO, Davidson JR, Stein DJ, Pedersen R, Musgnung J, Tian XW, Ahmed S, Baldwin DS: A pooled analysis of gender and trauma-type effects on responsiveness to treatment of PTSD with venlafaxine extended release or placebo. *J Clin Psychiatry* 2008; 69: 1529-39.

[100] Seedat S, Warwick J, van Heerdeen B, Hugo C, Zungu-Dirwayi N, Van Kradenburg J, Stein DJ: Single photon emission computed tomography in posttraumatic stress disorder before and after treatment with a selective serotonin reuptake inhibitor. *J Affect Disord* 2004; 80: 45-53.

[101] Bossini L, Tavanti M, Lombardelli A, Calossi S, Polizzotto NR, Galli R, Vatti G, Pieraccini F, Castrogiovanni P: Changes in hippocampal volume in patients with post-traumatic stress disorder after sertraline treatment. *J Clin Psychopharmacol* 2007; 27: 233-5.

[102] Ravindran LN, Stein MB: Pharmacotherapy of PTSD: *Premises, Principles, and Priorities. Brain Research,* Mar 27, 2009 [epub ahead of print].

In: Neurobiology of Post-Traumatic Stress Disorder
Editors: L. Sher and A. Vilens, pp. 291-308
ISBN: 978-1-61470-593-2

Chapter XVIII

The Role of NMDA Receptors in the Pathophysiology and Treatment of Anxiety Syndromes

Viviane Labrie and John C. Roder
Samuel Lunenfeld Research Institute, Mount Sinai Hospital, Toronto, Ontario, Canada
University of Toronto, Toronto, Ontario, Canada

Abstract

Aberrant glutamatergic transmission has been implicated in the pathophysiology of anxiety disorders. Alterations in NMDA receptor (NMDAR) expression, function, and binding are induced in animal models of fear and stress. NMDARs have a central role in mediating the acquisition and extinction of learned fear responses. Consequently, this has led to investigations examining the therapeutic utility of compounds targeting the NMDAR for the treatment of anxiety syndromes. High-affinity open channel blockers of the NMDAR promote anxiolytic effect in preclinical rodent studies; however the psychotomimetic effects induced by these antagonists have hampered their applicability in clinical settings. Instead, low-affinity NMDAR inhibitors and agonists of the NMDAR glycine binding site have demonstrated therapeutic potential without significant adverse effects. Furthermore, NMDAR glycine site activators have been shown to be effective in accelerating fear extinction in animal studies, and recent clinical trials have supported their efficacy in enhancing exposure therapy in patient with anxiety syndromes. Thus, an improved understanding of the molecular mechanisms underlying stress, fear, and extinction has led to the discovery of highly promising treatment interventions involving the NMDAR.

Introduction

Anxiety disorders are a highly prevalent class of psychiatric conditions that include posttraumatic stress syndrome, obsessive-compulsive disorder, social anxiety, and panic disorder. These disorders are often chronic, causing significant disability, functional impairment, and a decrease in quality of life [1, 2]. Current treatment strategies involve pharmacotherapy and/or psychotherapy, however, a substantial proportion of patients with anxiety syndromes are poorly responsive to conventional therapies [3-5]. In addition, many pharmaceutical treatments are associated with troublesome side effects that can limit tolerability and compliance [3, 4]. Consequently, further understanding the molecular mechanisms underlying anxiety and fear responses can provide insight for the development of novel treatment approaches.

The pathogenesis of anxiety involves a complex interplay of multiple genes, neurotransmitter systems, and environmental factors. Growing evidence supports that the glutamatergic system has an important role in the biological mechanism of fear and stress responses that contribute to the development of pathological anxiety conditions [6-11]. Glutamatergic neurotransmission is mediated by several receptor subtypes, including the *N*-methyl-D-aspartate receptor (NMDAR). In this chapter, we explore the involvement of NMDARs in the pathophysiology and treatment of anxiety disorders. NMDARs are abundantly expressed in brain regions mediating fear and anxiety, including the amygdala, hippocampus, and prefrontal cortex [12], and show changes in function and binding in response to stress [6-9]. Activation of NMDARs is centrally involved in the synaptoplastic processes mediating learning and memory formation [13] that are relevant to the onset and subsequent symptomatology of many anxiety disorders, such as posttraumatic stress syndrome. Moreover, fear learning, fear extinction, and behavioral flexibility are disrupted following inhibition of NMDARs [14-18]. This has recently led to the investigation of novel treatments targeting the NMDAR for the amelioration of several anxiety disorders, and therapeutic effects have been observed [19, 20]. In addition, activators of the NMDAR have been applied to exposure-based psychotherapy, a treatment strategy that involves the extinction of pathological behaviors and memories. Clinical trials in patients with anxiety disorders have demonstrated that compounds modulating NMDAR function are capable of enhancing the effects of exposure therapy [21-23]. Thus, the mapping of core pathways, neuroanatomical regions, and neurotransmitters involved in anxiety and fear extinction indicate an involvement of NMDARs. This has prompted the development of novel NMDAR-based treatment interventions that demonstrate promising effects for the improvement of anxiety disorders.

The NMDAR: Structure and Modulatory Sites

NMDARs in the central nervous system (CNS) are heteromeric protein complexes that consist of at least one NR1 subunit together with various combinations of NR2 and/or NR3 subunits [24]. The *Grin1* gene is alternatively spliced producing eight NR1 isoforms that associate with different constellations of NR2 (NR2A-D) and NR3 subunits (NR3A and NR3B) to form a multitude of different NMDAR receptors with distinct biophysical properties [25] and specific patterns of expression during development and in the mature mammalian CNS [26]. The complexity of NMDARs is further augmented through post-translational modifications, that include phosphorylation, glycosylation, and ubiquitination, affecting cellular distribution and function of the receptor [27].

At resting membrane potential, the pore of the NMDAR channel is blocked by magnesium, and this block can be removed by membrane depolarization mediated by α-amino-3-hydroxyl-5-methyl-4-isoxazole-propionate (AMPA) receptors [27]. The NMDAR also contains a glutamate recognition site on the NR2 subunit, and a glycine or D-serine modulatory site on the NR1 subunit [28, 29]. The glycine/D-serine site on the NMDAR must be occupied for glutamate to activate the receptor [29]. The unique property of being both voltage-dependent and ligand-gated gives the NMDAR the ability to act as a coincidence detector for presynaptic activity (glutamate release) and post-synaptic activity (adequate depolarization of the post-synaptic membrane). Once activated, the NMDAR channel permits the influx of calcium, which stimulates intracellular signaling cascades that can subsequently affect synaptic plasticity and gene transcription [30]. NMDAR-dependent forms of synaptic plasticity play an essential role in many types of learning and memory formation [13, 31].

NMDARs are present throughout the brain, and are principally neuronal, though they can also be expressed on astrocytes. Beyond the glutamate and glycine binding site, they contain several regulatory sites sensitive to polyamines, Zn^{2+}, protons, and glutathione [24, 27]. Additionally, the NMDAR can also be modulated by several artificially-derived inhibitory compounds, including phencyclidine (PCP), MK-801, and ketamine, which are high-affinity open channel blockers [24, 27].

The Involvement of the NMDAR in the Pathophysiology of Anxiety

The glutamatergic system is thought to have an important role in the pathogenesis of anxiety. Animal tests involving fear and stress promote an increase in NR1 subunit expression in the ventral tegmental area [6] and elevate NR1 and NR2B mRNA levels in the hippocampus [7, 32]. Exposure to stress also induces a robust increase in glutamate release in brain areas, such as the hippocampus, prefrontal cortex, and striatum [10, 33], and produces alterations in the function or binding of the NMDAR and its glycine site in rodents [8, 9, 34, 35]. Accordingly, altered levels of glutamate and NMDAR binding and expression have been observed in the brains of depressed patients and suicide victims [36-39, but see 40, 41]. Stress-induced glutamatergic activation induces neurotoxic effects in the hippocampus that are facilitated by the interaction of glucocorticoids and glutamate [42, 43] and involves NMDAR and other glutamate receptor subtypes [44-46]. The neurotoxic effects of stress have been proposed to underlie the decrease in hippocampal volume associated with human posttraumatic stress disorder (PTSD) [47]. Furthermore, aberrant NMDAR activation alongside dysfunctional GABAergic tone in the dorsal periaqueductal gray [48], amygdala [49], or dorsomedial hypothalamus [50] has been implicated in the etiology of panic conditions. Rodents with a chronic inhibition of GABA synthesis display an enhancement in glutamatergic excitation that through NMDAR activation leads to the development of panic-like states and increased anxiety [50]. Thus, abnormal NMDAR function is present under conditions that promote anxious behaviors and may significantly contribute to the pathophysiology of several anxiety disorders.

Stimulation of NMDARs is also crucial for the formation and retention of fear memories. Fear conditioning paradigms in which a previously neutral stimuli is paired with an aversive outcome have shown that NMDAR activation in the amygdala mediates fear learning and memory formation [51-53]. Infusion of the competitive NMDAR antagonist DL-2-amino-5-phosphonavaleric acid (APV) into the amygdala blocks learning of conditioned fear [15, 54-56]. Hippocampal NMDARs have also been shown to be important in encoding fear memories involving contextual information or a trace interval (temporal gap) [57-59]. Strong NMDAR activation is capable of producing long-term potentiation (LTP), a form of synaptic plasticity that is a major cellular mechanism underlying learning and memory [13]. Induction of NMDAR-dependent LTP within the amygdala and in efferent pathways of the amygdala has been shown to cause lasting increases in fearfulness in animals exposed to traumatic stress [60]. Consequently, NMDARs have an important role in establishing and maintaining fear-based memories that in turn modulate anxiety behaviors and stress responses.

Alterations in glutamatergic neurotransmission have also been implicated in the pathophysiology of obsessive-compulsive disorder (OCD) and other psychiatric illnesses characterized by cognitive inflexibility and persistent repetitive behaviors. Neuroimaging studies of patients with OCD observe elevations in normal glucose metabolism in the orbitofrontal cortex, caudate nucleus, and thalamus [61], which normalize in response to effective pharmacological and behavioral therapy treatments [62]. The glucose metabolic signal is considered to primarily reflect glutamatergic excitatory transmission [63]. In accordance, glutamate concentrations have been found to be abnormally elevated in the cerebrospinal fluid of OCD patients [64]. Magnetic resonance spectroscopy studies have also demonstrated glutamatergic dysfunctions in individuals with OCD, with an increase in glutamate levels in the caudate of patients that is reversed with successful treatment interventions [65-68]. Furthermore, genetic variation in the NMDA-NR2B receptor has been associated with susceptibility to OCD [69, 70]. A transgenic mouse model of OCD and Tourette's syndrome has been developed [71]. These mice exhibit a chronic potentiation of cortical and limbic dopamine D1 neurons that induces glutamatergic signaling [71]. In this mouse model administration of the NMDAR antagonist MK-801 produced an exacerbation of OCD-like perseverative behaviors and stereotypies [74], which may be related to the capacity of NMDAR inhibitors to indirectly increase presynaptic glutamate efflux in frontocortical brain regions [72, 73]. NMDAR inhibition has also been shown to impair cognitive flexibility and cause persistent behaviors in numerous preclinical studies and in humans [16, 75-79]. Thus, aberrant NMDAR-mediated glutamatergic transmission is capable of inducing maladaptive repetitive behaviors relevant to OCD.

Targeting NMDARs to Produce Anxiolytic Effects

Pharmacological reduction of NMDAR function using competitive, non-competitive, and glycine site antagonists diminishes anxiety-like behaviors in preclinical rodent studies [11, 80]. NMDAR inhibitors reduce anxiety in both conditioned behavioral models, such as Vogel's conflict test, and unconditioned behavioral models of anxiety, that include the elevated plus-maze, novel open field, ultra-sonic vocalization test, and acoustic startle task [80-84]. Blockade of the NMDAR also prevents fear conditioning in rats and mice [54, 55, 57]. Recently, the NMDAR antagonist ketamine has been shown to reverse the physiological effects of chronic stress exposure in rats [85]. Acute and chronic ketamine treatments reversed the stress-induced increase in adrenal gland weight, promoted the regain of body weight, and normalized circulating glucocorticoid levels [85]. Anxiolytic-like effects have also been observed following genetic inactivation of the NMDA-NR2A subunit in mice [86] and in the $Grin1^{D481N}$ mice that carry a point mutation in the NR1 subunit resulting in a five-fold reduction in NMDAR glycine site affinity [87, 88]. Decreased anxiety in the NR2A knock-out and $Grin1^{D481N}$ mice was observed in unconditioned behavioral models, including the novel open field, elevated plus-maze, and light-dark exploration test [86-88]. Diminished anxiety in the $Grin1^{D481N}$ mice was reversible by direct and indirect agonists of the NMDAR glycine site [87]. Studies in rats and humans examining individual differences in anxiety levels indicate that occupancy of the NMDAR glycine site and/or glutamatergic activity may be reduced in naturally less anxious subjects, compared to highly anxious subjects [89, 90]. Together, these studies indicate that reduced NMDAR activation may promote anxiolytic-like behaviors.

The potential clinical utility of NMDAR antagonists for the treatment of anxiety disorders has been greatly limited by their propensity to elicit unfavorable psychotomimetic side effects [91, 92]. Studies in healthy humans have found that high-affinity NMDAR channel blockers, such as PCP and ketamine, produce dissociative states that involve disturbances in identity and perception [91, 92]. Moreover, schizophrenia-like symptoms, including psychosis and cognitive impairments have been described [91, 92]. Interestingly, the dissociative perceptual alterations induced by NMDAR antagonists are remarkably similar to the dissociative symptoms evoked by severe stress exposure, as experienced by patients that develop PTSD [93]. Thus, potent NMDAR antagonism produces adverse effects relevant to psychotic disorders and extreme stress, and is not considered to be a promising therapeutic approach.

In contrast, memantine, a low-affinity NMDAR channel blocker with rapid-off rate kinetics [94], has recently been demonstrated to alleviate anxiety symptoms in clinical trials [19, 95]. Memantine blocks the sustained activation of the NMDAR in the presence of high concentrations of glutamate as seen in pathological conditions, but rapidly leaves the receptor upon transient physiological activation by low concentrations of glutamate [96]. Memantine has been approved for use in patients with Alzheimer's disease and is considered to be a well-tolerated compound with minimal side effects [97]. In an open-label trial with 10 OCD patients, memantine was administered either as monotherapy or in conjunction with conventional medications [19]. Individuals with OCD experienced a significant reduction in symptoms over a 12-week period, and no serious adverse effects were observed [19]. In the same study, memantine was not found to be effective in ameliorating symptoms in patients with generalized anxiety disorder [19]. However, memantine significantly reversed anxiety symptoms in patients with major depressive disorder and cormorbid alcohol dependence in a double-blind, 26-week clinical trial [95]. Although these positive findings are considered preliminary, they warrant larger, placebo-controlled investigations.

A pilot clinical trial with the direct NMDAR glycine site agonist D-serine has also shown promising therapeutic effects for the treatment of chronic PTSD [20]. In a double-blind, placebo-controlled trial D-serine ameliorated anxiety symptoms, general PTSD symptomatology, and decreased the re-experiencing of traumatic memories over a 6-week treatment period [20]. D-serine was well-tolerated and devoid of significant side effects [20], which is consistent with NMDAR glycine site activators exhibiting minimal adverse effects in clinical trials with schizophrenia patients [98-100]. D-cycloserine, a partial agonist of the NMDAR glycine site, has also demonstrated beneficial effects in individuals with chronic PTSD, including improvements in numbing, avoidance, and anxiety symptoms; however these results were confounded by a relatively large placebo-effect demonstrating similar improvements [101]. In PTSD patients, D-cycloserine also reduced perseverative errors in the Wisconsin Card Sorting Task, a widely used measure of cognitive flexibility [101]. Similarly, glycine, an activator of the NMDAR glycine site, has demonstrated a capacity to diminished persistent behaviors related to anxiety. Glycine was shown to provide some therapeutic benefit in patients with OCD, although this study was limited by a high drop-out rate [102]. Consistent with these findings are animal studies that have shown that enhanced NMDAR glycine site function improves cognitive flexibility in wild-type mice [16, 103] and reverses behavioral perseveration induced by genetic and pharmacological manipulation [75, 79, 104].

Modulators of the NMDAR Glycine Site Facilitate Extinction and Exposure Therapy

The most commonly used treatment strategies for the majority of anxiety syndromes, including PTSD, social anxiety disorder, OCD, generalized anxiety disorder, and panic disorder, involve pharmacotherapy or psychotherapy. Exposure therapy is one of the most effective forms of psychotherapy and is based on the concept of extinction, whereby repeated and prolonged exposures to a fear-inducing stimulus without consequence in a controlled environment diminishes uncontrolled fear responses underlying the anxiety disorder. Attempts to combine pharmacotherapy and exposure therapy has traditionally led to disappointing results [105, 106], which is likely due to the propensity of anxiolytics to interfere with the processes of exposure therapy. However, an alternate approach has recently emerged. Rather than reducing the symptoms of anxiety, a pharmacotherapy that enhances the learning occurring during exposure therapy has the potential to improve the effectiveness of the exposure procedures.

Extinction learning paradigms in rodents are a close correlate to exposure therapy in humans. Extinction is considered to be a form of inhibitory learning that acts to suppress fear responses [107-109]. Like other types of learning, extinction has been shown to depend on glutamate NMDAR-associated signaling pathways, gene expression, and protein synthesis [107-109]. Indeed, infusions the NMDAR antagonist APV into the amygdala dose-dependently blocks the extinction of a fear-based memory [14, 15]. In response to these findings, NMDAR activators were examined in fear conditioning protocols to determine their efficacy in ameliorating extinction learning. Growing evidence indicates that stimulation of the NMDAR glycine site with the partial agonist D-cycloserine is capable of enhancing the extinction of a fear memory [110-112]. D-cycloserine administered systemically or injected directly into the amygdala was found to dose-dependently augment the extinction of conditioned startle [110] or freezing responses [111]. Administration of the NMDAR glycine site antagonist HA-966 blocked the extinction enhancement by D-cycloserine [110].

Subsequent studies examining mice given exogenous D-serine treatments or with a genetic inactivation of the D-serine catabolic enzyme D-amino acid oxidase (DAO) have demonstrated that elevated cerebral D-serine levels accelerate extinction in a contextual fear conditioning task and in a test of spatial memory [103].

Interestingly, D-cycloserine has been found to enable generalized extinction, such that diminished fear responses are exhibited for extinguished and non-extinguished stimuli [113]. Animals trained to fear two different cues (a light and tone stimulus) were given extinction training sessions for only one of the cues [113]. Animals that received D-cycloserine during the extinction training session demonstrated a greater attenuation of fear responses to both cues on test day [113]. This implies that administration of D-cycloserine in exposure therapy may be capable of generalizing extinction to other associated cues even if they are not directly treated in therapy.

In clinical settings, D-cycloserine has been shown to enhance exposure-based psychotherapy for a number of anxiety disorders. The first study examined individuals with a fear of heights (acrophobia). Single doses of placebo or D-cycloserine were administered prior to 2 sessions of virtual reality exposure therapy, in which participants were exposed to a simulated environment with elevated heights [21]. In comparison to placebo, D-cycloserine was found to substantially improve the effects of exposure therapy, resulting in a larger reduction of overall acrophobic symptoms at 1 week and 3 months following treatment [21]. In addition, participants that received D-cycloserine demonstrated greater decreases in skin conductance fluctuations, a physiological measure of anxiety, during the virtual reality exposures, and reported a greater improvement compared to placebo in general measures of real-world acrophobia symptoms [21]. Subsequent placebo-controlled studies in patients with social phobia [22, 114] and OCD [23, 115] have confirmed the ability of D-cycloserine to augment the effects of exposure therapy. In studies with patients with social anxiety disorders, participants received placebo or D-cycloserine before public speaking exposure sessions [22, 114]. Patients treated with D-cycloserine and exposure therapy were found to have significantly less social anxiety, demonstrating improvements immediately after treatment and 1-month after the last treatment session in measures of social fear, avoidance, dysfunctional cognitions, and quality of life functioning [22, 114]. D-cycloserine also enhanced exposure therapy in patients with OCD, reducing the number of exposure sessions required to improve symptoms and decreasing obsession-related distress in comparison to placebo-treated patients [23, 115]. However, although D-cycloserine accelerated the effects of exposure therapy in OCD, improvements beyond the placebo-treated group were not apparent post-treatment nor at a 1- or 3-month follow-up exam [23, 115]. Also, D-cycloserine was not effective in enhancing exposure procedures in individuals with a non-clinical fear of spiders [116] and in healthy individuals exposed to a differential shock conditioning paradigm [117]. This suggests that the ameliorative effects of D-cycloserine may be limited to clinical populations, as extinction procedures alone may be sufficient to produce maximal improvements in non-clinical samples.

Molecular and Cellular Mechanisms Mediating Extinction

How does stimulation of the NMDAR glycine site accelerate extinction learning? A recent study has demonstrated that facilitation of extinction by D-cycloserine involves phosphatidylinositol 3-kinase (PI-3K) and mitogen-activated protein kinase (MAPK)-dependent signaling cascades and requires de novo protein synthesis within the amygdala [118]. Infusions of a PI-3K or MAPK inhibitor in the basolateral nucleus of the amygdala prevented the ameliorative effects of D-cycloserine on extinction in a potentiated startle task in rats [118]. Similarly, intra-amygdala administration of a transcription and translation inhibitor abolished the D-cycloserine-induced increase in extinction learning [118].

Studies in rodents have demonstrated that NMDAR antagonists can block the retention of a learned extinction experience, suggesting that NMDARs are involved in the consolidation of extinction [15, 119]. Consolidation is a process that stabilizes new memory traces after the initial acquisition period. Administration of D-cycloserine after extinction training has been shown to facilitate extinction retention [111]. Increasing the delay of D-cycloserine administration following the extinction session produced a linear decrease in its facilatory effect [111]. This finding indicates that D-cycloserine may improve extinction by augmenting consolidation processes.

Compounds targeting the NMDAR glycine site may facilitate extinction by regulating synaptic plasticity. NMDAR activation is critically involved in the synaptoplastic processes underlying the acquisition of novel behaviors [13] and has a role in mediating behavioral alterations in response to changing environmental conditions [16-18]. Compelling evidence indicates that NMDAR-dependent long-term depression (LTD), a form of synaptic plasticity, may contribute to extinction and behavioral flexibility. Exogenous application of the NMDAR glycine site agonist D-serine has been shown to strongly enhance NMDA-NR2B receptor-dependent LTD in rodent hippocampal slices [16, 120]. Elevated D-serine levels were found to improve extinction learning in a contextual fear conditioning procedure and in a spatial memory task [103]. Augmented D-serine levels also enhanced the ability to learn novel platform locations in a spatial memory task, indicating that D-serine improves adaptive learning [16, 103]. Mice that overexpress NR2B receptors similarly exhibit better learning and memory in tasks that include extinction of fear associations [31]. Conversely, blockade of NR2B receptors has been reported to inhibit hippocampal LTD in the adult mouse and has been found to induce persistent behaviors in a spatial memory and serial reaction time task [16, 121]. Inhibition of NR2B receptors also disrupts the acquisition and retention of fear extinction [18]. Transgenic mice that exhibit deficits in hippocampal LTD demonstrate impairments in spatial reversal learning and working memory, suggesting that loss of LTD results in behavioral inflexiblity [17, 122, 123]. In addition, animals lacking LTD demonstrate a deficiency in fear extinction acquisition and retention [124]. Finally, the propensity for LTD has been shown to increase with exposure to a novel environments and when salient environmental cues are relocated [125, 126]. Thus, NMDAR-mediated LTD appears to have an important role in extinction learning and adaptive behaviors. Activators of the NMDAR glycine site (i.e. D-serine, D-cycloserine) may augment extinction by enhancing the propensity for LTD.

These studies demonstrating the involvement of signaling cascades, protein synthesis, consolidation, and LTD are a first step in uncovering the neurocellular mechanisms contributing to the facilitation of extinction by NMDAR glycine site agonists. Further exploration will provide a substantial contribution to our understanding of the molecular processes involved and may reveal novel drug targets for the enhancement of exposure therapy.

Conclusion

Investigations into the biological mechanisms of fear, stress, and anxiety have supported an important contribution of the glutamate system in the pathophysiology and treatment of anxiety disorders. Alterations in NMDAR levels and activation are demonstrated in preclinical models of stress [6-9] and in individuals experiencing severe stress [36-39]. NMDARs also have a central role in fear learning and retention [127], and can modulate maladaptive repetitive behaviors relevant to anxiety syndromes [16, 17]. Pharmacological and genetic animal studies have demonstrated that potent NMDAR inactivation promotes anxiolytic effects [11, 80, 86, 87]; however high-affinity NMDAR antagonists have also been shown to induce psychotic and cognitive disturbances in animals and humans [91, 92, 128], impeding their therapeutic utility. In contrast, preliminary clinical trials with memantine, a low-affinity NMDAR antagonist with fast offset kinetics, have demonstrated ameliorative effects for anxiety symptoms [19, 95]. Agonists of the NMDAR glycine site have also demonstrated therapeutic potential for treatment of PTSD [20]. In animal studies, NMDARs were shown to be highly involved in mediating fear extinction. Extinction of learned fear responses were abolished following NMDAR inhibition [14, 15], but enhanced by D-cycloserine, a partial agonist of the NMDAR glycine site [110-112]. Since fear extinction in rodents is similar to exposure-based psychotherapy in humans, translational research from preclinical to clinical studies was initiated to examine the therapeutic effects of D-cycloserine [112]. D-cycloserine was found to facilitate the effects of exposure therapy for a number of anxiety syndromes, including phobias, social anxiety, and OCD [21-23]. Further, research will be necessary to determine the ideal conditions required for D-cycloserine to produce maximal facilitation of exposure therapy. In addition, examining the potential for D-cycloserine to augment effects of exposure therapy in a greater number of anxiety disorders, including as PTSD, will also be necessary. Moreover, D-serine and glycine transporter inhibitors are more effective activators of the NMDAR glycine site and may be superior to D-cycloserine in facilitating exposure procedures, although this has yet to be investigated in clinical trials. Overall, treatments targeting NMDARs have demonstrated promising therapeutic effects for the amelioration of many anxiety syndromes. This supports the potential success of treatments based on mechanistic insight: discoveries on the neurocircuitry and neurotransmitters involved in fear and extinction led to the development of novel therapies involving the NMDAR for the treatment of anxiety symptoms in the clinical setting.

Acknowledgments

V.L. is supported by a Natural Sciences and Engineering Research Council (NSERC, Canada) studentship. J.C.R. is a Canada Research Council (CRC) chair. The author's studies are supported by the CIHR (GMH-79044).

References

[1] Lochner C, Mogotsi M, du Toit PL, Kaminer D, Niehaus DJ,Stein DJ: Quality of life in anxiety disorders: a comparison of obsessive-compulsive disorder, social anxiety disorder, and panic disorder. *Psychopathology* 2003; 36:255-262.

[2] Seedat S, Lochner C, Vythilingum B,Stein DJ: Disability and quality of life in post-traumatic stress disorder: impact of drug treatment. *Pharmacoeconomics* 2006; 24:989-998.

[3] Denys D,de Geus F: Predictors of pharmacotherapy response in anxiety disorders. *Curr Psychiatry Rep* 2005; 7:252-257.

[4] Zhang W,Davidson JR: Post-traumatic stress disorder: an evaluation of existing pharmacotherapies and new strategies. *Expert Opin Pharmacother* 2007; 8:1861-1870.

[5] Pollack MH, Otto MW, Roy-Byrne PP, Coplan JD, Rothbaum BO, Simon NM,Gorman JM: Novel treatment approaches for refractory anxiety disorders. *Depress Anxiety* 2008; 25:467-476.

[6] Fitzgerald LW, Ortiz J, Hamedani AG,Nestler EJ: Drugs of abuse and stress increase the expression of GluR1 and NMDAR1 glutamate receptor subunits in the rat ventral tegmental area: common adaptations among cross-sensitizing agents. *J Neurosci* 1996; 16:274-282.

[7] Yamamoto S, Morinobu S, Fuchikami M, Kurata A, Kozuru T,Yamawaki S: Effects of single prolonged stress and D-cycloserine on contextual fear extinction and hippocampal NMDA receptor expression in a rat model of PTSD. *Neuropsychopharmacol* 2008; 33:2108-2116.

[8] Harvey BH, Oosthuizen F, Brand L, Wegener G,Stein DJ: Stress-restress evokes sustained iNOS activity and altered GABA levels and NMDA receptors in rat hippocampus. *Psychopharmacol* 2004; 175:494-502.

[9] Nowak G, Ossowska G, Jopek R,Papp M: Strychnine-insensitive glycine/NMDA sites are altered in two stress models of depression. *Pol J Pharmacol* 1998; 50:365-369.

[10] Bagley J,Moghaddam B: Temporal dynamics of glutamate efflux in the prefrontal cortex and in the hippocampus following repeated stress: effects of pretreatment with saline or diazepam. *Neurosci* 1997; 77:65-73.

[11] Bergink V, van Megen HJ,Westenberg HG: Glutamate and anxiety. *Eur Neuropsychopharmacol* 2004; 14:175-183.

[12] Mcdonald AJ: Glutamate and aspartate immunoreactive neurons of the rat basolateral amygdala: colocalization of excitatory amino acids and projections to the limbic circuit. *J Comp Neurol* 1996; 365:367-379.

[13] Whitlock JR, Heynen AJ, Shuler MG,Bear MF: Learning induces long-term potentiation in the hippocampus. *Science* 2006; 313:1093-1097.

[14] Falls WA, Miserendino MJ,Davis M: Extinction of fear-potentiated startle: blockade by infusion of an NMDA antagonist into the amygdala. *J Neurosci* 1992; 12:854-863.
[15] Lee H,Kim JJ: Amygdalar NMDA receptors are critical for new fear learning in previously fear-conditioned rats. *J Neurosci* 1998; 18:8444-84454.
[16] Duffy S, Labrie V,Roder JC: D-serine augments NMDA-NR2B receptor-dependent hippocampal long-term depression and spatial reversal learning. *Neuropsychopharmacol* 2008; 33:1004-1018.
[17] Nicholls RE, Alarcon JM, Malleret G, Carroll RC, Grody M, Vronskaya S,Kandel ER: Transgenic mice lacking NMDAR-dependent LTD exhibit deficits in behavioral flexibility. *Neuron* 2008; 58:104-117.
[18] Sotres-Bayon F, Bush DE,LeDoux JE: Acquisition of fear extinction requires activation of NR2B-containing NMDA receptors in the lateral amygdala. *Neuropsychopharmacol* 2007; 32:1929-1940.
[19] Feusner JD, Kerwin L, Saxena S,Bystritsky A: Differential Efficacy of Memantine for Obsessive-Compulsive Disorder vs. Generalized Anxiety Disorder: An Open-Label Trial. *Psychopharmacol* 2009; 42:81-93.
[20] Heresco-Levy U, Vass A, Bloch B, Wolosker H, Dumin E, Balan L, Deutsch L,Kremer I: Pilot controlled trial of d-serine for the treatment of post-traumatic stress disorder. *Int J Neuropsychopharmacol* 2009; 15:1-8.
[21] Ressler KJ, Rothbaum BO, Tannenbaum L, Anderson P, Graap K, Zimand E, Hodges L,Davis M: Cognitive enhancers as adjuncts to psychotherapy: use of D-cycloserine in phobic individuals to facilitate extinction of fear. *Arch Gen Psychiatry* 2004; 61:1136-1144.
[22] Hofmann SG, Meuret AE, Smits JA, Simon NM, Pollack MH, Eisenmenger K, Shiekh M,Otto MW: Augmentation of exposure therapy with D-cycloserine for social anxiety disorder. *Arch Gen Psychiatry* 2006; 63:298-304.
[23] Kushner MG, Kim SW, Donahue C, Thuras P, Adson D, Kotlyar M, McCabe J, Peterson J,Foa EB: D-cycloserine augmented exposure therapy for obsessive-compulsive disorder. *Biol Psychiatry* 2007; 62:835-838.
[24] Cull-Candy S, Brickely S,Farrant M: NMDA receptor subunits: diversity, development and disease. *Curr Op Neurobiol* 2001; 11:327–335.
[25] Monyer H, Burnashev N, Laurie DJ, Sakmann B,Seeburg PH: Developmental and regional expression in the rat brain and functional properties of four NMDA receptors. *Neuron* 1994; 12:529–540.
[26] Dumas TC: Developmental regulation of cognitive abilities: modified composition of a molecular switch turns on associative learning. *Prog Neurobiol* 2005; 76:189–211.
[27] Dingledine R, Borges K, Bowie D,Traynelis SF: The glutamate receptor ion channels. *Pharmacol Rev* 1999; 51:7-61.
[28] Johnson JW,Ascher P: Glycine potentiates the NMDA response in cultured mouse brain neurons. *Nature* 1987; 325:529-531.
[29] Clements JD,Westbrook GL: Activation kinetics reveal the number of glutamate and glycine binding sites on the *N*-methyl-D-aspartate receptor. *Neuron* 1991; 7:605-613.
[30] Bading H, Ginty DD,Greenberg ME: Regulation of gene expression in hippocampal neurons by distinct calcium signaling pathways. *Science* 1993; 260:181-186.
[31] Tang YP, Shimizu E, Dube GR, Rampon C, Kerchner GA, Zhuo M, Liu G,Tsien JZ: Genetic enhancement of learning and memory in mice. *Nature* 1999; 401:63-69.

[32] Bartanusz V, Aubry JM, Pagliusi S, Jezova D, Baffi J,Kiss JZ: Stress-induced changes in messenger RNA levels of N-methyl-D-aspartate and AMPA receptor subunits in selected regions of the rat hippocampus and hypothalamus. *Neurosci* 1995; 66:247-252.

[33] Ho YJ, Chang YC, Liu TM, Tai MY, Wong CS,Tsai YF: Striatal glutamate release during novelty exposure-induced hyperactivity in olfactory bulbectomized rats. *Neurosci Lett* 2000; 287:117-120.

[34] Nowak G, Redmond A, McNamara M,Paul IA: Swim stress increases the potency of glycine at the N-methyl-D-aspartate receptor complex. *J Neurochem* 1995; 64:925-927.

[35] Yoneda Y, Han D,Ogita K: Preferential induction by stress of the N-methyl-D-aspartate recognition domain in discrete structures of rat brain. *J Neurochem* 1994; 63:1863-1871.

[36] Pfleiderer B, Michael N, Erfurth A, Ohrmann P, Hohmann U, Wolgast M, Fiebich M, Arolt V,Heindel W: Effective electroconvulsive therapy reverses glutamate/glutamine deficit in the left anterior cingulum of unipolar depressed patients. *Psychiatr Res* 2003; 122:185-192.

[37] Nowak G, Ordway GA,Paul IA: Alterations in the N-methyl-D-aspartate (NMDA) receptor complex in the frontal cortex of suicide victims. *Brain Res* 1995; 675:157-164.

[38] Auer DP, Pütz B, Kraft E, Lipinski B, Schill J,Holsboer F: Reduced glutamate in the anterior cingulate cortex in depression: an in vivo proton magnetic resonance spectroscopy study. *Biol Psychiatry* 2000; 47:305-313.

[39] Law AJ,Deakin JF: Asymmetrical reductions of hippocampal NMDAR1 glutamate receptor mRNA in the psychoses. *Neuroreport* 2001; 12:2971-2974.

[40] Palmer AM, Burns MA, Arango V,Mann JJ: Similar effects of glycine, zinc and an oxidizing agent on [3H]dizocilpine binding to the N-methyl-D-aspartate receptor in neocortical tissue from suicide victims and controls. *J Neural Trans Gen Sect* 1994; 96:1-8.

[41] Holemans S, De Paermentier F, Horton RW, Crompton MR, Katona CL,Maloteaux JM: NMDA glutamatergic receptors, labelled with [3H]MK-801, in brain samples from drug-free depressed suicides. *Brain Res* 1993; 616:138-143.

[42] You JM, Yun SJ, Nam KN, Kang C, Won R,Lee EH: Mechanism of glucocorticoid-induced oxidative stress in rat hippocampal slice cultures. *Can J Physiol Pharmacol* 2009; 87:440-447.

[43] Stein-Behrens BA, Lin WJ,Sapolsky RM: Physiological elevations of glucocorticoids potentiate glutamate accumulation in the hippocampus. *J Neurochem* 1994; 63:596-602.

[44] Armanini MP, Hutchins C, Stein BA,Sapolsky RM: Glucocorticoid endangerment of hippocampal neurons is NMDA-receptor dependent. *Brain Res* 1990; 532:7-12.

[45] Lu J, Goula D, Sousa N,Almeida OF: Ionotropic and metabotropic glutamate receptor mediation of glucocorticoid-induced apoptosis in hippocampal cells and the neuroprotective role of synaptic N-methyl-D-aspartate receptors. *Neurosci* 2003; 121:123-131.

[46] Elliott EM,Sapolsky RM: Corticosterone enhances kainic acid-induced calcium elevation in cultured hippocampal neurons. *J Neurochem* 1992; 59:1033-1040.

[47] Bremner JD, Randall P, Scott TM, Bronen RA, Seibyl JP, Southwick SM, Delaney RC, McCarthy G, Charney DS,Innis RB: MRI-based measurement of hippocampal volume

in patients with combat-related posttraumatic stress disorder. *Am J Psychiatry* 1995; 152:973-981.
[48] Jenck F, Moreau JL,Martin JR: Dorsal periaqueductal gray-induced aversion as a simulation of panic anxiety: elements of face and predictive validity. *Psychiatry Res* 1995; 57:181-191.
[49] Sajdyk TJ,Shekhar A: Sodium lactate elicits anxiety in rats after repeated GABA receptor blockade in the basolateral amygdala. *Eur J Pharmacol* 2000; 394:265-273.
[50] Johnson PL,Shekhar A: Panic-prone state induced in rats with GABA dysfunction in the dorsomedial hypothalamus is mediated by NMDA receptors. *J Neurosci* 2006; 26:7093-7104.
[51] Walker DL,Davis M: The role of amygdala glutamate receptors in fear learning, fear-potentiated startle, and extinction. *Pharmacol Biochem Behav* 2002; 71:
[52] LeDoux JE: Emotional memory: in search of systems and synapses. *Ann N Y Acad Sci* 1993; 702:149-157.
[53] McGaugh JL, Cahill L,Roozendaal B: Involvement of the amygdala in memory storage: interaction with other brain systems. *Proc Natl Acad Sci USA* 1996; 93:13508-13514.
[54] Miserendino MJ, Sananes CB, Melia KR,Davis M: Blocking of acquisition but not expression of conditioned fear-potentiated startle by NMDA antagonists in the amygdala. *Nature* 1990; 345:716-718.
[55] Campeau S, Miserendino MJ,Davis M: Intra-amygdala infusion of the N-methyl-D-aspartate receptor antagonist AP5 blocks acquisition but not expression of fear-potentiated startle to an auditory conditioned stimulus. *Behav Neurosci* 1992; 106:569-574.
[56] Kim M,McGaugh JL: Effects of intra-amygdala injections of NMDA receptor antagonists on acquisition and retention of inhibitory avoidance. *Brain Res* 1992; 585:35-48.
[57] Misane I, Tovote P, Meyer M, Spiess J, Ogren SO,Stiedl O: Time-dependent involvement of the dorsal hippocampus in trace fear conditioning in mice. *Hippocampus* 2005; 15:418-426.
[58] Quinn JJ, Loya F, Ma QD,Fanselow MS: Dorsal hippocampus NMDA receptors differentially mediate trace and contextual fear conditioning. *Hippocampus* 2005; 15:665-674.
[59] Stiedl O, Birkenfeld K, Palve M,Spiess J: Impairment of conditioned contextual fear of C57BL/6J mice by intracerebral injections of the NMDA receptor antagonist APV. *Behav Brain Res* 2000; 116:157-168.
[60] Adamec R: Transmitter systems involved in neural plasticity underlying increased anxiety and defense--implications for understanding anxiety following traumatic stress *Neurosci Biobehav Rev* 1997; 21:755-765.
[61] Saxena S, Brody AL, Schwartz JM,Baxter LR: Neuroimaging and frontal-subcortical circuitry in obsessive-compulsive disorder. *Br J Psychiatry* 1998; 173 (suppl. 35):26-37.
[62] Baxter LR, Schwartz JM, Bergman KS, Zzuba MP, Guze BH, Mazziotta JC, Alazraki A, Selin CE, Ferng HK, Munford P,Phelps ME: Caudate glucose metabolic rate changes with both drug and behavior therapy for obsessive-compulsive disorder. *Arch Gen Psychiatry* 1992; 49:681-689.

[63] Sestini S: The neural basis of functional neuroimaging signal with positron and single-photon emission tomography. *Cell Mol Life Sci* 2007; 64:1778-1784.
[64] Chakrabarty K, Bhattacharyya S, Christopher R,Khanna S: Glutamatergic dysfunction in OCD. *Neuropsychopharmacol* 2005; 30:1735-1740.
[65] Moore GJ, MacMaster FP, Stewart C,Rosenberg DR: Case study: caudate glutamatergic changes with paroxetine therapy for pediatric obsessive-compulsive disorder. *J Am Acad Child Adolesc Psychiatry* 1998; 37:663-667.
[66] Bolton J, Moore GJ, MacMillan S, Stewart CM,Rosenberg DR: Case study: caudate glutamatergic changes with paroxetine persist after medication discontinuation in pediatric OCD. *J Am Acad Child Adolesc Psychiatry* 2001; 40:903-906.
[67] Rosenberg DR, MacMaster FP, Keshavan MS, Fitzgerald KD, Stewart CM,Moore GJ: Decrease in caudate glutamatergic concentrations in pediatric obsessive-compulsive disorder patients taking paroxetine. *J Am Acad Child Adolesc Psychiatry* 2000; 39:1096-1103.
[68] Rosenberg DR, Mirza Y, Russell A, Tang J, Smith JM, Banerjee SP, Bhandari R, Rose M, Ivey J, Boyd C,Moore GJ: Reduced anterior cingulate glutamatergic concentrations in childhood OCD and major depression versus healthy controls. *J Am Acad Child Adolesc Psychiatry* 2004; 43:1146-1153.
[69] Arnold PD, Rosenberg DR, Mundo E, Tharmalingam S, Kennedy JL,Richter MA: Association of a glutamate (NMDA) subunit receptor gene (GRIN2B) with obsessive-compulsive disorder: a preliminary stud. *Psychopharmacol* 2004; 184:530-538.
[70] Arnold PD, Macmaster FP, Richter MA, Hanna GL, Sicard T, Burroughs E, Mirza Y, Easter PC, Rose M, Kennedy JL,Rosenberg DR: Glutamate receptor gene (GRIN2B) associated with reduced anterior cingulate glutamatergic concentration in pediatric obsessive-compulsive disorder. *Psychiatr Res* 2009; 172:136-139.
[71] Nordstrom EJ,Burton FH: A transgenic model of comorbid Tourette's syndrome and obsessive-compulsive disorder circuitry. *Mol Psychiatry* 2002; 7:617-625, 524.
[72] Moghaddam B, Adams B, Verma A,Daly D: Activation of glutamatergic neurotransmission by ketamine: a novel step in the pathway from NMDA receptor blockade to dopaminergic and cognitive disruptions associated with the prefrontal cortex. *J Neurosci* 1997; 17:2921-2927.
[73] Jackson ME, Homayoun H,Moghaddam B: NMDA receptor hypofunction produces concomitant firing rate potentiation and burst activity reduction in the prefrontal cortex. *Proc Natl Acad Sci USA* 2004; 101:8467-8472.
[74] McGrath MJ, Campbell KM, Parks CR,Burton FH: Glutamatergic drugs exacerbate symptomatic behavior in a transgenic model of comorbid Tourette's syndrome and obsessive-compulsive disorder. *Brain Res* 2000; 877:23-30.
[75] Andersen JD,Pouzet B: Spatial memory deficits induced by perinatal treatment of rats with PCP and reversal effect of D-serine. *Neuropsychopharmacol* 2004; 29:1080-1090.
[76] Krystal JH, Bennett A, Abi-Saab D, Belger A, Karper LP, D'Souza DC, Lipschitz D, Abi-Dargham A,Charney DS: Dissociation of ketamine effects on rule acquisition and rule implementation: possible relevance to NMDA receptor contributions to executive cognitive functions. *Biol Psychiatry* 2000; 47:137-143.
[77] van der Meulen JA, Bilbija L, Joosten RN, de Bruin JP,Feenstra MG: The NMDA-receptor antagonist MK-801 selectively disrupts reversal learning in rats. *Neuroreport* 2003; 14:2225-2228.

[78] Stefani MR,Moghaddam B: Systemic and prefrontal cortical NMDA receptor blockade differentially affect discrimination learning and set-shift ability in rats. *Behav Neurosci* 2005; 119:420-428.

[79] Lipina T, Labrie V, Weiner I,Roder J: Modulators of the glycine site on NMDA receptors, D-serine and ALX-5407, display similar beneficial effects to clozapine in mouse models of schizophrenia. *Psychopharmacol* 2005; 179:54-67.

[80] Kotlinska J,Liljequist S: A characterization of anxiolytic-like actions induced by the novel NMDA/glycine site antagonist, L-701,324. *Psychopharmacol* 1998; 135:175-181.

[81] Dunn RW, Flanagan DM, Martin LL, Kerman LL, Woods AT, Camacho F, Wilmot CA, Cornfeldt ML, Effland RC, Wood PL,Corbett R: Stereoselective R-(+) enantiomer of HA-966 displays anxiolytic effects in rodents. *Eur J Pharmacol* 1992; 214:207-214.

[82] Corbett R,Dunn RW: Effects of HA-966 on conflict, social interaction, and plus maze behaviors. *Drug Dev Res* 1991; 24:201-205.

[83] Wiley JL, Cristello AF,Balster RL: Effects of site-selective NMDA receptor antagonists in an elevated plus-maze model of anxiety in mice. *Eur J Pharmacol* 1995; 294:101-107.

[84] Anthony EW,Nevins ME: Anxiolytic-like effects of N-methyl-D-aspartate-associated glycine receptor ligands in the rat potentiated startle test. *Eur J Pharmacol* 1993; 250:317-324.

[85] Garcia LS, Comim CM, Valvassori SS, Réus GZ, Stertz L, Kapczinski F, Gavioli EC,Quevedo J: Ketamine treatment reverses behavioral and physiological alterations induced by chronic mild stress in rats. *Prog Neuropsychopharmacol Biol Psychiatry* 2009; 33:450-455.

[86] Boyce-Rustay JM,Holmes A: Genetic inactivation of the NMDA receptor NR2A subunit has anxiolytic- and antidepressant-like effects in mice. *Neuropsychopharmacol* 2006; 31:2405-2414.

[87] Labrie V, Clapcote SJ,Roder JC: Mutant mice with reduced NMDA-NR1 glycine affinity or lack of D-amino acid oxidase function exhibit altered anxiety-like behaviors. *Pharmacol Biochem Behav* 2008; 91:610-620.

[88] Kew JNC, Koester A, Moreau JL, Jenck F, Quagazzal AM, Mutel V, Richards JG, Trube G, Fischer G, Montkowski A, Hundt W, Reinscheid RK, Pauly-Evers M, Kemp JA,Bluethmann H: Functional consequences of reduction in NMDA receptor glycine affinity in mice carrying targeted point mutations in the glycine binding site. *J Neurosci* 2000; 20:4037-4049.

[89] Ho YJ, Hsu LS, Wang CF, Hsu WY, Lai TJ, Hsu CC,Tsai YF: Behavioral effects of d-cycloserine in rats: the role of anxiety level. *Brain Res* 2005; 1043:179-185.

[90] Tsang SW, Vinters HV, Cummings JL, Wong PT, Chen CP,Lai MK: Alterations in NMDA receptor subunit densities and ligand binding to glycine recognition sites are associated with chronic anxiety in Alzheimer's disease. *Neurobiol Aging* 2008; 29:1524-1532.

[91] Javitt DC,Zukin SR: Recent advances in the phencyclidine model of schizophrenia. *Am J Psychiatry* 1991; 148:1301-1308.

[92] Krystal JH, Karper LP, Seibyl JP, Freeman GK, Delaney R, Bremner JD, Heninger GR, Bowers MBJ,Charney DS: Subanesthetic effects of the noncompetitive NMDA

antagonist, ketamine, in humans. Psychotomimetic, perceptual, cognitive, and neuroendocrine responses. *Arch Gen Psychiatry* 1994; 51:199-214.

[93] Krystal JH, D'Souza DC, Petrakis IL, Belger A, Berman RM, Charney DS, Abi-Saab W,Madonick S: NMDA agonists and antagonists as probes of glutamatergic dysfunction and pharmacotherapies in neuropsychiatric disorders. *Harv Rev Psychiatry* 1999; 7:125-143.

[94] Rogawski MA: Low affinity channel blocking (uncompetitive) NMDA receptor antagonists as therapeutic agents--toward an understanding of their favorable tolerability. *Amino Acids* 2000; 19:133-149.

[95] Muhonen LH, Lönnqvist J, Juva K,Alho H: Double-blind, randomized comparison of memantine and escitalopram for the treatment of major depressive disorder comorbid with alcohol dependence. *J Clin Psychiatry* 2008; 69:392-399.

[96] Rammes G, Danysz W,Parsons CG: Pharmacodynamics of memantine: an update. *Curr Neuropharmacol* 2008; 6:55-78.

[97] Farlow MR, Graham SM,Alva G: Memantine for the treatment of Alzheimer's disease: tolerability and safety data from clinical trials. *Drug Saf* 2008; 31:577-585.

[98] Tsai G, Yang P, Chung L-C, Lange N,Coyle JT: D-serine added to antipsychotics for the treatment of schizophrenia. *Biol Psychiatry* 1998; 44:1081-1089.

[99] Lane HY, Chang YC, Liu YC, Chiu CC,Tsai GE: Sarcosine or D-serine add-on treatment for acute exacerbation of schizophrenia: a randomized, double-blind, placebo-controlled study. *Arch Gen Psychiatry* 2005; 62:1196-1204.

[100] Heresco-Levy U, Javitt DC, Ebstein R, Vass A, Lichtenberg P, Bar G, Catinari S,Ermilov M: D-serine efficacy as add-on pharmacotherapy to risperidone and olanzapine for treatment-refractory schizophrenia. *Biol Psychiatry* 2005; 57:577-585.

[101] Heresco-Levy U, Kremer I, Javitt DC, Goichman R, Reshef A, Blanaru M,Cohen T: Pilot-controlled trial of D-cycloserine for the treatment of post-traumatic stress disorder. *Int J Neuropsychopharmacol* 2002; 5:301-307.

[102] Greenberg WM, Benedict MM, Doerfer J, Perrin M, Panek L, Cleveland WL,Javitt DC: Adjunctive glycine in the treatment of obsessive-compulsive disorder in adults. *J Psychiatr Res* 2009; 43:664-670.

[103] Labrie V, Duffy S, Wei W, Barger S, Baker GB,Roder JC: Genetic inactivation of D-amino acid oxidase enhances extinction and reversal learning in mice. *Learn Mem* 2008; 16:28-37.

[104] Labrie V, Lipina T,Roder JC: Mice with reduced NMDA receptor glycine affinity model some of the negative and cognitive symptoms of schizophrenia. *Psychopharmacol* 2008; 200:217-230.

[105] Black DW: Efficacy of combined pharmacotherapy and psychotherapy versus monotherapy in the treatment of anxiety disorders. *CNS Spectr* 2006; 11:29-33.

[106] Pull CB: Combined pharmacotherapy and cognitive-behavioural therapy for anxiety disorders. Curr Opin Psychiatry 2007; 20:30-35.

[107] Davis M, Ressler K, Rothbaum BO,Richardson R: Effects of D-cycloserine on extinction: translation from preclinical to clinical work. *Biol Psychiatry* 2006; 60:369-375.

[108] Cammarota M, Bevilaqua LR, Barros DM, Vianna MR, Izquierdo LA, Medina JH,Izquierdo I: Retrieval and the extinction of memory. *Cell Mol Neurobiol* 2005; 25:465-474.

[109] Lin CH, Yeh SH, Lu HY,Gean PW: The similarities and diversities of signal pathways leading to consolidation of conditioning and consolidation of extinction of fear memory. *J Neurosci* 2003; 23:8310-8317.

[110] Walker DL, Ressler KJ, Lu KT,Davis M: Facilitation of conditioned fear extinction by systemic administration or intra-amygdala infusions of D-cycloserine as assessed with fear-potentiated startle in rats. *J Neurosci* 2002; 22:2343-2351.

[111] Ledgerwood L, Richardson R,Cranney J: Effects of D-cycloserine on extinction of conditioned freezing. *Behav Neurosci* 2003; 117:341-349.

[112] Norberg MM, Krystal JH,Tolin DF: A meta-analysis of D-cycloserine and the facilitation of fear extinction and exposure therapy. *Biol Psychiatry* 2008; 63:1118-1126.

[113] Ledgerwood L, Richardson R,Cranney J: D-cycloserine facilitates extinction of learned fear: effects on reacquisition and generalized extinction. *Biol Psychiatry* 2005; 57:841-847.

[114] Guastella AJ, Richardson R, Lovibond PF, Rapee RM, Gaston JE, Mitchell P,Dadds MR: A randomized controlled trial of D-cycloserine enhancement of exposure therapy for social anxiety disorder. *Biol Psychiatry* 2008; 63:544-549.

[115] Wilhelm S, Buhlmann U, Tolin DF, Meunier SA, Pearlson GD, Reese HE, Cannistraro P, Jenike MA,Rauch SL: Augmentation of behavior therapy with D-cycloserine for obsessive-compulsive disorder. *Am J Psychiatry* 2008; 165:335-341.

[116] Guastella AJ, Dadds MR, Lovibond PF, Mitchell P,Richardson R: A randomized controlled trial of the effect of D-cycloserine on exposure therapy for spider fear. *J Psychiatr Res* 2007; 41:466-471.

[117] Guastella AJ, Lovibond PF, Dadds MR, Mitchell P,Richardson R: A randomized controlled trial of the effect of D-cycloserine on extinction and fear conditioning in humans. *Behav Res Ther* 2007; 45:663-672.

[118] Yang YL,Lu KT: Facilitation of conditioned fear extinction by d-cycloserine is mediated by mitogen-activated protein kinase and phosphatidylinositol 3-kinase cascades and requires de novo protein synthesis in basolateral nucleus of amygdala. *Neurosci* 2005; 134:247-260.

[119] Santini E, Muller RU,Quirk GJ: Consolidation of extinction learning involves transfer from NMDA-independent to NMDA-dependent memory. *J Neurosci* 2001; 21:9009-9017.

[120] Zhang Z, Gong N, Wang W, Xu L,Xu TL: Bell-shaped D-serine actions on hippocampal long-term depression and spatial memory retrieval. *Cereb Cortex* 2008; 18:2391-2401.

[121] Higgins GA, Enderlin M, Haman M,Fletcher PJ: The 5-HT2A receptor antagonist M100, 907 attenuates motor and 'impulsive-type' behaviours produced by NMDA receptor antagonism. *Psychopharmacol* 2003; 170:309-319.

[122] Morice E, Billard JM, Denis C, Mathieu F, Betancur C, Epelbaum J, Giros B,Nosten-Bertrand M: Parallel loss of hippocampal LTD and cognitive flexibility in a genetic model of hyperdopaminergia. *Neuropsychopharmacol* 2007; 32:2108-2116.

[123] Zeng H, Chattarji S, Barbarosie M, Rondi-Reig L, Philpot BD, Miyakawa T, Bear MF,Tonegawa S: Forebrain-specific calcineurin knockout selectively impairs bidirectional synaptic plasticity and working/episodic-like memory. *Cell* 2001; 107:617-629.

[124] Dalton GL, Wang YT, Floresco SB,Phillips AG: Disruption of AMPA receptor endocytosis impairs the extinction, but not acquisition of learned fear. *Neuropsychopharmacol* 2007; 33:2416-2426.

[125] Kemp A,Manahan-Vaughan D: Hippocampal long-term depression and long-term potentiation encode different aspects of novelty acquisition. *Proc Natl Acad Sci USA* 2004; 101:8192-8197.

[126] Manahan-Vaughan D,Braunewell KH: Novelty acquisition is associated with induction of hippocampal long-term depression. *Proc Natl Acad Sci USA* 1999; 96:8739-8744.

[127] Walker DL,Davis M: Are fear memories made and maintained by the same NMDA receptor-dependent mechanisms? *Neuron* 2004; 41:680-682.

[128] Bubeníková-Valesová V, Horácek J, Vrajová M,Höschl C: Models of schizophrenia in humans and animals based on inhibition of NMDA receptors. *Neurosci Biobehav Rev* 2008; 32:1014-1023.

In: Neurobiology of Post-Traumatic Stress Disorder
Editors: L. Sher and A. Vilens, pp. 309-330
ISBN: 978-1-61470-593-2

Chapter XIX

Extinction of Fear Conditioning: Relevance of Recent Findings for the Treatment of Post-traumatic Stress Disorder

Lia R. M. Bevilaqua, M. Cammarota, J. H. Medina, I. Izquierdo
Pontificia Universidade Catolica de Rio Grande do Sul, Porto Alegre, Brasil (LRMB, MC, II)
and Universidad de Buenos Aires, Buenos Aires, Argentina (JHM)

Abstract

Extinction is the learned inhibition of the expression of previously acquired memories. It is particularly useful for the treatment of fear motivated disorders, particularly PTSD. It relies on the interlinked electrophysiological and molecular activity of the ventromedial prefrontal cortex, the hippocampus, the lateral or basolateral amygdala and the entorhinal cortex. For some memories it may involve in addition other brain sites, like the insular cortex.

Introduction: Retrieval, Extinction and Reconsolidation

Retrieval of a consolidated memory by presenting the learned stimulus but not the associated reinforcement leads to two opposing processes [1]: extinction, which weakens expression of that memory [2], and reconsolidation, which strengthens it [3,4]. Extinction is directly observable; reconsolidation must be inferred from observations that post-retrieval treatments reduce further retrieval in subsequent sessions [4,5]. Here we will concentrate on extinction because of its widespread use in the treatment of PTSD. For other recent reviews on extinction, see [1,6-8].

Extinction increases with repetition of the unreinforced retrieval and can be initiated at almost any time after training [hours, days, years [2,9,10]. Reconsolidation is not gradual: it is triggered by a single retrieval trial, but only when this is carried out a few hours or days after the (last) training session [11]. This might be due to the fact that both at the hippocampal molecular level, and at the systems level, retrieval is different 3-24 h and, say, 31 days after training. At 3 h from inhibitory avoidance training, retrieval requires hippocampal AMPA and metabotropic receptors, as well as ERK and PKA activity [12,13]. At 24 h, it requires, in addition, the concomitant participation of the amygdala and the entorhinal, parietal and cingulate cortex [14,15]. One month later, retrieval requires hippocampal ERK and PKA activity and glutamate metabotropic but not ionotropic receptors [12], and the involvement of entorhinal and parietal cortex but perhaps not the amygdala [16]. The pharmacology of long-term extinction and short-term memory reactivation are different; which has been taken to underlie the fact that both engage different post-retrieval processes [17]. Indeed, extinction has been suggested to start several seconds after CS offset at least in the crab [18], whereas reconsolidation process appears to be initiated immediately upon CS presentation. Contrarily to reconsolidation, extinction is easier and quicker for old and/or not-too-well-remembered memories (eg., [9]).

Therefore, extinction and reconsolidation are different processes resulting from the labilization of the trace by unreinforced retrieval [4,19,20]. To be sure, there are similarities between the two processes: dependence on retrieval, dependence on context, [21-24], requirement of ERKs [25] and transcription and/or translation in hippocampus [26] and amygdala [27]. However, for the most part, extinction and reconsolidation rely on different sets of brain structures [8,28-30], on different cell types within the same brain structure (the amygdala, [31]), and on different molecular processes within the same structure (the vmPFC, [32]).

Extinction obviously is a form of learning, as attested by the existence of time-dependent spontaneous recovery [33,34], by the fact that responses can be relearned more quickly after extinction than they were in the original training [2,9], and very particularly by the existence of retrograde amnesia for extinction [35]. It is possible that in 17-day-old rats, in which the hippocampal system is immature [36] extinction may involve unlearning as well [37]. Extinction uses different interactions among prefrontal regions in the 17-day old rat than in older animals [38]; and it has been suggested that the role of the amygdala in extinction is different in the young and the mature rat [37].

In contrast to extinction, reconsolidation is not a form of learning; it is just one possible, and major, consequence of the first of a series of retrieval sessions [3,5,26].

In avoidance tasks in which the response requirement is to inhibit a response, extinction is of course more difficult than in active avoidance tasks. This is probably because in the former there is superposition of the requirements of avoidance and extinction learning: response omission is at the same time the correct response and signals the beginning of extinction [9]. After intense or prolonged extinction, reinstatement of the original learning requires the activation of the transcription factor CREB (cAMP-responsive element binding protein) in hippocampus and amygdala [39] and gene transcription and protein synthesis in the hippocampus [40].

Rarely has it been shown that memory can effectively be updated or enhanced by postretrieval treatments or information [41]. In 1974, decades before the word "reconsolidation" was coined, Loftus and Palmer reported that leading questions at the time of retrieval can strongly and lastingly modify previously acquired memories [42]. Subjects were shown a short film of a car accident, and then asked at what speed the cars were going when they contacted, hit, bumped, collided or smashed with each other. Depending on which of these words was used in the question, the answers ranged from 30 mph (contacted) to 40.5 mph (smashed), and some of the latter reported having seen blood in the scene, which had not been the case. Clearly, memory was modified by the introduction of new information at the time of retention testing; i.e, information from the movie reconsolidated with that of the leading question.

Both extinction and reconsolidation require gene transcription and protein synthesis in specific brain regions and/or specific brain cells. The regions and the cells vary with the task [5,25,28,30,39,43-45].

Extinction of Learned Fear

PTSD consists of the exaggerated persistence and repeated retrieval of fearful and stressful memories, and is now viewed by many as a deficit of extinction [8,20,46-49]. Both fear reactivity and fear recovery phenotypes have in fact been suggested [8,50]. The extinction of learned fear serves the adaptive purpose of preventing repeated or otherwise unwanted retrieval of fearsome or otherwise stressful memories [20]. Accordingly, extinction is widely used as a treatment of PTSD, often under the name of "exposure therapy" [6,7,10]. In extinction the original memory is preserved [51,52], but its expression is inhibited. Testimony to this are both the fact that there is spontaneous recovery of the extinguished response, which is a function of time [33,34], and that its relearning is easier than the original learning [2,9,33,51]. Preservation of the fear memory is useful for having it available when needed, in order to avoid similar harmful or fearsome situations in the future. Inhibition of its retrieval helps the individual not to remember such experiences more often than needed.

There are other risk factors for PTSD aside from deficient extinction [8]: maladaptive (catastrophic) appraisal [53], a history of exposure to stressors and/or corticosterone [54], a propensity to develop panic attacks [55], which is indeed associated with enhanced resistance to extinction [56]. Context-dependent extinction recall is impaired in schizophrenia, a disease with impaired function of the anterolateral prefrontal cortex and the temporal lobe [57], and therefore involving a difficulty to perceive or recall contexts.

Conditioned Fear and PTSD

Clearly, a reduction of reconsolidation and an enhancement of extinction of conditioned fear can be useful for the treatment of PTSD. But extinguished fear responses can recur spontaneously [34], after a change of context (renewal, [21-23,58]), and particularly if the reinforcement is reintroduced [33,40,59,60]. The hippocampus, entorhinal cortex and fornix are critical for the control of extinction by context [22,23].

The extension of animal models of fear conditioning to humans advocated by many in recent years [50,58,61-63] has already yielded important results (see [6-8]). The study of conditioned fear in humans with or without a background of anxiety has confirmed the findings on animal fear conditioning, both with regard to the effect of extinction [64] and to that of drugs upon it (MK-801; D-cycloserine, see below) [58,64].

Short- versus Long-delay Extinction

The initiation of extinction shortly after fear learning is usually viewed as less efficacious than the introduction of extinction at later times (i.e., days), both for animal and for human fear conditioning [65-67] and appetitive conditioning as well [66]. Recent studies in humans suggest that this is not necessarily so [60]. Davis and his group furnished evidence that different neural mechanisms are recruited depending on the temporal delay of fear extinction [68]. In rats, early extinction actually produces a short-lasting suppression of conditioned fear [65,66], that according to some is [67] and according to others is not [51] context dependent. It is accompanied by a decrease of pCREB in the amygdala [69]. Some have suggested that short-term extinction may have mechanisms different from those of long-term extinction [68]; others have not [70]. There is no strong evidence in humans for the widespread belief that if extinction is carried out shortly after training the chances for spontaneous recovery are reduced [71].

Massed and Spaced Extinction Trials; Changes of Context

There is general agreement that spaced trials are in the long run more effective than massed trials for the extinction of fear conditioning [59]. Compound stimulus presentation deepens extinction and reduces spontaneous recovery and reconditioning in rats and pigeons [72]. Behavioral effects of extinction are indeed highly specific to the stimulus constellation that has gone through the extinction procedure, and this may have clinical implications [73]. Extinction can be carried out in the same context in which fear was learned, or in different contexts, that may be presented consecutively or not; this does not make extinction learning less context specific or weaken fear expression or renewal [74].

In psychiatric practice, spaced rather than massed extinction trials are used in the therapy of PTSD. Results with the latter are less reliable and there may be recurrence of the traumatic memory (spontaneous recovery, renewal) [6-8].

Brain Systems Involved in Fear Retrieval

Abundant evidence indicates that both extinction [75] and reconsolidation [4] directly derive from retrieval; i.e., the various molecular processes that accompany retrieval gate or initiate both extinction and reconsolidation. In humans, episodic retrieval involves the hippocampus [12,13,76] in association with other brain structures. Fear retrieval is usually measured by the remembrance of episodes, rarely by fear recall, both in animals and in humans (see below).

The systems needed for the retrieval of fear conditioning in rats have been studied in one-trial avoidance in rats. In fact, retrieval of fear conditioning requires not only the hippocampus but also the BLA, entorhinal, parietal and anterior cingulate cortex (ACC) [12,13]. In CA1, entorhinal and parietal cortex and in the ACC it involves other-than-NMDA glutamate receptors plus protein kinase A, and ERK activation. Fear retrieval in animals and humans requires β-noradrenergic receptors in the BLA: it is blocked by locally infused or systemically administered propranolol or other β-blockers [77] and this is strongly modulated by circulating glucocorticoids, that being the reason why anxiety or stress impair retrieval; in humans this effect of glucocorticoids is accompanied by reduced blood flow in the medial temporal lobe [78].

The role of various neurotransmitter receptors in retrieval of inhibitory avoidance has been studied in CA1, entorhinal and parietal cortex, ACC and the BLA by local infusion of specific antagonists. AMPA and mGlu receptors play a role in all areas analysed [14,16]. In CA1, mGlu2s, mGlu3s, mGlu4s and mGlu6s are necessary for retrieval, whereas mGlu1s play an inhibitory role [79]. Dopaminergic D1, β-noradrenergic and muscarinic cholinergic receptors are necessary for retrieval in the four areas; 5HT1A receptors play instead an inhibitory role [15]; of these, in the BLA, only β-adrenoceptors appear to be involved in the retrieval of this task [15,80].

The amnesic effect of posttraining propranolol or other β-blockers has indeed been extensively studied in laboratory animals, both by systemic and by intra-amygdala or intrahippocampal administration [77,80]. In humans, propranolol administered shortly after a traumatic experience prevents the consolidation of that experience and thus effectively erases it [81-83]. Similar effects have been described with the anesthetic sevoflurane [84] or the anxiolytic drug, alprazolam [85]. There are advocates and opponents of the possible application of β-blockers to the treatment of PTSD. The advocates favor the erasure or non-consolidation of traumatic memories [81-84]. The opponents cite the fact that while β-blockers or other post-experience amnesic agents will cancel the consolidation of those memories, they will also cancel that of all other memories acquired shortly before or after them, which were in the process of being consolidated at about the same time. Among the memories that would also be erased are those of the eventual escape from, or termination (relief) of, the traumatic event, which could be used in therapy. Still others object the erasure of any memory; they believe that deleting any memory would be detrimental to our personality. A counterargument against this is that it is hardly likely that the persistence of the memory of torture or rape would be of any benefit to anyone.

Whatever the philosophical position about the ethics or the convenience of erasing any memory, the treatment is usually not practical since people do not normally carry propranolol pills with them, let alone anesthetics, in prevention of the eventuality of a traumatic experience (an accident, a crime, a terrorist attack, an earthquake).

Post-retrieval Propranolol

A recent variant of this type of treatment that holds promise is the post-retrieval administration of propranolol, which should cancel reconsolidation of traumatic or other memories, and thus be beneficial for PTSD [86,87]. The treatment has been described as very effective in canceling the memory that generates the PTSD, without affecting others. Indeed, propranolol has been reported to inhibit the reconsolidation of several types of memory [3,88,89], including that of cocaine [90,91], particularly when given into the lateral amygdala [88,89], and it has been predicted that this effect could be of use in the treatment of PTSD [89]. It should be borne in mind that reconsolidation is most effective when presented hours or a few days, but not weeks, after the original training [10,92]. The same should hold for the effect of post-retrieval propranolol.

An important variant of the retrieval-mediated models for PTSD treatment has been recently advocated by LeDoux and his group [19]. It consists of an additional retention trial preceding a formal short-term extinction procedure, which, by rendering the trace more labile, makes the subsequent extinction more effective. The procedure actually makes short-term extinction as effective as long-term extinction [66,67], and would be very convenient for PTSD patients inasmuch as the memories that haunt them would cease to be expressed shortly after the initiation of the treatment, so they would not need to retrieve them again and again over successive therapy sessions over many days.

An acceleration of the rate of extinction through the use of D-cycloserine (see below) has been proposed as an important adjunct for the psychotherapy of PTSD [64,93].

Brain Areas Involved in Fear Extinction: Ventromedial Prefrontal Cortex (vmPFC)

The vmPFC is linked to the amygdala and hippocampus, and plays a role of its own [12] in fear motivated learning. Abundant recent evidence has shown quite unequivocally that the vmPFC is important for extinction. Early lesion studies attributed the role to the medial PFC, but soon it was realized that the key portion of it for extinction is the ventromedial portion (vmPFC) [94]. The recall of extinction is hindered by the intra-vmPFC infusion of protein synthesis inhibitors [95], ERK inhibitors [96], and NMDA receptor antagonists [97,98]. Unit activity in vmPFC neurons in response to the CS change during extinction [99]. Fear conditioning and extinction specifically modify the intrinsic excitability of infralimbic prefrontal cortex pyramidal cells: it is lower during acquisition, and higher at the time of extinction [100].

Thickness of the medial orbitocortical prefrontal cortex is greater in humans with good retention of extinction [101], which has suggested that an atrophy or underdevelopment of the vmPFC might be at the root of PTSD, and that PTSD might be viewed as due to a deficit of extinction [20,63,64,101]. The view of reduced extinction as a key vulnerability factor for PTSD has been endorsed by biochemical studies in mice [102] and by behavioral observations in humans [46].

Brain Areas Involved in Fear Extinction: The Hippocampus

Many believe that the site(s) involved in memory formation (which he called "formation of temporary connections") also participate in extinction. We now know, since patient H.M. [103], that memories are processed initially in the temporal lobe, particularly in the hippocampus and amygdala [80] and stored later in other parts of the brain [16,103]. Extinction, as mentioned, involves and requires the vmPFC (see above). Does it also involve and require other brain regions as well, such as the hippocampus and amygdala?

Much recent research indicates that it does, and that this depends on the task. A reduction of hippocampal size has been reported in PTSD [104-106]. Conversely, increased hippocampal blood flow was observed using PET techniques in such patients, with changes reported also in mPFC and amygdala [106,107].

In relation to extinction, the hippocampus has been best studied in rodents, which have a large dorsal extension easily amenable to cannulae and electrode implantation. The dorsal CA1 region of the hippocampus plays a role in extinction of one-trial fear avoidance. Inhibition of protein synthesis [28] or gene expression [29] in this region shortly before or right after the first of a series of extinction trials blocks extinction over the next trials. The inhibitors given alone in the absence of a retrieval session have no effect on performance in subsequent retrieval trials [28]. The same can be obtained with intra-CA1 infusions of AP5, an antagonist of glutamate NMDA receptors, of the calcium-calmodulin kinase inhibitor (CaMKII), KN62, of the cAMP-dependent protein kinase (PKA) inhibitor Rp-cAMPs, or of the extracellularly regulated kinase pathway (ERK) inhibitor PD098059 ([44]; see also [108]), of pre-extinction muscimol [109] or with post-extinction hypothermia [35]. NMDA receptors, CaMKII, PKA and ERKs are necessary for the consolidation of this and many other tasks [80]. Posttraining hypothermia has long been known to cause time-dependent retrograde amnesia [110]. These findings indicate that extinction is a form of learning that consolidates using molecular mechanisms similar to those of most other learnings.

Important recent evidence showed that segregated populations of principal CA1 neurons are in charge of conditioning and extinction of fear conditioning [11]. These experiments frontally collide with the old-fashioned notion that the same cells that acquire and consolidate should be involved in extinction (see [111-113]). Tronson et al. [111] analysed by immunofluorescence cells positive to the early gene, cFos, and to ERK, and observed predominant nuclear activation of either protein during conditioning and extinction. Selective immunotoxin-induced lesions of medial septal muscarinic cholinergic cells selectively abolished ERK activation and extinction of fear conditioning but not cFos and conditioning. In the extinction of fear, ERK phosphorylation in CA1 plays a key role in the prediction of error [114]. The prediction of error depends on shock expectancy and is crucial in the extinction of shock avoidance responses, in which the omission of the unconditioned footshock signals both a correct response to the original training and the initiation of extinction, thereby tending to make the latter slower [9].

Concerning possible roles of hippocampal ERK in extinction, it was recently shown that active Rap2 transgenic mice show reduced ERK signaling, fewer dendritic spines, enhanced long-term depression and impaired extinction both of a spatial task and of fear conditioning [102]. The Rap family antagonizes Ras, the GTPase that initiates ERK activity.

Retrieval activates the NF-kappa-B pathway, which is necessary for reconsolidation [115,116]; apparently extinction entails the inhibition of that pathway [117]. Extinction can be facilitated by inhibition of the anchoring of PKA to A-kinase anchoring proteins in mouse hippocampus [118].

Brain Areas Involved in Fear Extinction: BLA and/or LA

It has been known now for quite some time that the lateral amygdale (LA) (see [8,118]) and/or the basolateral amygdale (BLA) [77] is involved crucially in the perception of, reactions to, and learning of, emotional memory. A role of the LA or BLA in the extinction of such memory was proposed by Davis and his group [6,120], who showed it to be blocked by the local infusion of the NMDA receptor antagonist, AP5, by Le Doux and his collaborators [8], and by Quirk and his associates, who, early on, correlated unit activity measured in the LA with conditioning and extinction of fear conditioning [121,122]. The latter recently devised a network model of acquisition and extinction of conditioned fear associations in the LA [123]. Concerning the LA vs. BLA controversy [119,120,121-123], perhaps the recent histochemical work of Wilson and Murphy [124] has seriously tipped the balance in favor of a role of the LA, at least in contextual fear conditioning. These authors used the fos-tau-lacZ mouse to map functional activated expression in neurons right after contextual fear learning and found a discrete population in the LA in which this happens. .

The central nucleus of the amygdala has been suggested by most authors [31,69] as the main output station for conditioned fear responses. It is active in the acquisition but probably not the extinction of such responses [113]. It is controlled by inhibitory projections from the so-called intercalated amygdala neurons (ITC), which, if selectively lesioned by a toxin that selectively targets micro-opioid receptors, seriously hinder extinction; thus, ITCs might well be the crucial cells in the amygdala that regulate extinction; they are indeed innervated by BLA projections [31].

Davis et al. [93] showed that the intra-amygdala infusion of D-cycloserine, a partial agonist at the glycine site of the NMDA receptor enhances, and that of ERK inhibitors depressed extinction of a task well-known to be mediated by that brain region. Davis and his group worked extensively in the fear-potentiated startle paradigm, a simple task whose underlying circuitry has been reliably demonstrated by this group and serves as a model for fear. The effect of D-cycloserine on extinction of conditioned fear is hindered by inhibitors of ERK and phosphatidylinositol 3-kinase (PI-3K) activity as well as by actinomycin D and anisomycin given into the BLA [125] and by proteasome inhibitors [126], which suggests a more complex mechanism than just an influence of the amino acid on the glycine site. D-cycloserine has been tested clinically as an adjunct to extinction therapy in PTSD [64,93].

Protein synthesis inhibitors or β-noradrenergic blockers given into the central amygdaloid nucleus block acquisition but not extinction of conditioned taste aversion, the opposite is true when these substances are given into the BLA [113]. This further illustrates the involvement of different cells in each of the two behavioral processes, as the experiments of Tronson et al. [111] have demonstrated for subpopulations of CA1 cells in another task. Berman and Dudai [112] had reported that these same substances given into the insular cortex block extinction of that task, as they had been previously shown to block its acquisition. Interestingly, they found that ERK inhibitors affect consolidation but not extinction of this task in the insula; which is the opposite of what Tronson et al. [111] found in the hippocampus with conditioned fear. Extinction of one-trial avoidance is also blocked by infusions of inhibitors of gene expression or protein synthesis and of an ERK inhibitor, but not by a PKA inhibitor, into the BLA [29]. ERK activation is also required for the extinction of auditory fear conditioning [127] and fear-potentiated startle [128].

Calcineurin in the amygdala has been recently investigated in relation to the extinction of conditioned fear; increases of this enzyme and of its phosphorylated form have been described in the amygdala during extinction of conditioned fear, and its inhibitors block the extinction [129]. A decrease of pCREB has also been described during extinction of conditioned fear in the amygdala [69,130], an effect that can be hindered by inhibitors of PI-3K [129]. This has been taken as an indication of unlearning, but it can also be viewed as an indication of inhibited retrieval, which is more in accordance with the most widely accepted views about extinction [33,34]. Retrieval of fear responses is indeed followed by a synthesis of pCREB [131].

The participation of NMDA receptors in the BLA in extinction has been ascertained in the potentiated startle task (see above). Sotres-Bayon et al. [132], through selective NR2B blockade, demonstrated that NMDA receptors are required for the initial acquisition rather than for the retention of fear extinction. This finding correlates with those using more classic NMDA blockers in hippocampus [44] and entorhinal cortex [133]. Interestingly, as in many other forms of learning, BDNF signaling in amygdala is required for the consolidation but not the acquisition of extinction [134].

Most data so far point to an active role of the LA or BLA in generating extinction. It has been suggested that depotentiation of thalamic input synapses onto the LA is the basis of fear extinction; extinction is indeed accompanied by reversal of the conditioning-induced enhancement of surface expression of AMPA receptor subunits in synaptosomes from LA [135]. Depotentiation is of course an active process; so this finding may not be incompatible with the others commented above.

Amygdala hyperfunction has been described in fMRI studies in humans with phobic fear, and found to normalize after exposure therapy [136].

The mechanisms involved in extinction control by the BLA involve muscarinic cholinergic synapses [137], as do those in the hippocampus [111].

For a summing up of much of the evidence on the role of the amygdala in the extinction of conditioned fear, see [138].

Brain Areas Involved in Fear Extinction: The Entorhinal Cortex

The entorhinal cortex links the amygdala and hippocampus to each other and also links both with the neocortex through afferent and efferent pathways. It was recently shown that it also plays a role in the extinction of one-trial avoidance alongside that of the other two structures: Infusion of AP5, or of the CaMKII inhibitor, KN93, or of the protein synthesis inhibitor, anisomycin, but not of an ERK inhibitor into the entorhinal cortex right before the first of 4 extinction sessions in that task inhibits extinction altogether [133].

Entorhinal lesions, which effectively isolate the hippocampus from the extra-temporal neocortex, interfere with the context-dependence of both retrieval and extinction; a similar effect is obtained with fornix lesions [139]. This probably due to the fact that the hippocampus, which no doubt plays a key role in context dependency [22,23] must be connected to all of its major afferent and efferent projections in order to manifest its role.

Brain Areas Involved in Fear Extinction: The Anterior Cingulate Cortex

Recently the anterior cingulate cortex (ACC) was found to play a role in the consolidation [140] and retrieval [14] of one-trial inhibitory avoidance.

fMRI studies in humans has shown that the ACC plays a role in fear expression [140]. The ACC is anatomically and functionally linked to the mPFC, vmPFC, amygdala and hippocampus (see references in [14,140,141]).

In an important study of combat-exposed twins, of which one developed PTSD and the other did not, a substantial reduction of ACC gray matter volume was found in the former [142]. Rostral ACC volume predicts treatment response for extinction therapy of PTSD [143]. Reduced N-acetylaspartate levels in the ACC have been reported in PTSD by human imaging techniques, and claimed to be a more sensitive marker than brain volume [144]. In infant rats, functional maturation of the ACC, mPFC and orbitofrontal cortex measured by fluorodeoxyglucose uptake was found to be associated with the maturation of extinction behavior [38].

Are there Circuits that Include Both the vmPFC and Amygdala or Hippocampus Involved in the Regulation of Extinction?

As commented above, the vmPFC, ACC, hippocampus, and LA or BLA are all involved in the generation or regulation of extinction, and, in the case of conditioned taste aversion, the insular cortex is crucial. There is, in fact, fMRI evidence for a role in extinction of circuits involving the vmPFC and the amygdala [119,145], the vmPFC and the hippocampus [146-148], and ACC, hippocampus and amygdala [149]. There have been recent excellent fMRI studies suggesting dissociable roles for the hippocampus and the amygdala in human extinction [149,150], or between the BLA and the mPFC; the latter is, and the former is not, apparently involved in reextinction [151]. Double dissociations between the role of NR2B-containing NMDA receptors in extinction [98], and for sensitivity to inhibition by the local infusion of a dopamine D1 receptor blocker, SCH23390 [152], have been reported for vmPFC and amygdala in rats. In opposition to this, a similar impairment of c-Fos and zif-268 expression in mPFC and amygdala have been described in mice with resistance to extinction [153].

Long-standing anatomical, clinical and experimental evidence certainly favors the view of circuits including vmPFC, hippocampus and LA or BLA in extinction, as there are for the encoding and consolidation of fear motivated memory [154]. In patients with PTSD, fMRI studies show increased regional blood flow in bilateral hippocampus and right amygdala relative to controls, and symptom severity is associated with both measures [155].

Care must be taken in interpreting fMRI data in relation to their bearing upon what is already well known from anatomical and molecular studies about the involvement of the LA or BLA, the hippocampus and the ACC in relation to both retrieval and extinction of fear memories. Differently from them, the vmPFC is believed to play a role in extinction but not necessarily in retrieval [20,97-99]. In some cases, a reciprocal relationship and opposing association between mPFC or vmPFC and amygdala or hippocampus has been described (eg. 145,155]. In others, a CS-evoked positive correlation of human vmPFC and hippocampus activity develops after extinction and is context dependent expressed in extinction, but not in conditioning [146].

Projections of the mPFC to the amygdala have been studied by careful electrophysiological methods by Likhtik and his coworkers [156]. Projections of the mPFC to the central amygdala, which are the ones that have been most usually proposed to be inhibitory and to mediate extinction, are very sparse; the main projections are to the BLA, and are excitatory. Therefore, the mechanisms whereby the mPFC (or vmPFC) controls the role of the amygdala in extinction are, at best, obscure; the BLA should play a role in accordance with, and of the same sign as, that of the vmPFC. The central amygdala is linked to the BLA by ITCs, which have been suggested to be crucially involved in extinction [31,156,157].

In the common and widely used inbred mouse strain 129S1 impaired fear extinction and cortico-amygdala circuit abnormalities (a reduced expression of c-Fos and zif-268 in both regions) have been described [158]. The possible usefulness of this mouse strain as a model for extinction has not been further explored.

fMRI does not demonstrate mechanisms but simply shows changes in relative blood flow; it is important to be clear as to what these changes are compared against, and whether they reflect retrieval, extinction, or just resting or steady states in PTSD patients relative to controls, for example. A recent paper by Prickaerts and his group [159] is sobering in relation to how wide can fMRI data interpretations be in terms of whether a given brain region is involved in this or that aspect of extinction and/or retrieval. They found that phosphodiesterase inhibitors enhance object memory independently of local cerebral blood flow, measured by much more sensitive methods than fMRI: quantitative iodoantiphyrine and 2-deoxyglucose autoradiography. Phosphodiesterase inhibitors enhance the formation of many types of memory, in general through an influence on cyclic nucleotide mediated enzymatic activities [80].

Rauch and coworkers [62] have proposed that PTSD involves exaggerated amygdala responses reflecting high reactivity to fear, deficient hippocampal function mediating deficits in the appreciation of safe contexts, and deficient vmPFC function mediating deficits in extinction. But it must be remembered that others have shown a concurrent role of both the BLA and the hippocampus [6,28,29,44,113], as well as of the entorhinal cortex that links them with each other and with the PFC [8,133], in the extinction of fear memories.

Importantly, in PTSD there are altered fMRI responses in the amygdala and vmPFC both for trauma-specific responses, and also for that of emotionally salient but trauma-unrelated responses [160]. This supports the concept of a general extinction failure in, and perhaps as a basis of, PTSD commented above [62].

There is recent literature not directly related to extinction that point to a variety of functional interconnections between the entorhinal cortex and the mPFC[161], the ACC and amygdala dependent fear [162], and the relations of the mPFC and hippocampus with the accumbens nucleus [20,163,164]. It has been recently shown in humans that dopaminergic mechanisms determine whether memories will persist beyond 2-3 days through an action presumably in the hippocampus [165]. Dopamine modulates hippocampal-prefrontal interactions in order to drive behaviors guided by memories [166]. Possible roads to the further study of extinction mechanisms might lie among these references [8,20].

Conclusion

Extinction is a form of learning that consists of the gradual inhibition of retrieval. It is itself triggered by retrieval and relies on gene expression, protein synthesis and various molecular processes in the ventromedial prefrontal cortex, hippocampus, lateral or basolateral amygdala and entorhinal cortex, acting in concert. Extinction of fear motivated memories is particularly useful in the therapy of PTSD and other forms of learned fear.

References

[1] Eisenhardt D, Menzel R: Extinction learning, reconsolidation and the internal reinforcement hypothesis. *Neurobiol Learn Mem* 2007; 87:167-173.

[2] Pavlov IP: Conditioned Reflexes: *An Investigation of the Physiological Activity of the Cerebral Cortex*. London, Oxford University Press, 1927. 1927

[3] Sara SJ: Strengthening the shaky trace through retrieval. *Nature Rev Neurosci* 2000; 1:212-213.

[4] Nader K: Memory traces unbound: *Trends Neurosci* 2003; 26:65-72.

[5] Nader K, Schafe GE, LeDoux JE: Fear memories require protein synthesis in the amygdala for reconsolidation after retrieval. *Nature* 2000; 406: 722-726.

[6] Myers KM, Davis M: Mechanisms of fear extinction. Mol Psychiat 2007; 12:120-150.

[7] Quirk GJ, Mueller D: Neural mechanisms of extinction learning and retrieval. *Neuropsychopharmacol* 2008; 33:56-72.

[8] Yehuda R, Le Doux J: Response variation following trauma: a translational neuroscience approach to understanding PTSD. *Neuron* 2007; 56:19-32.

[9] Izquierdo I, Wyrwicka W, Sierra G, Segundo JP. Établissement d' un reflex conditionné de trace pendant le sommeil natural chez le Chat. *Actual Neurophysiol* 1965; 6:277-296.

[10] Beckett WS: Post-traumatic stress disorder. *New Eng J Med* 2002; 346:130-132.

[11] Milekic MH, Alberini C: Temporally graded requirement for protein synthesis following memory reactivation. *Neuron* 2002; 36:521-525.

[12] Izquierdo LA, Barros DM, Medina JH, Izquierdo I: Novelty enhances retrieval of one-trial avoidance learning in rats 1 or 31 days after training unless the hippocampus is

inactivated by different receptor antagonists and enzyme inhibitors. *Behav Brain Res.* 2000 Dec 20;117(1-2):215-20.

[13] Viola H, Furman M, Izquierdo LA, Alonso M, Barros DM, de Souza MM, Izquierdo I, Medina JH: Phosphorylated cAMP response element-binding protein as a molecular marker of memory processing in rat hippocampus: effect of novelty. *J Neurosci* 2000; 20: RC112.

[14] Barros DM, Izquierdo LA, Mello e Souza T, Ardenghi P, Pereira P, Medina JH, Izquierdo I: Molecular signaling pathways in the cerebral cortex are required for retrieval of one-trial avoidance learning in rats. *Behav Brain Res* 2000; 183-192.

[15] Barros DM, Mello e Souza T, De David T, Choi H, Aguzzoli A, Madche C, Ardenghi P, Medina JH, Izquierdo I: Simultaneous modulation of retrieval by dopaminergic D(1), beta-noradrenergic, serotonergic-1A and cholinergic muscarinic receptors in cortical structures of the rat. *Behav Brain Res* 2001; 124:1-7.

[16] Izquierdo I, Quillfeldt JA, Zanatta M, Quevedo J, Schaeffer E, Schmitz PK, Medina JH: Sequential involvement of hippocampus and amygdala, entorhinal cortex and parietal cortex in the formation and expression of memory for inhibitory avoidance in rats. *Eur J Neurosci* 1997; 9: 786-793.

[17] Lee JL, Milton AL, Everitt BJ: Reconsolidation and extinction of conditioned fear: inhibition and potentiation. *J Neurosci* 2006; 26:10051-10056.

[18] Pérez-Cuesta LM, Hepp Y, Pedreira ME, Maldonado H: Memory is not extinguished along with CS presentation but within a few seconds after CS-offset. *Learn Mem* 2007; 14:101-108.

[19] Monfils MH, Cowansage KK, Klann E, LeDoux JE: Extinction-reconsolidation boundaries: key to persistent attenuation of fear memories. *Science* 2009; 324:951-955.

[20] Peters J, Kalivas PW, Quirk GJ: Extinction circuits for fear and addiction overlap in prefrontal cortex. *Learn Mem* 2009; 16: 279-288.

[21] Bouton ME: Context and behavioral processes in extinction. *Learn Mem* 2004; 11: 485-494.

[22] Ji J, Maren S: Hippocampal involvement in contextual modification of fear extinction. *Hippocampus* 2007; 17:749-758.

[23] Ji J, Maren S: Differential roles for hippocampal areas CA1 and CA3 in the contextual encoding and retrieval of extinguished fear. *Learn Mem* 2008; 15:244-251.

[24] Maroun M, Akirav I: Arousal and stress effects on consolidation and reconsolidation of recognition memory. *Neurophyschpharmacol* 2008; 33: 394-405.

[25] Rossato JI, Bevilaqua LR, Lima RH, Medina JH, Izquierdo I, Cammarota M: On the participation of Hippocampal p38 mitogen-activated protein kinase in extinction and reacquisition. *Neurosci* 2006; 143:15-23.

[26] Rossato JI, Bevilaqua LR, Myskiw JC, Medina JH, Izquierdo I, Cammarota M: On the role of hippocampal protein synthesis in the consolidation and reconsolidation of object recognition memory. *Learn Mem* 2007; 14:36-46.

[27] Duvarci S, Nader K, LeDoux JE: De novo mRNA synthesis is required for both consolidation and reconsolidation of fear memories in the amygdala. *Learn Mem* 2008; 15:747-755.

[28] Vianna MR, Szapiro G, McGaugh JL, Medina JH, Izquierdo I: Retrieval of memory for fear-motivated training initiates extinction requiring protein synthesis in the rat hippocampus. *Proc Natl Acad Sci USA* 2001; 98: 12251-12254.

[29] Vianna MR, Coitinho AS, Izquierdo I: Role of the hippocampus and amygdala in the extinction of fear-motivated learning. *Curr Neurovasc Res* 2004; 1:55-609.

[30] Lima RH, Rossato JI, Furini CR, Bevilaqua LR, Izquierdo I, Cammarota M: Infusion of protein synthesis inhibitors in the entorhinal cortex blocks consolidation but not reconsolidation of object recognition memory. *Neurobiol Learn Mem* 2009; 91:466-472.

[31] 31.Likhtik E, Popa D, Apergis-Schoute J, Fidacaro GA, Paré D: Amygdala intercalated neurons are required for expression of fear extinction. *Nature* 2008; 454:642-645.

[32] 32.Maroun M, Akirav I: Differential involvement of dopamine D1 receptor and MEK signaling pathway in the ventromedial prefrontal cortex in consolidation and reconsolidation of recognition memory. *Learn Mem* 2009; 16:243-247.

[33] 33 Rescorla RA: Retraining of extinguished Pavlovian stimuli. *J Exp Psychol Anim Behav Process* 2001; 27:115-124.

[34] 34 Rescorla RA: Spontaneous recovery. *Lean Mem* 2004; 11:501-509.

[35] 35 Briggs JF, Riccio DC: Retrograde amnesia for extinction: similarities with amnesia for original acquisition memories. *Learn Behav* 2007; 35:131-140.

[36] 36 Altman J, Brunner RL, Bayer SA: The hippocampus and behavioral maturation. *Behav Biol* 1973; 8:557-596.

[37] Kim JH, Richardson R: The effect of temporary amygdala inactivation and reextinction of fear in the developing rat: unlearning as a potential as a potential mechanism for extinction early in development. *J Neurosci* 2008; 28:1282-1290.

[38] Nair HP, Berndt JD, Barrett D, Gonzalez-Lima F: Maturation of extinction behavior in infant rats: large-scale regional interactions with medial prefrontal cortex, Orbitofrontal cortex, and anterior cingulate cortex. *J Neurosci* 2001; 21:4400-4407.

[39] Mamiya N, Fukushima H, Suzuki A, Matsuyama Z, Homma S, Frankland PW, Kida S: Brain region-specific gene expression required for reconsolidation and extinction of contextual fear memory. *J Neurosci* 2009; 29:402-413.

[40] Cammarota M, Bevilaqua LR, Kerr D, Medina JH, Izquierdo I: Inhibition of mRNA and protein synthesis in the CA1 region of the dorsal hippocampus blocks reinstatement of an extinguished conditioned response. *J Neurosci* 2003; 23:737-741.

[41] Hupbach A, Hardt O, Gomez R, Nadel L: The dynamics of memory: context-dependent updating. *Learn Mem* 2008; 15:574-579.

[42] Loftus EF, Palmer JC: Reconstruction of automobile destruction: An example of the interaction between language and memory. *J Verb Learn Verb Behav* 1974; 13:585-589.

[43] Kida S, Josselyn SA, de Ortiz SP, Kogan JH, Chevere I, Masushige S, Silva AJ: CREB required for the stability of new and reactivated gear memories. *Nat NNeurosci* 2002; 5:348-355.

[44] Szapiro G, Vianna MR, McGaugh JL, Medina JH, Izquierdo I: The role of NMDA glutamate receptors, PKA, MAPK and CAMKII in the hippocampus in the extinction of conditioned fear. *Hippocampus* 2003; 13: 53-58.

[45] Alberini C: Transcription factors in long-term memory and synaptic plasticity. *Physiol Rev* 2009; 89: 121-145.

[46] Guthrie RM, Bryant RA: Extinction learning before trauma and subsequent posttraumatic stress. *Psychosom Med* 2006; 68:307-311.

[47] Wessa M, Flor H: Failure of extinction of fear responses in posttraumatic stress disorder: evidence from second-order conditioning. *Am J Psychiat* 2007; 164:1684-1692.

[48] Milad MR, Orr SP, Lasko NB, Chang Y, Rauch SL, Pitman RK: Presence and acquired origin of reduced recall for fear extinction in PTSD: results of a twin study. *J Psychiat Res* 2008; 42:515-520.

[49] Jovanovic T, Norrholm SD, Fennell JE, Keyes M, Fiallos AM, Myers KM, Davis M, Duncan CJ: Posttraumatic stress disorder may be associated with fear inhibition: relation to symptom severity. *Psychiat Res* 2009; 167:151-160.

[50] Bush DE, Sotres-Bayon F, LeDoux JE: Individual differences in fear: isolating fear reactivity and fear recovery phenotypes. *J Trauma Stress* 2007; 20:413-422.

[51] Alvarez RP, Biggs A, Chen G, Pine DS, Grillon C: Contextual fear conditioning in humans: cortico-hippocampal and amygdala contributions. *J Neurosci* 2008: 28:6211-6219.

[52] Lattal KM, Stafford JM: What does it take to demonstrate memory erasure? Theoretical comment on Norrholm et al. (2008): *Behav Neurosci* 2008; 122:1186-1190.

[53] Bryant RA, Guthrie RM: Maladaptive appraisals as a risk factor for posttraumatic stress: a study of trainee firefighters. *Psychol Sci* 2005; 16:749-752.

[54] Gourley SL, Ketves AT, Olausson P, Taylor JR: A history of corticosterone exposure regulates fear extinction and cortical NR2B, GluR2/3, and BDNF. *Neuropsychopharmacol* 2009; 34: 707-716.

[55] Hinton DE, Hofmann SG, Pitman RK, Pollack MH, Barlow DH: The panic attack-posttraumatic stress disorder model: applicability to orthostatic panic among Cambodian refugees. *Cogn Behav Ther* 2008; 37:101-116.

[56] Michael T, Blechert J, Vriends N, Margraf J, Wilhelm FH: Fear conditioning in panic disorder: Enhanced resistance to extinction. *J Abnorm Psychol* 2007; 116:612-617.

[57] Holt DJ, Lebron-Milad K, Milad MR, Rauch SL, Pitman RK, Orr SP, Cassidy BS, Walsh JP, Goff DC: Extinction memory is impaired in schizophrenia. *Biol Psychiat* 2009; 65:455-463.

[58] Hermans D, Craske MG, Mineka S, Lovibond PF: Extinction in human fear conditioning. *Biol Psychiat* 2006; 60:361-368.

[59] Leung HT, Bailey GK, Laurent V, Westbrook RF: Rapid reaquisition of fear to a completely extinguished context is replaced by transient impairment with additional extinction training. *J Exp Psychol Anim Behav Process* 2007; 33:299-313.

[60] Schiller D, Levy I, Niv Y, LeDoux JE, Phelps EA: From fear to safety and back: reversal of fear in the human brain. *J Neurosci* 2008; 28:11517-11525.

[61] Delgado MR, Olsson A, Phelps EA: Extending animal models of fear conditioning to humans. *Biol Psychol* 2006; 73:39-48.

[62] Rauch SL, Shin LM, Phelps EA. Neurocircuitry models of posttraumatic stress disorder and extinction: human neuroimaging research – past, present and future. *Biol Psychiat* 2006; 60:376-382.

[63] Milad MR, Rauch SL, Pitman RK, Quirk GJ: Fear extinction in rats: Implications for human brain imaging and anxiety disorders. *Biol Psychol* 2006; 73:61-71;

[64] Norberg MM, Krystal JH, Tolin DF: A meta-analysis of D-cycloserine and the facilitation of fear extinction and exposure therapy. *Biol Psychiat* 2008; 63:1118-1126

[65] Maren S, Chang CH: Recent fear is resistant to extinction. *Proc Natl Acad Sci USA* 2006; 103:18020-18025.

[66] Woods AM, Bouton ME: Immediate extinction causes a less durable loss of performance than delayed extinction following either fear or appetitive conditioning. *Lean Mem* 2008; 15:909-920.

[67] Chang CH, Maren S: Early extinction after fear conditioning yields a context-independent and short-term suppression of conditional freezing in rats. *Learn Mem* 2009; 16:62-68.

[68] Myers KM, Ressler KJ, Davis M: Different mechanisms of fear extinction dependent on length of time since fear acquisition. *Learn Mem* 2006; 13:216-223.

[69] Izumi T, Inoue T, Kato A, Kitaichi Y, Nakagawa S, Koyama T: Changes in amygdala neural activity that occur with the extinction of context-dependent conditioned fear stress. *Pharmacol Biochem Behav* 2008; 90:297-304.

[70] Cammarota M, Bevilaqua LR, Rossato JI, Ramirez M, Medina JH, Izquierdo I: Relationship between short- and long-term memory and short- and long-term extinction. *Neurobiol Learn Mem* 2005; 84:25-32.

[71] Norrholm SD, Vervliet B, Jovanovic T, Boshoven W, Myers KM, Davis M, Rothbaum B, Duncan EJ: Timing of extinction relative to acquisition: a parametric analysis of fear extinction in humans. *Behav Neurosci* 2008; 122:1016-1030.

[72] Rescorla RA: Deepened extinction from compound stimulus presentation. *J Exp Psychol Anim Behav Process* 2006; 32:135-144.

[73] Vervliet B, Vansteenwetgen D, Hermans D, Eelen P: Concurrent excitors limit the extinction of conditioned fear in humans. *Behav Res Ther* 2007; 45:375-383.

[74] Bouton ME, Westbrook RF, Corcoran KA, Maren S: Contextual and temporal modulation of extinction: behavioral and biological mechanisms. *Biol Psychiat* 2006: 60:352-360.

[75] Cammarota M, Barros DM, Vianna MR, Bevilaqua LR, Coitinho A, Szapiro G, Izquierdo LA, Medina JH, Izquierdo I: The transition from memory retrieval to extinction. *An Acad Bras Cienc* 2004; 76:573-582.

[76] Eldridge LL, Engel SA, Zeineh MM, Bookheimer SY, Knowlton BJ: A dissociation of encoding and retrieval processes ijn the human hippocampus. *J Neurosci* 2005; 25:3280-3286.

[77] McGaugh JL: The amygdala modulates the consolidation of memories of emotionally arousing experiences. *Annu Rev Neurosci* 2004; 27:1-28.

[78] De Quervain DJ, Henke K, Aerni A, Terrier V, McGaugh JL, Berthold T, Nitsch RM, Back A. Roozendaal B, Hock C: Glucocorticoid-induced impairment of declarative memory retrieval is associated with reduced blood flor in the medial temporal lobe. Eur J Neurosci 2003; 17:1296-1302.

[79] Szapiro G, Barros DM, Ardenghi P, Vianna MR, Choi H, Silva T, Medina JH, Izquierdo I: Facilitation and inhibition of retrieval in two aversive tasks in rats by intrahippocampal infusion of agonists of specific glutamate metabotropic receptor subtypes, *Psychopharmacology* 2001; 156:297-401.

[80] Izquierdo I, Bevilaqua LRM, Rossato JI, Bonin JS, Medina JH, Cammarota M. Different molecular cascades in different sites of the brain control consolidation. *Trends Neurosci* 2006; 29:496-505.

[81] Pitman RK, Sanders KM, Zusman RM, Healy AR, Cheema F, Lasko NB, Cahill L, Orr SP: Pilot study of secondary prevention of posttraumatic stress disorder with propranolol. *Biol Psychiat* 2002; 51:189-192.
[82] Pitman RK, Delahanty DL: Conceptually driven pharmacologic approaches to acute trauma. *CNS Spectr* 2005; 10:99-106.
[83] Gruber RP, Roberts C, Schooler W, Pitman RK: Preventing postsurgical dissatisfaction syndrome after rhinoplasty with propranolol: a pilot study. *Plast reconstr Surg* 2009; 123:1072-1078.
[84] Alkire MT, Gruver R, Miller J, McReynolds JR, Hahm EL, Cahill L: Neuroimagining analysis of an anesthetic gas that blocks human emotional memory. *Proc Natl Acad Sci USA* 2008; 105:1722-1727.
[85] Matar MA, Zohar J, Kaplan Z, Cohen H: Alprazolam treatment immediately after stress exposure interferes with the normal HPA-stress response and increases vulnerability to subsequent stress in an animal model of PTSD. *Eur Neuropsychopharmacol* 2009; 19:283-295.
[86] Brunet A, Orr SP, Tremblay J, Robertson K, Nader K, Pitman RK: Effect of post-retrieval propranolol on psychophysiologic responding during subsequent script-friven traumatic imagery in post-traumatic stress disorder. *J Psychiat Res* 42:503-506.
[87] Kindt M, Soeter M, Vervliet B: Beyond extinction: erasing human fear responses sand preventing the return of fear. *Nat Neurosci* 2009; 12:256-258.
[88] Debiec J, LeDoux JE: Disruption of reconsolidation but not consolidation of auditory fear conditioning by noradrenergic blockade in the amygdala. *Neurosci* 2004; 129:267-272.
[89] Debiec J, LeDoux JE: Noradrenergic signalling in the amygdala contributes to the reconsolidation of fear memory: treatment implications for PTSD. *Ann NY Acad Sci* 2006; 1071:521-524.
[90] Milton AL, Lee JL, Everitt BJ: Reconsolidation of appetitive memories for both natural and drug reinforcement is dependent on (beta)-adrenergic receptors. *Learn Mem* 2008; 15:88-92.
[91] Fricks-Gleason AN, Marshall JF: Post-retrieval beta-adrenergic receptor blockade: effects on extinction and reconsolidation of cocaine-cue memories. *Learn Mem* 2008; 15:643-648.
[92] Tronson NC, Taylor JR: Molecular mechanisms of memory reconsolidation. *Nat Rev Neurosci* 2007; 8:262-275.
[93] Davis M, Ressler K, Rothbaum BO, Richardson E: Effects of D-cycloserine on extinction: translation from preclinical to clinical work. *Biol Psychiat* 2006; 60:369-375.
[94] MorganMA, LeDoux JE: Differential contribution of dorsal and ventral medial prefrontal cortex to the acquisition and extinction of conditioned fear in rats. *Behav Neurosci* 1995; 109: 681–688.
[95] Santini E, Ge H, Ren K, Peña de Ortiz S, Quirk GJ: Consolidation of fear extinction requires protein synthesis in the medial prefrontal cortex. *J Neurosci* 2004; 24:5704-5710.
[96] Hugues S, Deschaux O, Garcia R: Postextinction infusion of a mitogen-activated protein kinase inhibitor into the medial prefrontal cortex impairs memory of the extinction of conditioned fear. *Learn Mem* 2004; 11:540-543.

[97] Burgos-Robles A, Vidal-Gozález I, Santini E, Quirk GJ: Consolidation of fear extinction requires NMDA receptor-dependent bursting in the ventromedial prefrontal cortex. *Neuron* 2007; 53:871-880.

[98] Sotres-Bayon, F., Diaz-Mataix, L., Bush, D.E., LeDoux, J.E. Dissociable roles for the ventromedial prefrontal cortex and amygdala in fear extinction: NR2B contribution. *Cereb. Cortex* 2008; 19: 474–482.

[99] Milad MR, Quirk GJ: Neurons in medial prefrontal cortex signal memory for fear extinction. *Nature* 2002; 420:7-74.

[100] Santini E, Quirk GJ, Porter JT: Fear conditioning and extinction modify the intrinsic excitability of infralimbic neurons. *J Neurosci 2008*; 28: 4028-4036.

[101] Rauch SL, Milad MR, Orr SP, Quinn BT, Fischl B, Pitman RK: Orbitofrontal thickness, retention of fear extinction and extraversion. *Neuroreport* 2005; 16:1909-1912.

[102] Ryu J, Futai K, Feliu M, Weinberg E, Sheng M: Constitutively active Rap2 transgenic mice display fewer dendritic spines, reduced extracellular signal-regulated kinase signaling, enhanced long-term depression, and impaired spatial learning and fear extinction. *J Neurosci* 2008; 28:8178-8188.

[103] Squire LR: The legacy of patient H.M. to neuroscience. *Neuron* 2009; 61:6-9.

[104] Lindauer RJ, Olff M, van Meijel EP, Carlier IV, Gersons BP: Cortisol, learning, memory, and attention in relation to smaller hippocampal volume in police officers with posttraumatic stress disorder. *Biol Psychiat* 2006; 59:171-177.

[105] Lindauer RJ, Vlieger EJ, Jalink M, Olff M, Carlier IV, Majoie CB, den Heeten GJ, Gersons BP: Smaller hippocampal volume in Dutch police officers with posttraumatic stress disorder. *Biol Psychiat* 2004; 56:356-363.

[106] Shin LM, Shin PS, Heckers S, Krangel TS, Macklin ML, Orr SP, Lasko N, Segal E, Makris N, Richert K, Levering J, Schacter DL, Alpert NM, Fischman AJ, Pitman RK, Raugh SL: Hippocampal function in posttraumatic stress disorder. *Hippocampus* 2004; 14:292-300.

[107] Shin LM, Orr SP, Carson MA, Rauch SL, Macklin ML, Lasko NB, Peters PM, Metzger LJ, Dougherty DD, Cannistraro PA, Alpert NM, Fischman AJ, Pitman RK: Regional cerebral blood flow in the amygdala and medial prefrontal cortex during traumatic imagery in male and female Vietnam veterans with PTSD. *Arch Gen Psychiat* 2004; 61:168-176.

[108] Fischer A, Radulovacki M, Schrick C, Sananbenesi F, Godovac-Zimmermann J, Radulovacki J: Hippocampal Mek/Erk signaling mediates extinction of contextual freezing behavior. *Neurobiol Learn Mem* 2007; 87:149-158.

[109] Corcoran KA, Desmond TJ, Frey KA, Maren S . Hippocampal inactivation disrupts the acquisition and contextual encoding of fear extinction. *J Neurosci* 2005; 25:8978-8987.

[110] Jensen RA, Riccio D: Effects of prior experience upon retrograde amnesia produced by hypothermia. *Physiol Behav* 1970; 5:1291-1294.

[111] Tronson NC, Schrick C, Guzman YF, Huh KH, Srivastava DP, Penzes P, Guedea AL, Gao C, Radulovic J: Segregated populations of hippocampal principal CA1 neurons mediating conditioning and extinction of contextual fear. *J Neurosci* 2009; 29:3387-3394.

[112] Berman DE, Dudai Y: Memory extinction, learning anew, and learning the new: dissociations in the molecular machinery of learning in cortex. *Science 2001*:291:2417-2419.

[113] Bahar A, Dorfman N, Dudai Y: Amygdalar circuits required for either consolidation or extinction of taste aversion memory are not requird for reconsoilidation. *Eur J Neurosci* 2004; 19:1115-1118.

[114] Huh KH, Guzman YF, Tronson NC, Guedea AL, Goao C, Radulovic J: Hippocampal Erk mechanisms linking prediction error to fear extinction: roles of shock expectancy and contextual aversive valence. *Learn Mem* 2009; 16:273-278.

[115] Lubin FD, Sweatt JD:. The IkappaB kinase regulates chromatin structure during reconsolidation of conditioned fear memories. *Neuron* 2007; 55:942-957.

[116] Boccia M, Freudenthal R, Blake M, de la Fuente V, Acosta G, Baratti C, Romano A: Activation of hippocampal nuclear factor-kappa B by retrieval is required for memory reconsolidation. *J Neurosci.* 2007; 27:13436-13445.

[117] Merlo E, Romano A: Memory extinction entails the inhibition of the transcription factor NF-kappaB. *PloS ONE* 2008; 3: e3687.

[118] Nijholt IM, Ostroveanu A, Scheper WA, Penke B, Luiten PG, Van der Zee EA, Eisel UL. Inhibition of PKA anchoring to A-kinase anchoring proteins impairs consolidation and facilitates extinction of contextual fear memories. *Neurobiol Learn Mem.* 2008 Jul;90(1):223-229.

[119] Delgado MR, Nearing KT, LeDoux JE, Phelps EA: Neural circuitry underlying the regulation of conditioned fear and its relation to extinction. *Neuron* 2008; 59:829-838.

[120] Falls WA, Miserendino MJ, Davis M: Extinction of fear-potentiated startle: blockade by infusion of an NMDA antagonist into the amygdala. *J Neurosci.* 1992; 12: 854-863.

[121] Quirk GJ, Armony JL, LeDoux JE: Fear conditioning enhances different temporal components of tone-evoked spike trains in auditory cortex and lateral amygdala. *Neuron.* 1997; 19: 613-624.

[122] Maren S, Quirk GJ: Neuronal signaling of fear memory. *Nat Rev Neurosci* 2004; 5:844-852.

[123] Li G, Nair SS, Quirk GJ: A biologically realistic network model of acquisition and extinction of conditioned fear associations in lateral amygdala neurons. *J Neurophysiol* 2009; 101:1629-1646.

[124] Wilson WM, Murphy M: A discrete population of neurons in the lateral amygdala is specifically activated by contextual fear conditioning. *Learn Mem* 2009; 16:357-361.

[125] Yang YL, Lu KT. Facilitation of conditioned fear extinction by d-cycloserine is mediated by mitogen-activated protein kinase and phosphatidylinositol 3-kinase cascades and requires de novo protein synthesis in basolateral nucleus of amygdala. *Neuroscience* 2005;134(1):247-260.

[126] Mao SC, Lin HC, Gean PW: Augmentation of fear extinction by infusion of glycine transporter blocker into the amygdala. *Mol Pharmacology* 2009; EPub ahead f Print May 1.

[127] Herry C, Trifilieff P, Micheau J, Lüthi A, Mons N: Extinction of auditory fear conditioning requires MAPK/ERK activation in the basolateral amygdala. Eur *J Neurosci;* 2006; 24: 261-269.

[128] Lu KT, Walker DL, Davis M: Mitogen-activated protein kinase cascade in the basolateral nucleus of amygdala is involved in extinction of fear-potentiated startle. *J Neurosci* 2001; 21:RC162.
[129] Lin CH, Yeh SH, Leu TH, Chang WC, Wang ST, Gean PW: Identification of Calcineurin as a key signal in the extinction of fear memory. *J Neurosci* 2003a; 23:1574-1579.
[130] Lin CH, Yeh SH, Lu HY, Gean PW: The similarities and diversities of signal pathways leading to consolidation of conditioning and consolidation of extinction of fear memory. *J Neurosci* 2003; 23:8310-8317.
[131] Szapiro G, Izquierdo LA, Alonso M, Barros D, Paratcha G, Ardenghi P, Pereira P, Medina JH, Izquierdo I: Molecular mechanisms of memory retrieval. *Neurochem Res* 2002; 27:1491-1498.
[132] Sotres-Bayon F, Bush DE, LeDoux JE: Acquisition of fear extinction requires activation of NR2B-containing NMDA receptors in lateral amygdala. *Neuropsychopharmacology* 2007; 32:1929-1940.
[133] Bevilaqua LR, Bonini, JS, Rossatto JI, Izquierdo LA, Cammarota M, Izquierdo I: The entorhinal cortex plays a role in extinction. *Neurobiol Learn Mem* 2006; 86:192-197.
[134] Chhatwal JP, Stanek-Rattiner L, Davis nM, Ressler KJ: Amygdala BDNF signaling is required for consolidation but not encoding of extinction. *Nat Neurosci* 2006; 9:870-872.
[135] Kim J, Lee S, Park K, Hong I, Song B, Son G, Park H, Kim WR, Park E, Choe HK, Kim H, Lee C, Sun W, Kim K, Shin KS, Choi S: Amygdala depotentiation and fear extinction. *Proc Natl Acad Sci USA* 2007; 104:20955-20960.
[136] Goossens L, Sunaert S, Peeters R, Griez EJ, Schruers KR: Amygdala hyperfunction in phobic fear normalizes after exposure. *Biol Psychiat* 2007; 62:1119-1125.
[137] Boccia MM, Blake MG, Baratti CM, McGaugh JL: Involvement of the basolateral amygdala in muscarinic cholinergic modulation of extinction memory consolidation. *Neurobiol Learn Mem* 2009; 91:93-97.
[138] Barad M, Gean PW, Lutz B: The role of the amygdala in the extinction of conditioned fear. *Biol Ppsychiat* 2006; 60:3222-328.
[139] Ji J, Maren S: Lesions of the entorhinal cortex or fornix disrupt the context-dependence of fear extinction in rats. *Behav Brain Res.* 2008 Dec 12;194(2):201-6
[140] Malin EL, Ibrahim DY, Tu JW, McGaugh JL: Involvement of the rostral anterior cingulate cortex in consolidation of inhibitory avoidance memory: interaction with the basolateral amygdala. *Neurobiol Learn Mem* 2007; 87:295-302.
[141] Milad MR, Quirk GJ, Pitman RK, Orr SP, FISCHL B, Rauch SL: A role for the human anterior cingulate cortex in fear expression. *Biol Psychiat* 2007; 62:1191-1194.
[142] Kasai K, Yamasue H, Gilbertson MW, Shenton ME, Rauch SL, Pitman RK: Evidence for acquired pregenual anterior cingulate gray matter loss from a twin study of combat-related posttraumatic stress disorder. *Biol Psychiat* 2008; 63:550-556.
[143] Bryant RA, Felmingham K, Whitford TJ, Kemp A, Hughes G, Peduto A, Williams LM: Rostral anterior cingulate volume predicts treatment response to cognitive-behavioural therapy for posttraumatic stress dirorder. *J Psychiatry Neurosci* 2008; 33:142-146.

[144] Schuff N, Neylan TC, Fox-Bosetti S, Lenoci N, Samuelson KW, Studholme C, Kornak J, Marmar CR, Weiner MW. Abnormal N-acetylaspartate in hippocampus and anterior cingulate in posttraumatic stress disorder. *Psychiat Res* 2008; 162:147-157.

[145] Phelps EA, Delgado MR, Nearing KI, LeDoux JE: Extinction learning in humans: role of the amygdala and vmPFC. *Neuron* 2004; 43:897-905.et al., 2008;

[146] Kalisch R, Korenfeld E, Stephan KE, Weiskopf N, Seymour B, Dolan RJ: Context-dependent human extinction memory is mediated by a ventromedial prefrontal and hippocampal network. *J Neurosci* 2006; 26:9503-9511.

[147] Corcoran KA, Quirk GJ: Recalling safety: cooperative functions of the ventromedial prefrontal cortex and the hippocampus in extinction. *CNS Spectr* 2007; 12:200-206.

[148] Milad MR, Quirk GJ, Pitman RK, Orr SP, Fischl B,Rauch SL: Recall of fear extinction in humans activates the ventromedial prefrontal cortex and hippocampus in concert. *Biol Psychiat* 2007; 62:446-454.

[149] Lang S, Kroll A, Lipinski SJ, Wessa M, Ridder S, Christmann C, Scchad LR, Flor H: Context conditioning and extinction in humans: differential contribution of the hippocampus, amygdala and prefrontal cortex. *Eur J Neurosci* 2009. 29:823-832.

[150] Marschner A, Kalisch R, Vervliet B, Vansteenwegen D, Büchel C: Dissociable roles for hippocampus and the amygdala in human cued versus context fear conditioning. *J Neurosci* 2008; 28:9030-9036.

[151] Laurent V, Westbrook RF: Distinct contributions of the basolateral amygdala and the medial prefrontal cortex to learning and relearning extinction of context conditioned fear. *Learn Mem* 2008; 15: 657-666.

[152] Hikind N, Maroun M: Microinfusion of the D1 receptor antagonist, SCH23390 into the IL but not the BLA impairs consolidation of extinction of auditory fear conditioning. *Neurobiol Learn Mem* 2008; 90:217-222.

[153] Herry C, Mons N: Resistance to extinction is associated with impaired immediate early gene induction in medial prefrontal cortex and amygdala. *Eur JNeurosci* 2004; 20:781-790.

[154] Izquierdo LA, Barros DM, da Costa JC, Furini C, Zinn C, Cammarota M, Bevilaqua LR, Izquierdo |I : A link between role of two prefrontal areas in immediate memory and in long-term memory consolidation. *Neurobiol Learn Mem* 2007; 88:160-188.

[155] Shin LM, Rauch SL, Pitman RK: Amygdala, medial prefrontal cortex, and hippocampal function in PTSD. *Ann N Y Acad Sci* 2006; 1071:67-79.

[156] Likhtik E, Pelletier JG, Paz R, Paré D: Prefrontal control of the amygdala. *J Neurosci* 2005; 25:7429-7437.

[157] Likhtik E, Pelletier JG, Popescu AT, Paré D: Identification of basolateral amygdala projection cells and interneurons using extracellular recordings. *J Neurophysiol* 2006; 96:3257-3265.

[158] Hefner K, Whittle N, Juhasz J, Norcross M, Karlsson RM, Saksida LM, Bussey TJ, Singewald N, Holmes A: Impaired fear extinction learning and cortico-amygdala circuit abnormalities in a common genetic mouse strain. *J Neurosci* 2008; 28: 8074-8085.

[159] Rutten K, van Donkelaar EL, Ferrington L, Blokland A, Bollen E, Steinbusch HWM, Kelly PAT, Prickaerts JHHJ: Phosphodiesterase inhibitors enhance object memory independent of cerebral blood flow and glucose utilization in rats. *Neuropharmacology* 2009; 34:1914-1925.

[160] Phan KL, Britton JC, Taylor SF, Fig LM, Liberzon I: Corticolimbic blood flow during nontraumatic emotional processing in posttraumatic stress disorder. *Arch Gen Psychiat* 2006; 63:184-192.

[161] Valenti O, Grace AA: Entorhinal cortex inhibits medial prefrontal cortex and modulates the activity states of electrophysiologically characterized pyramidal neurons in vivo. *Cereb Cortex* 2009; 19:658-674.

[162] Bissière S, Plachta N, Hoyer D, McAllister KH, Olpe HR, Grace AA, Cryan JF: The rostral anterior cingulate cortex modulates the efficiency of amygdala-dependent fear learning. *Biol Psychiat* 2008; 63:821-831.

[163] McGinty VB, Grace AA: Activity-dependent depression of medial prefrontal cortex inputs to accumbens neurons by the basolateral amygdala. *Neurosci* 2009; EPub ahead of print.

[164] Belujon P, Grace AA: Critical role of the prefrontal cortex in the regulation of hippocampus-accumbens information flow. *J Neurosci* 2008; 28:9797-9805.

[165] Izquierdo I, Bevilaqua LR, Rossato JI, Lima RH, Medina JH, Cammarota M: Age-dependent and age-independent human memory persistence is enhanced by delayed posttraining methylphenidate administration. *Proc Natl Acad Sci USA* 2008; 105:19504-19507.

[166] Goto Y, Grace AA: Dopamine modulation of hippocampal-prefrontal cortical interaction drives memory-guided behavior. *Cereb Cortex* 2008; 18:1407-1414.

In: Neurobiology of Post-Traumatic Stress Disorder
Editors: L. Sher and A. Vilens, pp. 331-344
ISBN: 978-1-61470-593-2

Chapter XX

A Preliminary Look at the Potential Role of the Endocannabinoid System in the Management of PTSD

George A. Fraser
Consultant psychiatrist, Canadian Forces Health Services Centre (Ottawa), Mental Health Department.

Abstract

The newly discovered endocannabinoid system (ECS) and the ongoing research into its role in the central nervous system (CNS) has important ramifications in the understanding and treatment of posttraumatic stress disorder (PTSD). While the connection with this system and PTSD is in its early stages, and only one small clinical study involving patients with PTSD being treated with an endocannabinoid agonist has been published to date, [1] nonetheless emerging investigations into the endocannabinoid system are promising and warrant a dialogue and speculation about the potential benefits of endocannabinoid ligands in the treatment of PTSD. This chapter will consider three areas related to the endocannabinoid system and how it may be connected to the field of posttraumatic stress disorder. First will be an overview of the recent discovery of endocannabinoid system. Next, research studies pointing to potentially interesting relationships between cannabinoids, endocannabinoids and PTSD will be discussed. Finally, clinical issues and questions arising out of actual field experience using nabilone, a synthetic cannabinoid, will be presented. This latter area will be of interest to clinicians who may consider using a synthetic cannabinoid for the first time with a PTSD patient. It will be based on the above mentioned clinical trial which used a synthetic cannabinoid in managing treatment resistant PTSD nightmares which was carried out by this author.

Introduction

The Endocannabinoid System

The recent discovery of the endocannabinoid system (ECS) came about with research into the site of action on the most famous of all cannabinoids, *Cannabis sativa*, best known as marijuana. The many effects of this plant have been known for literally thousands of years. These include euphoria, anxiolysis and even hallucinogenic effects. It was only in 1964 that the major psychoactive constituent of cannabis, delta-9-tetrahydrocannabinol (THC) was discovered and later synthesized. [2]

The next major step came in 1988 when a radiolabelled synthetic cannabinoid was found to bind to brain membrane. [3] Within a few years two major cannabinoid receptors had been identified – CB1 and CB2. [3,4]

CB1 receptors are primarily located in the brain, found in areas such as the hippocampus, amygdala, basal ganglia, hypothalamus, cerebellum, areas of the cerebral cortex and the nucleus accumbens. CB2 receptors are primarily located in the reproductive and immune systems. [5]

As it seems unlikely that nature provided endogenous cannabinoid receptors solely as convenient central nervous system (CNS) landing sites for marijuana, the search for natural cannabinoid ligands went ahead. By 1996, two major endocannabinoids had been identified. These two ligands are *N*-arachidonoylethanolamine [AEA] and 2-arachidonoylglycerol [2-AG]). Since then, others have been discovered. It appears that AEA and 2-AG are synthesized and released on demand. [6] Interestingly, they are released by the postsynaptic neuron. This phenomenon has been noted by Dr. Stephen Stahl who writes, "not all neurotransmission is classic or anterograde or from top to bottom - namely, presynaptic to postsynaptic. Postsynaptic neurons may also communicate with presynaptic neurons from the bottom to the top via retrograde neurotransmission." [7] The CB1 receptors are on the presynaptic neurons and respond to retrograde diffusion of the endocannabinoids from adjacent postsynaptic neurons.

In overview, it is now know that there are at least two cannabinoid receptors in the brain. From the standpoint of PTSD, the CB1 receptors that are primarily present in the CNS should be the focus of future study. The CB receptors and endogenous ligands form what has been called the endocannabinoid system. Any ligand which interacts with the endogenous cannabinoid receptors, whether endogenous (i.e. AEA and 2-AG) or exogenous (i.e. cannabis, nabilone), is classified as a cannabinoid.

Some writers have lightly referred to the endocannabinoids as "the brain's own marijuana." While such a comment can be understood, the danger is that such a comment may incorrectly associate all cannabinoids with marijuana and the medical risks associated with smoked marijuana. When discussing the use of the synthetic cannabinoid, nabilone, in the clinical trial, it was very important to clarify for patients that nabilone was not marijuana. [1] As well, those patients were told that nabilone did not give the euphoric effects, the psychomimetic effects of marijuana, nor did it have any known 'street value'.

Before looking at the possible CNS interactions which may explain the mechanisms of exogenous cannabinoids, and more specifically the synthetic cannabinoid analogue, nabilone, it is important to point out to the reader why a trial of nabilone for PTSD nightmares was conceived. This author had been somewhat frustrated for a number of years because of treatment resistant nightmares experienced by some patients being treated with conventional medications used in treatment for PTSD, despite good response to many other symptoms. By chance, a publication of a single case study involving PTSD and nabilone was noted by this writer. In that study nabilone was successfully used by a psychiatrist to replace a patient's self prescribed smoked marijuana being used for relief of PTSD symptoms. [8] In that report, an incidental note mentioned that nightmares had been reduced in that patient while on nabilone. Consequently the author of this current chapter decided to try nabilone with three patients who were struggling with distressing persistent nightmares related to their PTSD. These nightmares had not responded to various trials of medications. The responses and comments made by these three patients resulting from the addition of nabilone were so positive that it was decided to use nabilone in a more formal trial with future patients presenting with PTSD nightmares which similarly had not responded to currently available hypnotics, antidepressants and atypical antipsychotics being used to treat their symptoms.

There had been reports in the literature of the successful use of an antihypertensive, prazosin in managing PTSD nightmares. [9, 10] Prazosin had earlier been unsuccessfully tried in some of the cases in the current author's study, but in retrospect, not in adequate doses. Nothing seemed to work in stopping the nightmares at that time.

The clinical trial using nabilone for PTSD nightmares took place over a two year period and was subsequently published in 2009. [1] Those who participated had reached a therapeutic impasse with various trials of previous medications and were so distressed by their persistent nightmares that they were quite happy to volunteer for the trial, The results will be discussed in more detail later but will be introduced here by repeating the abstract of that publication. [1]

"This is the report of an open label clinical trial to evaluate the effects of nabilone, an endocannabinoid receptor agonist, on treatment-resistant nightmares in patients diagnosed with posttraumatic stress disorder (PTSD). Charts of 47 patients diagnosed with PTSD and having continuing nightmares in spite of conventional antidepressants and hypnotics were reviewed after adjunctive treatment with nabilone was initiated. These patients had been referred to a psychiatric specialist outpatient clinic between 2004 and 2006. The majority (72%) of patients receiving nabilone experienced either cessation of nightmares or a significant reduction in nightmare intensity. Subjective improvement in sleep time, the quality of sleep, and the reduction of daytime flashbacks and night sweats were also noted by some patients. The results of this study indicate the potential benefits of nabilone, a synthetic cannabinoid, in patients with PTSD experiencing poor control of nightmares with standard pharmacotherapy. This is the first report of the use of nabilone (Cesamet®, Valeant Canada Ltd., Montreal, Canada) for the management of treatment resistant nightmares in PTSD."

Cannabinoids and their Role in Fear, Memory and Sleep

At this early stage in the literature on the ECS and the relationship with PTSD, very little clinical information is available. However, there are a number of studies looking at the effects of cannabinoids in animals. As well, experience with Δ-9 THC and the CNS offers some interesting observations on the potential effects of cannabinoids in anxiety and stress. Consequently, it is possible to review some early research findings which can lead us to hypothesize why nabilone stopped the PTSD nightmares. As well, speculation can be ventured which may support the belief that the study of the ECS may turn out to be a rewarding area in the understanding and future therapy of PTSD.

PTSD and the Brain

The central features of PTSD suggest dysfunction of several brain structures involved in memory, fear, arousal, and the stress response. Brain imaging studies of PTSD patients have shown amygdala hyper-responsivity [11-16], blunted response in the medial prefrontal cortex [17, 18], altered perfusion in limbic and frontal regions, and loss of volume in the anterior cingulate. [19, 20] Also noted has been a decrease in hippocampal volume in those with PTSD. [21] In a study of women with memories of childhood sexual abuse, Bremner and colleagues observed increased amygdala activation during fear conditioning and attenuated activation of the medial prefrontal cortex during extinction compared non-PTSD subjects. [22]

The areas of the brain mentioned above are rich in CB1 receptors, so it is reasonable to assume that the ECS has a role to play in many of these functions. The anecdotal reports of widespread use of cannabis among war veterans with PTSD also suggests a therapeutic effect, as many report fewer memory flashbacks, better sleep, and relief from anxiety when they used marijuana.

Effects of Cannabinoids on Memory and Anxiety

Pharmacological efforts at treating PTSD have had limited success though new approaches continue. In 1994, investigators at the University of California discovered that propranolol impaired memory for a disturbing story, but not a neutral one, suggesting that enhanced memory associated with emotional experiences involves the beta-adrenergic system. [23] Reinforcing these findings was a study by Pitman and colleagues, who discovered that propranolol prevented PTSD in patients if given with a few hours after a traumatic event. [24] Being a beta adrenergic antagonist, it appears that propranolol interferes with adrenaline which is involved in consolidation of fearful memories within the amygdala in the immediate aftermath of traumatic events.

In recent years, the CB1 receptor has been found to have a central role in the extinction of aversive memories. In work performed at the Max Planck Institute of Psychiatry, CB1-deficient knockout mice subjected to auditory fear-conditioning tests demonstrated normal memory acquisition and consolidation, but were unable to extinguish the memory. [25] Similar results were seen in wild-type mice who were treated with a CB1 antagonist, demonstrating that CB1 is required for memory extinction. During the extinction protocol, levels of the endocannabinoids anandamide and 2-arachidonylglycerol were raised in the basolateral amygdala complex of both normal and knockout mice, a region known to control extinction of aversive memories. Moreover, the investigators found that upon low-frequency stimulation of amydala neurons, the release of GABA (gamma-aminobutyric acid) was completely absent in CB1-deficient mice and was blocked in normal mice given the CB1 antagonist. These findings suggest that endocannabinoids reduce GABA release in the basolateral amygdala, aiding the extinguishing of the fear-conditioned response.

Hippocampal neurogenesis may be yet another mechanism by which fear and anxiety is reduced through the ECS. Jiang and colleagues discovered that cannabinoids and endocannabinoids promote embryonic and adult hippocampus neurogenesis by acting on CB1 receptors, producing anxiolytic and antidepressant-like effects. [26] Interestingly, cannabis can promote adult hippocampal neurogeneration following chronic administration. Opiates, alcohol, nicotine, and cocaine all suppress hippocampal neurogenesis. [27-30]

Azad and colleagues discovered that inhibitory GABAergic transmission induces retrograde endocannabinoid signaling via activation of the adenylyl cyclase–protein kinase A pathway and release of anandamide. [31] Their findings indicate that anandamide decreases the activity of inhibitory interneurons in the amygdala. Reflecting on the findings of their earlier study [25], the authors suggested that inhibitory CB1-dependent GABAergic transmission might enhance the excitability of specific neuronal circuits which are instrumental in the formation of new memory traces. In other words, the creation of new memory traces may be a mechanism by which fear-conditioned memory is extinguished. [32]

Subsequent work by a Brazilian team further clarified the role of cannabinoids in memory extinction. [33] Rats subjected to a fear conditioning and extinction test were treated with a cannabinoid agonist (WIN 55,212) or antagonist (SR 147778) 24 hours or 30 days after conditioning. The agonist facilitated the extinction of contextual fear memory at both time points, whereas the antagonist prevented it.

Cannabinoids and Sleep

Sleep disturbances are core features of PTSD, namely insomnia and nightmares. As well, nocturnal anxiety attacks, leg movements, and sleep apnea, may also be present. [34, 35] Several pharmacological approaches have been used to treat PTSD-related sleep disorders; the most promising to date are prazosin [36] and olanzapine. [37] However, cannabis and TCH are known to facilitate initiation of sleep, increase stage 4 sleep, and reduce REM sleep [38], so it is logical to assume that cannabinoids may play a role in improving sleep architecture in patients with PTSD. A Mexican group reported that the endocannabinoid anandamide increased adenosine in the basal forebrain and increased sleep. [39] The soporific effects were found to be mediated by the CB1 receptor.

Hypotheses

The investigation of PTSD and cannabinoids is still relatively recent, yet at this stage studies present a diverse set of observations that give clues to explain how cannabinoids, and nabilone in particular, might reduce nightmares.

To summarize our knowledge to date, high levels of CB1 expression are found in the amygdala, hippocampus, anterior cingulate, and prefrontal cortex; all areas involved in emotion and emotional regulation. Endocannabinoids may be generated in the amygdala during anxiety and might regulate emotional states by influencing amygdala outputs. The ECS may be part of a negative feedback system that limits anxiety. [40]

Chronic stress produced by PTSD may exhaust the ECS, leading to disruptions in mood control and sleep. Nightmares might be a combination of disrupted sleep structure combined with anxiety breakthrough caused by ECS exhaustion. Nabilone may stabilize CB1 receptors in the exhausted ECS, thus normalizing the ECS to a certain degree.

At this point, we are left with more questions than answers. Does nabilone enhance the inhibition of fearful memory through its action on the limbic forebrain? Does it have an effect on the extinction of fear in memories of trauma, thus preventing their breakthrough in dreams? Is the effect of nabilone on nightmares due to a general anxiolytic effect, or does nabilone have an overall effect on sleep architecture, improving it perhaps so that disturbing emotions are not triggered during sleep? If so, how might they do that? Only further research can answer these questions. Patients in the trial often commented, "the nabilone stops the nightmares but not the normal dreams." That has led this author to speculate that nightmares have an origin in the amygdala thus are suppressed by nabilone, whereas it does not effect normal dreaming which originates elsewhere in the CNS.

Clinical Issues Related to Using a Synthetic Cannabinoid for the Treatment of Treatment Resistant PTSD Nightmares

Using a medication (nabilone) for the first time for a condition which was 'off label' (PTSD nightmares) and for which there was no anecdotal clinical experience to draw upon presented a challenging situation. The distress caused by terrifying nightmares as described by the patients, and their associated fear of going to sleep, was enough to overcome the initial hesitancy to prescribe a medication which possibly might help but had not been tried in nightmare management before. The idea of adding a trial with nabilone was initially discussed with three patients being treated for PTSD at that time who had distressing PTSD nightmares which had not responded to various medication trials. They were told that as a synthetic cannabinoid, nabilone had similarities in composition to marijuana. 'Off label' usage was explained, and they were told they would be the first to volunteer to try this medication to see if it had an effect on their nightmares. They were all quite eager and agreed, hoping at last that something could abate these terrible nightmares, though perhaps expecting that it would fail like so many medications before.

Fortunately these three initial patients all had rapid cessation of their nightmares. Their profuse praise for the positive effects was unexpected and so sincere that it appeared that there was something special happening here. Comments like, "Now I wake up without having to deal with the memory of the nightmares" or "Now I am not afraid to go to sleep" pointed to the clinical importance of targeting nightmares. The enthusiasm of these three patients inspired the decision to go ahead with an open label clinical trial on subsequent patients. Records would be kept comparing nightmare frequency and intensity before and during therapy with nabilone. More details are available in the already mentioned published study

[1]. Over the next two years records were kept on 47 subsequent patients who volunteered for the trial.

The author will now discuss issues that were presented during the trial as well as questions asked by colleagues when results of this clinical trial were later presented at medical meetings and conferences. Issues and questions will be highlighted in bold print in question format and then answered. Keep in mind that this paper relates to a relatively small clinical study. However there already exists a wealth of excellent articles available which have been written in the field of endocannabinoids and the potential role in PTSD and anxiety. It is hoped that the clinical results of this trial using a cannabinoid in patients suffering PTSD will dovetail with research findings and animal studies and will support speculations about the potential benefits of the medical use of cannabinoids.

Isn't Nabilone just Marijuana?

No! While it is a synthetic analogue of THC, it is not identical. It is not THC. It does not give a positive urine screen for THC. It does not give the euphoria of THC . There is no history of "street value" in spite of being available in Canada for over 25 years.

It is classified as a cannabinoid only because it acts on the body's natural cannabinoid receptors, but it is not marijuana. This is a very important issue to have clarified for both patients and physicians. Even at this time of writing there are many colleagues who are not yet aware of the recent major developments and understanding of the endocannabinoid system which is why it is important to point out the difference between 'cannabinoid' and 'cannabis'..

Is Cesamet (nabilone) a Controlled Substance in Canada and the United States?

Cesamet is classified as a narcotic in Canada and is a Class 2 drug under the U.S. Controlled Substances Act. What that means in Canada is that while any physician can prescribe it, refills cannot be added to that script. A new prescription must be written for refills. Otherwise there are no restrictions.

Do Patients Develop Dependency or Tolerance?

To date, there is no conclusive scientific evidence to support or refute the dependence or tolerance associated with nabilone. There is support showing that tolerance to the side-effects of nabilone develop quickly without a reduction in the therapeutic effect. [41]

What needs to be considered in an Overdose of Nabilone?

The danger of an overdose with cannabinoids is extremely low as there are very few CB receptors in the cardiorespiratory centers of the brain stem. [42]

What about Interactions of Nabilone with other Psychotropics being Used in Patients with PTSD?

As expected, there is no literature available to answer this question at this time. The patients in the study had previously tried various medications for their nightmares. It was not uncommon for them to be on a combination of psychotropic medications. Some were on a single medication, others were on antidepressants such as of serotonin selective reuptake inhibitors (SSRIs) , serotonin norepinephrine reuptake inhibitors (SNRIs), and norepinephrine and dopamine reuptake inhibitors (NDRIs). As well, these medications were often in addition to atypical antipsychotic agents, hypnotics and anticonvulsants (being used as mood stabilizers).

Not wanting to disrupt a regimen that had already been established and not wanting to confuse dose reduction effects with side effects related to the introduction of nabilone, a decision was made to hold medications at their current level and to cautiously introduce nabilone while carefully monitoring for adverse reactions. This decision worked out well and there were no major reactions with the addition of nabilone. On the other hand, considering that 28% of the trial patients did discontinue due to unpleasant side effects, it is possible that some were related to drug interactions.

While there were no major adverse reactions to the addition of nabilone in the study, at one subsequent medical talk, a family physician who was using the tricyclic antidepressant, amitriptyline, for depression (150 mg) reported adding nabilone (1 mg), because of comorbid chronic neuropathic pain, had caused an adverse reaction. Such an adverse interaction had not been noted at seminars given by pain specialists who had often used nabilone with amitriptyline. Pain specialists in Canada have been using nabilone for neuropathic pain for years with good success. In fact, prescriptions for nabilone written in Canada are by far more often prescribed for pain management than for nausea of chemotherapy.. This writer is indebted to these pain specialists. Much that was learned about the clinical experiences with nabilone was learned from attending seminars given by these pain specialists who had been using cannabinoids for many years. When the patient of that family doctor had the apparent adverse reaction (nabilone with amitriptyline), the doctor ordered a blood level of amitriptyline and found it to be above acceptable levels. The nabilone was discontinued and the adverse effects stopped. This adverse reaction was discussed with a number of pain specialists who said they experienced no such difficulties when combining nabilone and amitriptyline. However the doses used by pain specialists are generally in the lower ranges i.e. 10-25 mg. amitriptyline at bedtime as opposed to the higher doses when it is used for depression. This single case prompted this writer to contact the medical department of the related pharmaceutical company (Valeant, Canada). Their reply pointed out that nabilone is highly bound to plasma protein and could displace other drugs. Since both nabilone and amitriptyline are substrates of the same microsomal isoenzymes CYP2C9 and CYP3A4, consequently there could be problems with this combination.

Other potential problems could arise if nabilone was used in conjunction with alcohol, benzodiazepines or psychomimetic substances (including marijuana). Medical conditions that could possibly present problems would be cardiovascular or liver disease. As well a combination of nabilone with medications used for cardiac conditions (i.e. antihypertensives, beta blockers would warrant caution.)

The reader is to be reminded that these observations are based on a small clinical study and more rigorous clinical trials and experience will be needed before one can determine the safety of nabilone in conjunction with psychotropic medications and comorbid medical conditions.

What Side Effects were Experienced by Patients Taking Nabilone in the Study?

The following side effects were reported in this study of 47 patients. Because the study focused mainly on changes in nightmares, percentages of side effects were not recorded. No side effect was significantly distressing. Those patients stopped the medication resulting in rapid return to pre-nabilone functioning. Others, with mild side effects quickly adjusted to these and were able to continue taking nabilone.

Side effects noted, in no special order, include depersonalization, derealization, dizziness, dry mouth, drowsiness/somnolescence, nausea, constipation, increased appetite and a 'stoned' feeling. One case reported morning urinary frequency. One case with borderline low blood pressure experienced postural hypotension. Another single case reported paranoid ideation: "I became paranoid and felt everyone was watching me. It was the same as when I used marijuana as a teenager." This quickly dissipated when she stopped the nabilone. Patients were not admitted to the study if they were currently using marijuana or other psychomimetics. As well, they were not accepted if they had an experience of an adverse event in the past with marijuana. This patient forgot her earlier experience until it recurred with nabilone. She was sorry she had to discontinue nabilone as it had quickly stopped her chronic nightmares.

What Dose of Nabilone do you Begin with?

There were no guidelines to follow in reference to the dose needed for PTSD nightmares. However the initial dose for nausea control recommended by the company was 1.0 mg once or twice daily with a maximum of 6.0 mg. Based on that, and the fact that peak plasma levels were attained within two hours, a decision was made to start with 1 milligram (1.0 mg) one hour before bedtime. That dose was too high for a number of patients but was the only capsule strength available at initiation of the study. Fortunately, instructions for diluting for lower dosages are available and can be prepared by any pharmacy. Early into the study the 0.5 mg strength became available, also in capsule form. More recently the 0.25 mg capsule became available in Canada. Response rates in the study varied from 0.2 mg. to 5.0 mg. with the most common dose being 0.5 mg. Presently, this author recommends a starting dose of 0.25 mg (if that strength is not available then the 0.5 mg capsule) and titration up or down every three to seven days depending on availability of therapists to monitor response. Increases are initially in 0.25 mg increments, but if higher doses are required, a 0.5 mg increment would be appropriate. In most cases, the response was no change in nightmares until the therapeutic dose was reached, at which time the nightmares stopped that night. If the starting dose happened to be the right one for that patient, the nightmares would stop the first or second night of taking the medication. Tolerance was not noted, and that dose was maintained without breakthrough nightmares.

Though available for over 25 years in Canada, nabilone was only introduced into the USA in 2006 and is only available there presently at the 1.0 mg capsule strength. A decision could be made to begin with that higher strength with a caution given to the patient that this may be too strong and there may be a need to arrange for a dilution to a 0.25 or 0.5 mg dose. Conversely, one could begin with a dilution of 0.25 mg. from the onset and adjust accordingly.

How long do Patients Need to Be on Nabilone?

It is really too early to say. Patients who forgot their medication in the first couple of months noted a return of nightmares on the first or second night. One patient who has been successfully on nabilone for four months went on vacation and forgot his medications. While the nightmares resumed immediately, they were no longer at the same intensity and he decided to wait before recommencing nabilone and soon realized he no longer needed to use the drug. Others were able to stop the medication without major recurrence after six months to one year. Some likely will need the medication on a long-term basis. This author left private practice at his previous trauma clinic one year after the study was completed. Patients had been dispersed to various therapists so statistics are not available as to when these patients stopped or whether they are still on nabilone. The author recommended patients continue with other aspects of their PTSD therapy i.e. Cognitive Behavior Therapy (CBT), Prolonged Exposure Therapy (PE) [43], Eye Movement Desensitization and Reprocessing (EMDR), [44] Imagery Rehearsal Therapy (IRT) [45], medications and other therapies while on nabilone. With this ongoing therapy, which hopefully also dealt with nightmare content, it is hoped that nabilone would eventually be able to be discontinued. It is suspected that single incident trauma may need a shorter duration of medication compared to those with complex PTSD and many years of nightmares. Patients are instructed to attempt to get off nabilone every six months.

Two other Cannabinoids Are Available by Prescription. Could they be Useful in Managing PTSD Nightmares as well? What about Marijuana?

Probably, however their use is not recommended. Sativex® (delta-9-tetrahydrocannabinol-cannabidiol) is a buccal spray indicated as adjunctive treatment for the symptomatic relief of neuropathic pain in multiple sclerosis in adults. Marinol® (dronabinol) has a delta-9-tetrahydrocannabinol structure and like nabilone is listed as an antiemetic for cancer chemotherapy. Unlike Cesamet (nabilone), both Sativex and Marinol have the actual THC structure (nabilone is only a synthetic analogue) and both come with warnings of abuse potential and psychological dependency. They both will test positive in urine tests for THC, nabilone will not. Marijuana, which contains various other compounds, is known to be carcinogenic and certainly fails any recommendation for treatment for any aspect of PTSD therapy in this writer's opinion. Indeed, persons who may be using marijuana for self-treatment for PTSD would be well served by asking their doctor to consider a trial of nabilone to replace marijuana and eventually withdraw the nabilone (they should be forewarned that they will not get any euphoria from the nabilone). Of these, only nabilone would be considered appropriate in the therapy of PTSD nightmares (as well it may benefit other PTSD symptoms - but this will need to await future studies).

What Should be Done Next in Reference to Using Cannabinoids in Treating PTSD?

Certainly a replication of this study on the effects on PTSD nightmares would be useful. As well, a larger study looking at all PTSD symptoms and how they may respond to nabilone would be recommended. These trials could include sleep studies (polysomnography) to better understand the effect on the stages of sleep (and to rule out sleep apnea which is frequently noted in military personnel diagnosed with PTSD in the author's current work with the Canadian military's Operational Trauma and Stress Support Center.) Polysomnography could clarify if awakenings were due to nightmares, night terrors, nocturnal panic attacks, sleep apnea or other causes. It would also be interesting to observe which sleep event is associated with the profuse sweating while asleep which is commented on by many with PTSD. As well, a trial comparing nabilone and prazosin would be an interesting research project.

Conclusion

Knowledge of the endocannabinoid system has grown exponentially over the past decade. Cannabinoid receptors are present in the human body and located in areas relevant to Posttraumatic Stress Disorder (PTSD) i.e. amygdala, hippocampus and prefrontal cortex. Pre-clinical and clinical evidence points to the importance of the endocannabinoid system in the field of PTSD and the potential role of approved cannabinoids in treating PTSD nightmares and as well as other symptoms. Positive clinical experience and research findings warrant further evaluation and controlled clinical trials to explore the role of the endocannabinoid system and its relation to anxiety disorders including PTSD.

Acknowledgments

The author thanks Bruce Wilson and Bonnie Durrett for their valuable help in preparing this chapter.

References

[1] 1. Fraser G: The use of a synthetic cannabinoid in the management of treatment resistant nightmares in posttraumatic stress disorder. *CNS Neurosci Ther* 2009; 15:84-88.

[2] Gaoni Y, Mechoulam R: Isolation, structure and partial synthesis of an active constituent of hashish. *J Am Chem Soc* 1964; 86:1646-7.

[3] Devane WA, Dysarz FA 3rd, Johnson MR, Melvin LS, Howlett AC: Determination and characterization of a cannabinoid receptor in rat brain. *Mol Pharmacol* 1988; 34:605-13.

[4] Matsuda LA, Lolait SJ, Brownstein MJ, Young AC, Bonner TI: Structure of a cannabinoid receptor and functional expression of the cloned CDNA. *Nature* 1990; 346:561-4.

[5] Munro S, Thomas KL, Abu-Shaar M: Molecular characterization of a peripheral receptor for cannabinoids. *Nature* 1993;365:61-5.

[6] Lynch M: Preclinical science regarding cannabinoids as analgesics: An overview. *Pain Res Manage.* 2005; 10(Suppl A): 7A-14A.

[7] Stahl SM: Stahl's Essential Pharmacology - Neuroscientific Basis and Practical Applications. 3rd edition. Cambridge, England: Cambridge University Press 57; 2008.

[8] Bartolucci G. Nabilone and posttraumatic stress disorder. *Four Zero One Pharma. Aug.* 2004.

[9] Taylor HR, Freeman MK, Cates ME: Prazosin for treatment of nightmares related to posttraumatic stress disorder. *Am J Health Syst Pharm* 2008; 65:716-22.

[10] Dierks MR, Jordan JK, Sheehan AH: Prazosin treatment of nightmares related to posttraumatic stress disorder. *Ann Pharmacother* 2007; 41:1013-7.

[11] Lanius RA, Williamson PC, Densmore M, Boksman K, Gupta MA, Neufeld RW, Gati JS, Menon RS: Neural correlates of traumatic memories in posttraumatic stress disorder: a functional MRI investigation. *Am J Psychiatry* 2001; 158:1920–1922.

[12] Lanius RA, Williamson PC, Hopper J, Densmore M, Boksman K, Gupta MA, Neufeld RW, Gati JS, Menon RS: Recall of emotional states in posttraumatic stress disorder: an fMRI investigation. *Biol Psychiatry* 2003; 53:204–210.

[13] Lanius RA, Williamson PC, Boksman K, Densmore M, Gupta M, Neufeld RW, Gati JS, Menon RS: Brain activation during script-driven imagery induced dissociative responses in PTSD: a functional magnetic resonance imaging investigation. *Biol* Psychiatry 2002; 52:305–311.

[14] Rauch SL, Whalen PJ, Shin LM, McInerney SC, Macklin ML, Lasko NB, Orr SP, Pitman RK: Exaggerated amygdala response to masked facial stimuli in posttraumatic stress disorder: a functional MRI study. *Biol Psychiatry* 2000; 47:769–776.

[15] Rauch SL, van der Kolk BA, Fisler RE, Alpert NM, Orr SP, Savage CR, Fischman AJ, Jenike MA, Pitman RK: A symptom provocation study of posttraumatic stress disorder using positron emission tomography and script-driven imagery. *Arch Gen Psychiatry* 1996; 53:380–387.

[16] Shin LM, Wright CI, Cannistraro PA, Wedig MM, McMullin K, Martis B, Macklin ML, Lasko NB, Cavanagh SR, Krangel TS, Orr SP, Pitman RK, Whalen PJ, Rauch SL: A functional magnetic resonance imaging study of amygdala and medial prefrontal

cortex responses to overtly presented fearful faces in posttraumatic stress disorder. *Arch Gen Psychiatry* 2005; 62:273–281.

[17] Bremner JD, Staib LH, Kaloupek D, Southwick SM, Soufer R, Charney DS: Neural correlates of exposure to traumatic pictures and sound in Vietnam combat veterans with and without posttraumatic stress disorder: a positron emission tomography study. *Biol Psychiatry* 1999; 45:806–816.

[18] Shin LM, Whalen PJ, Pitman RK, Bush G, Macklin ML, Lasko NB, Orr SP, McInerney SC, Rauch SL: An fMRI study of anterior cingulate function in posttraumatic stress disorder. *Biol Psychiatry* 2001; 50:932–942.

[19] Woodward SH, Kaloupek DG, Streeter CC, Martinez C, Schaer M, Eliez S: Decreased anterior cingulate volume in combat-related PTSD. *Biol Psychiatry* 2006; 59:582–587.

[20] Kitayama N, Quinn S, Bremner JD: Smaller volume of anterior cingulate cortex in abuse-related posttraumatic stress disorder. *J Affect Disord* 2006; 90:171–174.

[21] Bremner JD, Randall P, Scott TM, Bronen RA, Seibyl JP, Southwick SM, Delaney RC, McCarthy G, Charney DS, Innis RB: MRI-based measurement of hippocampal volume in patients with combat-related posttraumatic stress disorder. *Am J Psychiatry* 1995;152:973-81.

[22] Bremner JD, Vermetten E, Schmahl C, Vaccarino V, Vythilingam M, Afzal N, Grillon C, Charney DS: Positron emission tomographic imaging of neural correlates of a fear acquisition and extinction paradigm in women with childhood sexual-abuse-related post-traumatic stress disorder. *Psychol Med* 2005; 35:791–806.

[23] Cahill L, Prins B, Weber M, McGaugh JL: Beta-adrenergic activation and memory for emotional events. *Nature* 1994; 371:702-4.

[24] Pitman RK, Sanders KM, Zusman RM, Healy AR, Cheema F, Lasko NB, Cahill L, Orr SP: Pilot study of secondary prevention of posttraumatic stress disorder with propranolol. *Biol Psychiatry* 2002; 51:189-92.

[25] Marsicano G, Wotjak CT, Azad SC, Bisogno T, Rammes G, Cascio MG, Hermann H, Tang J, Hofmann C, Zieglgänsberger W, Di Marzo V, Lutz B: The endogenous cannabinoid system controls extinction of aversive memories. *Nature.* 2002; 418:530-4.

[26] Jiang W, Zhang Y, Xiao L, Van Cleemput J, Ji SP, Bai G, Zhang X: Cannabinoids promote embryonic and adult hippocampus neurogenesis and produce anxiolytic- and antidepressant-like effects. *J Clin Invest* 2005; 115:3104-16.

[27] Eisch, AJ, Barrot M, Schad CA, Self DW. Nestler EJ: Opiates inhibit neurogenesis in the adult rat hippocampus. *Proc Natl Acad Sci U S A* 2000;97: 7579–7584.

[28] Nixon K, Crews FT: Binge ethanol exposure decreases neurogenesis in adult rat hippocampus. *J Neurochem* 2000; 83:1087–1093.

[29] Abrous DN, Adriani W, Montaron MF, Aurousseau C, Rougon G, Le Moal M, Piazza PV: Nicotine self-administration impairs hippocampal plasticity. *J Neurosci* 2002; 22:3656–3662.

[30] Yamaguchi M, Suzuki T, Seki T, Namba T, Juan R, Arai H, Hori T, Asada T: Repetitive cocaine administration decreases neurogenesis in adult rat hippocampus. *Ann N Y Acad Sci* 2004; 1025:351–362.

[31] Azad SC, Monory K, Marsicano G, Cravatt BF, Lutz B, Zieglgänsberger W, Rammes G: signaling. *J Neurosci* 2004; 24:9953-61.

[32] Suzuki A, Josselyn SA, Frankland PW, Masushige S, Silva AJ, Kida S: Memory reconsolidation and extinction have distinct temporal and biochemical signatures. *J Neurosci* 2004;24:4787-95.

[33] Pamplona FA, Prediger RD, Pandolfo P, Takahashi RN: The cannabinoid receptor agonist WIN 55,212-2 facilitates the extinction of contextual fear memory and spatial memory in rats. *Psychopharmacology (Berl).* 2006; 188:641-9.

[34] Germain A, Hall M, Krakow B, Shear MK, Buysse DJ: A brief sleep scale for posttraumatic stress disorder: Pittsburgh Sleep Quality Index Addendum for PTSD. *J Anxiety Disord* 2005; 19:233-244.

[35] Krakow B, Melendrez D, Pedersen B, Johnston L, Hollifield M, Germain A, Koss M, Warner TD, Schrader R: Complex insomnia: insomnia and sleep-disordered breathing in a consecutive series of crime victims with nightmares and PTSD. *Biol Psychiatry* 2001; 49:948–953.

[36] Thompson CE, Taylor FB, McFall ME, Barnes RF, Raskind MA: Nonnightmare distressed awakenings in veterans with posttraumatic stress disorder: response to prazosin. *J Trauma Stress.* 2008;21:417-20.

[37] States JH, St Dennis CD: Chronic Sleep Disruption and the Reexperiencing Cluster of Posttraumatic Stress Disorder Symptoms Are Improved by Olanzapine: Brief Review of the Literature and a Case-Based Series. *Prim Care Companion J Clin Psychiatry* 2003; 5:74-79.

[38] Schierenbeck T, Riemann D, Berger M, Hornyak M: Effect of illicit recreational drugs upon sleep: cocaine, ecstasy and marijuana. *Sleep Med Rev.* 2008; 12:381-9.

[39] Murillo-Rodriguez E, Blanco-Centurion C, Sanchez C, Piomelli D, Shiromani PJ: Anandamide enhances extracellular levels of adenosine and induces sleep: an in vivo microdialysis study. *Sleep* 2003; 26:943-7.

[40] Gaetani S, Cuomo V, Piomelli D. Anandamide hydrolysis: a new target for anti-anxiety drugs? *Trends Mol Med* 2003 ;9:474-8.

[41] Lemberger L, Rowe: Clinical pharmacology of nabilone, a cannabinol derivative. *Clin Pharmacol Ther 1975*;18:720-6.

[42] Walker JM, Huang SM: Cannabinoid analgesia. Pharmacol Ther 2002;95: 127-35.

[43] Foa E, Riggs D, Massie E,Yarczower M: The impact of fear activation on the efficacy of exposure treatment for PTSD. *Behavior Therapy* 1995;26:487-499.

[44] Spates CR. Koch E, Cusack K, Pagoto S, Waller S: Eye Movement Desensitization and Reprocessing. In *Effective Treatment for PTSD: guidelines from the International Society for Traumatic Stress Studies.* Edit by Edna Foa, Terrence Keane, Matthew Friedman, Judith A. Cohen. 2nd edition, 2009 The Guilford Press. New York. pp. 279-305.

[45] Krakow B, Hollifield M, Johnston L, Koss M, Schrader R, Warner TD, Tandberg D, Lauriello J, McBride L, Cutchen L, Cheng D, Emmons S, Germain A, Melendrez D, Sandoval D, Prince H: Imagery rehearsal therapy for chronic nightmares in sexual assault survivors with posttraumatic stress disorder: a randomized controlled trial. *JAMA* 2001;286:537-45.

In: Neurobiology of Post-Traumatic Stress Disorder
Editors: L. Sher and A. Vilens, pp. 345-354
ISBN: 978-1-61470-593-2

Chapter XXI

The Use of Anesthesia in an Animal Model of Posttraumatic Stress Disorder

***Vinuta Rau*[1], *Michael S. Fanselow*[2] *and Edmond I Eger II*[1]**
[1]University of California, San Francisco, California, USA
[2]University of California, Los Angeles, California, USA

Abstract

The experience of surgery can produce both physical and psychological stress. Awareness during anesthesia is a major concern of patients preoperatively. While its occurrence is relatively rare, awareness increases the likelihood of negative post-surgical outcomes, including Posttraumatic Stress Disorder (PTSD). By using an animal model of PTSD called stress-enhanced fear learning (SEFL), we have studied how inhaled anesthetics suppress PTSD-like symptomotology. Sufficient concentrations of the inhaled anesthetics isoflurane and nitrous oxide suppress PTSD-like symptomotology, as measured by SEFL. However, their mechanism of action is unclear. We have examined the role of *N*-methyl-D-aspartate receptors (NMDA-R) previously, and we examine the role of γ-aminobutyric acid type A receptors (GABAA-R) in this chapter. The results collectively show that neither receptor system alone suppresses SEFL, suggesting that multiple receptor systems underlie the ability of inhaled anesthetics to suppress SEFL.

Anesthesia, Awareness and PTSD

Undergoing surgery is both a physically and psychologically stressful event. In addition to fear of the unknown and loss of control, patients cite fear of experiencing awareness during anesthesia as one of their major worries about surgery preoperatively [1]. While highly popularized, the likelihood of experiencing awareness during anesthesia is relatively rare. A recent study reports the incidence of awareness during anesthesia at 0.13%, which translates into an expected number of 26,000 cases each year [2]. These patients are at increased risk of developing negative sequellae like Posttraumatic Stress Disorder (PTSD) subsequently. In fact, it has been reported that more than half of patients who experience awareness during anesthesia go on to develop PTSD [3]. These individuals show the hallmark signs of PTSD; re-experiencing aspects of the trauma in various ways, avoiding things associated with the trauma, and increased arousal [4]. For example, patients report fear, intrusive thoughts and flashbacks when seeing the blue color of scrubs, when watching hospital-themed television shows or when smelling rubbing alcohol [5].

There are several reasons why an individual may experience awareness during anesthesia. Light anesthesia; due to the type of surgery (*e.g.* Cesarean section), hypovolemia or minimal cardiac reserve; increased anesthetic requirement due to age, smoking or drug use; or anesthetic machine misuse or malfunction are the most common [6]. Awareness during anesthesia is not necessary for negative sequellae to develop after surgery. In fact, being able to recall the details of the traumatic event is not necessary for PTSD to occur [7-9]. For example, Osterman and van der Kolk (1998) describe a patient who "following painful awareness ... developed a sudden, intense inexplicable aversion to jazz. She had no memory of music being played in the operating room until almost 5 years following her awareness when she recalled that one of her favorite jazz pieces was played during her surgery" [5]. Thus, despite whether or not there are conscious memories of the events of surgery, patients who experience adverse events during this time are susceptible to negative psychological outcomes like PTSD.

Stress-Enhanced Fear Learning, an Animal Model of PTSD

Because it is difficult to study this important clinical problem prospectively in patients, we have utilized an animal model of PTSD based on the principles of Pavlovian fear conditioning to better understand how inhaled anesthetics can best suppress negative psychological outcomes. In a typical fear conditioning procedure, rodents are taught to associate an aversive stimulus, such as a shock, with a neutral one, such as a tone or the environment (*i.e.*, the context) in which the shock is received. After pairing the aversive and neutral cues, the previously neutral cue will elicit a fear response in the absence of the aversive event. Rodents will show their innate fear response, freezing or immobility, in response to the tone or context. Freezing serves as an index of fear learning that can be quantified [10, 11].

Typically, freezing is proportional to the number and intensity of shocks an animal receives, up to some asymptotic level. However, if a rat is pre-exposed to a stressful event prior to fear conditioning, the rat will show an enhanced acquisition of new fear learning or stress-enhanced fear learning (SEFL). In a typical SEFL experiment, rats are placed into a context (Context A) and given a series of 15 footshocks, which simulates a stressful or "traumatic" event. The following day, they are placed in a second different context (Context B) and given a single footshock, a reminder of the original stressful event. Over the following two days, the animals are tested for their freezing response in each context in the absence of shock. When animals are placed back in the 15-shock context, they show a fear response proportional to the number of shocks received there (about 80% freezing) [12]. Animals not pre-exposed to the footshock stressor show 50% freezing in the 1-shock context while animals pre-exposed to the 15-shock stressor show an enhancement that is almost double that of non-pre-shocked animals, 90% freezing [12].

SEFL represents a permanent change in the neurobiological mechanisms that mediate fear conditioning such that a reminder of the original stressful event produces a greatly exaggerated fear response [12]. This is evidenced by the fact that attempts to eliminate fear to the 15-shock context through extinction or pharmacological means does not suppress SEFL [12]. That is, rats that have no memory of the traumatic event, the 15-shock stressor, still show SEFL in the single shock context. SEFL is akin to symptomotology that PTSD sufferers show in which reminders of the traumatic event produce a disproportionately strong emotional response. Thus, like PTSD, SEFL is characterized by exaggerated responses to a stressful event, a facilitated development of new fear responses and resistance to extinction (*i.e.*, exposure therapy).

In a series of experiments, we assessed how two inhaled anesthetics, isoflurane and nitrous oxide, suppressed SEFL [13]. In the first experiment, rats were given various concentrations of isoflurane (0%, 0.2%, 0.4%, 0.6%, 0.8% or 1%) during pre-exposure to the stressor. We found that isoflurane at concentrations of 0.4% and higher suppressed conditional fear to the 15-shock context and SEFL. We found a similar outcome with nitrous oxide at a concentration of 70-80%. These results demonstrate that a sufficient concentration of an inhaled anesthetic can suppress SEFL. However, if a low concentration is given, for example 0.2% isoflurane, there is an incomplete suppression of the 15-shock context and SEFL still occurs.

Generally, it is thought that inhaled anesthetics, like isoflurane and nitrous oxide, suppress learning and memory by increasing inhibition and decreasing excitation [14]. The neurotransmitter γ-aminobutyric acid (GABA) mediates most neural inhibition in the mammalian central nervous system through its action on GABA type A receptors (GABAA-R), while glutamate mediates most neural excitation in the nervous system though its action on *N*-methyl-D-aspartate receptors (NMDA-R). Isoflurane enhances the functional effect of GABA on GABAA-R and inhibits the functional effect of glutamate on NMDA-R, while nitrous oxide has little effect on GABAA-R and works primarily by functional inhibition of NMDA-R [14, 15]. Because both isoflurane and nitrous oxide affect multiple receptor targets, we cannot isolate the molecular site of action at which isoflurane and nitrous oxide are producing their effects. We have shown previously that delivery of an NMDA-R antagonist ((2R)-amino-5-phosphonovaleric acid or APV) into the ventricles of the brain suppresses fear conditioned to the 15-shock context, but does not suppress SEFL [12]. Because of APV's failure to suppress SEFL, we thought it likely that inhibition of GABAA-R is responsible for the behavioral suppression produced by isoflurane and nitrous oxide. Midazolam is a GABAA-R agonist that is often administered during anesthesia because of its anxiolytic and amnestic properties. Midazolam has been shown to suppress the acquisition of conditional fear [16, 17], but its effect on SEFL is unknown. In the following two experiments, we examined whether suppressing GABAA-R by administering midazolam during stress pre-exposure would, like isoflurane and nitrous oxide, suppress SEFL.

The Effect of Midazolam on SEFL

Materials and Methods

The protocol for these studies was approved by the University of California San Francisco Institutional Animal Care and Use Committee. Naïve adult male Long Evans rats (Harlan; Indianapolis, IN) were used. The rats were housed 2-3 per cage (42 cm long x 25 cm wide x 20 cm high), and food and water were available *ad libitum*. The animals were kept on a 12:12-hr light-dark cycle, with lights coming on at 6:00 am, and all experimental procedures took place during the light cycle. All rats were handled daily for approximately 20 s each for a week before each experiment.

Fear conditioning took place in four identical acrylic training chambers (27 cm L x 24.5 cm W x 20 cm H). The lid to each chamber had acrylic sidewall guides that fit inside the chamber and an 8 cm diameter hole fitted with a rubber stopper that was transiently removed to place rats inside the chamber. The chambers were used in two different procedures, and they were altered between procedures to maximize the rats' ability to differentiate between them. Pre-exposure to the stressor took place in Context A, and training to the single shock took place in Context B. The contexts differed in their texture, lighting, odor, ambient noise and grid floor. In Context A, the walls were wiped with a pine-scented cleaner (5% Pine Scented Disinfectant, Midland, Inc., Sweetwater, TN) before and after each session. The grid floor used to deliver shocks was composed of 14 stainless steel bars, each bar 5 mm in diameter, all bars spaced 13 mm from center to center. These floors were connected to a shock delivery system (San Diego Instruments, San Diego, CA). Overhead fluorescent room lights were left on, and a squirrel fan supplied background noise (65 db).

In Context B, rubber mats with suction backing covered the clear acrylic walls and ceiling. These mats were wiped with acetic acid (1%, Fisher Scientific; Fair Lawn, NJ) before and after each session. The grid floors used to deliver a single shock in this context were composed of 31 stainless steel bars, each 3 mm in diameter, spaced 7 mm center to center, which were also wired to the shock generator. Context B appeared dark to the rats, being lit by a single 30 W red bulb placed at the opposite end of the room. Before each set of animals entered the room, an aerosol (Amphyl Disinfectant Deodorant Spray, Reckitt Benckiser Inc.; Wayne, NJ) was sprayed in the room. A compact disc player played a disc of white noise (65 db). In both Context A and B, the front walls of each chamber were clear and allowed a video camera and corresponding VCR to record the animals' behavior for analysis.

Experiment 1

Seventy-two rats were used in this study, and the number of rats/group ranged from 8-16. On day 1, rats were transported in their home cages from their housing area to a small dark room in the laboratory. Sets of four rats at a time were transported from the dark room to the study room and weighed. Each animal was then given an intraperitoneal injection of midazolam (0.375, 0.75, 1.5, 3, or 6 mg/kg) or a vehicle (0 mg/kg). Midazolam was dissolved in sterile saline, and the injection volume used was 1 kg/ml sterile saline. Animals receiving the vehicle were given a saline injection of the same volume. Injections were given 40 min prior to pre-exposure to stress.

Rats received either 0 or 4 shocks (1 mA, 1 s shocks with a variable intershock interval of 240-480 s) during Context A pre-exposure. Previous studies have shown that pre-exposure to 4 or 15 shocks produces a similar amount of SEFL, both short- and long-term [18]. Four shocks were used as a stressor in this experiment because the session duration of 23 min (vs. 93 min for the 15 shock protocol) was more amenable to the short duration of action of midazolam. Animals receiving no shock during pre-exposure were placed in Context A for the same time as shocked animals. After the pre-exposure, the animals were removed from the chambers and placed in their home cages. The rats were returned to their housing area upon completion of the last group.

The following day, the animals were transported in cages that were the same size as the home cage, but outfitted with opaque plastic dividers such that each cage housed 3 rats separated from one another. These cages were placed in the laboratory housing area, now illuminated with a fluorescent bulb. Rats in each squad were then transferred to the Context B test area in separate plastic pots (23 cm high x 25 cm diameter) containing woodchip bedding and covered with opaque lids. In Context B, the rats were given a single shock (1 mA, 1 s) 192 s after placement in the chamber and then were removed from the chamber after an additional 32 s. On day 3, animals were returned to Context B for an 8 min 32 s test of conditional fear (the context test). On day 4, rats were returned to Context A and given another 8 min 32 s context test. The design and procedure for this experiment is outlined in Table 1. Freezing, defined as the absence of all movement except that necessary for respiration, was measured during these context tests [19]. Each animal's behavior was scored every 8 s during the observation period, and a percentage of freezing calculated from these observations.

Analysis of variance (ANOVA) was used to analyze freezing scores, and Fisher's PLSD was used for preplanned group comparisons.

Table 1. The design and procedure for Experiment 1

Group	*n*	DAY 1 Context A		DAY 2 Context B	DAY 3 Context B	DAY 4 Context A
		Midazolam Dose (mg/kg)	Pre-exposure Treatment			
1	12	0	0 shocks	1 Shock	Context Test	Context Test
2	12	0	4 shocks			
3	8	0.375	4 shocks			
4	8	0.75	4 shocks			
5	8	1.5	4 shocks			
6	8	3.0	4 shocks			
7	8	6.0	4 shocks			

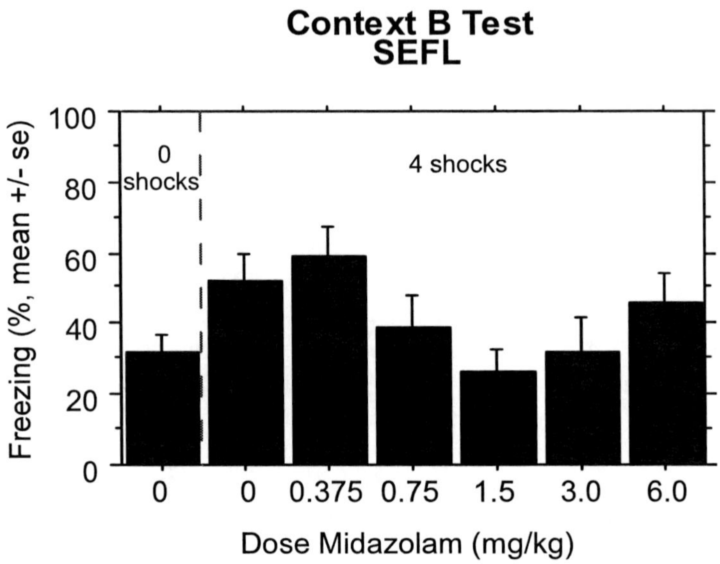

Figure 1. Experiment 1, Context B Freezing. Midazolam produced a "U"-shaped suppression of SEFL. Rats receiving 4 shocks during pre-exposure showed SEFL in comparison to non-shocked control rats. Shock pre-exposed rats treated with 1.5 mg/kg midazolam showed a significant suppression of SEFL compared to shock pre-exposed rats receiving 0 mg/kg midazolam.

Results

Context B Freezing

Midazolam produced a "U"-shaped suppression of SEFL with its lowest and highest doses producing less suppression than the mid-range dose, 1.5 mg/kg, which produced the most robust suppression (Figure 1). The overall one-way ANOVA with treatment as the factor (with 7 levels) revealed a significant effect of treatment, $F(6,65) = 2.42$ ($P < 0.05$) on Context B freezing. *A priori* comparisons between groups showed SEFL between vehicle treated groups. That is, rats receiving 4 shocks during pre-exposure showed enhanced freezing compared to non-shocked controls ($P < 0.05$). Shock pre-exposed rats treated with 1.5 mg/kg midazolam showed a significant suppression of SEFL compared to shock pre-exposed rats receiving 0 mg/kg midazolam ($P < 0.05$). Midazolam doses of 1.5 mg/kg and higher were statistically indistinguishable from non-shocked control rats.

Context A Freezing

Midazolam at doses of 0.75 mg/kg and higher produced suppression of conditional fear in Context A (Figure 2). The overall one-way ANOVA revealed a significant effect of treatment, $F(6,65) = 11.67$ ($P < 0.0001$) on Context A freezing. *A priori* comparisons confirmed that with the exception of rats receiving 0.375 mg/kg midazolam, all midazolam-treated rats showed statistically suppressed freezing levels compared to shocked rats receiving vehicle injections ($P < 0.05$) and were statistically similar to non-shocked control rats.

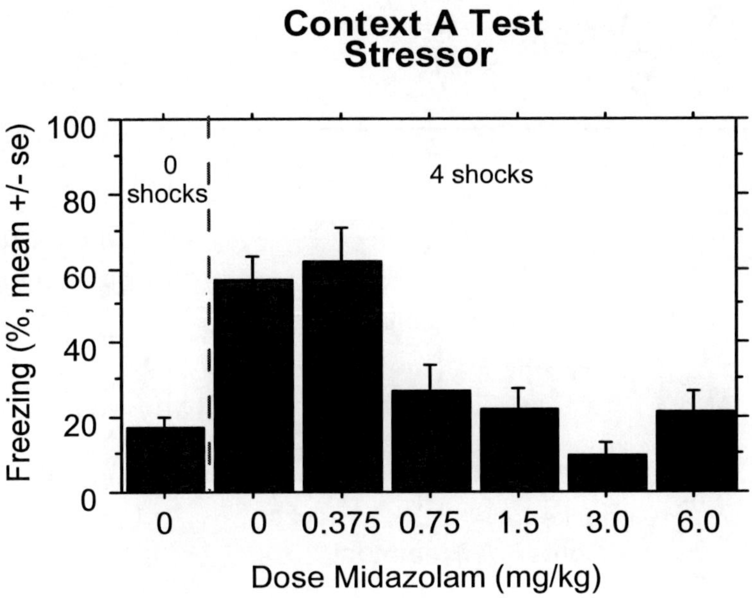

Figure 2. Experiment 1, Context A Freezing. Midazolam at doses of 0.75 mg/kg and higher and produced suppression of conditional fear.

These results show that a sufficient dose of midazolam (1.5 mg/kg) produces suppression of SEFL and a suppression of fear conditioned to Context A. However, in order for this to be a reliable effect, it must be long lasting. PTSD is characterized by its symptoms having a prolonged duration [4]. In order for SEFL to be a reliable model of PTSD symptomotology, it must also produce a long lasting effect. We have shown previously that rats pre-exposed to the shock stressor and given the single shock 90 days later still show SEFL [18]. Likewise, isoflurane produces a long lasting suppression of SEFL. Rats treated with 0.6% isoflurane during shock pre-exposure show a suppression of SEFL when the single shock is given 90 days later [13]. In the following experiment, we assessed whether midazolam also produced a long lasting suppression.

Experiment 2

We conducted a follow-up experiment to examine whether midazolam produced a long-lasting suppression of SEFL. Forty-eight rats were used in this experiment (12 rats/group). In this experiment, the procedure was slightly changed, and a 5-day period was inserted between the Context A shock pre-exposure and the Context B single shock. We tested one midazolam dose with this new procedure, 1.5 mg/kg, the dose that produced the most robust suppression of SEFL in Experiment 1. A two-factorial design was used here, with shock pre-exposure as one factor (0 or 4 shocks) and midazolam dose as the second factor (0 or 1.5 mg/kg).

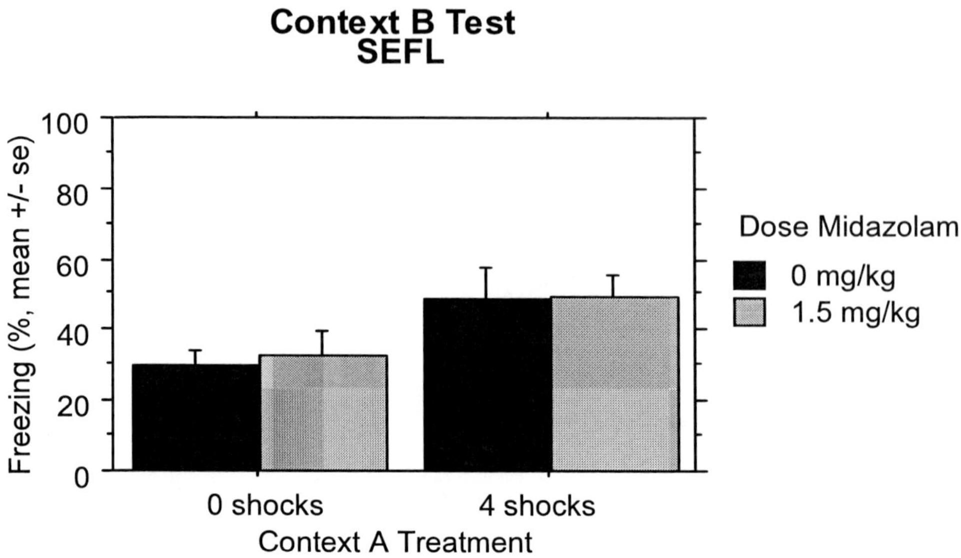

Figure 3. Experiment 2, Context B Freezing. When the 5-day waiting period was placed between Context A pre-exposure and the Context B single shock, midazolam no longer suppressed SEFL.

Context B Freezing

When the 5-day waiting period was placed between Context A pre-exposure and the Context B single shock, midazolam did not suppress SEFL (Figure 3). The two-way ANOVA conducted on these observations revealed a reliable effect of Context A treatment, $\underline{F}(1,44) = 6.85$ ($P < 0.05$) on Context B freezing, but there was no reliable main effect of midazolam treatment or interaction between factors.

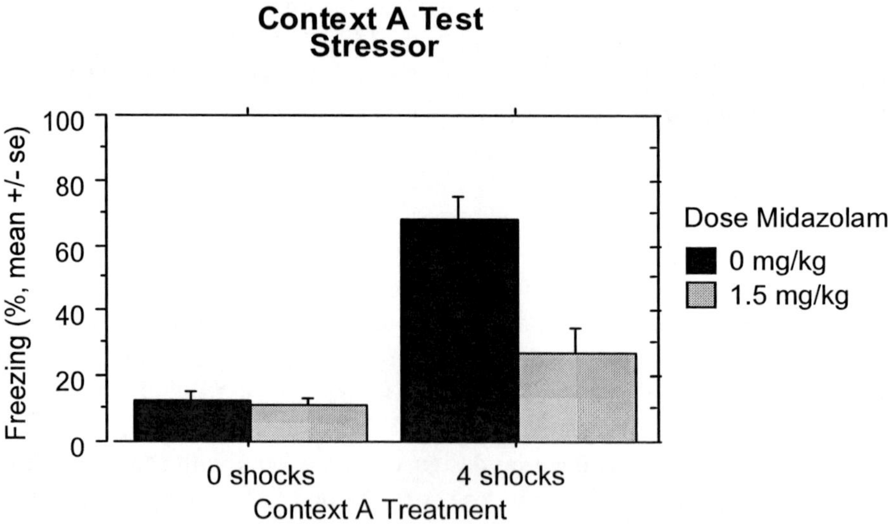

Figure 4. Experiment 2, Context A Freezing. Midazolam suppressed conditional fear in Context A, replicating our findings in Experiment 1.

Context A Freezing

Midazolam did suppress conditional fear in Context A (Figure 4), replicating our findings in Experiment 1. The two-way ANOVA revealed a significant effect of Context A treatment, $\underline{F}(1,44) = 44.06$ ($P < 0.0001$), a reliable effect of midazolam $\underline{F}(1,44) = 15.48$ ($P < 0.0005$), and a significant interaction of the two factors $\underline{F}(1,44) = 13.31$ ($P < 0.005$).

While it initially appeared that midazolam suppressed SEFL in Experiment 1, Experiment 2 showed that this was not a long lasting effect. It is possible that midazolam had a lingering effect during Context B training (on day 2) in Experiment 1 that was responsible for producing the suppression of SEFL in Context B. These results along with those from the APV study described earlier [12] suggest that perhaps combined antagonism of NMDA-R and agonism of GABAA-R are necessary for suppression of SEFL. This hypothesis would account for isoflurane and nitrous oxide's ability to suppress SEFL, but neither APV nor midazolam to do it alone. More research is necessary to directly test this hypothesis.

Conclusion

The onset of physical illness and treatment at the hospital can produce high levels of fear, anxiety, helplessness and a fear of dying, making PTSD an unfortunate outcome in many cases [20]. While awareness during anesthesia accounts for only a small number of these cases, it lends itself to the study of improved outcomes because amnestic agents are being given during the traumatic event of surgery. One of the goals of anesthesia during surgery is to produce a total suppression of memory formation. Unfortunately, we do not fully understand how inhaled anesthetics produce amnesia. Therefore, we must focus on developing a better understanding to prevent negative sequellae from developing after unfortunate incidents like awareness. Using an animal model of PTSD such as SEFL can help towards this understanding. Our research has shown that sufficient concentrations of inhaled anesthetics such as isoflurane or nitrous oxide will suppress memories of the traumatic event as well as future PTSD-like symptomotology. However, much more research is necessary to further understand the molecular basis of these effects. Our previous work and the experiments described in this chapter with midazolam suggest that there may not be just one target and that instead a simultaneous effect on multiple receptor targets is crucial for suppression of SEFL. Our future research efforts will continue to examine this idea. Also, as mentioned previously, negative sequellae may develop due to insufficient anesthesia, which may be necessary due to the type of surgery or vulnerabilities of the patient. Here we need to further study what anesthetic adjuvants may be given with reduced concentrations of anesthesia to fully block memory formation during anesthesia. The use of the SEFL model will help us to develop a better understanding of which inhaled anesthetics and anesthetic adjuvants best suppress PTSD-like symptomotology and the mechanisms by which they work.

References

[1] Sandlin D: A closer look at bispectral index monitoring. *J Perianesth Nurs* 2001; 16(6):420-2.

[2] Sebel PS, Bowdle TA, Ghoneim MM, Rampil IJ, Padilla RE, Gan TJ, Domino KB: *The incidence of awareness during anesthesia: a multicenter United States study. Anesth Analg* 2004; 99(3):833-9.

[3] Osterman JE, Hopper J, Heran WJ, Keane TM, van der Kolk BA: Awareness under anesthesia and the development of posttraumatic stress disorder. *Gen Hosp Psychiatry* 2001; 23(4):198-204.

[4] American Psychiatric Association: *Diagnostic and Statistical Manual of Mental Disorders*. Washington, DC, American Psychiatric Press, 1994.

[5] Osterman JE, van der Kolk BA: Awareness during anesthesia and posttraumatic stress disorder. *Gen Hosp Psychiatry* 1998; 20(5):274-81.

[6] Ghoneim MM: Awareness during anesthesia. Anesthesiology 2000; 92(2):597-602.

[7] Bryant RA: Posttraumatic stress disorder and traumatic brain injury: can they co-exist? *Clin Psychol Rev* 2001; 21(6):931-48.

[8] Krikorian R, Layton BS: Implicit memory in posttraumatic stress disorder with amnesia for the traumatic event. *J Neuropsychiatry Clin Neurosci* 1998; 10(3):359-62.

[9] Layton BS, Krikorian R, Dori G, Martin GA, Wardi K: Posttraumatic stress disorder with amnesia following asphyxiation. *Ann N Y Acad Sci* 2006; 1071:488-90.

[10] Bolles R, Collier A: The effect of predictive cues on freezing in rats. *Anim. Lear. Behav.* 1976; 4(1A):6-8.

[11] Fanselow MS: Neural organization of the defensive behavior system responsible for fear. *Psychonomic Bulletin and Review* 1994; 1:429-438.

[12] Rau V, DeCola JP, Fanselow MS: Stress-induced enhancement of fear learning: an animal model of posttraumatic stress disorder. *Neurosci Biobehav Rev* 2005; 29(8):1207-23.

[13] Rau V, Oh I, Laster M, Eger EI, 2nd, Fanselow MS: Isoflurane suppresses stress-enhanced fear learning in a rodent model of post-traumatic stress disorder. *Anesthesiology* 2009; 110(3):487-95.

[14] Campagna JA, Miller KW, Forman SA: Mechanisms of actions of inhaled anesthetics. *N Engl J Med* 2003; 348(21):2110-24.

[15] Grasshoff C, Drexler B, Rudolph U, Antkowiak B: Anaesthetic drugs: linking molecular actions to clinical effects. *Curr Pharm Des* 2006; 12(28):3665-79.

[16] Fanselow MS, Helmstetter FJ: Conditional analgesia, defensive freezing, and benzodiazepines. *Behav Neurosci* 1988; 102(2):233-43.

[17] Pain L, Launoy A, Fouquet N, Oberling P: Mechanisms of action of midazolam on expression of contextual fear in rats. *Br J Anaesth* 2002; 89(4):614-21.

[18] Rau V, Fanselow MS: Exposure to a stressor produces a long lasting enhancement of fear learning in rats. *Stress* 2009; 12(2):125-33.

[19] Fanselow MS: Conditioned and unconditional components of post-shock freezing. *Pavlov J Biol Sci* 1980; 15(4):177-82.

[20] Tedstone JE, Tarrier N: Posttraumatic stress disorder following medical illness and treatment. *Clin Psychol Rev* 2003; 23(3):409-48.

In: Neurobiology of Post-Traumatic Stress Disorder
Editors: L. Sher and A. Vilens, pp. 355-361
ISBN: 978-1-61470-593-2

Chapter XXII

Translational Neuroscience and Treatment Development for Posttraumatic *Stress Disorder*

Leo Sher
Columbia University and New York State Psychiatric Institute,
New York, New York, USA

Abstract

This chapter reports on the symposium, "Translational neuroscience may inform treatment development for posttraumatic stress disorder" (Chair – Leo Sher, M.D., Co-Chair – Alexander Neumeister, M.D.) that was held on June 28, 2009 during the 9th World Congress of Biological Psychiatry (June 28-July 2, 2009, Palace of Congresses (Le Palais des Congrès de Paris), Paris, France). Compelling animal and human data indicate that stress compromises neurotrophic functions producing neural atrophy and reducing neuroplasticity resulting in dysfunction in brain circuitry associated with PTSD. This problem has been identified as the most important priority for studies of the neurobiology of PTSD. This symposium was the first translational interdisciplinary attack on this problem, from molecular neuroscience to treatment and targets major gaps in our understanding of the neurobiology, neural circuitry, phenomenology, and treatment of PTSD. Four research works were presented at the symposium: "Early postnatal stress: Impact on neural circuits and relevance to psychophysiology" by Machiko Matsumoto (Japan), "Long term alterations in brain and neurobiology in PTSD" by Eric Vermetten (Netherlands), "Mechanisms of resilience: Lessons from PET imaging" by Alexander Neumeister (USA) and "Neurobiology of suicidal behavior in PTSD and implications for suicide prevention" by Leo Sher (USA).

Introduction

The 9th World Congress of Biological Psychiatry was held on June 28-July 2, 2009, at the Palace of Congresses (Le Palais des Congrès de Paris) in Paris, France. The Congress consisted of an opening lecture, 5 plenary lectures, 4 special lectures, 88 symposia, 10 debates, 10 treatment guidelines sessions, 14 workshops, 36 free communications, 5 young scientist award sessions, and 51 guided poster tours with over 1020 posters. Most research presentations were from the USA, Germany, France, UK, and Japan. A lot of presentations were from Switzerland, Netherlands, Canada, Italy, Israel, Australia, Brazil, Austria, India, Sweden, Russia, Spain, Greece, and Argentina.

Many symposia including "Translational neuroscience may inform treatment development for posttraumatic stress disorder," "Suicide prevention: Focusing on modifiable risk factors," "Recent advances in the neurobiology of suicide," "Psychiatric disorders and thyroid axis functioning across lifecycles," "New findings on the neurobiological basis of language abnormalities in schizophrenia," "Impulse disorders - categorical versus dimensional approaches: State of the evidence," "New approaches to diagnostics and treatment in personality disorders," "Genetic strategies in the search for genes in schizophrenia," "Neurosciences of suicidal behaviour: New targets for prevention?" "Are biological measures useful in predicting antipsychotic treatment response in schizophrenia?" "Biological markers and clinical features in depression: A difficult puzzle," "Neurobiological correlates of dissociation and influence on information processing in stress-related disorders," "Aß Immunotherapy of Alzheimer's disease," "EEG and MEG response to mismatch stimuli as an index of schizophrenia neurobiology and genetics," "New strategies and tools in the field of transcranial magnetic stimulation (TMS)," "Social cognition, social hormones and psychophathology," "Neural circuitry and mood disorders: Relationship to course of illness and treatment outcome," "Recent advances in neuropharmacology of stress" were met with a significant interest by Congress participants. The Congress Program and Abstracts have been published online at http://www.wfsbp-congress.org/.

The new leadership of the World Federation of Societies of Biological Psychiatry (WFSBP) was elected on July 1, 2009. Prof. Florence Thibaut (France) is the new President of the WFSBP. Prof. Robertas Bunevicius (Lithuania) and Prof. Constantin R. Soldatos (Greece) are the new Vice-Presidents of the WFSBP. Prof. Norbert Müller (Germany) is the Secretary-Treasurer. Prof. Yoshio Hirayasu (Japan) and Prof. Jorge Tellez-Vargas (Colombia) are Associate Secretary-Treasurers.

The 10th World Congress of Biological Psychiatry will be held in Prague, Czech Republic, on May 29-June 3, 2011. The WFSBP is also sponsoring the 3rd International Forum "Innovation in Psychiatry" (INNOPSY 2009) (Milan, Italy, November 18-20, 2009).

The Symposium

The symposium, "Translational neuroscience may inform treatment development for posttraumatic stress disorder" (Chair – Leo Sher, M.D., Co-Chair – Alexander Neumeister, M.D.) was held on June 28, 2009. Four research works were presented at the symposium: "Early postnatal stress: Impact on neural circuits and relevance to psychophysiology" by

Machiko Matsumoto (Japan), "Long term alterations in brain and neurobiology in PTSD" by Eric Vermetten (Netherlands), "Mechanisms of resilience: Lessons from PET imaging" by Alexander Neumeister (USA) and "Neurobiology of suicidal behavior in PTSD and implications for suicide prevention" by Leo Sher (USA).

Compelling animal and human data indicate that stress compromises neurotrophic functions producing neural atrophy and reducing neuroplasticity resulting in dysfunction in brain circuitry associated with PTSD [1,2]. This problem has been identified as the most important priority for studies of the neurobiology of PTSD. This symposium was the first translational interdisciplinary attack on this problem, from molecular neuroscience to treatment and targets major gaps in our understanding of the neurobiology, neural circuitry, phenomenology, and treatment of PTSD.

Early Postnatal Stress: Impact on Neural Circuits and Relevance to Psychophysiology

In her talk, Dr. Matsumoto has suggested that traumatic events during early life may precipitate long-lasting alterations in the functional properties underlying emotional expression that are attributable to the pathophysiology of stress-related disorders [3]. She summarized the recent work of her group elucidating whether early postnatal stress alters the neural circuits associated with fear memory, using electrophysiological approaches combined with behavioral analysis. Rats exposed to footshock (FS) stress during the second (2W) or the third (3W) postnatal week were subjected to contextual fear conditioning (CFC), an animal model of anxiety, at the postadolescent period (10 to 12 weeks). Fear-related freezing behavior during exposure to CFC was markedly attenuated in the 2W-FS group, presumably due to disturbance of the retention and/or consolidation of fear memory, an effect that was attributable to synaptic transmission in the hippocampal CA1 field. Aversive stress exposure at the third postnatal week (3W-FS) impaired extinction of context-dependent fear memory evaluating by expression of freezing, which appears to be associated with dysfunction of hippocampal-medial prefrontal cortex circuits. Thus, the altered behavior observed at the postadolescent period seemed to be the result of neurodevelopmental perturbations elicited by early life stress, in turn, a "critical period" exists in development of neural system underlying emotional expression. Furthermore, the decreased attenuation of freezing behavior observed in 3W-FS was abolished by repeated treatment with either the partial N-methyl- D-aspartate receptor agonist D-cycloserine or the serotonin 1A receptor agonist tandospirone. Dr. Matsumoto and her colleagues have concluded that considering similarity between fear extinction training in animals and exposure psychotherapy in humans, these pharmacological interventions may have clinical significance in the treatment of anxiety-related disorders, such as PTSD. In other words, a better understanding of the neural mechanisms of extinction deficits caused by early postnatal stress could help to clarify the pathophysiology of PTSD in which fear extinction is compromised.

Long Term Alterations in Brain and Neurobiology in PTSD

Dr. Vermetten has stated that brain areas involved in the stress response include the medial prefrontal cortex, hippocampus and amygdale [4]. Neurohormonal systems that act on brain areas to modulate PTSD symptoms and memory include glucocorticoids and norepinephrine. Dysfunction of these brain areas is responsible for symptoms of PTSD. Brain imaging studies show that PTSD patients have increased amygdala reactivity during fear acquisition. Other studies show smaller hippocampal volume. A failure of medial prefrontal/anterior cingulate activation with re-experiencing of the trauma is hypothesized to represent a neural correlate of the failure of extinction seen in PTSD.

Dr. Vermetten emphasized that the brain has the capacity for plasticity in the aftermath of traumatic stress. Antidepressant treatments and changes in environment can reverse the effects of stress on hippocampal neurogenesis. In humans with PTSD paroxetine increases hippocampal volume and improves memory function in conjunction with improving PTSD symptoms. Phenytoin, which blocks the effects of stress on the hippocampus in animal studies, also increases hippocampal volume in PTSD patients. Dr. Vermetten has noted that future studies should use brain imaging and neurobiology to assess plasticity in PTSD, preferrably in true prospective designs, or in relation to treatment. These can include both functional neuroimaging and neuroreceptor imaging to track the course of change during treatment, or to predict which traumatized individuals will develop (chronic) PTSD. The information from such studies will provide valuable information that will guide the development of new treatments.

Mechanisms of Resilience: Lessons from PET Imaging

Dr. Neumeister has suggested that there is an increasing body of research on psychosocial factors associated with resistance to severe trauma [5]. The neurobiological mechanisms that contribute to the ability of many people to adapt with minimal life disruption after the exposure to sever trauma is yet not fully understood. Even though animal studies suggest an important role for the serotonin (5-HT) type 1B receptor in the response to severe inescapable stress it is not clear at all whether such models constitute relevant models for stress response in humans. The selective 5-HT1B receptor radioligand, [11C]P943, a selective, high affinity 5-HT1B receptor antagonist, permits for the first time in vivo assessment of central 5-HT1B receptor binding using positron emission tomography (PET). Dr. Neumeister and his colleagues studied healthy controls without trauma exposure and healthy controls who had experienced severe trauma but did not develop psychiatric symptoms. Following a transmission scan, [11C]P943 was injected by pump over one min and high resolution research tomograph (HRRT) list mode data were acquired for 120 minutes. Dr. Neumeister and his associates have demonstrated that trauma-exposed healthy controls relative to non-traumatized controls show a downregulation of 5-HT1B receptor expression in a cortico-striatal-limbic circuit that has been consistently implicated in PTSD. Compensatory downregulation of 5-HT1B receptors among trauma-exposed healthy controls leads to improved 5-HT transmission which allows the chemical signal (5-HT) to restore the affective/anxiety circuitry and prevents symptom development after severe trauma. Serotonergic and non-serotonergic mechanisms mediate these effects.

Neurobiology of Suicidal Behavior in PTSD and Implications for Suicide Prevention

Dr. Sher has reviewed and discussed the role of neurobiological factors in the pathophysiology of suicidal behavior in PTSD [6]. A comprehensive review of the scientific literature on the neurobiology of suicidal behavior in PTSD and the role of biological treatments in suicide prevention in patients with PTSD has been carried out. A new concept related to the comorbidity of PTSD and major depressive disorder (MDD) has been developed. Multiple neurobiological factors including serotonin, glutamate, gamma-aminobutyric acid, norepinephrine, neuropeptide Y, corticotrophin-releasing hormone, dopamine, opioids, and thyroid hormones are involved in the pathophysiology of PTSD. PTSD is associated with suicidal behavior. Serotonergic and hypothalamic-pituitary adrenal (HPA) axis abnormalities are associated with both PTSD and suicidal behavior and may underlie both conditions.

PTSD is frequently comorbid with MDD [7]. Dr. Sher has previously proposed that some or all individuals diagnosed with comorbid PTSD and MDD have a separate psychobiological condition that can be termed "post-traumatic mood disorder" (PTMD) [8]. Dr. Sher's theory was based on the fact that a significant number of studies suggested that patients suffering from comorbid PTSD and MDD differed clinically and biologically from individuals with PTSD alone or MDD alone. Individuals with comorbid PTSD and MDD are characterized by greater severity of symptoms, increased suicidality and the higher level of impairment in social and occupational functioning compared to individuals with PTSD alone or MDD alone. Neurobiological evidence supporting the concept of PTMD includes the findings from neuroendocrine challenge, cerebrospinal fluid, neuroimaging, sleep and other studies. This increased suicidality may be related to neurobiological differences between PTMD and PTSD alone or MDD alone. Dr. Sher has also proposed a model of suicidal behavior in war veterans with PTMD [9]. The model consists of the following components: 1) genetic factors; 2) prenatal development; 3) biological and psychosocial influences from birth to mobilization/deployment; 4) mobilization/pre-deployment stress; 5) combat stress, traumatic brain injury, and physical injury; 6) post-deployment stress; 7) biological and psychosocial influences after the deployment; 8) trigger (precipitant) of a suicidal act; and 9) suicidal act. The first four components determine vulnerability to combat stress. The first seven components determine predisposition to suicidal behavior, a key element that differentiates PTMD patients who are at high risk from those at lower risk. Suicidal behavior in PTMD can be attributed to the coincidence of a trigger with a predisposition for suicidal behavior.

Dr. Sher has suggested that prevention of suicidal behavior in individuals with PTSD should include biological treatments of both PTSD and comorbid conditions. Medications affecting the serotonergic system may reduce symptoms of PTSD, treat comorbid conditions, such as MDD, and reduce suicidality. Medications affecting the HPA system might potentially have a beneficial effect on individuals with PTSD and suicidal behavior. Suicidal behavior in patients with PTSD is an underestimated and understudied problem. Studies of the neurobiology, treatment and prevention of suicidal ideation and behavior in PTSD are merited.

Conclusion

Over a lifetime, exposure to traumatic events occurs at fairly high levels. Research conducted during and after military and combat service, terrorism, sexual and physical assault, motor vehicle accidents and natural disasters reveals significant psychiatric consequences of trauma exposure [10-14]. PTSD is among the most common conditions associated with trauma exposure yet the majority of exposed and acutely symptomatic persons naturally recover functioning. For the significant minority that does not recover, PTSD and related co-morbidity can be quite debilitating. The importance of finding ways to prevent PTSD is underscored by recent reports of high prevalence among veterans of ongoing and past military conflicts, survivors of disasters, terrorist attacks, shooting incidents and all too common interpersonal violence and accidents. Until the processes involved in the adjustment to trauma and the development of symptoms are fully understood, decisions about treatment will be based on existing empirical evidence and theories, which have heuristic value but do not provide an integrated and comprehensive view of trauma-related psychopathology. Future translational studies of PTSD may lead to better understanding of the pathophysiology of PTSD and identification of targets for therapeutic drugs.

References

[1] Francati V, Vermetten E, Bremner JD. Functional neuroimaging studies in posttraumatic stress disorder: review of current methods and findings. *Depress Anxiety* 2007;24(3):202-18.

[2] Reagan LP, Grillo CA, Piroli GG. The As and Ds of stress: metabolic, morphological and behavioral consequences. *Eur J Pharmacol* 2008;585(1):64-75.

[3] Matsumoto M. Early postnatal stress: Impact on neural circuits and relevance to psychophysiology. In: *9th World Congress of Biological Psychiatry,* Paris, France, June 28 – July 2, 2009. Abstracts, p. 28.

[4] Vermetten E. Long term alterations in brain and neurobiology in PTSD. In: *9th World Congress of Biological Psychiatry,* Paris, France, June 28 – July 2, 2009. Abstracts, p. 28.

[5] Neumeister A. Mechanisms of resilience: Lessons from PET imaging. In: *9th World Congress of Biological Psychiatry,* Paris, France, June 28 – July 2, 2009. Abstracts, p. 28.

[6] Sher L. Neurobiology of suicidal behavior in posttraumatic stress disorder and implications for suicide prevention. In: *9th World Congress of Biological Psychiatry,* Paris, France, June 28 – July 2, 2009. Abstracts, pp. 28-29.

[7] Sher L. Suicide in war veterans: the role of comorbidity of PTSD and depression. *Exp Rev Neurother* 2009;9(7):921-923.

[8] Sher L. The concept of post-traumatic mood disorder. *Med Hypoth* 2005;65(2):205-210.

[9] Sher L. A model of suicidal behavior in war veterans with posttraumatic mood disorder. *Med Hypoth* 2009;73(2):215-219.

[10] Sher L, Vilens A. (editors). *War and Suicide.* Hauppauge, New York: Nova Science Publishers, in press.

[11] Sher L, Vilens A. (editors). *Suicide in the Military.* Hauppauge, New York: Nova Science Publishers, in press.

[12] Sher L, Vilens A. (editors). *Terror and Suicide. Hauppauge,* New York: Nova Science Publishers, in press.

[13] Friedman MJ, Keane TM, (Editor), Resick PA (editors). *Handbook of PTSD: Science and Practice.* New York: Guilford Press, 2007.

[14] Yehuda R. *Treating Trauma Survivors With PTSD.* Washington, DC: American Psychiatric Publishing, Inc., 2002.

Index

A

B

C

D

E

F

G

H

I

N

O

P

Q

R

S

T

U

V

W

Y

Z